Lecture Notes in Artificial Intelligence 1397

Subseries of Lecture Notes in Computer Science
Edited by J. G. Carbonell and J. Siekmann

Lecture Notes in Computer Science
Edited by G. Goos, J. Hartmanis and J. van Leeuwen

Springer
*Berlin
Heidelberg
New York
Barcelona
Budapest
Hong Kong
London
Milan
Paris
Santa Clara
Singapore
Tokyo*

Harrie de Swart (Ed.)

Automated Reasoning with Analytic Tableaux and Related Methods

International Conference, TABLEAUX'98
Oisterwijk, The Netherlands, May 5-8, 1998
Proceedings

Springer

Series Editors

Jaime G. Carbonell, Carnegie Mellon University, Pittsburgh, PA, USA
Jörg Siekmann, University of Saarland, Saarbrücken, Germany

Volume Editor

Harrie de Swart
Tilburg University, Faculty of Philosophy
P.O. Box 90153, 5000 LE Tilburg, The Netherlands
E-mail: H.C.M.deSwart@kub.nl

Cataloging-in-Publication Data applied for

Die Deutsche Bibliothek - CIP-Einheitsaufnahme

Automated reasoning with analytic tableaux and related methods
: international conference ; tableaux '98, Oisterwijk, The Netherlands,
May 5 - 8, 1998 ; proceedings / [International Conference Tableaux].
Harrie de Swart (ed.). - Berlin ; Heidelberg ; New York ; Barcelona ;
Budapest ; Hong Kong ; London ; Milan ; Paris ; Santa Clara ;
Singapore ; Tokyo : Springer, 1998
  (Lecture notes in computer science ; Vol. 1397 : Lecture notes in
  artificial intelligence)
ISBN 3-540-64406-7

CR Subject Classification (1991): F.4.1, I.2.3

ISBN 3-540-64406-7 Springer-Verlag Berlin Heidelberg New York

This work is subject to copyright. All rights are reserved, whether the whole or part of the material is concerned, specifically the rights of translation, reprinting, re-use of illustrations, recitation, broadcasting, reproduction on microfilms or in any other way, and storage in data banks. Duplication of this publication or parts thereof is permitted only under the provisions of the German Copyright Law of September 9, 1965, in its current version, and permission for use must always be obtained from Springer-Verlag. Violations are liable for prosecution under the German Copyright Law.

© Springer-Verlag Berlin Heidelberg 1998
Printed in Germany

Typesetting: Camera ready by author
SPIN 10637003   06/3142 – 5 4 3 2 1 0   Printed on acid-free paper

This volume contains the papers presented at TABLEAUX'98, the International Conference on Analytic Tableaux and Related Methods, held on May 5-8, 1998 in Oisterwijk (conference centre Boschoord), near Tilburg, The Netherlands.

This conference was a continuation of international workshops/conferences on Theorem Proving with Analytic Tableaux and Related Methods held in Lautenbach near Karlsruhe (1992), Marseille (1993), Abingdon near Oxford (1994), St. Goar near Koblenz (1995), Terrasini near Palermo (1996), and Pont-à-Mousson near Nancy (1997).

Tableau methods have been found to be a convenient formalism for automating deduction in various non-standard logics as well as in classical logic. Areas of application include verification of software and computer systems, deductive databases, knowledge representation and its required inference engines, and system diagnosis.

The conference brought together researchers interested in all aspects – theoretical foundations, implementation techniques, systems development, and applications – of the mechanization of reasoning with tableaux and related methods.

From the 34 papers submitted, 17 original *research papers* and 3 original *system descriptions* were selected by the program committee for presentation at the conference and for inclusion in these proceedings, together with the *invited lectures*. Abstracts of the *tutorials* have also been included. These proceedings also contain the summary of the *comparison* of theorem provers for modal propositional logics, as part of the Tableaux'98 conference, together with the contributions of the persons who participated in this comparison.

As before, Tableaux'98 attracted interest from many parts of the world with papers from many countries.

Acknowledgements I would like to thank Michael Franssen for his support in handling the many files, Kirsten van den Hoven for creating and maintaining the Tableaux'98 web page, Jozef Pijnenburg for his invaluable help in preparing the final manuscript, and all other people without whose help this conference would not have been possible: the authors who submitted papers, the speakers, the organizers of the tutorials and the comparison, the members of the program committee, the referees, the secretarial office of the Faculty of Philosophy of Tilburg University, and, last but not least, the sponsors. They made organizing the Tableaux'98 conference a pleasant experience.

February 1998   Harrie de Swart

Previous Tableaux Workshops/Conferences

1992 Lautenbach, Germany
1994 Abingdon, England
1996 Terrasini, Italy

1993 Marseille, France
1995 St. Goar, Germany
1997 Pont-à-Mousson, France

Invited Speakers

N.G. de Bruijn   Eindhoven University of Technology
A. Bundy   University of Edinburgh
E. Clarke   Carnegie Mellon University

Program Chair

H.C.M. de Swart   Tilburg University

Program Committee

M. d'Agostino   Ferrara University, Italy
K. Broda   Imperial College, London, U.K.
R. Dyckhoff   St. Andrews University, U.K.
C. Fermüller   TU Wien, Austria
M. Fitting   CUNY, New York City, U.S.A.
U. Furbach   Koblenz University, Germany
D. Galmiche   LORIA, Nancy, France
R. Goré   Australian National University
J. Goubault-Larrecq   GIE Dyade, France
R. Hähnle   Karlsruhe University, Germany
R. Hasegawa   Kyushu University, Japan
R. Letz   TU Munich, Germany
U. Moscato   Milan University, Italy
N. Murray   SUNY at Albany, U.S.A.
N. Olivetti   Torino University, Italy
D. Pearce   DFKI, Saarbrücken, Germany
J. Posegga   Deutsche Telekom, Germany
E. Rosenthal   University of New Haven, U.S.A.
P. Schmitt   Karlsruhe University, Germany
C. Schwind   LIM-CNRS, Marseille, France
H. de Swart   Tilburg University, The Netherlands
P. Voda   Comenius University, Slovakia

Referees

Each submitted paper was refereed by three members of the Program Committee. In some cases, they consulted specialists who were not on the committee. We gratefully mention their names:

W. Ahrendt	M. Baaz
D. Basin	P. Baumgartner
B. Beckert	D. Cantone
L. Cholvy	S. Costantini
U. Egly	M. Ferrari
C. Kreitz	D. Larchey-Wendling
P. Miglioli	M. Ornaghi
T. Schaub	R. Sebastiani
V. Sofronie-Stokkermans	S. Schwendimann
F. Stolzenburg	W. Veldman
C. Weidenbach	A. Zanardo

Sponsoring Institutions

Faculty of Philosophy, Tilburg University

SUN Microsystems Nederland B.V.

PTT Telecom

Dutch Academy of Sciences

Research Group "Logic and Informationsystems" of Eindhoven and Tilburg Universities

Compulog

Dutch Graduate School in Logic

Position Papers

The conference program included the presentation of seven position papers. Informal proceedings containing these papers appeared as the internal scientific report "Position Papers Tableaux'98" of the Faculty of Philosophy, Tilburg University, P.O. Box 90153, NL 5000 LE, Tilburg, The Netherlands:

A Tableau Calculus and a Cut-Free Sequent Calculus for Dummett's Predicate Logic
Alessandro Avellone, Mauro Ferrari, Pierangelo Miglioli, Ugo Moscato

Modal Tableaux with Propagation Rules and Structural Rules
Marcos A. Castilho, Luis Farinas del Cerro, Olivier Gasquet, Andreas Herzig

Fibred Modal Tableaux
Dov M. Gabbay, Guido Governatori

Sequent Calculi and Free Variable Tableaux for Non-bivalent Logics
Reinhard Muskens

Using Binary Decision Diagrams to Discover Functional Dependencies and Independencies
Sanjai Rayadurgam, Duminda Wijesekara, Mats Heimdahl

Comprehension Schemata in Tableaux
Benjamin Shults

The Tableaux Method with Propositional Constant Substitutions
George Tagviashvili

Table of Contents

Extended Abstracts of Invited Lectures

Philosophical Aspects of Computerized Verification of Mathematics 1
   N.G. de Bruijn

A Science of Reasoning 10
   Alan Bundy

Model Checking: Historical Perspective and Example 18
   Edmund M. Clarke, Sergey Berezin

Comparison

Comparison of Theorem Provers for Modal Logics -
Introduction and Summary.. 25
   Peter Balsiger, Alain Heuerding

FaCT and DLP 27
   Ian Horrocks, Peter F. Patel-Schneider

Prover KT4 31
   Michel Levy

LeanK 2.0 33
   Bernhard Beckert, Rajeev Goré

Logics Workbench 1.0 35
   Peter Balsiger, Alain Heuerding, Stefan Schwendimann

Optimised Functional Translation and Resolution 36
   Ullrich Hustadt, Renate A. Schmidt, Christoph Weidenbach

Benchmark Evaluation of □KE 38
   Jeremy Pitt

Abstracts of the Tutorials

Implementation of Propositional Temporal Logics Using BDDs 40
   Geert L.J.M. Janssen

Computer Programming as Mathematics in a Programming Language and
Proof System CL 42
   Ján Komara, Paul J. Voda

Contributed Research Papers

A Tableau Calculus for Multimodal Logics and some (Un)decidability
Results 44
   Matteo Baldoni, Laura Giordano, Alberto Martelli

Hyper Tableau - The Next Generation. 60
   Peter Baumgartner

Fibring Semantic Tableaux 77
   Bernhard Beckert, Dov Gabbay

A Tableau Calculus for Quantifier-Free Set Theoretic Formulae 93
  Bernhard Beckert, Ulrike Hartmer
A Tableau Method for Interval Temporal Logic with Projection 108
  Howard Bowman, Simon Thompson
Bounded Model Search in Linear Temporal Logic and Its Application to
Planning 124
  Serenella Cerrito, Marta Cialdea Mayer
On Proof Complexity of Circumscription 141
  Uwe Egly, Hans Tompits
Tableaux for Finite-Valued Logics with Arbitrary Distribution Modalities . 156
  Christian G. Fermüller, Herbert Langsteiner
Some Remarks on Completeness, Connection Graph Resolution and
Link Deletion 172
  Reiner Hähnle, Neil V. Murray, Erik Rosenthal
Simplification and Backjumping in Modal Tableau 187
  Ullrich Hustadt, Renate A. Schmidt
Free Variable Tableaux for a Logic with Term Declarations 202
  P.J. Martín, A. Gavilanes, J. Leach
Simplification: A General Constraint Propagation Technique for
Propositional and Modal Tableaux 217
  Fabio Massacci
A Tableaux Calculus for Ambiguous Quantification 232
  Christof Monz, Maarten de Rijke
From Kripke Models to Algebraic Counter-Valuations 247
  Sara Negri, Jan von Plato
Deleting Redundancy in Proof Reconstruction 262
  Stephan Schmitt, Christoph Kreitz
A New One-Pass Tableau Calculus for PLTL 277
  Stephan Schwendimann
Decision Procedures for Intuitionistic Propositional Logic by Program
Extraction 292
  Klaus Weich

Contributed System Descriptions

The FaCT System 307
  Ian Horrocks
Implementation of Proof Search in the Imperative Programming
Language Pizza 313
  Christian Urban
P-SETHEO: Strategy Parallelism in Automated Theorem Proving 320
  Andreas Wolf

Author Index 325

Philosophical Aspects of Computerized Verification of Mathematics

N.G. de Bruijn

Department of Mathematics and Computing Science
Eindhoven University of Technology
PO Box 513, 5600MB Eindhoven, The Netherlands
wsdwnb@win.tue.nl

Abstract. This invited lecture discusses various philosophical aspects of computerized verification of mathematics. Particular attention is given to the influences of type-theoretical verification systems. The paper is halfway between a full paper and an extended abstract. The reason is that an extensive text of a very similar lecture (Venice, 1995) is to be published in [7].

1 Introduction

Computer aided verification has philosophical aspects. The design of a verification system is a product of the designer's view on mathematics, and, on the other hand, the development and the usage of a verification system may reshape one's philosophy. That word will be used lightheartedly here. It is not taken as serious professional philosophy, but just as meditation about the way one does one's job.

What used to be called philosophy of mathematics in the past was for a large part subject-oriented. Most people characterized mathematics by its *subject matter*, classifying it as the science of *space* and *number*. From the verification system's point of view, however, subject matter is irrelevant. Verification involves the rules of mathematical reasoning, not the subject.

The author's philosophy is definitely anti-platonistic. Mathematical language is so perfect that it can talk coherently about things that do not exist at all (such discussions might even lead to *proofs* of their non-existence). So one should not claim any kind of platonistic existence of things on the sole ground that one has talked about them in the same style as about the real world.

Some of the points of view displayed in this lecture are matters of taste, but most of them were imposed by the task of letting a machine follow what we say, a machine *without* any knowledge of our mathematical culture and without any knowledge of physical laws. The author's ideas can be carried back to his design of the Automath system in the late 1960's, with quite some mutual inspiration between the way to look upon mathematics on the one hand and the structure and usage of the verification system on the other. See [6] for philosophical items concerning Automath, and [1,2,8] for general information about the Automath project.

Such *type-theoretical* verification systems call for new attitudes: throughout the 20-th century most mathematicians had been trained to think in terms of *untyped* sets.

2 Predominance of language

During the 20-th century the role played by the *language* of mathematics became more and more prominent. After millennia of mathematics one has finally reached a level of understanding that can be physically represented. Mankind managed to disentangle the intricate mixture of language, metalanguage and interpretation, isolating a body of formal, abstract mathematics that machines can verify completely. Machines handle language, and nothing but language.

Philosophically, the fact that mathematics can now completely be checked by machines should not be underestimated.

If a computer has to check mathematics, one has to feed it with texts and to request it to check whether those texts obey the rules of the game. The rules interpret correctness of a piece of text in the light of what was accepted earlier: it is the matter of correctness of a complete *book*.

In our relation with the machine there is language and nothing but language. Machines have no idea about *meaning* in the usual philosophical sense. But they can handle "meaning" adequately if we take the word in the sense of a mapping from a language system into another one.

The language of mathematics is a living organism. There have been many cases in the past where mathematicians began talking metalanguage, making and proving statements about the mathematical text itself. Subsequently they admitted such arguments as legitimate language. An example that took about the whole 19-th century to mature, is the acceptance of a *function* as a mathematical object.

Such transitions change the borderline between language and metalanguage. A recent case is the paradigm of "proofs as objects", that has by no means settled yet in the general mathematical community.

3 Levels of activity

The world of mathematics can be subdivided in several ways; here we mainly consider the levels of formalization involved with the full spectrum ranging from discovery to complete formal verification. It is like a *conveyor belt* where people are processing a mathematical product which starts as a vague idea, is transformed from stage to stage, and ends as a finished completely formal and impeccable text.

This conveyor belt may employ various kinds of people. Towards the end of the belt, the workers have to be more meticulous, more bureaucratic, and hardly need to "understand" what they are working on.

Do there always have to be different people at different stations along the belt? One can learn from the area of mathematics publishing, which rapidly changed over the last few decades. Instead of using assistants, typists and professional printing people, many creative mathematicians are now desk-top publishers themselves.

If the theorem proving community and the proof checking community do their work well, one might get a similar situation for the conveyor belt. The belt might be handled by the creative mathematician all by himself. Automated theorem provers may do useful work at several places of the belt, in particular at the far end (the proof checking station), where it may be left to a machine to fill all sorts of trivial little gaps.

4 Teaching mathematics

Teaching is an essential part of the mathematician's trade, mainly because of the central role of proving. To prove means to explain and to convince.

But is teaching always completely convincing? It has often been said that mathematics is taught by *intimidation*, and learned by *imitation*. There is certainly some truth in this, in particular as far as it concerns the structure and basic rules of mathematics. Teaching hardly ever specified what definitions, assumptions, axioms, theorems, variables and proofs really are.

It may be quite true that it is better not to burden beginners with such questions. But would mathematics teachers be able to explain these fundamentals later? Would they be able to explain them to themselves? A verification system forces us to be quite clear about those structural items, and may therefore have a positive influence on mathematics teaching. If we fail to explain the basic structure of mathematics to a machine then it is an illusion to think that we can explain it, without intimidation, to a student.

5 Influences of system efficiency

The development of a verification system requires quite some efficiency in order to cope with limited resources, in particular with restricted hardware and software. This has a danger: the urge for efficiency can lead to a structure one might deplore later on. But there can be positive effects too. The need for efficiency may reveal similarities, suggesting ways to treat similar things alike, making them even more similar than they were.

This is illustrated by the discovery of the "proofs as objects" paradigm. If we want a machine to accept a particular application of a theorem, we have to feed it with object expressions for the variables as well as with proofs (or references to proofs) for the updated assumptions, and these two kinds of things are interwoven. In both cases we have similar dependency on parameters and similar substitution mechanism. At the start, attitudes with respect to objects and proofs are very different, but this changes under the attempts to draw full profit from the similarity. This gives two innovations at the same time. First, proofs are treated the same way as objects, and proof classes the same way as object types. Secondly, we notice that the proof classes depend on parameters, so they require *dependent types*, and the natural step is to allow dependent types for object types too.

In the matter of efficiency, the most essential thing is preventing exponential growth and even quadratic growth of the time a machine needs for checking what we write. In order to keep in pace with the ordinary presentation of mathematics, we have to require linear time. It means *feasibility*.

6 Influences of systematic notation

If we want a machine to digest our mathematical material, we have to revise our notational habits occasionally. Poor notation can obstruct insight and development, good notation can be a stimulus for discovery, in particular since it can promote ideas on the metalanguage level.

Lambda notation is an example. If it had not existed before, it would have been discovered at once in the first attempts to get mathematical contact with a machine.

7 Natural deduction

One of the pillars of Automath was what is sometimes called Fitch style of natural deduction. It fully deserves the name "natural" since it was used by mathematicians long before it was ever formally described: presentation of mathematics in the form of nested blocks, where blocks are opened either by making an assumption or by introducing a (possibly typed) variable.

Natural deduction follows the way people reasoned *before* it was tried to explain logic by means of an algebra of truth values. Such Boolean logic is metatheory of classical reasoning. It does not *show* what that reasoning is. It is silly that education in elementary logic so often takes truth values as point of departure.

8 Types

Having types is not a new idea either. On the contrary: it was the standard idea before the doctrine "everything is a set" emerged. In that doctrine "set" means "untyped set".

Most mathematicians may still think in terms of types today, even when *preaching* that everything is a set.

One can get a feeling for the meaning of "type" by inspecting English sentences containing the word group "is a". The sentence "The capital of Italy is a big city" expresses that "the capital of Italy" has the type "big city". On the left of the group "is a" we have a string of words that has the form of a *name*: an accurate description of something, describing it uniquely; on the right we have a *substantive group* or a single substantive.

With this natural language interpretation in mind, one can find all sorts of opportunities to introduce types, thereby enriching the scope of formal mathematics. Once a verification system is ready for attaching types to things, types can be used in many ways, even simultaneously. An expression representing a particular proof of some proposition Q can be given the type "proof of Q", an expression describing the construction of the centre of a circle c can be given the type "construction for the centre of c". This amalgamates several worlds: the world of objects, the one of proofs and the one of geometrical constructions; in each one of these worlds things from the other worlds can play the role of parameters (see [3,5]).

Types play the same role as substantives in natural language. They have always been used on a large scale in mathematics, but there was no tradition to express them in terms of symbols. Since antiquity one used symbols and composite expressions as *names* of objects, Leibniz and Boole began to use symbols to represent *sentences*, but for *substantives* such a thing was not done. Yet it can be profitable. In particular one can handle dependent substantives, similar to dependent types.

A description of the syntax of natural mathematical language on the basis of sentences, substantives and names can be found in [4].

9 Proofs as objects

The "proofs as objects" paradigm has philosophical consequences: it is a quite revolutionary shift of the borderline between language and metalanguage. The sentence "p is a proof of Q" (where Q is some proposition and p is a proof) used to be metalanguage, but in a type-oriented verification system it belongs to the mathematical language itself, just like "$x + y$ is a rational number". The word group "proof of Q" is used in the same way as the group "rational number"; both word groups can be called *types*.

This principle that proofs can be treated the same way as objects has been given various names, like "formulas as types" and "propositions as types".

10 Constructivism

Different people may give different definitions of constructivism. For some it is the rejection of the axiom of choice; others will say that the use of classical logic is non-constructive already.

Constructivity is a point of view that accepts a particular language and a particular set of axioms, but refuses language extensions and further axioms or further primitive notions.

It is unreasonable to ban extra axioms if one is liberal about language extensions at the same time. One can easily fool oneself: if one allows quantification over all propositions, it is possible to introduce the proposition that all propositions are true, and to use that as a definition of falsum. That gives the falsum rule as a theorem, and one does not have to call it an axiom any more. Nevertheless one has introduced the same non-constructivity, and possibly more, in a disguised form.

11 The scope of mathematics

Mathematics is *not* just the study of numbers and geometrical figures, as it used to be said in the past. And it is *not* just set theory.

Verification systems have a different point of view. For them, mathematics is anything they can verify. This is flexible, of course, since languages can be extended, in particular by internalizing pieces of metalanguage.

Typed lambda calculus is capable of handling almost anything we call mathematics. It can handle types of objects, of proofs, of geometrical constructions, of computer programs, and whatever else might come up in the abstract sciences. Therefore it seems attractive to claim that *mathematics is anything that can be handled by typed lambda calculus*.

But one can go beyond this. Lambda calculus can be extended in many ways. One can have extensions which allow using large pieces of text (like *theories*) as objects.

12 Platonism

Mathematics seems to talk about *things*, but do these really exist? We have no way to find out, and worse: we have no way to express what we mean by existence. Yet many people have that funny feeling of having to choose between existence and non-existence.

One might call it mathematical *platonism* to consider the mathematician as a journalist, and anti-platonism to consider him as a novelist.

Verification systems definitely put an end to platonism. The only things these systems deal with are language texts. These texts themselves have a certain physical existence: they can be represented by means of ink on paper and by electric or magnetic charges in computer hardware. But the things *discussed by* the texts usually lack physical existence.

It is instructive to compare a mathematics verification system with a machine that simply verifies whether a given list of chess moves represents a legitimate game. Any sensible program achieves this by building the chess board positions in the machine's memory, checking whether in each one of those positions the next move of the list is admitted by the rules of chess, and updating the board accordingly. The list of moves *talks* about positions, and *both* the list and the positions are physically represented. In a mathematics verification system this is different. Even where mathematical objects *might* be represented physically, the system does not do it. The mathematical text is judged by its internal coherence, irrespective of interpretations or meanings.

Platonists and anti-platonist are alike if it comes to forming mental images, possibly suggesting connections with older statements with similar images. It is both helpful and stimulating; it does not matter whether it is considered as fact or as fiction.

As long as no *use* is made of the claimed existence of mathematical objects, platonism does no harm at all; it is just irrelevant.

But platonism is certainly confusing for beginning mathematics students in the matter of the existence quantifier, where the word "existence" has a different meaning.

Platonism has left its traces in the anti-platonistic world too. The idea of negation somehow acknowledges the platonistic idea that there is truth beyond provability. All mathematicians who handle classical logic agree that any statement is either true or false. Their confidence was not even shaken by the discovery of undecidable propositions.

The old question whether mathematical situations are *discovered* or *created* is entirely a platonist's problem.

13 Mathematics and the real world

The physicist E. Wigner made a famous statement about the incredible and undeserved success of mathematics in explaining the real world. Mathematicians may find it even more remarkable that it holds in spite of the fact that a large part of that successful mathematics has always been immature, incomplete, unfinished and partially incorrect.

But if we believe that there is a final correct mathematical description of the entire physical reality, Wigner's observation gets into a new light under the claim that all correct mathematics can be physically represented, and that formal mathematical reasoning can be a very clear *part* of that real world. Needless to say, the human brain was a part of that real world all the time, but by no means a very clear part.

14 Changing roles

There can be changes in *attitudes* about mathematics under the influence of the proofs-as-objects system.

There is the old chicken-and-egg problem of mathematics and logic: is mathematics based on logic or is logic a branch of mathematics? If a verification system handles proofs as objects, the difference between logic and mathematics vanishes almost entirely. Logical fundamentals have the same form as mathematical axioms. Logical derivation rules can be derived, sometimes using mathematics. Such rules can be applied, both to logic and to mathematics.

Traditionally, logic is just the part of mathematics that is not taught explicitly at the time it should, but studied later as a kind of metatheory. Verification systems can change our attitudes in this matter.

The attitudes of today's mathematicians might be vaguely sketched by the roles of what shall here be called M, P, F and L.

M is the body of all mathematical truth: mathematical objects and their relations.

P is the real physical world. All sorts of events in P seem to be reflected by objects and relations in M.

F is a formal system, meticulously expressing the foundations for the establishment of truth in M. For most people, it contains logic and the foundations of formal set theory.

L is the discussion language in which we think, write and talk. It is a mixture of words and formulas, and usually *not* very formal.

The standard attitude seems to be that L *talks about* M, and that F provides the *authority*. But F does not provide rules for handling the physical world. The relation between M and P is discussed in physics.

Under the influence of verification systems the roles can change. It may even go so far that F's authority will be taken over by some purified form of L. M lives on as a kind of imagined reality, giving inspiration and motivation when we talk L. And if we need mathematics in studying the real world, L can talk directly about P, without interference by M. So L can serve as a language for physics too.

F becomes *metatheory*, not necessary for L's authority, but possibly useful for studying the limits of what L can achieve. As a mathematical theory, F is just one of the many mathematical subjects that F can handle.

In this picture the mathematician becomes a formalist, but there is nothing against it. Formalism is the essence of mathematics, as all non-mathematicians know. Over the years the word "formalist" got a negative connotation, it even became an insult. Wrongly: the emotional dividing line should not be drawn between non-formalism and formalism, but between formalism and *bureaucratic formalism*. A bureaucratic formalist is someone who applies rules meticulously without higher motives like "imagination", "meaning", "sense" and "beauty". In that sense the machine is a bureaucratic formalist, and we ourselves try to avoid becoming one.

15 Absolute safety?

Formal verification does not lead to *absolute* safety. Who checks the checker and the checking program? The program may be compared to its specifications, but who checks the specifications? The only thing we can do is to require the specifications and checkers to be exceedingly simple and transparent.

For a part of the problem there is an obvious way out: one can take a verification program itself as the definition of correctness. That is about the same as what seems to have been the general norm in the past: a thing was correct if and only if meticulous human verification found no flaws. But there is an important difference: in the human case there used to be no explicitly stated rules for such verification.

One has to be modest, but nevertheless it can be claimed that a simple and transparent computerized verification system can check human-made mathematics much more dependable than humans themselves could ever do.

16 What is a proof?

Verification systems insist on formal proofs. It is reasonable to ask whether this is really the only thing there is. What one often expects from a proof is that it gives *confidence*. As an example we imagine some theorem t and we imagine that we have a metatheoretical proof q for the statement that there *exists* a piece of text r which our verification system will accept as a proof for t. This gives confidence, yet we do not have a proof for t, as long as we do not actually "have" r.

We cannot claim that with the acceptance of a verification system we have finally settled the question about what a proof is, but anyway we have raised the discussion to a higher level.

References

1. N.G. de Bruijn, A survey of the project Automath. In: *To H.B. Curry: Essays in combinatory logic, lambda calculus and formalism*, ed. J.P. Seldin and J.R. Hindley, Academic Press 1980, pp. 579-606. Reprinted in: Nederpelt [8].
2. N.G. de Bruijn, Checking mathematics with computer assistance. In *Notices American Mathematical Society, vol 38(1), Jan. 1991, pp 8-15*.
3. N.G. de Bruijn, Formalization of constructivity in Automath. In: Nederpelt [8].
4. N.G. de Bruijn, The Mathematical Vernacular, a language for mathematics with typed sets. In: Nederpelt [8].
5. N.G. de Bruijn, On the roles of types in mathematics. In: *The Curry-Howard Isomorphism*, ed. Ph. de Groote. *Cahiers du Centre de Logique, vol. 8, pp. 27-54*. Academia, Louvain-la-Neuve (Belgique) 1995.
6. N.G. de Bruijn, Reflections on Automath. In: Nederpelt [8].
7. N.G. de Bruijn, Type-theoretical checking and philosophy of mathematics. In: "*Twenty-Five Years of Constructive Type Theory*", to be published by Oxford University Press.
8. Nederpelt, R.P., Geuvers, J.H. and de Vrijer, R.C. (editors), *Selected Papers on Automath, Studies in Logic, vol. 133*. North-Holland 1994.

A Science of Reasoning (Extended Abstract) *

Alan Bundy

Department of Artificial Intelligence,
University of Edinburgh,
Edinburgh, EH1 1HN, Scotland.
bundy@edinburgh.ac.uk, Tel: 44-131-650-2716.

Abstract. How can we understand reasoning in general and mathematical proofs in particular? It is argued that a high-level understanding of proofs is needed to complement the low-level understanding provided by Logic. A role for computation is proposed to provide this high-level understanding, namely by the association of *proof plans* with proofs. Criteria are given for assessing the association of a proof plan with a proof.

1 Motivation: The Understanding of Mathematical Proofs

We argue that Logic[1] is not enough to understand reasoning. It provides only a low-level, step by step understanding, whereas a high-level, strategic understanding is also required. Many commonly observed phenomena of reasoning cannot be explained without such a high-level understanding. Furthermore, automatic reasoning is impractical without a high-level understanding.

We propose a science of reasoning which provides both a low- and a high-level understanding of reasoning. It combines Logic with the concept of *proof plans*, [Bundy, 1988]. We illustrate this with examples from mathematical reasoning, but it is intended that the science should eventually apply to all kinds of reasoning.

2 The Need for Higher-Level Explanations

A proof in a logic is a partially ordered set of formulae where each formula in the set is either an axiom or is derived from earlier formulae in the set by a rule of inference. Each mathematical theory defines what it means to be a formula, an axiom or a rule of inference. Thus Logic provides a low-level explanation of a mathematical proof. It explains the proof as a sequence of steps and shows how each step follows from previous ones by a set of rules. Its concerns are limited to the soundness of the proof, and to the truth of proposed conjectures in models of logical theories.

---

* The research reported in this paper was supported by EPSRC grant GR/L/11724. I would like to thank two anonymous referees and other members of the mathematical reasoning group at Edinburgh for feedback, especially Richard Boulton, Mitch Harris, Colin Phillips, Frank van Harmelen and Toby Walsh. The full version of this extended abstract appeared in [Bundy, 1991].
[1] We adopt the convention of using uncapitalised 'logic' for the various mathematical theories and capitalised 'Logic' for the discipline in which these logics are studied.

While Logic provides an explanation of how the steps of a proof fit together, it is inadequate to explain many common observations about mathematical proofs.

- Mathematicians distinguish between understanding each step of a proof and understanding the whole proof.
- Mathematicians recognise families of proofs which contain common structure.
- Mathematicians distinguish between 'interesting' and 'standard' steps of a proof.
- Mathematicians describe proofs to each other at different levels of detail. Their high level descriptions contain only very brief summaries of standard steps but give more detail about the interesting ones.
- Mathematicians use their experience of previously encountered proofs to help them discover new proofs.
- Mathematicians often have an intuition that a conjecture is true, but this intuition is fallible.
- Students of mathematics, presented with the same proofs, learn from them with varying degrees of success.

3 Common Structure in Proofs

Several researchers in automatic theorem proving have identified common structure in families of proofs. For instance,

- [Bundy & Welham, 1981] describes the common structure in solutions to symbolic equations. This common structure was implemented in a process of *meta-level inference* which guided the search for solutions to equations.
- [Bundy et al, 1988] describes the common structure in inductive theorems about natural numbers, lists, *etc*. This common structure was implemented as an *inductive proof plan* which was used to guide the search for proofs of such theorems.
- [Bledsoe et al, 1972] describes the common structure in theorems about limits of functions in analysis. This common structure was implemented as the *limit heuristic* and used to guide the search for proofs of such theorems.
- [Wos & McCune, 1988] describes the common structure in attempts to find fixed-points combinators. This common structure was implemented as the *kernel method* and used to guide the search for such fixed-points.
- [Polya, 1965] describes the common structure in ruler and compass constructions. This common structure was implemented by [Funt, 1973] and used to guide the search for such constructions.
- [Huang et al, 1995] and [Gow, 1997] describe the common structure in diagonalization proofs. This common structure has been implemented (twice) as a diagonalization proof plan and used to guide the proofs of a variety of theorems.

4 Proof Plans

Common structure in proofs can be captured in *proof plans*. These proof plans are represented by three kinds of computational object:

Tactics: are computer programs which construct part of a proof by applying rules of inference in a theorem proving system. A simple tactic might apply only a single rule of inference; a compound tactic will be defined in terms of simpler tactics and might construct a whole proof.

Methods: are logical specifications of tactics. In particular, a method describes the preconditions for the use of a tactic and the effects of using it. These preconditions and effects are syntactic properties of the logical expressions manipulated by the tactic and are expressed in a meta-logic.

Critics: capture common patterns of failure of methods and suggest patches to the partial proof. A critic is associated with a method and is similar in structure, except that its preconditions describe a situation in which the method fails. Instead of effects it has instructions on how to patch the failed proof plan.

A proof planner uses the methods to construct a customised tactic for the current conjecture. It combines general-purpose tactics so that the effects of earlier ones achieve the preconditions of later ones. The specifications of the tactics, which the methods provide, enable the proof planner to conduct meta-level inference which matches problems to the tactics which are best placed to solve them. The customised tactic constructed by the proof planner is input to a theorem prover which uses it to try to prove the conjecture.

In general, a complete specification of a tactic is not obtainable. So the proof planning process is fallible. Using critics we may be able to recover from an initial failure. For instance, the preconditions of a method may succeed, but those of one its sub-methods fail. In this case a critic associated with the sub-method may suggest a patch to the proof plan. Each sub-method may have several associated critics, corresponding to different failure patterns of its preconditions. Each critic will suggest a different way to patch the proof. So the form of proof failure suggests an appropriate patch. This productive use of failure via critics is made possible by proof planning and is one of its most powerful features, [Ireland & Bundy, 1996].

Proof plans have been implemented at Edinburgh in the *Oyster-CLAM* system, [Bundy *et al*, 1990], and at Saarbrücken in the Ωmega system, [Benzmüller *et al*, 1997]. *Oyster* is a theorem prover for Intuitionist Type Theory. *CLAM* is a plan formation program which has access to a number of general-purpose methods and critics for inductive proofs. *CLAM* constructs a special-purpose tactic for each conjecture by reasoning with its methods and critics. This specialised tactic is then executed by *Oyster*, constructing a proof. Ωmega works in a similar way. The search for a proof plan at the meta-level is considerably cheaper than the search for a proof at the object-level. This makes proof plans a practical solution to the problems of search control in automatic theorem proving.

5 The High-Level Understanding of Proofs

Thus a high-level explanation of a proof of a conjecture is obtained by associating a proof plan with it. The tactic of this proof plan must construct the proof. The method of this proof plan must describe both the preconditions which made this tactic appropriate for proving this conjecture and the effects of this tactic's application on the conjecture. It must also describe the role of each sub-tactic in achieving the preconditions of later sub-tactics and the final effects of the whole tactic.

In fact, this association provides a multi-level explanation. The proof plan associated with the whole proof provides the top-level explanation. The immediate sub-tactics and sub-methods of this proof plan provide a medium-level explanation of the major sub-proofs. The tactics and methods associated with individual rules of inference provide a bottom-level explanation, which is similar to that already provided by Logic.

The general-purpose tactics and methods which we will use to build proof plans, and the association of proof plans with proofs will constitute the theories of our science of reasoning. This extends the way in which logical theories and the association of logical proofs with real proofs and arguments, constitute the theories of Logic (especially Philosophical Logic). Just as Logic also has meta-theories about the properties of and relations between logical theories, we may also be able to develop such meta-theories about proof plans.

6 What is the Nature of our Science of Reasoning?

Before we can dignify this proposed study of the structure of proofs with the epithet *science* we must address a fundamental problem about the nature of such a science. Traditional sciences like Physics and Chemistry study physical objects and the way they interact. The subject of our proposed science is proof plans. But proof plans are not physical objects. If they can be said to exist at all it is in the minds of mathematicians proving theorems, teachers explaining proofs and students understanding them. Physicists assume that the electrons in the apple I am eating as I write are essentially the same as the electrons in some distant star. But proof plans will differ from mind to mind and from time to time. There will be billions of such proof plans. Are we doomed merely to catalogue them all? Given the difficulty of discovering the nature of even one such proof plan, what a difficult and ultimately pointless task this would be. We would prefer to narrow our focus on a few representative proof plans. But on what basis could these few be chosen?

Fortunately, this is not a new problem. It is one faced by all human sciences to some extent and it is one that has been solved before. Consider the science of Linguistics. In Linguistics the theories are grammars and the association of grammatical structure with utterances. Linguists do not try to form different grammars for each person, but try to form a grammar for each language, capturing the commonality between different users of that language. They try to make these grammars as parsimonious as possible, so that they capture the maximum amount of generality within and between languages. Linguists do not claim that everyone or anyone has these target grammars stored in their head — nor, indeed, that anyone has a grammar at all — only that they *specify* the grammatical sentences of the language.

Another example is Logic itself. Again judged by the arguments people produce, the logical laws differ between minds and vary over time. Logicians do not try to capture this variety, but confine themselves to a few logics which specify 'correct' arguments. As with grammatical sentences, correct arguments are identified by initial observation of arguments actually used and consultation with experts to decide which of these are correct.

I place our proposed science of reasoning between Linguistics and Logic. Proof plans are more universal than grammatical rules, but it is possible to associate different, equally appropriate proof plans with the same proof. The study of proof plans appeals both to an empirical study of the way in which mathematicians structure their proofs and to reflection on the use of logical laws to put together proofs out of parts.

Thus there are strong precedents for a science that takes mental objects as its domain of study and tames the wide diversity of exemplars by imposing a normative explanation informed by reflection and empirical study. It only remains to propose criteria for associating proof plans with proofs that will enable us to prefer one proof plan to another. This we can do by appealing to general scientific principles. Our proposals are given in the next section.

7 Criteria for Assessing Proof Plans

If there were no criteria for the association of proof plans with proofs, then we could carry out our programme by associating with each proof an *ad hoc* tactic consisting of the concatenation of the rules of inference required to reproduce it, and constructing an *ad hoc* method in a similar way. This would not go beyond the existing logical explanation.

The only assessment criterion we have proposed so far is *correctness*, *i.e.* that the tactic of the proof plan associated with a proof will construct that proof when executed. We now discuss some other possible criteria.

- *Psychological Validity*: a proof plan gets more credit if there is experimental evidence that all, most or some mathematicians producing or studying proofs also structured a proof in the way suggested by some proof plan. This criterion is only applicable if we are trying to model human reasoning, but it can be suggestive even when we are not.
- *Expectancy*: a proof plan gets more credit if it provides some basis for predicting whether it will succeed.
- *Generality*: a proof plan gets credit from the number of proofs or sub-proofs with which it is associated and for which it accounts.
- *Prescriptiveness*: a proof plan gets more credit the less search its tactic generates and the more it prescribes exactly what rules of inference to apply.
- *Simplicity*: a proof plan gets more credit for being succinctly stated.
- *Efficiency*: a proof plan gets more credit when its tactic is computationally efficient.
- *Parsimony*: the overall theory gets more credit the fewer general-purpose proof plans are required to account for some collection of proofs.

Initially, a proof plan may be suggested by its author's intuition of how s/he proved some theorems, perhaps augmented by more or less formal studies of other mathematicians. The criteria of correctness, expectancy, generality, prescriptiveness, simplicity, efficiency and parsimony can then be used to generalise and refine it.

8 The Role of the Computer

So far we have not involved the computer in this methodological discussion. One might expect it to play a central role. In fact, computers have no role in the *theory*, but play an important *practical* role. *Computation* plays a central role in the theory, because the tactics are procedures and they are part of the theory of our science of reasoning. It is not, strictly speaking, necessary to implement these tactics on a computer, since they can be executed by hand. However, in practice, it is highly convenient. It makes the process of checking that the tactics meet the criteria of §7 both more efficient and less error prone. Machine execution is convenient:

- for speeding up correctness testing, especially when the proof plans are long, or involve a lot of search, or when a large collection of conjectures is to be tested;
- to automate the gathering of statistics, *e.g.* on size of search space, execution time, *etc*;
- to ensure that a tactic has been accurately executed; and
- to demonstrate to other researchers that the checking has been done by a disinterested party.

In this way the computer can assist the rapid prototyping and checking of hypothesised proof plans. Furthermore, in its 'disinterested party' role, the computer acts as a sceptical colleague, providing a second opinion on the merits of hypothesised proof plans that can serve as a source of inspiration. Unexpected positive and negative results can cause one to revise ones current preconceptions.

9 The Relation to Automatic Theorem Proving

Although our science of reasoning might find application in the building of high performance, automatic theorem provers, the two activities are not co-extensive. They differ both in their motivation and their methodology.

I take the conventional motivation of automatic theorem proving to be the building of theorem provers which are empirically successful, without any necessity to understand why. The methodology is implied by this motivation. The theorem prover is applied to a random selection of theorems. Unsuccessful search spaces are studied in a shallow way and crude heuristics are added which will prune losing branches and prefer winning ones. This process is repeated until the law of diminishing returns makes further repetitions not worth pursuing. The result is fast progress in the short term, but eventual deadlock as different proofs pull the heuristics in different directions. This description is something of a caricature. No ATP researchers embody it in its pure form, but aspects of it can be found in the motivation and methodology of all of us, to a greater or lesser extent.

Automatic theorem provers based on proof plans make slower initial progress. Initial proof plans have poor generality, and so few theorems can be proved. The motivation of understanding proofs mitigates against crude, general heuristics with low prescriptiveness and no expectancy. The 'accidental' proof of a theorem is interpreted as a fault caused by low prescriptiveness, rather than a lucky break. However, there is no eventual deadlock to block the indefinite improvement of the theorem prover's performance. If two or more proof plans fit a theorem then either they represent legitimate alternatives both of which deserve attempting or they point to a lack of prescriptiveness in the preconditions which further proof analysis should correct.

Thus, we expect a science of reasoning will help us build better automatic theorem proving programs in the long term, although probably not in the short term.

10 Conclusion

In this paper we have proposed a methodology for reaching a multi-level understanding of mathematical proofs as part of a science of reasoning. The theories of this science consist of a collection of general-purpose proof plans, and the association of special-purpose proof plans with particular proofs. Each proof plan consists of a tactic and a method which partially specifies it. Special-purpose proof plans can be constructed by a process of plan formation which entails reasoning with the methods of the general-purpose proof plans and critics which provide standard patches for commonly occurring failure patterns.

Ideas for new proof plans can be found by analysing mathematical proofs using our intuitions about their structure and, possibly, psychological experiments on third party mathematicians. Initial proof plans are then designed which capture this structure. These initial proof plans are then refined to improve their expectancy, generality, prescriptiveness, simplicity, efficiency and parsimony. Scientific judgement is used to find a balance between these sometimes opposing criteria. Computers can be used as a workhorse, as a disinterested party to check the criteria and as a source of inspiration.

Proof planning can be applied to automatic theorem proving as a heuristic technique for proof search. It may also be used in interactive theorem proving to improve the communication with the user. The proof can be automatically divided into manageable chunks and the relationships between these chunks can be described in terms of the preconditions and effects of the tactics. Lastly, proof plans may find some role in mathematical education as a basis for structuring the proof and describing the process of proof discovery.

The design of general-purpose proof plans and their association with particular proofs is an activity of scientific theory formation that can be judged by normal scientific criteria. It requires deep analysis of mathematical proofs, rigour in the design of tactics and their methods, and judgement in the selection of those general-purpose proof plans with real staying power. Our science of reasoning is normative, empirical and reflective. In these respects it resembles other human sciences like Linguistics and Logic. Indeed it includes parts of Logic as a sub-science.

Personal Note

For many years I have regarded myself as a researcher in automatic theorem proving. However, by analysing the methodology I have pursued in practice, I now realise that my real motivation is the building of a science of reasoning in the form outlined above. Now that I have identified, explicitly, the science in which I have been implicitly engaged, I intend to pursue it with renewed vigour. I invite you to join me.

References

[Benzmüller et al, 1997] Benzmüller, C., Cheikhrouhou, L., Fehrer, D, Fiedler, A., Huang, X., Kerber, M., Kohlhase, K., Meirer, A, Melis, E., Schaarschmidt, W., Siekmann, J. and Sorge, V. (1997). Ωmega: Towards a mathematical assistant. In McCune, W., (ed.), *14th Conference on Automated Deduction*, pages 252-255. Springer-Verlag.

[Bledsoe et al, 1972] Bledsoe, W. W., Boyer, R. S. and Henneman, W. H. (1972). Computer proofs of limit theorems. *Artificial Intelligence*, 3:27-60.

[Bundy & Welham, 1981] Bundy, A. and Welham, B. (1981). Using meta-level inference for selective application of multiple rewrite rules in algebraic manipulation. *Artificial Intelligence*, 16(2):189-212. Also available from Edinburgh as DAI Research Paper 121.

[Bundy, 1988] Bundy, Alan. (1988). The use of explicit plans to guide inductive proofs. In Lusk, R. and Overbeek, R., (eds.), *9th Conference on Automated Deduction*, pages 111-120. Springer-Verlag. Longer version available from Edinburgh as DAI Research Paper No. 349.

[Bundy, 1991] Bundy, Alan. (1991). A science of reasoning. In Lassez, J.-L. and Plotkin, G., (eds.), *Computational Logic: Essays in Honor of Alan Robinson*, pages 178-198. MIT Press. Also available from Edinburgh as DAI Research Paper 445.

[Bundy et al, 1988] Bundy, A., van Harmelen, F., Hesketh, J. and Smaill, A. (1988). Experiments with proof plans for induction. Research Paper 413, Dept. of Artificial Intelligence, University of Edinburgh, Appeared in Journal of Automated Reasoning, 7, 1991.

[Bundy et al, 1990] Bundy, A., van Harmelen, F., Horn, C. and Smaill, A. (1990). The Oyster-Clam system. In Stickel, M. E., (ed.), *10th International Conference on Automated Deduction*, pages 647-648. Springer-Verlag. Lecture Notes in Artificial Intelligence No. 449. Also available from Edinburgh as DAI Research Paper 507.

[Funt, 1973] Funt, B. V. (October 1973). A procedural approach to constructions in Euclidean geometry. Unpublished M.Sc. thesis, University of British Columbia.

[Gow, 1997] Gow, J. (1997). *The Diagonalization Method in Automatic Proof*. Undergraduate project dissertation, Dept of Artificial Intelligence, University of Edinburgh.

[Huang et al, 1995] Huang, X., Kerber, M. and Cheikhrouhou. (1995). Adapting the diagonalization method by reformulations. In Levy, A. and Nayak, P., (eds.), *Proc. of the Symposium on Abstraction, Reformulation and Approximation (SARA-95)*, pages 78-85. Ville d'Esterel, Canada.

[Ireland & Bundy, 1996] Ireland, A. and Bundy, A. (1996). Productive use of failure in inductive proof. *Journal of Automated Reasoning*, 16(1-2):79-111. Also available as DAI Research Paper No 716, Dept. of Artificial Intelligence, Edinburgh.

[Polya, 1965] Polya, G. (1965). *Mathematical discovery*. John Wiley & Sons, Inc, Two volumes.

[Wos & McCune, 1988] Wos, L. and McCune, W. (1988). Searching for fixed point combinators by using automated theorem proving: a preliminary report. Technical Report ANL-88-10, Argonne National Laboratory.

Model Checking: Historical Perspective and Example (Extended Abstract) *

Edmund M. Clarke and Sergey Berezin

Carnegie Mellon University — USA

Abstract. Model checking is an automatic verification technique for finite state concurrent systems such as sequential circuit designs and communication protocols. Specifications are expressed in propositional temporal logic. An exhaustive search of the global state transition graph or system model is used to determine if the specification is true or not. If the specification is not satisfied, a counterexample execution trace is generated if possible. By encoding the model using Binary Decision Diagrams (BDDs) it is possible to search extremely large state spaces with as many as 10^{120} reachable states. In this paper we describe the theory underlying this technique and outline its historical development. We demonstrate the power of model checking to find subtle errors by verifying the Space Shuttle *Three-Engines-Out Contingency Guidance Protocol*.

1 Introduction

Logical errors found late in the design phase are an extremely important problem for both circuit designers and programmers. During the past few years, researchers at Carnegie Mellon University have developed an alternative approach to verification called *temporal logic model checking* [10, 11]. In this approach specifications are expressed in a propositional temporal logic, and circuit designs and protocols are modeled as state-transition systems. An efficient search procedure is used to determine automatically if the specifications are satisfied by the transition systems.

Model checking has several important advantages over mechanical theorem provers or proof checkers for verification of circuits and protocols. The most important is that the procedure is completely automatic. Typically, the user provides a high level representation of the model and the specification to be checked. The model checking algorithm will either terminate with the answer *true*, indicating that the model satisfies the specification, or give a counterexample execution that shows why the formula is not satisfied. The counterexamples

---

* This research is sponsored by the the Semiconductor Research Corporation (SRC) under Contract No. 97-DJ-294, the National Science Foundation (NSF) under Grant No. CCR-9505472, and the Defense Advanced Research Projects Agency (DARPA) under Contract No. DABT63-96-C-0071. Any opinions, findings and conclusions or recommendations expressed in this material are those of the authors and do not necessarily reflect the views of SRC, NSF, DARPA, or the United States Government.

are particularly important in finding subtle errors in complex transition systems. The procedure is also quite fast, and usually produces an answer in a matter of minutes or even seconds. Partial specifications can be checked, so it is unnecessary to specify the circuit completely before useful information can be obtained regarding its correctness. Finally, the logic used for specifications can directly express many of the properties that are needed for reasoning about concurrent systems.

The main disadvantage of this technique is the state explosion which can occur if the system being verified has many components that can make transitions in parallel. Recently, however, the size of the transition systems that can be verified by model checking techniques has increased dramatically. The initial breakthrough was made in the fall of 1987 by McMillan, who was then a graduate student at Carnegie Mellon. He realized that using an explicit representation for transition relations severely limited the size of the circuits and protocols that could be verified. He argued that larger systems could be handled if transition relations were represented implicitly with *ordered binary decision diagrams* (OBDDs) [6]. By using the original model checking algorithm with the new representation for transition relations, he was able to verify some examples that had more than 10^{20} states [9, 21]. He made this observation independently of the work by Coudert, et. al. [12] and Pixley [23–25] on using OBDDs to check equivalence of deterministic finite-state machines. Since then, various refinements of the OBDD-based techniques by other researchers at Carnegie Mellon have pushed the state count up to more than 10^{120} [7].

2 Temporal Logic Model Checking

Pnueli [26] was the first to use temporal logic for reasoning about the concurrent programs. His approach involved proving properties of the program under consideration from a set of axioms that described the behavior of the individual statements in the program. The introduction of temporal logic model checking algorithms in the early 1980's allowed this type of reasoning to be automated. Since checking that a single model satisfies a formula is much easier than proving the validity of a formula for all models, it was possible to implement this technique very efficiently. The first algorithm was developed by Clarke and Emerson in [10]. Their algorithm was polynomial in both the size of the model determined by the program under consideration and in the length of its specification in Computational Tree Logic (CTL). They also showed how *fairness* could be handled without changing the complexity of the algorithm. This was an important step since the correctness of many concurrent programs depends on some type of fairness assumption; for example, absence of starvation in a mutual exclusion algorithm may depend on the assumption that each process makes progress infinitely often.

At roughly the same time Quielle and Sifakis [27] gave a model checking algorithm for a similar branching-time logic, but they did not analyze its complexity or show how to handle an interesting notion of fairness. Later Clarke,

Emerson, and Sistla [11] devised an improved algorithm that was linear in the product of the length of the formula and in the size of the global state graph. Sistla and Clarke [28] also analyzed the model checking problem for a variety of other temporal logics and showed, in particular, that for linear temporal logic the problem was PSPACE complete.

A number of papers demonstrated how the temporal logic model checking procedure could be used for verifying network protocols and sequential circuits ([2], [3], [4], [5], [11], [15], [22]). Early model checking systems were able to check state-transition graphs with between 10^4 and 10^5 states at a rate of about 100 states per second. In spite of these limitations, model checking systems were used successfully to find previously unknown errors in several published circuit designs.

Alternative techniques for verifying concurrent systems were proposed by a number of other researchers. The approach developed by Kurshan [17, 18] was based on checking *inclusion* between two automata. The first machine represented the system that was being verified; the second represented its specification. Automata on infinite tapes (ω-automata) were used in order to handle fairness. Pnueli and Lichtenstein [20] reanalyzed the complexity of checking linear-time formulas and discovered that although the complexity appears exponential in the length of the formula, it is linear in the size of the global state graph. Based on this observation, they argued that the high complexity of linear-time model checking might still be acceptable for short formulas. Emerson and Lei [16] extended their result to show that formulas of the logic CTL*, which combines both branching-time and linear-time operators, could be checked with essentially the same complexity as formulas of linear temporal logic. Vardi and Wolper [29] showed how the model checking problem could be formulated in terms of automata, thus relating the model checking approach to the work of Kurshan.

3 New Implementations

In the original implementation of the model checking algorithm, transition relations were represented explicitly by adjacency lists. For concurrent systems with small numbers of processes, the number of states was usually fairly small, and the approach was often quite practical. Recent implementations [9, 21] use the same basic algorithm; however, transition relations are represented implicitly by *ordered binary decision diagrams* (OBDDs) [6]. OBDDs provide a canonical form for boolean formulas that is often substantially more compact than conjunctive or disjunctive normal form, and very efficient algorithms have been developed for manipulating them. Because this representation captures some of the regularity in the state space determined by circuits and protocols, it is possible to verify systems with an extremely large number of states—many orders of magnitude larger than could be handled by the original algorithm.

The implicit representation is quite natural for modeling sequential circuits and protocols. Each state is encoded by an assignment of boolean values to the set of state variables associated with the circuit or protocol. The transition relation can, therefore, be expressed as a boolean formula in terms of two sets of variables, one set encoding the old state and the other encoding the new. This formula is then represented by a binary decision diagram. The model checking algorithm is based on computing fixed points of *predicate transformers* that are obtained from the transition relation. The fixed points are sets of states that represent various temporal properties of the concurrent system. In the new implementations, both the predicate transformers and the fixed points are represented with OBDDs. Thus, it is possible to avoid explicitly constructing the state graph of the concurrent system.

The model checking system that McMillan developed as part of his Ph.D. thesis is called SMV [21]. It is based on a language for describing hierarchical finite-state concurrent systems. Programs in the language can be annotated by specifications expressed in temporal logic. The model checker extracts a transition system from a program in the SMV language and uses an OBDD-based search algorithm to determine whether the system satisfies its specifications. If the transition system does not satisfy some specification, the verifier will produce an execution trace that shows why the specification is false. The SMV system has been distributed widely, and a large number of examples have now been verified with it. These examples provide convincing evidence that SMV can be used to debug real industrial designs.

4 Related Verification Techniques

A number of other researchers have independently discovered that OBDDs can be used to represent large state-transition systems. Coudert, Berthet, and Madre [12] have developed an algorithm for showing equivalence between two deterministic finite-state automata by performing a breadth first search of the state space of the product automata. They use OBDDs to represent the transition functions of the two automata in their algorithm. Similar algorithms have been developed by Pixley [23–25]. In addition, several groups including Bose and Fisher [1], Pixley [23], and Coudert, et. al. [13] have experimented with model checking algorithms that use OBDDs. Although the results of McMillan's experiments [8, 9] were not published until the summer of 1990, his work is referenced by Bose and Fisher in their 1989 paper [1].

5 Example: Space Shuttle Digital Autopilot

We illustrate the power of model checking to find subtle errors by considering a protocol used by the Space Shuttle. We discuss the verification of the *Three-Engines-Out Contingency Guidance Requirements* using the SMV model checker. The example describes what should be done in a situation where all of the three main engines of the Space Shuttle fail during the ascent. The main task of the *Space Shuttle Digital Autopilot* is to separate the shuttle from the external tank.

This task has many different input parameters, and it is important to make sure that all possible cases and input values are taken into account.

The Digital Autopilot chooses one of the six contingency regions depending on the current flight conditions. Each region uses different maneuvers for separating from the external tank. This involves computing a guidance *quaternion*. Usually, the region is chosen once at the beginning of the contingency and is maintained until separation occurs. However, under certain conditions a change of region is allowed. In this case, it is necessary to recompute the quaternion and certain other output values. Using SMV we were able to find a counterexample in the program for this task. We discovered that when a transition between regions occurs, the autopilot system may fail to recompute the quaternion and cause the wrong maneuver to be made. The guidance program consists of about 1200 lines of SMV code. The number of reachable states is $2 \cdot 10^{14}$, and it takes 60 seconds to verify 40 CTL formulas.

Specifically, the error occurs when a change is made from region 2 to region 1. Region 2 is selected initially if the Shuttle is descending and the dynamic pressure is not safe for attitude independent separation. In this region it is necessary to consider the position of the craft relative to its velocity vector, and the quaternion computed in this region is supposed to minimize the *angle of attack* and the *side slip*. However, if the side slip is too big and the dynamic pressure builds up too quickly, meaning that we do not have enough time to perform the maneuver, then the program performs the transition to region 1 — an attitude independent emergency separation.

In this mode, in contrast to region 2, the current values of the angle of attack and the side slip must be frozen, and the tank will separate as soon as the *angle rates* become relatively small. A special flag called `Freeze_flag` is set to indicate this maneuver. However, the quaternion from region 2 is not recomputed and causes the space shuttle to rotate. This violates the condition that the angle of attack and sideslip should be frozen. Since the part of the specifications we possessed does not indicate whether the `Freeze_flag` has a precedence over the quaternion or not, this situation may lead to an incorrect behavior of the Space Shuttle in a critical situation.

The same example was also verified by Judith Crow at SRI [14] using an explicit state model checker called Murϕ. She had to abstract away many variables to avoid the state explosion problem, and her model was not as complete as ours. She found a similar error in the transition from region 2 to region 1, but for a different variable, which turned out to be correct in our model. Instead, the error shows up in the quaternion, which she didn't consider.

References

1. S. Bose and A. L. Fisher. Automatic verification of synchronous circuits using symbolic logic simulation and temporal logic. In L. Claesen, editor, *Proceedings of the IMEC-IFIP International Workshop on Applied Formal Methods for Correct VLSI Design*, November 1989.

2. M. C. Browne and E. M. Clarke. Sml: A high level language for the design and verification of finite state machines. In *IFIP WG 10.2 International Working Conference from HDL Descriptions to Guaranteed Correct Circuit Designs, Grenoble, France*. IFIP, September 1986.
3. M. C. Browne, E. M. Clarke, and D. Dill. Checking the correctness of sequential circuits. In *Proceedings of the 1985 International Conference on Computer Design*, Port Chester, New York, October 1985. IEEE.
4. M. C. Browne, E. M. Clarke, and D. Dill. Automatic circuit verification using temporal logic: Two new examples. In *Formal Aspects of VLSI Design*. Elsevier Science Publishers (North Holland), 1986.
5. M. C. Browne, E. M. Clarke, D. L. Dill, and B. Mishra. Automatic verification of sequential circuits using temporal logic. *IEEE Transactions on Computers*, C-35(12):1035–1044, December 1986.
6. R. E. Bryant. Graph-based algorithms for boolean function manipulation. *IEEE Transactions on Computers*, C-35(8):677–691, August 1986.
7. J. R. Burch, E. M. Clarke, and D. E. Long. Symbolic model checking with partitioned transition relations. In A. Halaas and P. B. Denyer, editors, *Proceedings of the 1991 International Conference on Very Large Scale Integration*, August 1991. Winner of the Sidney Michaelson Best Paper Award.
8. J. R. Burch, E. M. Clarke, K. L. McMillan, and D. L. Dill. Sequential circuit verification using symbolic model checking. In *Proceedings of the 27th ACM/IEEE Design Automation Conference*, pages 46–51. IEEE Computer Society Press, June 1990.
9. J. R. Burch, E. M. Clarke, K. L. McMillan, D. L. Dill, and L. J. Hwang. Symbolic model checking: 10^{20} states and beyond. *Information and Computation*, 98(2):142–170, June 1992.
10. E. M. Clarke and E. A. Emerson. Synthesis of synchronization skeletons for branching time temporal logic. In D. Kozen, editor, *Logic of Programs: Workshop, Yorktown Heights, NY, May 1981*, volume 131 of *Lecture Notes in Computer Science*. Springer-Verlag, 1981.
11. E. M. Clarke, E. A. Emerson, and A. P. Sistla. Automatic verification of finite-state concurrent systems using temporal logic specifications. *ACM Transactions on Programming Languages and Systems*, 8(2):244–263, April 1986.
12. O. Coudert, C. Berthet, and J. C. Madre. Verification of synchronous sequential machines based on symbolic execution. In J. Sifakis, editor, *Proceedings of the 1989 International Workshop on Automatic Verification Methods for Finite State Systems, Grenoble, France*, volume 407 of *Lecture Notes in Computer Science*. Springer-Verlag, June 1989.
13. O. Coudert, J. C. Madre, and C. Berthet. Verifying temporal properties of sequential machines without building their state diagrams. In Kurshan and Clarke [19].
14. Judith Crow. Finite-state analysis of space shuttle contingency guidance requirements. Technical Report NASA Contractor Report 4741, SRI International, Menlo Park, CA, May 1996.
15. D. L. Dill and E. M. Clarke. Automatic verification of asynchronous circuits using temporal logic. *IEE Proceedings*, Part E 133(5), 1986.
16. E.A. Emerson and Chin Laung Lei. Modalities for model checking: Branching time strikes back. *Twelfth Symposium on Principles of Programming Languages*, New Orleans, La., January 1985.
17. Z. Har'El and R. P. Kurshan. Software for analytical development of communications protocols. *AT&T Technical Journal*, 69(1):45–59, Jan.–Feb. 1990.

18. R. P. Kurshan. Analysis of discrete event coordination. In J. W. de Bakker, W.-P. de Roever, and G. Rozenberg, editors, *Proceedings of the REX Workshop on Stepwise Refinement of Distributed Systems, Models, Formalisms, Correctness*, volume 430 of *Lecture Notes in Computer Science*, pages 414–453. Springer-Verlag, May 1989.
19. R. P. Kurshan and E. M. Clarke, editors. *Proceedings of the 1990 Workshop on Computer-Aided Verification*. Springer-Verlag, June 1990.
20. O. Lichtenstein and A. Pnueli. Checking that finite state concurrent programs satisfy their linear specification. In *Proceedings of the Twelfth Annual ACM Symposium on Principles of Programming Languages*, pages 97–107. Association for Computing Machinery, January 1985.
21. K. L. McMillan. *Symbolic Model Checking: An Approach to the State Explosion Problem*. PhD thesis, Carnegie Mellon University, 1992.
22. B. Mishra and E.M. Clarke. Hierarchical verification of asynchronous circuits using temporal logic. *Theoretical Computer Science*, 38:269–291, 1985.
23. C. Pixley. A computational theory and implementation of sequential hardware equivalence. In R. Kurshan and E. Clarke, editors, *Proc. CAV Workshop (also DIMACS Tech. Report 90-31)*, Rutgers University, NJ, June 1990.
24. C. Pixley, G. Beihl, and E. Pacas-Skewes. Automatic derivation of FSM specification to implementation encoding. In *Proceedings of the International Conference on Computer Desgin*, pages 245–249, Cambridge, MA, October 1991.
25. C. Pixley, S.-W. Jeong, and G. D. Hachtel. Exact calculation of synchronization sequences based on binary decision diagrams. In *Proceedings of the 29th Design Automation Conference*, pages 620–623, June 1992.
26. A. Pnueli. A temporal logic of concurrent programs. *Theoretical Computer Science*, 13:45–60, 1981.
27. J.P. Quielle and J. Sifakis. Specification and verification of concurrent systems in CESAR. In *Proceedings of the Fifth International Symposium in Programming*, 1982.
28. A. P. Sistla and E.M. Clarke. Complexity of propositional temporal logics. *Journal of the ACM*, 32(3):733–749, July 1986.
29. M. Y. Vardi and P. Wolper. An automata-theoretic approach to automatic program verification. In *Proceedings of the First Annual Symposium on Logic in Computer Science*. IEEE Computer Society Press, June 1986.

Comparison of Theorem Provers for Modal Logics – Introduction and Summary

Peter Balsiger and Alain Heuerding*

IAM, University of Bern, Switzerland

Abstract. The Tableaux 98 conference included a comparison of automated theorem provers for some modal logics. Our aim was to make the existing provers better known and to show what possibilities they offer. This comparison included benchmarks for the propositional modal logics K, KT and S4. Although efficiency is an important aspect, depending on the intended application other qualities can be as important, such as portability, construction of counter-models, user-friendliness, or small size.
We first discuss our aims in more detail, explain the applied benchmark method, and finally give a short summary of the results. The submissions of the participants follow in alphabetic order.

During the last years, there has been considerable progress in the area of theorem provers for modal logics, and various methods have been proposed and implemented. One aim of this comparison is to make the available provers known to a wider audience. Others can profit from these experiences, such that they can come up with 'better' provers instead of reinventing the wheel.

The apostrophes above already indicate that it is not at all clear when a prover is better than another one. Efficiency is probably the criterion that comes first to ones mind, and efficiency was indeed an important part of this comparison. But first we would like to show that there are many other aspects that can be as important.

To achieve good portability and maintainability is certainly harder for a large prover than for a prover with few lines of code. Moreover, every optimisation makes a prover more error-prone. If an application does not require a very efficient prover, a small prover might therefore be preferable to a fast and intricate one.

Although we concentrated on three logics in this comparison, it is of course an advantage if the same prover can deal with other logics as well, e.g. with further propositional modal logics or with extensions of K. Again this can be a trade-off: If we use a specialised prover for each logic, it can be tuned, but then each prover has to be checked for errors separately.

A decision procedure is preferable to a non-complete prover. Sometimes the user would like even more information, e.g. be a proof if the formula is provable and otherwise a counter-model. Provided that the model is not too large, then the

---
* Work supported by the Swiss National Science Foundation, 21-43197.95

latter can be helpful to understand why a formula is not provable. For educational purposes also the user-friendliness is an important issue.

All submissions include the results of the respective prover on a set of benchmark formulas. There is a large number of modal logics. We confined ourselves to the three propositional modal logics K, KT, S4, as they seem to be among the most widespread ones. Such a restriction is always somewhat arbitrary.

For each of the logics K, KT, S4 there were nine provable and nine unprovable parametrised formulas (with names ending in 'p' and 'n', respectively). Let $A(n)$ be one of these parametrised formulas. The participants had to decide for which n they could decide in less than 100 seconds whether the formula is provable or not. Example: The number 6 for the parametrised formula k_branch_p means that the prover returned the correct result for the formula $k\_branch\_p(6)$ in less than 100 seconds, and that it took more than 100 seconds to compute the result for $k\_branch\_p(7)$. (The formulas will still be available after the conference via http://lwbwww.unibe.ch:8080/LWBinfo.html; there you will also find a technical report concerning these benchmark formulas.) We did not try to standardise the hardware, but because of the form of the benchmark this should not influence the results too much.

One property of the benchmark method is that the result consists of relatively few numbers, but it is still not easy to decide whether one prover is faster than another. Since the scalable formulas have different characteristics, a prover can be very fast in one case and slow in another, and often there is a considerable difference between provable and unprovable formulas.

The results show that the time when it was hard for a prover to decide whether a formula like $\Box(\Box(p \to \Box p) \to p) \to (\Diamond \Box p) \to p$ is provable in S4 is over. Just to give an impression we display the formula $k\_lin\_p(3)$, which proved to be easy for all the participants.

$\neg(\Box((p_1 \land \Box p_1 \land p_1 \to p_2) \lor (\neg p_1 \to \neg(\Box p_2 \land p_2))) \land \Box(\Box(p_1 \land \Box p_1 \land p_1 \to p_2) \lor (\neg p_1 \to \neg(\Box p_2 \land p_2))) \land \Box((p_1 \land \Box p_1 \land p_1 \to p_2) \lor \Box(\neg p_1 \to \neg(\Box p_2 \land p_2))) \to \Box(p_1 \land \Box p_1 \land p_1 \to p_2) \lor \Box(\neg p_1 \to \neg(\Box p_2 \land p_2))) \lor (\Box(p_3 \land \Box p_3 \to p_3) \lor \Box(p_3 \land \Box p_3 \to p_3)) \lor (\neg(\Box((p_2 \land \Box p_2 \land p_2 \to p_3) \lor (\neg p_2 \to \neg(\Box p_3 \land p_3))) \land \Box(\Box(p_2 \land \Box p_2 \land p_2 \to p_3) \lor (\neg p_2 \to \neg(\Box p_3 \land p_3))) \land \Box((p_2 \land \Box p_2 \land p_2 \to p_3) \lor \Box(\neg p_2 \to \neg(\Box p_3 \land p_3))) \to \Box(p_2 \land \Box p_2 \land p_2 \to p_3) \lor \Box(\neg p_2 \to \neg(\Box p_3 \land p_3))) \lor \neg(\Box((p_3 \land \Box p_3 \land p_3 \to p_4) \lor (\neg p_3 \to \neg(\Box p_4 \land p_4))) \land \Box(\Box(p_3 \land \Box p_3 \land p_3 \to p_4) \lor (\neg p_3 \to \neg(\Box p_4 \land p_4))) \land \Box((p_3 \land \Box p_3 \land p_3 \to p_4) \lor \Box(\neg p_3 \to \neg(\Box p_4 \land p_4))) \to \Box(p_3 \land \Box p_3 \land p_3 \to p_4) \lor \Box(\neg p_3 \to \neg(\Box p_4 \land p_4))))$

Another conclusion from the comparison is that there is no method that clearly outperforms all the others. It seems that at the moment not the method alone decides whether or not a prover is fast, but first and foremost the way it is implemented and the optimisations that are used.

The good results of some provers should not prevent people from implementing a new prover; as we have outlined above, efficiency is only one aspect. However, we think that a prover, provided that is not designed for a specific class of formulas, should not be announced as 'very fast' or 'state-of-the-art' or ..., unless it can compete with the participants of this comparison.

Last but not least we would like to thank Roy Dyckhoff for his help.

FaCT and DLP

Ian Horrocks[1,2] and Peter F. Patel-Schneider[3]

[1] Medical Informatics Group, Department of Computer Science,
University of Manchester, Manchester M13 9PL, UK
horrocks@cs.man.ac.uk
[2] IRST, Istituto per la Ricerca Scientifica e Tecnologica, I-38050 Povo TN, ITALY
[3] Bell Labs Research, Murray Hill, NJ, U.S.A.
pfps@research.bell-labs.com

FaCT: The tests were performed using FaCT version 1.2. FaCT is a description logic classifier whose description language is a superset of $\mathbf{K4_{(m)}}$ and whose subsumption reasoning uses a sound and complete tableaux algorithm. FaCT employs a wide range of optimisations, in particular a form of dependency directed backtracking called *backjumping* which can significantly reduce the size of the search space [5]. The FaCT algorithm does not support **KT** and **S4** explicitly, but FaCT includes a preprocessing and encoding optimisation which is also able to apply the standard embedding of **KT** and **S4** in **K** and **K4** respectively: the time taken for preprocessing and embedding is included in the results. Programming language: Common Lisp (compiled).

DLP: The ideas in FaCT are being incorporated into a new generation of Description Logic systems. Initial experiments in this effort have resulted in a modal prover for a superset of $\mathbf{K4_{(m)}}$, which has provisionally been called DLP. The DLP prover has control over several options, including backjumping and caching partial results. Both of these mechanisms have proved to be very useful in the benchmarks, with caching being the more powerful. As an experimental prover, there are essentially no user amenities in DLP, but the final Description Logic system will have a full user interface and other amenities. Programming language: ML (compiled).

The other provers: For comparative purposes the tests for **K** and **KT** were repeated using three other available provers: Crack version 1.0 beta 15 [3], KSAT [4] and Kris [2,1]. Crack and Kris are also description logic classifiers which use sound and complete tableaux algorithms while KSAT is a $\mathbf{K_{(m)}}$ prover which uses an algorithm based on propositional satisfiability (SAT) testing. None of these systems supports transitive relations so they could not be used for **S4**. The **KT** tests were performed by using the standard embedding of **KT** in **K**: the time taken for the embedding is not included in the results for these systems.

All three systems are programmed in Common Lisp (compiled). It should be pointed out that neither Crack nor Kris are intended as stand-alone **K** provers and for many classes of formula a significant improvement in their performance could be achieved by preprocessing and encoding large formulae, a technique which is used by both FaCT and KSAT. Both Crack and Kris support much richer logics than **K** (for example Crack can reason about converse relations) and can also reason about nominals (individuals).

Availability: The sources for FaCT are available from the first authors home page: *http://www.cs.man.ac.uk/ horrocks*; the DLP prover is currently under development, but the benchmark version and full timing results are also available from the same location. Contacts for information about the other systems are:

Crack — Enrico Franconi, *franconi@irst.itc.it*;
KSAT — Roberto Sebastiani, *rseba@irst.itc.it*;
Kris   — H.-J. Burckert, *hjb@dfki.uni-sb.de*.

Advantages: FaCT has been tested using several Common Lisps including GNU Lisp and should thus be highly portable. As well as **K**, **KT** and **S4** it can also deal with **K4**. The implemented logic is significantly more expressive than **S4**: it includes support for a hierarchy of multiple modalities (roles), functional roles and global axioms. DLP should also be highly portable: the ML compiler runs on a variety of platforms and is freely available from several sites, including *http://cm.bell-labs.com/cm/cs/what/smlnj*.

Hardware and Software: For DLP: SPARC clone; main memory 132MB; 150 MHz Ross RT626 CPU; SML-NJ compiler, version 109.32. For the other provers: Sun Ultra 1; main memory 32MB; 147 MHz CPU; Solaris; Allegro CL 4.3.

Results: To demonstrate the effectiveness of the backjumping optimisation the tests were also performed using FaCT with backjumping disabled: the resulting prover is referred to as FaCT*. The results of the tests are given in Tables 1, 2 and 3. Both FaCT and DLP performed reasonably well with all classes of **K** and **KT** formula, trivially solving most of the **K** formulae, and in the case of DLP many of the **KT** and **S4** formulae.

FaCT and DLP significantly outperformed all the other provers, and in many cases they also exhibited a completely different qualitative performance. For example, with *k_dum_p* the other provers all show an exponential increase in solution times with increasing formula size, whereas the times taken by FaCT and DLP increase very little for larger formulae (and FaCT is already 2,000 times faster for the largest formula solved by another system).

The results for FaCT* demonstrate that backjumping accounts for a significant proportion of FaCT's performance advantage over the other systems, particularly with respect to provable formulae, and experiments with DLP suggest that caching is even more effective. However the performance of FaCT* still compares favourably with that of the other systems and it still exhibits a

Table 1. Results for **K**, **KT** and **S4**

Formulae	FaCT		FaCT*		DLP		Crack		KSAT		Kris	
	p	n	p	n	p	n	p	n	p	n	p	n
k_branch_	6	4	3	3	13	11	2	1	8	8	3	3
k_d4_	>20	8	15	8	>20	>20	2	3	8	5	8	6
k_dum_	>20	>20	15	>20	>20	>20	3	>20	11	>20	15	>20
k_grz_	>20	>20	8	>20	>20	>20	1	>20	17	>20	13	>20
k_lin_	>20	>20	7	>20	>20	>20	5	2	>20	3	6	9
k_path_	7	6	5	5	>20	>20	2	6	4	8	3	11
k_ph_	6	7	5	6	6	8	2	3	5	5	4	5
k_poly_	>20	>20	>20	>20	>20	>20	>20	>20	13	12	11	>20
k_t4p_	>20	>20	>20	>20	>20	>20	1	1	10	18	7	5

Table 2. Results for **KT**

Formulae	FaCT		FaCT*		DLP		Crack		KSAT		Kris	
	p	n	p	n	p	n	p	n	p	n	p	n
kt_45_	>20	>20	13	>20	>20	>20	0	0	5	5	4	3
kt_branch_	6	4	3	3	16	11	2	2	8	7	3	3
kt_dum_	11	>20	8	>20	>20	>20	0	1	7	12	3	14
kt_grz_	>20	>20	5	>20	>20	>20	0	0	9	>20	0	5
kt_md_	4	5	3	5	3	>20	2	4	2	4	3	4
kt_path_	5	3	2	2	6	>20	1	5	2	5	1	13
kt_ph_	6	7	4	5	7	18	2	2	4	5	3	3
kt_poly_	>20	7	>20	6	6	6	1	1	1	2	2	2
kt_t4p_	4	2	1	1	3	>20	0	1	1	1	1	7

Table 3. Results for **S4**

Formulae	FaCT		FaCT*		DLP	
	p	n	p	n	p	n
s4_45_	>20	>20	8	>20	>20	>20
s4_branch_	4	4	2	2	10	8
s4_grz_	2	>20	0	>20	9	>20
s4_ipc_	5	4	4	4	10	>20
s4_md_	8	4	3	4	3	>20
s4_path_	2	1	2	1	3	>20
s4_ph_	5	4	4	3	7	18
s4_s5_	>20	2	2	2	3	>20
s4_t4p_	5	3	1	1	>20	>20

different qualitative performance in some cases (e.g. k_lin_p). DLP is more effective with non-provable formulae, and for some classes of provable formulae it is outperformed by FaCT; this phenomenon is the subject of continuing research.

A well engineered C code implementation of KSAT is now available, and has been observed to outperform the Lisp version by a significant margin (as much as 100 times). It is likely that significant improvements to the performance of FaCT and DLP could also be achieved by employing more sophisticated software engineering.

References

1. F. Baader, E. Franconi, B. Hollunder, B. Nebel, and H.-J. Profitlich. An empirical analysis of optimization techniques for terminological representation systems. In B. Nebel, C. Rich, and W. Swartout, editors, *Principals of Knowledge Representation and Reasoning: Proceedings of the Third International Conference (KR'92)*, pages 270–281. Morgan-Kaufmann, 1992. Also available as DFKI RR-93-03.
2. F. Baader and B. Hollunder. A terminological knowledge representation system with complete inference algorithms. In *Processing declarative knowledge: International workshop PDK'91*, number 567 in Lecture Notes in Artificial Intelligence, pages 67–86, Berlin, 1991. Springer-Verlag.
3. P. Bresciani, E. Franconi, and S. Tessaris. Implementing and testing expressive description logics: a preliminary report. In Gerard Ellis, Robert A. Levinson, Andrew Fall, and Veronica Dahl, editors, *Knowledge Retrieval, Use and Storage for Efficiency: Proceedings of the First International KRUSE Symposium*, pages 28–39, 1995.
4. F. Giunchiglia and R. Sebastiani. A SAT-based decision procedure for $\mathcal{ALC}$. In L. C. Aiello, J. Doyle, and S. Shapiro, editors, *Principals of Knowledge Representation and Reasoning: Proceedings of the Fifth International Conference (KR'96)*, pages 304–314. Morgan Kaufmann, November 1996.
5. I. Horrocks. *Optimising Tableaux Decision Procedures for Description Logics*. PhD thesis, University of Manchester, 1997.

Prover KT4

Michel Levy

Laboratoire L.S.R.,
B.P.72, 38041 St Martin d'Hères, France

1 Prover

The proposed prover implements a decision procedure for the propositionnal logic KT4. It is based on a paper of Laurent Catach (see [1]).
The code can be loaded at the adress :
ftp://ftp.imag.fr/pub/PLIAGE/Michel.Levy/prover.bin
The program runs under the operating system Solaris. It is written in the Ocaml language (see [2]) and is compiled by the native-code compiler **ocamlopt**.

2 Algorithms

An assumption is a pair state, formula whose intuitive meaning is : I assume that the formula is true in its associated state.
  Each time you apply a rule to an assumption,this assumption receives a mark. A tableau is a list of assumptions and a relation between the states.

2.1 not modal rules

Let be A and B two formulae and e a state. When we apply a rule to the not marked assumption **e : (A or B)** we create two copies of the tableau containing the assumption, one copy receiving the new assumption **e : A** and another the new assumption **e : B**.

2.2 box rule

Let R the relation associated with a tableau containing the not marked assumption **e : (box A)**. When we apply a rule to this assumption, we add to the tableau the assumptions **f : A** for every state f such that e R* f, where R* is the transitive and reflexive closure of the R relation.

2.3 dia rule

In a given tableau, it's possible to apply a rule to an assumption of the form **e : (dia A)** only **if no other** rule can be applied. Let us consider such a tableau and such an assumption, R being the relation associated with the tableau.
  Let E the following set of formulae : B is member of E if and only if B is A or if the tableau has a already marked assumption **f : (dia B)** with f R* e. We have two cases :

1. We say that a formula B is in the state g, if the tableau contains an assumption **g** : **B**. If there exists a state g such that E is included in the the set of the formulae in the state g, we add the edge (e, g) to the relation r.
2. If such a state does not exist, we add a new state g, the edge (e,g) and every assumption **g** : **B** such that B is member of E.

3 Advantages of the prover

1. The source code, written in Ocaml [2], is short (less than 860 lines). So it's easy to maintain the prover.
2. Not only the prover test the validity of the formulae, but for a not-valid formula, it gives an counter-model of the formula.

4 Results

With the exception of the s4_md_n and s4_md_p formulae, my prover is less efficient that the LWB-prover. This exception is easy to explain : my prover reduce the modalities using the identities valid in KT4 : box box = box, dia dia = dia, dia box dia box = dia box. I give the results with the LWB-presentation, each filename of the benchmark is followed by the number of formulae proved (or disproved) in less that 100 seconds.

s4_45_p	1	s4_45_n	6
s4_branch_p	2	s4_branch_n	3
s4_grz_p	0	s4_grz_n	17
s4_ipc_p	5	s4_ipc_n	8
s4_md_p	21	s4_md_n	18
s4_path_p	1	s4_path_n	2
s4_ph_p	2	s4_ph_n	2
s4_s5_p	2	s4_s5_n	2
s4_t4p_p	0	s4_t4p_n	3

References

1. Catach,L. :
   TABLEAUX, A general Theorem Prover for Modal Logic.
   J.A.R.7(1991)
2. Leroy,X. :
   The Objective Caml system, release 1.03, Documentation and user's manual.
   I.N.R.I.A.(1996)

leanK 2.0

Bernhard Beckert[1] and Rajeev Goré[2]

[1] University of Karlsruhe, Institute for Logic, Complexity and Deduction Systems,
D-76128 Karlsruhe, Germany. E-mail: `beckert@ira.uka.de`
[2] Automated Reasoning Project, Australian National University,
Canberra, ACT, 0200, Australia. Email: `rpg@arp.anu.edu.au`

The Prover

leanK 2.0 implements an extension of the "Free Variable Tableaux for Propositional Modal Logics" reported by us in [1]. It performs depth first search and is based upon the original leanT^AP prover of Beckert and Posegga [2]. Formulae annotated with labels containing variables capture the universal and existential nature of the box and diamond modalities, respectively, with different variable bindings closing different branches. Prolog's built-in clause indexing scheme, unification facilities and built-in backtracking are used extensively.

In its new version, leanK's calculus includes additional methods for restriction the search space, which turn it into a decision procedure for the logics K, KT, and S4.

Availability

The source code for leanK is available at `http://i12www.ira.uka.de/modlean` on the *World Wide Web*.

Advantages

The main advantages of leanK are its modularity, its small size and its versatility. Minimal changes in the rules give provers for all the 15 basic normal modal logics. By sacrificing modularity we can obtain specialised (faster) provers for particular logics like K45D, G and Grz. It is easy to obtain an explicit counter-example from a failed proof attempt.

Programming Language, Operating System, Hardware

leanK is implemented in Prolog; we used SICStus Prolog 3, but leanK can easily be adapted to other Prolog dialects. For obtaining the results shown here, we used a Sun Ultra 1 Model 170 with 128 MB main memory, running under the Solaris operating system.

Results

The strength of leanK clearly is its small size and adaptability and not its performance. Nevertheless, leanK is able to solve at least a few formulae in most classes. With the exception of k_grz, leanK's performance for provable formulae is better than that for non-provable formulae.

Since leanK's calculus is better suited for serial logics, the results of the KD version for the provable K examples are shown as well.

	leanK -p	leanK _n	leanKD -p
k_branch	1	0	1
k_d4	1	1	> 20
k_dum	0	0	0
k_grz	0	> 20	2
k_lin	> 20	4	> 20
k_path	2	0	2
k_ph	3	1	2
k_poly	2	0	3
k_t4p	0	0	1

	leanKT -p	leanKT _n
kt_45	3	0
kt_branch	1	0
kt_dum	3	4
kt_grz	0	0
kt_md	3	2
kt_path	2	1
kt_ph	2	1
kt_poly	0	0
kt_t4p	> 20	0

	leanS4 -p	leanS4 _n
s4_45	0	0
s4_branch	0	0
s4_grz	0	0
s4_ipc	1	1
s4_md	2	2
s4_path	1	0
s4_ph	1	0
s4_s5	1	1
s4_t4p	0	0

References

1. Bernhard Beckert and Rajeev Goré. Free Variable Tableaux for Propositional Modal Logics. In *Proceedings, International Conference on Theorem Proving with Analytic Tableaux and Related Methods, Pont-a-Mousson, France*, pages 91–106. Springer, LNCS 1227, 1997.
2. Bernhard Beckert and Joachim Posegga. leanT^{AP}: Lean, Tableau-based Deduction. *Journal of Automated Reasoning*, Vol. 15, No. 3, pages 339–358, 1995.

Logics Workbench 1.0

Peter Balsiger, Alain Heuerding, and Stefan Schwendimann

IAM, University of Bern, Switzerland
lwb@iam.unibe.ch
http://lwbwww.unibe.ch:8080/LWBinfo.html

Prover: Logics Workbench (LWB), version 1.0. See the LWB home page for more information.

The LWB does backward proof search in two-sided sequent calculi. In the case of S4 we use a loop-check in order to ensure termination. With 'use-check' we cut off unnecessary branches generated by invertible rules with two premises; if e.g. in the proof of $\Delta \supset A, \Gamma$ the formula A is not 'used', then we know that $\Delta \supset A \wedge B, \Gamma$ is provable as well. Duplicate formulas are deleted. Structure sharing helps to reduce the copying of formulas and sequents. No heuristics. Programming language: C++ . Compiler: Sun C++ 4.0.1 . Operating system: Solaris 2.4 .

Availability: The binaries of the LWB 1.0 are available via the LWB home page (choose about the LWB, install the LWB).

You can also use the LWB 1.0 via WWW. Choose run a session via WWW on the LWB home page and type in your request.

Additional facilities of the prover: Graphical user interface, built-in programming language, progress indicator (a slider shows how the proof search is going on), trace of the proof search is available, various functions to convert formulas.

Hardware: Sun SPARCstation 5, main memory: 80MB, 1 CPU (70 MHz microSPARC II)

Timing: The timing includes parsing of the formulas and the construction of the corresponding data structure. The files loaded by the LWB have the following form: load(s4); timestart; provable(box p0 -> box box p0); timestop; quit;

Results:

class	p	n
k_branch_...	6	7
k_d4_...	8	6
k_dum_...	13	19
k_grz_...	7	13
k_lin_...	11	8
k_path_...	12	10
k_ph_...	4	8
k_poly_...	8	11
k_t4p_...	8	7

class	p	n
kt_45_...	5	4
kt_branch_...	5	6
kt_dum_...	5	10
kt_grz_...	6	> 20
kt_md_...	5	5
kt_path_...	10	9
kt_ph_...	4	8
kt_poly_...	14	2
kt_t4p_...	5	7

class	p	n
s4_45_...	3	5
s4_branch_...	11	7
s4_grz_...	9	> 20
s4_ipc_...	8	7
s4_md_...	8	6
s4_path_...	8	6
s4_ph_...	4	8
s4_s5_...	4	9
s4_t4p_...	9	12

Optimised Functional Translation and Resolution

Ullrich Hustadt[1], Renate A. Schmidt[1], and Christoph Weidenbach[2]

[1] Department of Computing, Manchester Metropolitan University,
Chester Street, Manchester M1 5GD, United Kingdom
{U.Hustadt, R.A.Schmidt}@doc.mmu.ac.uk

[2] Max-Planck-Institut für Informatik, Im Stadtwald, 66123 Saarbrücken, Germany
weidenb@mpi-sb.mpg.de

Prover: We facilitate modal theorem proving in a first-order resolution calculus implemented in SPASS Version 0.77 [4]. SPASS uses ordered resolution and ordered factoring, it supports splitting and branch condensing (splitting amounts to case analysis while branch condensing resembles branch pruning in the Logics Workbench), it has an extensive set of reduction rules including tautology deletion, subsumption and condensing, and it supports dynamic sort theories by additional inference and reduction rules.

The translation we use is the *optimised functional translation* [2]. It maps normal propositional modal logics into a class of *path logics*. Path logics are clausal logics over the language of the monadic fragment of sorted first-order logic with a special binary function symbol for defining accessibility. Clauses of path logics are restricted in that only Skolem terms which are constants may occur and the prefix stability property holds. Ordinary resolution without any refinement strategies is a decision procedure for the path logics associated with $K(m)$ and $KT(m)$ [3]. Our decision procedure for $S4$ uses an a priori term depth bound.

Availability: SPASS and a routine for the translation of modal formulae are available from http://www.mpi-sb.mpg.de/~hustadt/mdp

Advantages of the prover: SPASS is a fast and sophisticated state-of-the-art first-order theorem prover. Ordered inference rules and splitting are of particular importance when treating satisfiable formulae, while unit propagation and branch condensing are important for benchmarks based on randomly generated modal formulae [1].

Advantages of translation approaches: In its most general form the translation approach can deal with any complete, finitely axiomatizable, normal modal logic. Moreover, any first-order theorem prover can be used, that is, we may substitute SPASS with another theorem prover (not necessarily a resolution theorem prover). The relational and optimised functional translation approach are refinements of the general translation approach towards efficient modal theorem proving. The optimised functional translation is applicable to many propositional modal logics, including K, KT, and S4 and their multi-modal versions, but notably also to some second-order modal logics, like KM [2]. A general result shows that any first-order resolution theorem prover (with condensing) provides a decision procedure for a variety of modal logics [3].

Hardware: Sun Ultra 1 Model 170E (167 MHz UltraSPARC processor, 512 KB second level cache), 192 MB main memory.

Results: On classes of provable formulae (in the first column), the combination of the optimised functional translation approach and SPASS has little difficulty. Notable exceptions are the classes k_ph_p, kt_ph_p, $s4\_ph\_p$, k_branch_p and $s4\_ipc\_p$. Observe that SPASS solves more formulae in $s4\_branch\_p$ than in either kt_branch_p or k_branch_p. While for the basic modal logic, the classes of non-provable formulae (in the second column) are not harder than the classes of provable formulae, we see a noticeable difference between kt_dum_n, kt_poly_n, and kt_t4p_n and the corresponding classes of provable KT-formulae. However, the results are still acceptable. In contrast, for the classes $s4\_45\_n$, $s4\_grz\_n$, $s4\_s5\_n$, and $s4\_t4p\_n$ of non-provable S4-formulae the performance is unsatisfactory. We attribute this to using superposition and not E-unification, and to enforcing termination by an explicit term depth bound instead of a loop check. For the classes $s4\_45\_n$ and $s4\_s5\_n$ there are trivial satisfiability checks on the clause level which are not implemented in SPASS.

k_branch_p	9	k_branch_n	9
k_d4_p	> 20	k_d4_n	18
k_dum_p	> 20	k_dum_n	> 20
k_grz_p	> 20	k_grz_n	> 20
k_lin_p	> 20	k_lin_n	> 20
k_path_p	20	k_path_n	20
k_ph_p	6	k_ph_n	9
k_poly_p	16	k_poly_n	17
k_t4p_p	> 20	k_t4p_n	19
kt_45_p	17	kt_45_n	6
kt_branch_p	13	kt_branch_n	9
kt_dum_p	17	kt_dum_n	9
kt_grz_p	> 20	kt_grz_n	> 20
kt_md_p	16	kt_md_n	20
kt_path_p	> 20	kt_path_n	16
kt_ph_p	5	kt_ph_n	12
kt_poly_p	16	kt_poly_n	3
kt_t4p_p	> 20	kt_t4p_n	7
$s4\_45\_p$	9	$s4\_45\_n$	0
$s4\_branch\_p$	> 20	$s4\_branch\_n$	4
$s4\_grz\_p$	14	$s4\_grz\_n$	0
$s4\_ipc\_p$	6	$s4\_ipc\_n$	> 20
$s4\_md\_p$	9	$s4\_md\_n$	10
$s4\_path\_p$	15	$s4\_path\_n$	> 20
$s4\_ph\_p$	5	$s4\_ph\_n$	5
$s4\_s5\_p$	> 20	$s4\_s5\_n$	1
$s4\_t4p\_p$	11	$s4\_t4p\_n$	0

Table 1: Performance indices

References

1. U. Hustadt and R. A. Schmidt. On evaluating decision procedures for modal logics. In *Proc. IJCAI'97*, pages 202–207, 1997.
2. H. J. Ohlbach and R. A. Schmidt. Functional translation and second-order frame properties of modal logics. Res. Report MPI-I-95-2-002, MPI f. Informatik, Saarbrücken, 1995.
3. R. A. Schmidt. Resolution is a decision procedure for many propositional modal logics: Extended abstract. To appear in *Proc. AiML'96*, 1997.
4. C. Weidenbach, B. Gaede, and G. Rock. SPASS & FLOTTER version 0.42. In *Proc. CADE-13*, LNAI 1104, pages 141–145, 1996.

Benchmark Evaluation of □KE

Jeremy Pitt

Department of Electrical & Electronic Engineering,
Imperial College of Science Technology & Medicine, England
jvp@ee.ic.ac.uk   http://www-ics.ee.ic.ac.uk/

Prover: We used □KE (no version number), which was designed to be a generic theorem prover for the family of 15 normal model logics.

□KE is based on the calculus KE [1] and Fitting's prefixed tableaux [2], where the calculus is extended to include elimination rules for the □ and ◇ operators, and the prefixes are generalized to include variables. The type of any variable is determined by the logic and the extension to a prefix introduced by the □ and ◇ rules is also contingent on the logic. Prefixes must unify when applying KE's β and closure rules: although there are some extra side-conditions on the closure rule for certain logics, the unification algorithm remains constant for all logics. The paper [3] contains the details.

For all the logics, □KE uses depth first proof search. No optimizations are used. Programming language: ICL/ECRC ECL^iPSE^e Constraint Prolog 3.5.2 (compiled). Operating system: Solaris.

Availability: The sources are not available except via request from the author.

Advantages of the prover: □KE has no distinguishing features with regards to its size, speed, efficacity, correctness, portability, and maintenance.

The total size of all the code is approximately 52K, with an extra 27K if the modules for recording a proof tree and displaying the output in HTML or LaTeX are required. It is implemented in Prolog, with all the concomitant implications for speed and efficiency that this entails. The system is intended to provide coverage for all the normal modal logics, but there do appear to be types of problem for which the system is not best suited. The system is based on a calculus, as specified in [3], from which the (any) implementation could be developed: we do not have soundness and completeness proofs for this calculus (irrespective of any guarantee that these still hold in an implementation based on an incomplete inference engine). We would currently only commit ourselves to the intuitionistic statement that the system appears at best to be not unsound. Although □KE has been implemented in a "standard" programming language, one attempt at porting to a different Prolog platform was not quickly successful and was abandoned. The process of iterative and incremental development, which this system has undergone, would suggest, that in its current state, maintenance by any persons other than the current implementors would most likely be problematic. On the other hand, the system was aimed at covering a number of logics without loss of generality, and is intended for "real" applications rather than simply verifying a theoretical formulation. The utility of □KE is therefore threefold. Firstly, it handles the entire range of normal modal logics, using a generic decision procedure and unification algorithm, with logic-dependent side conditions on the

□, ◇, and closure rules, and very few other exceptions. Secondly, it achieves reasonable results on all three sets of benchmark tests which is suggestive that it can be used for non-trivial problems. Finally, it provides evidence that the approach based on the generalization of Fitting's prefixed tableau is sound, and valuable experience for future re-engineering.

Two other comments are worth making. Firstly, the proof (or non-proof) trees that □**KE** can be set to "dump" during runtime, or can construct for inspection afterwards, are reasonably perspicuous and a significant aid to explanation and discovery. A decent user interface would help considerably though. Secondly, these benchmarks only tested time, and not space. In previous work [4] we evaluated the first order theorem prover leanKE with another based on the tableau method: the results suggested that the simpler branching rules of the **KE** calculus made it more space-efficient, especially as the test problems became harder. It would be interesting to know if this also applies to the modal logic cases.

Hardware: Sun SPARC-5, main memory: 64MB, 110MHz microSPARC-2 CPU.

Results: The benchmark results for speed of execution are a more or less "normal distribution" that might be expected from an experimental system on 'unseen' problems, without any attempt to implement any optimizations. It appears to be moderate at some types of problem and rather less good at others. In nearly all cases the performance for the provable formulas was better then for the non-provable formulas, the exception being the Grz formulas. In general, it does better on K than on either KT or S4, and marginally better on KT than S4.

k_*_p		k_*_n		kt_*_p		kt_*_n		$s4\_*\_p$		$s4\_*\_n$	
branch	13	branch	3	45	14	45	2	45	8	45	0
d4	13	d4	3	branch	16	branch	15	branch	>20	branch	>20
dum	4	dum	4	dum	1	dum	1	grz	0	grz	>20
grz	3	grz	1	grz	0	grz	>20	ipc	6	ipc	4
lin	>20	lin	2	md	4	md	4	md	3	md	3
path	17	path	5	path	16	path	6	path	9	path	6
ph	4	ph	3	ph	4	ph	3	ph	4	ph	3
poly	17	poly	0	poly	0	poly	0	s5	1	s5	>20
t4p	0	t4p	3	t4p	7	t4p	17	t4p	3	t4p	1

Acknowledgements: Thanks are due to Professor Gerhard Jäger, Alain Heuerding, and other IAM personnel for the opportunity to work at the University of Berne, Switzerland, where some of this benchmarking work was undertaken.

References

1. M. D'Agostino and M. Mondadori. The Taming of the Cut. *Journal of Logic and Computation*, 4:285–319, 1994.
2. M. Fitting. Basic modal logic. In D. Gabbay, C. Hogger, & J. Robinson, eds., *Handbook of Logic in AI and Logic Programming, vol.1*, pp368–448. OUP, 1993.
3. J. Pitt and J. Cunningham. Distributed modal theorem proving with **KE**. In P. Miglioli, U. Moscato, D. Mundici, & M. Ornaghi, eds., *Theorem Proving with Analytic Tableaux and Related Methods*, LNAI1071, pp160–176. Springer-Verlag, 1996.
4. J. Pitt and J. Cunningham. Theorem proving and model building with the calculus **KE**. *Journal of the IGPL*, 4(1):129–150, 1996.

Implementation of Propositional Temporal Logics Using BDDs

G.L.J.M. Janssen

Eindhoven University of Technology
Department of Electrical Engineering, Room EH 9.26
P.O. Box 513, 5600 MB Eindhoven
E-mail: geert@ics.ele.tue.nl

1 Topic/Relevance

This tutorial intends to convey the step-by-step process by which computer programs for model checking and satisfiability testing for temporal logics may be derived from the theory. The idea is to demonstrate that it is very well possible to implement such a program in an efficient way without sacrificing a correct-by-construction approach. The tutorial will be fully self-contained, only a general knowledge of programming and propositional logic is assumed.

The proposed tutorial focuses on what might be coined "Implementing the theories". Unfortunately, too often an interesting approach or novel algorithm once published is not picked up by any user community. This is partly because no effort is spent in creating a state-of-the-art implementation in a readily accessible form. The intended audience are researchers in the field of theorem proving and any users of applications thereof. They will benefit from learning how theoretical results can be implemented in a well-written prototype program. Moreover, being a prototype should not be an excuse for not using advanced datastructures and algorithms. Therefore this tutorial takes an engineering approach to dealing with complicated issues such as reasoning in temporal logic, and shows that with a structured approach a powerful tool can be built that is able to handle real-life applications, for instance verification of sequential circuits.

2 Contents Outline

Below an outline is given of the tutorial contents. Much of the material on CTL model checking is based on the pioneering work in this area by Clarke et al.

1. **Introduction.** Presents the tutorial contents and sets its goals.
2. **Dags.** Directed acyclic graphs are the datastructure underlying BDDs. Dags will also be used to represent formulas. An efficient implementation based on a hash table and utilizing garbage-collection will be discussed.
3. **BDDs.** Explains what Binary Decision Diagrams are, how they can be efficiently implemented, and what their applications are.
4. **Kripke Structure Model.** Introduces the notion of a Kripke structure that is used as a model for the system we like to reason about.

5. **Computation Tree Logic.** Defines the syntax and semantics of a class of branching time logics.
6. **CTL Model Checking.** Shows how a model checker for CTL can be constructed in elegant way using a simple algorithm to calculate a fixed-point of a functional. Also, it is shown how BDDs can be exploited to represent the next-state relation and state-sets of the Kripke structure.
7. **Linear-time Temporal Logic.** Defines the syntax and semantics of the popular Manna/Pnueli propositional linear-time temporal logic.
8. **PTL Satisfiability Checking.** Shows how a satisfiability checker can be constructed for PTL. Again, BDDs will be used to represent the structure of the generated tableau.
9. **Diversion into μ-calculus.** Briefly explains Kozen's μ-calculus and shows how both CTL and PTL problems can be encoded in it. An implementation of μ-calculus is sketched.
10. **Summary and conclusions.** Summarizes the presented material and demonstrates the derived programs by some example runs and lists results of experiments.

The bibliography lists a number of articles and books on which the tutorial material will be based.

References

1. Z. Manna, A. Pnueli, "Verification of concurrent programs: the temporal framework," *The Correctness Problem in Computer Science*, eds. Robert S. Boyer, J. Strother Moore, International Lecture Series in Computer Science, Academic Press, New York, 1981.
2. J.R. Burch, E.M. Clarke, K.L. McMillan, "Symbolic Model Checking: 10^{20} States and Beyond," *International Workshop on Formal Methods in VLSI Design*, 1991.
3. E. Clarke, O. Grumberg, K. Hamaguchi, "Another Look at LTL Model Checking," *Proceeding 6-th Int. Conf. on Computer Aided Verification*, Springer-Verlag, LNCS 818, ed. David L. Dill, pp. 415–427.
4. Geert L.J.M. Janssen, "Hardware Verification using Temporal Logic: A Practical View," *Formal VLSI Correctness Verification, VLSI Design Methods-II*, Elsevier Science Publishers B.V. (North-Holland), ed. L.J.M. Claesen, IFIP, 1990, pp. 159–168.
5. Randal E. Bryant, "Graph-Based Algorithms for Boolean Function Manipulation," *IEEE Transactions on Computers*, Vol. C-35, Nr. 8, August 1986.
6. Karl S. Brace, Richard L. Rudell, Randal E. Bryant, "Efficient Implementation of a BDD Package," *Proceedings Design Automation Conference*, June 1990.
7. Kenneth L. McMillan, *Symbolic Model Checking*, Kluwer Academic Publishers, 1993.
8. E.A. Emerson, "Chapter 16: Temporal and Modal Logic," *Handbook of Theoretical Computer Science*, B: Formal Models and Semantics, ed. Jan van Leeuwen, Elsevier Science Publishers B.V., 1990, pp. 996–1072.
9. Rance Cleaveland, "Tableau-Based Model Checking in the Propositional Mu-Calculus," *Acta Informatica*, No. 27, 1990, pp. 725–747.

Computer Programming as Mathematics in a Programming Language and Proof System CL

Ján Komara and Paul J. Voda

Institute of Informatics, Comenius University Bratislava Slovakia.
E-mail: {komara,voda}@fmph.uniba.sk

CL (*Clausal Language*) is a computer programming language with mathematical syntax and a proof system based on Peano arithmetic which we have repeatedly used in the teaching of three (first and second year) undergraduate courses covering respectively *declarative programming, program verification,* and *program and abstract data specification.*

CL functions are over natural numbers, and yet CL has a look and feel of a modern functional language (higher-order functions are for the time being not covered). The coding of data structures into natural numbers is done via a pairing function which effectively identifies the domain of S-expressions of LISP with natural numbers. Recursion schemas available for the definition of CL functions are extremely programmer-friendly in that that they permit arbitrarily nested recursion where a previously defined measure of arguments goes down. By the well-known theorem of Tait on nested ordinal recursion (restricted in CL to ω) this does not lead outside of primitive recursive functions. Thus the Tait's theorem characterizes the CL programming language as being able to define exactly the unary primitive recursive functions (the effect of n-ary functions is achieved via pairing).

CL comes with its own proof system (intelligent proof checker) for proving properties of CL-defined functions such as the demonstration that they satisfy previously stated specifications. The proof system is also used for *proof obligations* where the user convinces the system that his recursively defined functions are properly introduced (they decrease arguments in certain measures).

The proof system is based on signed tableaux of Smullyan whose F-signed formulas are interpreted as goals to be proved and T-signed ones as assumptions. This permits a natural deduction style as used in mathematical practice and the proofs are easily described in English. It is amazing that CL seems to be the first system with such an obvious interpretation of signed tableaux.

The strength of the CL-proof system is characterized as a certain fragment of Peano Arithmetic. By the incompleteness result of Gödel, every formal system containing addition and multiplication admits only a fragment of recursive functions determined by its *proof* strength. Thus it seemed natural to us to choose that fragment of Peano arithmetic whose provably recursive functions are precisely the primitive recursive functions. This is the $I\Sigma_1$-*arithmetic* where induction axioms are restricted to Σ_1-formulas. Primitive recursive functions have very natural closure properties and certainly contain all feasibly computable functions.

Because CL uses strong recursion schemas for the definition of functions, its proof system requires a rich variety of induction schemas for proving their properties. The induction schemas are automatically derived from CL predicates characterizing data structures (such as lists, trees, tables) and amount to the shell principles of Boyer-Moore's system. Because CL has quantifiers, the induction schemas are extremely simply given in the form of Π_2-*rules* which by the well-known theorem of Parsons are reducible to Σ_1-induction axioms.

The domain of natural numbers is so well-known that the students have no problem understanding the meaning (semantics) of functions of CL and have a good intuition about their properties. This should be contrasted with similar systems with more complex and less intuitive domains (for instance PVS which is based on typed functionals). Our experience is that the students seem not only to understand but also enjoy CL.

A Tableau Calculus for Multimodal Logics and Some (Un)Decidability Results

Matteo Baldoni, Laura Giordano, and Alberto Martelli

Dipartimento di Informatica — Università degli Studi di Torino
Corso Svizzera, 185 — I-10149 Torino (Italy)
Tel. +39 11 74 29 111, Fax +39 11 75 16 03
E-mail: {baldoni,laura,mrt}@di.unito.it
URL: http://www.di.unito.it/~argo

Abstract. In this paper we present a *prefixed analytic tableau calculus* for a class of *normal multimodal logics* and we present some results about decidability and undecidability of this class. The class is characterized by axioms of the form $[t_1]\ldots[t_n]\varphi \supset [s_1]\ldots[s_m]\varphi$, called *inclusion axioms*, where the t_i's and s_j's are constants. This class of logics, called *grammar logics*, was introduced for the first time by Fariñas del Cerro and Penttonen to simulate the behaviour of grammars in modal logics, and includes some well-known modal systems. The prefixed tableau method is used to prove the *undecidability* of modal systems based on *unrestricted, context sensitive,* and *context free* grammars. Moreover, we show that the class of modal logics, based on *right-regular* grammars, are *decidable* by means of the *filtration methods*, by defining an extension of the Fischer-Ladner closure.

Keywords: Multimodal logics, Prefixed Tableaux methods, Decidability, Formal Grammars.

1 Introduction and motivations

Modal logics are widely used in artificial intelligence for representing *knowledge* and *beliefs* [19] together with other attitudes in *agent systems* like, for instance, *goals, intentions* and *obligations* [33]. Moreover, modal logics are well suited for representing *dynamic* aspects in *agent systems* and, in particular, to formalize reasoning about *actions* and *time*. Last but not least, modal logics are shown useful to extend logic programming languages with new features [31,13,4].

In this paper we focus on a class of *normal multimodal logics*, called *grammar logics*, which are characterized by a set of logical axioms of the form:

$$[t_1]\ldots[t_n]\varphi \supset [s_1]\ldots[s_m]\varphi \quad (n > 0; m \geq 0) \tag{1}$$

that we call *inclusion axiom*, where the t_i's and s_j's are modalities. This class includes some well-known modal systems such as K, $K4$, $S4$ and their multi-modal versions. Differently from other logics, such as those studied in [19], these systems can be *non-homogeneous* (i.e., every modal operator is not restricted to belong to the same system) and can contain some *interaction axioms* (i.e., every modal operator is not restricted to be independent from the others).

This class of logics has been introduced by Fariñas del Cerro and Penttonen in [11], where a method to define multimodal logics from *formal grammars* is presented, in such a way to simulate the behaviour of grammars. Given a formal grammar, a modality is associated to each terminal and nonterminal symbol, while, for each production rule of the form $t_1 \cdots t_n \to s_1 \cdots s_m$, an associated inclusion axiom $[t_1]\ldots[t_n]\varphi \supset [s_1]\ldots[s_m]\varphi$ is defined. In [11], it is shown that testing whether a word is generated by the formal grammar is equivalent to proving a theorem in the logic. Moreover, relying on this relation with formal grammars, an *undecidability* result for this class of multimodal logics is proved. However, in [11], *neither* a *proof method* is presented to deal with the class of grammar logics *nor (un)decidability* of restricted subclasses is studied.

In this paper, we develop an *analytic tableau calculus* for the class of *grammar logics*. The calculus is parametric with respect to each modal system in this class. In particular, it deals with *non-homogeneous* multimodal systems with arbitrary *interaction axioms* of the form (1).

The calculus is an extension of the one proposed in [26], which is closely related to the systems of prefixed tableaux presented in [14]. As a difference with [14], worlds are not represented by prefixes (which describe paths in the model from the initial world), but they are given an atomic name and the accessibility relationships among them are explicitly represented in a graph. The method is based on the idea of using the characterizing axioms of the logic as *"rewrite rules"* which create new paths among worlds in the counter-model construction.

Making use of the tableau calculus we prove the *undecidability* of the modal systems based on context sensitive and context-free grammars. Moreover, we show that the class of modal logics based on right regular grammars is *decidable*. We use the well-known *filtration methods* by defining an extension of the Fischer-Ladner closure for modal logics. This result is close to those that have been established for *propositional dynamic logic* [12, 20].

2 Grammar modal logics

Let us define a propositional multimodal language $\mathcal{L}$, containing the logical connectives $\wedge$, $\vee$, $\supset$, and $\neg$, a set of modal operators of the form $[t]$ and $\langle t \rangle$, where t belongs to a nonempty countable set MOD (the *alphabet of modalities*) and a nonempty countable set VAR of *propositional variables*. MOD and VAR are disjoint. The set of formulae of the languages are constructed as usual by means of the propositional variables, the connectives, and the modal operators.

We only consider *normal* modal logics, that is those ones whose axiomatization at least contains the axiom schemas for the classical propositional calculus, *modus ponens* and *necessitation* rules, and the axiom schema $K(t)$: $[t](\varphi \supset \psi) \supset ([t]\varphi \supset [t]\psi)$ for all modal operators. In particular, we focus on normal multimodal logics that are characterized by a set of axiom schemas of the form (1). We call these logics *grammar logics*. Let $\mathcal{A}$ be a set of inclusion axioms, we denote by $\mathcal{I}_\mathcal{L}^\mathcal{A}$ the grammar logic determined by the set $\mathcal{A}$ with $\mathcal{L}$ as underlying language, while we use $\mathcal{S}_\mathcal{L}^\mathcal{A}$ to denote its characterizing axiom systems (containing the axioms for normal modalities plus $\mathcal{A}$). As we will see, the inclusion axioms determine *inclusion properties* on the accessibility relations.

Some examples of grammar logics are the well-known modal systems K, T, $K4$, $S4$ [23], their multimodal versions K_n, T_n, $K4_n$, $S4_n$ [19], extensions of K_n and $S4_n$ with interaction axioms or with agent "any fool" in [16, 10, 3].

Example 1. (The friends puzzle) Peter is a friend of John, so if Peter knows that John knows something, then John knows that Peter knows that thing. That is, A_1: $[p][j]\varphi \supset [j][p]\varphi$, where $[p]$ and $[j]$ are modal operators of type $S4$ (i.e., A_2: $[p]\varphi \supset \varphi$, A_3: $[p]\varphi \supset [p][p]\varphi$, A_4: $[j]\varphi \supset \varphi$, and A_5: $[j]\varphi \supset [j][j]\varphi$) and they are used to denote what is known by Peter and John, respectively. Peter is married, so if Peter's wife knows something, then Peter knows the same thing, that is, A_6: $[wp]\varphi \supset [p]\varphi$ holds, where $[wp]$ is a modality of type $S4$ representing the knowledge of Peter's wife. John and Peter have an appointment, let us consider the following situation:

(1) $[p]time$ (3) $[wp]([p]time \supset [j]time)$
(2) $[p][j]place$ (4) $[p][j](place \wedge time \supset place)$

That is, (1) Peter knows the time of their appointment; (2) Peter also knows that John knows the place of their appointment. Moreover, (3) Peter's wife knows that if Peter knows the time of their appointment, then John knows that too; (4) Peter knows that if John knows the place and the time of their appointment, then John knows that he has an appointment. From this situation we will be able to prove $[j][p]appointment \wedge [p][j]appointment$, that is, each of the two friends knows that the other one knows that he has an appointment.

In order to define the meaning of a formula, we introduce the notion of *Kripke interpretation*. Formally, a Kripke interpretation M is a triple $(W, \{\mathcal{R}_t \mid t \in \text{MOD}\}, V)$, consisting of a non-empty set W of "*possible worlds*" and a set of *binary relations* $\mathcal{R}_t$ (one for each $t \in \text{MOD}$) on W, and a *valuation function* V, that is a mapping from $W \times \text{VAR}$ to the set $\{\mathbf{T}, \mathbf{F}\}$. We say that $\mathcal{R}_t$ is the *accessibility relation* of the modality $[t]$ and w' is *accessible* from w by means of $\mathcal{R}_t$ if $(w, w') \in \mathcal{R}_t$ (or $w\mathcal{R}_t w'$).

The meaning of a formula is given by means of a *satisfiability relation*, denoted by $\models$. Let $M = \langle W, \{\mathcal{R}_t \mid t \in \text{MOD}\}, V \rangle$ be a Kripke interpretation, w a world in W and φ a formula, then, we say that φ is *satisfiable in the Kripke interpretation M at w*, denoted by $M, w \models \varphi$, if the following conditions hold:

- $M, w \models \varphi$ and $\varphi \in \text{VAR}$ iff $V(w, \varphi) = \mathbf{T}$;
- $M, w \models \neg \varphi$ iff $M, w \not\models \varphi$;
- $M, w \models \varphi \wedge \psi$ iff $M, w \models \varphi$ and $M, w \models \psi$;
- $M, w \models \varphi \vee \psi$ iff $M, w \models \varphi$ or $M, w \models \psi$;
- $M, w \models \varphi \supset \psi$ iff $M, w \not\models \varphi$ or $M, w \models \psi$;
- $M, w \models [t]\varphi$ iff for all $w' \in W$ such that $(w, w') \in \mathcal{R}_t$, $M, w' \models \varphi$;
- $M, w \models \langle t \rangle \varphi$ iff there exists a $w' \in W$ such that $(w, w') \in \mathcal{R}_t$ and $M, w' \models \varphi$.

Let $\mathcal{M}_\mathcal{L}$ be the set of all Kripke interpretations, as defined above. For each grammar logic $\mathcal{I}_\mathcal{L}^\mathcal{A}$ we introduce a suitable notion of Kripke $\mathcal{A}$-interpretation, by adding some restriction on the accessibility relations. More precisely, let $M = (W, \{\mathcal{R}_t \mid t \in \text{MOD}\}, V)$ be a Kripke interpretation and let $\mathcal{A}$ be a set of inclusion axioms, we say M is a *Kripke $\mathcal{A}$-interpretation* if and only if for each axiom schema $[t_1][t_2]\ldots[t_n]\varphi \supset [s_1][s_2]\ldots[s_m]\varphi \in \mathcal{A}$, the following *inclusion property* on the accessibility relation holds:

$$\mathcal{R}_{t_1} \circ \mathcal{R}_{t_2} \circ \ldots \circ \mathcal{R}_{t_n} \supseteq \mathcal{R}_{s_1} \circ \mathcal{R}_{s_2} \circ \ldots \circ \mathcal{R}_{s_m} \quad (2)$$

where "$\circ$" means the relation composition $\mathcal{R}_t \circ \mathcal{R}_{t'} = \{(w, w'') \in W \times W \mid \exists w' \in W \text{ such that } (w, w') \in \mathcal{R}_t \text{ and } (w', w'') \in R_{t'}\}$[1].

The set of all Kripke $\mathcal{A}$-interpretations is denoted by $\mathcal{M}_\mathcal{L}^\mathcal{A}$ and it is a subset of $\mathcal{M}_\mathcal{L}$. Given a Kripke $\mathcal{A}$-intepretation $M = \langle W, \{\mathcal{R}_t \mid t \in \text{MOD}\}, V \rangle$ in $\mathcal{M}_\mathcal{L}^\mathcal{A}$, we say that a formula φ of $\mathcal{I}_\mathcal{L}^\mathcal{A}$ is *satisfiable in M* if $M, w \models_\mathcal{A} \varphi$ for some world $w \in W$. We say that φ is *valid in M* if $\neg\varphi$ is not satisfiable in M. Moreover, a formula φ is *satisfiable* if φ is $\mathcal{A}$-satisfiable in some Kripke $\mathcal{A}$-interpretation in $\mathcal{M}_\mathcal{L}^\mathcal{A}$ and $\mathcal{A}$-valid f it is valid in all Kripke $\mathcal{A}$-interpretations in $\mathcal{M}_\mathcal{L}^\mathcal{A}$ (in this case, we write $\models_\mathcal{A} \varphi$).

The axiom system $\mathcal{S}_\mathcal{L}^\mathcal{A}$ is *sound* and *complete* axiomatization with respect to $\mathcal{M}_\mathcal{L}^\mathcal{A}$ [2] (see also [11] for a subclass).

Due to the similarity between inclusion axioms and production rules in a grammar, we can associate to a given grammar a corresponding grammar logic.

A *grammar* is a quadruple $G = (V, T, P, S)$, where V and T are disjoint finite sets of *variables* and *terminals*, respectively. P is a finite set of *productions*, each production is of the form $\alpha \to \beta$, where the form of α and β depends on the *type* of grammar as follows[2]:

Production grammar form for different classes of languages

type-0	type-1	type-2	type-3
$\alpha \in (V \cup T)^*V(V \cup T)^*$	$\alpha \in (V \cup T)^*V(V \cup T)^*$	$\alpha \in V$	$\alpha \in V$
$\beta \in (V \cup T)^*$	$\beta \in (V \cup T)^+$	$\beta \in (V \cup T)^*$	$\beta = \sigma A$ or $\beta = \sigma$
	$|\beta| \leq |\alpha|$		$\sigma \in T^*, A \in V$

Finally, $S \in V$ is a special variable called the *start symbol* [21]. We say that the production $\alpha \to \beta$ is applied to the string $\gamma\alpha\delta$ to *directly derive* $\alpha\beta\delta$ in grammar G (written $\gamma\alpha\delta \Rightarrow_G \gamma\beta\delta$). The relation *derives*, $\Rightarrow_G^*$, is the reflexive, transitive closure of $\Rightarrow_G$. The *language generated* by a grammar G, denoted by $L(G)$ is the set of *words* $\{w \in T^* \mid S \Rightarrow_G^* w\}$.

Given a formal grammar $G = (V, T, P, S)$, we can associate to it a *grammar logic (based on G)* containing the modalities $\text{MOD} = V \cup T$ and characterized

---

[1] If $m = 0$ then we assume $\mathcal{R}_{s_1} \circ \mathcal{R}_{s_2} \circ \ldots \circ \mathcal{R}_{s_m} = I$, where I is the identity relation on W.

[2] We denote by "L^*" the Kleene closure of the language L (i.e. it denotes zero or more concatenation of L) and by "$+$" the positive closure of L (i.e. it denotes one or more concatenation of L) [21].

by the a axiom schema $[t_1]\ldots[t_n]\varphi \supset [s_1]\ldots[s_m]\varphi$, one for each production rule $t_1\cdots t_n \to s_1\cdots s_m \in P$, where the t_i's and s_j's are either in V or in T.

We will call *unrestricted, context sensitive, context-free,* and *right-regular* modal logic a grammar logic based on a type-0, type-1, type-2, and type-3 grammar, respectively.

3 A tableau calculus for grammar logics

Before introducing our tableau calculus, we need to define some notions. We define a *signed formula* Z as a formula prefixed by one of the two symbols **T** and **F** *(signs)*. For instance, if φ is a formula then, **T**φ and **F**φ are signed formulae.

Definition 1. *Let $\mathcal{L}$ be a propositional modal language and let $\mathcal{W}_C$ be a countable non-empty set of* constant world symbols *(or* prefixes*). A prefixed signed formula, $w : Z$, is a prefix $w \in \mathcal{W}_C$ followed by a signed formula Z.*

Intuitively, prefixes are used to name worlds, and a formula $w : \mathbf{T}\varphi$ ($w : \mathbf{F}\varphi$) on a branch of a tableau means that the formula φ is *true (false)* at the world w in the Kripke interpretation associated with that branch. We assume that $\mathcal{W}_C$ contains always at least the prefix i, that is interpreted as the *initial world*.

Definition 2. *Let $\mathcal{L}$ be a propositional modal language, an* accessibility relation formula $w \; \rho_t \; w'$, *where $t \in$ MOD, is a binary relation between prefixes of $\mathcal{W}_C$.*

We say that an accessibility relation formula $w \; \rho_t \; w'$ is *true* in a tableau branch if it belongs to that branch and, intuitively, this means that in the Kripke interpretation associated with that branch $(w, w') \in \mathcal{R}_t$ holds.

Remark 1. Using prefixed formulae is very common in modal theorem proving (see [17] for an historical introduction on the topic). We would like to mention the well-known *prefixed tableau systems* in [14] and the TABLEAUX system in [8]. In [14], differently than here and [26,8], a prefix is a sequence of integers which represents a world as a *path* from the initial world to it. As a result, instead of representing *explicitly* worlds and accessibility relations of a Kripke interpretation in a *graph*, by means of the accessibility relation formulae, [14] represents them by a set of paths, which can be considered as a *spanning tree* of the graph. Similar ideas are also used by other authors, such as the proposals in [25, 18, 32, 9].

In order to simplify the presentation of the calculus we use the well-known *uniform notation* for signed formulae [14] (see Fig. 1). In the following, we will often use α, β, ν^t, and π^t as formulae of the corresponding type.

A *tableau* is a *labeled tree* where each node consists of a *prefixed signed formula* or an *accessibility relation formula*. It is an attempt to build an interpretation in which a given formula is satisfiable. Starting from a formula φ, the interpretation is progressively constructed applying a set of *extension rules*, which reflect the semantics of the considered logic. At any stage, a branch of a tableau is a partial

α	α_1	α_2
$T(\varphi \wedge \psi)$	$T\varphi$	$T\psi$
$F(\varphi \vee \psi)$	$F\varphi$	$F\psi$
$F(\varphi \supset \psi)$	$T\varphi$	$F\psi$
$F(\neg\varphi)$	$T\varphi$	$T\varphi$

β	β_1	β_2
$F(\varphi \wedge \psi)$	$F\varphi$	$F\psi$
$T(\varphi \vee \psi)$	$T\varphi$	$T\psi$
$T(\varphi \supset \psi)$	$F\varphi$	$T\psi$
$T(\neg\varphi)$	$F\varphi$	$F\varphi$

ν^t	ν_0^t
$T([t]\varphi)$	$T\varphi$
$F(\langle t\rangle\varphi)$	$F\varphi$

π^t	π_0^t
$F([t]\varphi)$	$F\varphi$
$T(\langle t\rangle\varphi)$	$T\varphi$

Fig. 1. Uniform notation for propositional signed modal formulae.

description of an interpretation. In our case, the tableau method tries to build Kripke interpretations, one for each branch: the worlds are formed by the prefixes that appear on the branch, the accessibility relations for the modalities are given by means of the accessibility relation formulae, and the valuation function is given by means of the prefixed signed atomic formulae.

Now, we can present the set of extension rules. We say that a prefix w is *used* on a tableau branch if it occurs on the branch in some accessibility relation formula, otherwise we say that the prefix w is *new*.

Definition 3 ((Extension rules)). *Let $\mathcal{L}$ be a modal language and let $\mathcal{A}$ be a set of inclusion axioms, the extension rules for $\mathcal{I}_{\mathcal{L}}^{\mathcal{A}}$ are given in Fig. 2.*

$$\frac{w : \alpha}{\begin{array}{c}w : \alpha_1 \\ w : \alpha_2\end{array}} \; \alpha\text{-rule} \qquad \frac{w : \beta}{w : \beta_1 \mid w : \beta_2} \; \beta\text{-rule}$$

$$\frac{w : \nu^t \quad w \; \rho_t \; w'}{w' : \nu_0^t} \; \nu\text{-rule} \qquad \frac{w : \pi^t}{\begin{array}{c}w' : \pi_0^t \\ w \; \rho_t \; w'\end{array}} \; \pi\text{-rule}$$

where w' is *new* on the branch

$$\frac{w \; \rho_{s_1} \; w_1 \quad \cdots \quad w_{m-1} \; \rho_{s_m} \; w'}{\begin{array}{c}w \; \rho_{t_1} \; w'_1 \\ \vdots \\ w'_{n-1} \; \rho_{t_n} \; w'\end{array}} \; \rho\text{-rule}$$

where $w'_1, \ldots, w'_{n-1}$ are *new* on the branch
and $[t_1]\ldots[t_n]\varphi \supset [s_1]\ldots[s_m]\varphi \in \mathcal{A}$ ($n > 0$ and $m \geq 0$)

Fig. 2. Tableau rules for propositional inclusion modal logics.

The interpretation of the different kinds of extension rules is rather easy taking into account the possible-worlds semantics. The rules for the formula of type α and β are the usual ones of the classical calculus.

A formula of type ν^t is true at world w if ν_0^t is true in all worlds w' accessible from w by means of t. Therefore, if $w : \nu^t$ occurs on an open branch, we can add $w' : \nu_0^t$ to the end of that branch for any w' which is accessible from w by means of $\mathcal{R}_t$ (such that $w\ \rho_t\ w'$ is true on that branch).

A formula of type π^t is true at the world w if there exists a world w' accessible from w at which π_0^t is true. Therefore, if $w : \pi^t$ occurs on an open branch, we can add $w' : \pi_0^t$ to the end of that branch, provided w' is new and $w\ \rho_t\ w'$ is true on it.

The intuition behind ρ-rule is quite simple. Let us suppose, for instance, that $[t_1]\ldots[t_n]\varphi \supset [s_1]\ldots[s_m]\varphi \in \mathcal{A}$ is an axiom of our grammar logic $\mathcal{I}_\mathcal{L}^\mathcal{A}$. If $w\ \rho_{s_1}\ w_1,\ \ldots,\ w_{m-1}\ \rho_{s_m}\ w'$ are on a branch, then $(w, w_1) \in \mathcal{R}_{s_1},\ \ldots,$ $(w_{m-1}, w') \in \mathcal{R}_{s_m}$ in the Kripke interpretation associated with that branch. Since $[t_1]\ldots[t_n]\varphi \supset [s_1]\ldots[s_m]\varphi \in \mathcal{A}$ then, the corresponding inclusion property (2) must holds. Thus, we can add the formulae $w\ \rho_{t_1}\ w'_1,\ \ldots,\ w'_{n-1}\ \rho_{t_n}\ w'$ to that branch. Moreover, in the case of $m = 0$ we can always add the formulae $w\ \rho_{t_1}\ w'_1,\ \ldots,\ w'_{n-1}\ \rho_{t_n}\ w$, for every world constant w, provided that $w'_1,\ \ldots,\ w'_{n-1}$ are new on the branch.

Remark 2. It is worth noting that the ρ-rule works for the whole class of grammar logics. Nevertheless, the proposed tableau could be easily extended in order to deal with modal logics which are different than those we have considered. By introducing new rules, which operate on accessibility relation formulae, one could also deal with multimodal logics characterized by *serial*, *symmetric*, and *Euclidean* accessibility relations [2].

We say that a tableau branch is *closed* if it contains $w : \mathbf{T}\varphi$ and $w : \mathbf{F}\varphi$ for some formula φ. A tableau is *closed* if *every* branch in it is closed. Finally, let $\mathcal{L}$ be a modal language, $\mathcal{A}$ a set of inclusion axioms, and φ a formula. Then a closed tableau for $i : \mathbf{F}\varphi$ obtained by using the tableau rules of Fig. 2, is said to be a *proof* of φ.

Theorem 1. *Let $\mathcal{I}_\mathcal{L}^\mathcal{A}$ be grammar logic then, a formula φ of $\mathcal{L}$ has a tableau proof if and only if it is $\mathcal{A}$-valid.*

Due to space limitation we do not present here the proof of Theorem 1 but it follows the well-known guideline of [14, 25, 17] and it can be found in [2].

Example 2. In Figure 3 we have reported the proof of the first conjunct of the formula $[j][p]appointment \land [p][j]appointment$ of Example 1. We denote with "a" and "b" the two branches which are created by the β-rule at step 13., "c" and "d" the two ones created by the β-rule at step 14b., "e" and "f" the two ones created by the β-rule at step 17d. Moreover, to save space, we use "ap" instead of *appointment*, "tm" instead of *time*, and "pl" istead of *place*. The explanation: *1., 2., 3., and 4.*: formula (1), (2), (3), and (4); *5.*: goal, formula (5); *6. and 7.*: from 5., by π-rule; *8. and 9.*: from 6., by π-rule; *10. and 11.*: from 7. and 9., by A_1 and ρ-rule; *12.*: from 4. and 10., by ν-rule; *13.*: from 12. and 11., by ν-rule; *14a. and 14b*: from 13, by β-rule, branch "a" closes; *15c. and 15d.*: from 14b., by

β-rule; *16c.*: from 3. and 10., by ν-rule; *17c.*: from 16c. and 11., by ν-rule, branch "c" closes; *16d.*: from 10., by axiom A_6 and π-rule; *17d.*: from 2. and 16d., by ν-rule; *18e. and 18f*: from 17d., by β-rule; *19e.*: from 18e. and 11., by ν-rule, branch "e" closes; *19f. and 20f.*: from 18f., by π-rule; *21f.*: from 10. and 10f., by axiom A_3 and ρ-rule; *22f.*: from 1. and 21f., by ν-rule, branch "f" closes.

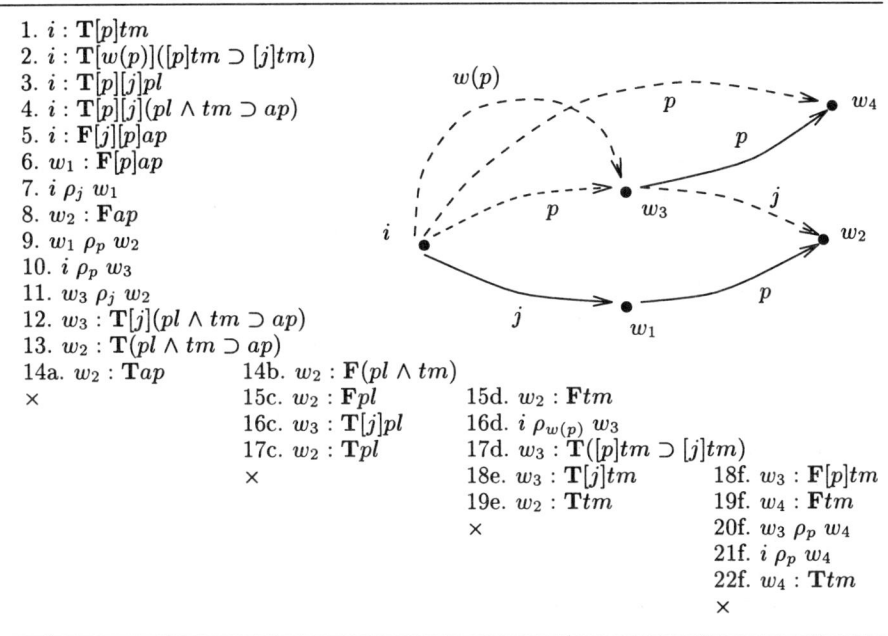

1. $i : \mathbf{T}[p]tm$
2. $i : \mathbf{T}[w(p)]([p]tm \supset [j]tm)$
3. $i : \mathbf{T}[p][j]pl$
4. $i : \mathbf{T}[p][j](pl \wedge tm \supset ap)$
5. $i : \mathbf{F}[j][p]ap$
6. $w_1 : \mathbf{F}[p]ap$
7. $i \, \rho_j \, w_1$
8. $w_2 : \mathbf{F}ap$
9. $w_1 \, \rho_p \, w_2$
10. $i \, \rho_p \, w_3$
11. $w_3 \, \rho_j \, w_2$
12. $w_3 : \mathbf{T}[j](pl \wedge tm \supset ap)$
13. $w_2 : \mathbf{T}(pl \wedge tm \supset ap)$

14a. $w_2 : \mathbf{T}ap$ 14b. $w_2 : \mathbf{F}(pl \wedge tm)$
 ×
                                 15c. $w_2 : \mathbf{F}pl$ 15d. $w_2 : \mathbf{F}tm$
                                 16c. $w_3 : \mathbf{T}[j]pl$ 16d. $i \, \rho_{w(p)} \, w_3$
                                 17c. $w_2 : \mathbf{T}pl$ 17d. $w_3 : \mathbf{T}([p]tm \supset [j]tm)$
                                  ×   18e. $w_3 : \mathbf{T}[j]tm$ 18f. $w_3 : \mathbf{F}[p]tm$
                                                                    19e. $w_2 : \mathbf{T}tm$ 19f. $w_4 : \mathbf{F}tm$
                                                                     ×   20f. $w_3 \, \rho_p \, w_4$
                                                                                                        21f. $i \, \rho_p \, w_4$
                                                                                                        22f. $w_4 : \mathbf{T}tm$
                                                                                                         ×

Fig. 3. ρ-rule as rewriting rule: counter-model construction of Example 1.

The ρ-rule can be regarded as a *rewriting* rule which creates new paths among worlds according to the inclusion properties of the grammar logic. In fact, given a tableau branch S, let w_0 and w_n two prefixes used on S, a *path* $\xi(w_0, w_n)$ is a collection $\{w_0 \, \rho_{t_1} \, w_1, w_1 \, \rho_{t_2} \, w_2, \ldots, w_{n-1} \, \rho_{t_n} \, w_n\}$ of accessibility relation formulae in S. We say that the path $\xi(w_0, w_n)$ *directly ρ-derives* the path $\xi'(w_0, w_n)$ if the path $\xi'(w_0, w_m)$ is obtained from $\xi(w_0, w_m)$ by means of the application of a ρ-rule to a subpath of $\xi(w_0, w_n)$. The relation *ρ-derive* is the reflexive, transitive closure of the relation *directly ρ-derive*. For instance, let us consider Fig. 3. Then, the path $\xi_1(i, w_2) = \{i \, \rho_j \, w_1, w_1 \, \rho_p \, w_2\}$ directly ρ-derives the path $\xi_2(i, w_2) = \{i \, \rho_p \, w_3, w_3 \, \rho_j \, w_2\}$, and ρ-derives the path $\xi_3(i, w_2) = \{i \, \rho_{wp} \, w_3, w_3 \, \rho_j \, w_2\}$.

For a path $\xi(w_0, w_n) = \{w_0 \, \rho_{t_1} \, w_1, \ldots, w_{n-1} \, \rho_{t_n} \, w_n\}$, we denote by $\overline{\xi}(w_0, w_n)$ the *word* $t_1 \cdots t_n$. It is worth noting that for a grammar logic $\mathcal{I}_{\mathcal{L}}^{\mathcal{A}}$ based on a grammar G, if $\xi(w_0, w_n)$ is a path occurring in a tableau branch, then, $\xi(w_0, w_n)$ ρ-derives a path $\xi'(w_0, w_n)$ *if and only if* $\overline{\xi'}(w_0, w_n) \Rightarrow_G^* \overline{\xi}(w_0, w_n)$.

4 Undecidability results for grammar logics

The tableau method developed in the previous section allows to generalize the correspondence between the membership problem for a given grammar and the validity problem in the corresponding grammar logic established by Fariñas del Cerro and Penttonen in [11].

Theorem 2. *Given a grammar $G = (V, T, P, S)$, let $\mathcal{I}_\mathcal{L}^A$ be the grammar logic based on G. Then, for any propositional variable p of $\mathcal{L}$, $\models_\mathcal{A} [S]p \supset [s_1] \ldots [s_m]p$ if and only if $S \Rightarrow_G^* s_1 \cdots s_m$, where the s_i's are in $V \cup T$.*

Proof. (*If*) Let us suppose that $\models_\mathcal{A} [S]p \supset [s_1] \ldots [s_m]p$, then, the tableau starting from $i : \mathbf{F}([S]p \supset [s_1] \ldots [s_m]p)$ closes. Now, by applying the β-rule we obtain: $i : \mathbf{T}[S]p$, $i : \mathbf{F}[s_1] \ldots [s_m]p$, and m times the π-rule: $w_1 : \mathbf{F}[s_2] \ldots [s_m]p$, $i\ \rho_{s_1}\ w_1$, ..., $w_m : \mathbf{F}p$, and $w_{m-1}\ \rho_{s_m}\ w_m$. Since, by hypothesis, the above tableau closes, the only way for this to happen is that after a finite number of applications of the ρ-rule we have the prefixed signed formula $w_m : \mathbf{T}p$ in the branch. This happens if the path $\xi(i, w_m) = \{i\ \rho_{s_1}\ w_1, \ldots, w_{m-1}\ \rho_{s_m}\ w_m\}$ ρ-derives the path $\xi'(i, w_m) = \{i\ \rho_S\ w_m\}$, that is, if there exits a derivation $\overline{\xi'}(i, w_m) \Rightarrow_G^* \overline{\xi}(i, w_m)$. (*Only if*) Assume $S \Rightarrow_G^* s_1 \cdots s_m$. Since a systematic attempt to prove $i : \mathbf{F}([S]p \supset [s_1] \ldots [s_m]p)$ generates a path $\xi(i, w_m) = \{i\ \rho_{s_1}\ w_1, \ldots, w_{m-1}\ \rho_{s_m}\ w_m\}$ and $\xi(i, w_m)$ ρ-derives the path $\xi'(i, w_m) = \{i\ \rho_S\ w_m\}$, after a finite number of steps the only branch of the tableau closes by $w_m : \mathbf{T}p$ and $w_m : \mathbf{F}p$.

It is well known that the problem of establishing if a word belongs to the language generated by an arbitrary type-0 grammar is undecidable [21]. Hence, we have the following corollary.

Corollary 1. *The validity problem for the class of grammar logics is undecidable.*

Indeed, this result has already been shown in [11]. However, Fariñas del Cerro and Penttonen do not prove Theorem 2 for the type-0 grammars but for a more restricted class of the grammar logics, that they call *Thue logics* because they are based on the *Thue systems* [6]. A Thue system is a type-0 grammars whose productions are *symmetric* and, thus, the Thue logics are grammar logics characterized by axiom schemas where the implication is replaced by the biimplication. In [11] the undecidability of grammar logics is proved by showing that the Thue logics are undecidable. In fact, since the membership problem for the Thue systems is *undecidable*, proving that a formula is a theorem of a Thue logic is also undecidable.[3]

---

[3] The Thue systems have also been used in [24] to define logics similar to those studied in [11], which, however, are not in the class on grammar logics since modalities enjoy some further properties like seriality and determinism. In [24] undecidability results are proved for this class of logics.

In [11] some problems are left open. In particular, it is not established whether more restricted classes of grammar logics, such as context sensitive, context-free, regular modal logics are decidable. In the following, we show that also the class of context sensitive and context-free modal logics are undecidable by reducing the solvability of the problem $L_1 \cap L_2 \neq \emptyset$ (where L_1 and L_2 are languages) to the satisfiability of formulas of context sensitive and context-free modal logics.

Theorem 3. Let $G_1 = (V_1, T_1, P_1, S_1)$ and $G_2 = (V_2, T_2, P_2, S_2)$ be two grammars such that $V_1 \cap V_2 = \emptyset$ and $T_1 = T_2 \neq \emptyset$. Then, there exists a grammar logic $\mathcal{I}_\mathcal{L}^A$ and a formula φ of $\mathcal{L}$ such that $\models_A \varphi$ if and only if $L(G_1) \cap L(G_2) \neq \emptyset$.

Proof. Let us define a grammar $G = (V, T, P, S)$, where $V = V_1 \cup V_2 \cup \{S\}$, $T = T_1 = T_2$, $P = P_1 \cup P_2 \cup \{S \to t, S \to S\,t \mid t \in T\}$, and $S \notin V_1$ and $S \notin V_2$. Then, we assume as $\mathcal{I}_\mathcal{L}^A$ the inclusion modal logic based on G and we consider the formula $\varphi_T(q) = \bigwedge_{t \in T}(\langle t \rangle q \wedge [S]\langle t \rangle q)$ where $q \in$ VAR. A tableau starting from $i : \mathbf{T}\varphi_T(q)$ is formed by only one branch that goes on forever. It is easy to see that for each word $x \in T^*$ the tableau branch contains a path $\xi(i, w)$ such that $\overline{\xi}(i, w) = x$. Now, let us define $\varphi = \varphi_T(q) \supset ([S_1]p \supset \langle S_2 \rangle p)$, where $p, q \in$ VAR and $p \neq q$. (*If*) Suppose that $\models_A \varphi$ then, the tableau starting from 1. $i : \mathbf{F}(\varphi_T(q) \supset ([S_1]p \supset \langle S_2 \rangle p))$ closes. Now, by applying twice the β-rule we obtain: 2. $i : \mathbf{T}\varphi_T(q)$, 3. $i : \mathbf{T}[S_1]p$, and 4. $i : \mathbf{F}\langle S_2 \rangle p$. Since the above tableau must close, the only way for this to happen is that after a finite number of steps we must have a pair of prefixed signed formulae $w : \mathbf{T}p$ and $w : \mathbf{F}p$, for some prefix w and, therefore, a path $\xi(i, w)$ that ρ-derives both the path $\xi_1(i, w) = \{i\ \rho_{S_1}\ w\}$ and the path $\xi_2(i, w) = \{i\ \rho_{S_2}\ w\}$. Thus, there is a derivation of $\overline{\xi}(i, w)$ both from $\overline{\xi_1}(i, w) = S_1$ and from $\overline{\xi_2}(i, w) = S_2$ ($S_1 \Rightarrow^*_G \overline{\xi}(i, w)$ and ($S_2 \Rightarrow^*_G \overline{\xi}(i, w)$), i.e. $\overline{\xi}(i, w) \in L(G_1) \cap L(G_2)$. (*Only if*) Assume that $S_1 \Rightarrow^*_{G_1} x$ and $S_2 \Rightarrow^*_{G_2} x$, for some $x \in T^*$. Since a systematic attempt to prove the formula $i : \mathbf{T}\varphi_T(q)$ can generate a path $\xi(i, w)$, for some prefix w, such that $\overline{\xi}(i, w) = y$, for any $y \in T^*$, after a finite number of steps we have a path $\xi'(i, w')$ such that $\overline{\xi'}(i, w') = x$. Thus, we have also the paths $\xi'_1(i, w') = \{i\ \rho_{S_1}\ w'\}$ and $\xi'_2(i, w') = \{i\ \rho_{S_2}\ w'\}$ by application of the ρ-rule for a finite number of times. This is enough to close the only branch of the tableau by $w' : \mathbf{T}p$ and $w' : \mathbf{F}p$.

It is well known that, given two arbitrary type-1 (type-2) grammars G_1 and G_2, it is undecidable if $L(G_1) \cap L(G_2) \neq \emptyset$ [21]. Hence, we have the following corollary.

Corollary 2. *The validity problem for the class of context sensitive and context-free modal logic is undecidable.*

5 A decidability result for grammar logics

In the previous section we have shown that it is not possible to supply a general decision procedure for the class of unrestricted, context sensitive and context-free modal logics. In this section, instead, we give a *decidability* result for *right regular grammar* logics, that is, those ones whose productions are of the form $A \to \sigma A'$, where A, A' are variables and σ a string of terminals.

Definition 4. Let $G = (V, T, P, S)$ be a right type-3 grammar and let A be a variable. Then, a derivation of a sentential form σX from A^4 is said to be non-recursive if and only if each variable of V appears in the derivation, apart from σX, at most once.

Proposition 1. Let $G = (V, T, P, S)$ be a right type-3 grammar, let A_0 be a variable and let $A_0 \Rightarrow_G^* \sigma_1 \cdots \sigma_n A_n \Rightarrow_G \sigma_1 \cdots \sigma_n \sigma_{n+1} A_{n+1}$ be a derivation, where either $A_{n+1} \in V$ or $A_{n+1} \in T$ and $A_i \to \sigma_{i+1} A_{i+1} \in P$, for $i = 0, \ldots, n$. Then, there exists a non-recursive derivation $A_0 \Rightarrow_G^* \sigma \sigma_{n+1} A_{n+1}$, for some $\sigma \in T^*$.

Proposition 2. Let $G = (V, T, P, S)$ be a right type-3 grammar. Then, the number of different non-recursive derivation by means of G is bounded by $\text{der}_G = |V| \cdot \sum_{i=1}^{|V|} n^i$, where n is the maximum number of production associated to a same variable of V.

The proofs of the proposition above are simple and they can be found in [2].

Let $G = (V, T, P, S)$ be a right type-3 grammar and $\mathcal{I}_\mathcal{L}^\mathcal{A}$ the regular inclusion modal logic based on G. Then, we define the *Fischer-Ladner closure* $FL(\varphi)$ of a formula φ of $\mathcal{L}$ (that only uses *existential modal operators, or,* and *negation*[5]) as follows:

- if $\psi \lor \psi' \in FL(\varphi)$ then $\psi \in FL(\varphi)$ and $\psi' \in FL(\varphi)$;
- if $\neg \psi \in FL(\varphi)$ then $\psi \in FL(\varphi)$;
- if $\langle t \rangle \psi \in FL(\varphi)$ and $t \in T$ then $\psi \in FL(\varphi)$;
- if $\langle A \rangle \psi \in FL(\varphi)$, $A \in V$, and there is a non-recursive derivation $A \Rightarrow_G^* t_1 \cdots t_n X$, where $t_1, \ldots, t_n \in T$ and either $X \in T \cup V$, then $\langle t_1 \rangle \ldots \langle t_n \rangle \langle X \rangle \psi \in FL(\varphi)$.

By Proposition 2 and the fact that φ has finite length, the Fischer-Ladner closure is finite for any formula of a right regular modal logic.

Consider a Kirpke $\mathcal{A}$-interpretation $M = \langle W, \{R_t \mid t \in \text{MOD}\}, V \rangle$ and a formula φ of $\mathcal{L}$, we define an *equivalence* relation $\equiv$ on state of W by: $w \equiv w'$ if and only if for all $\psi \in FL(\varphi)$ we have $M, w \models_\mathcal{A} \psi$ iff $M, w' \models_\mathcal{A} \psi$ (we use the notation $\overline{w}$ for this equivalence class). The *quotient* Kripke $\mathcal{A}$-interpretation $M^{FL(\varphi)} = \langle W^{FL(\varphi)}, \{\mathcal{R}_t^{FL(\varphi)} \mid t \in \text{MOD}\}, V^{FL(\varphi)} \rangle$ (the *filtration of M through* $FL(\varphi)$) is defined as follows:

- $W^{FL(\varphi)} = \{\overline{w} \mid w \in W\}$;
- $V^{FL(\varphi)}(\overline{w}, p) = V(w, p)$, for any $p \in \text{VAR}$ and $\overline{w} \in W^{FL(\varphi)}$;

---

[4] Note that, every sentential form derived from A has the form σX, where $\sigma \in T^*$ and either $X \in T$ or $X \in V$.

[5] Since all other connectives can be defined in terms of these, this is not a restrictive condition.

- $\mathcal{R}_t^{FL(\varphi)} \supseteq \{(\overline{w},\overline{w'}) \in W^{FL(\varphi)} \times W^{FL(\varphi)} \mid (w,w') \in \mathcal{R}_t\}$.

Moreover, $\mathcal{R}_t^{FL(\varphi)}$ is closed with respect to the inclusion axioms, that is, for each inclusion axiom $[t]\alpha \supset [s_1]\ldots[s_m]\alpha$ if $(\overline{w_0},\overline{w_1}) \in \mathcal{R}_{s_1}^{FL(\varphi)}$, ..., $(\overline{w_{m-1}},\overline{w_m}) \in \mathcal{R}_{s_m}^{FL(\varphi)}$ then the pair $(\overline{w_0},\overline{w_m})$ belongs to $\mathcal{R}_t^{FL(\varphi)}$.

The following lemma states that when we insert any extra binary relation between $\overline{w}$ and $\overline{w'}$ in a accessibility relation $\mathcal{R}_t^{FL(\varphi)}$ of $M^{FL(\varphi)}$, in order to satisfy the relative set of inclusion properties, it is not the case that there was any $\langle t\rangle\psi \in FL(\varphi)$ which was true at w while ψ itself was false at w' [22].

Lemma 1. *For all* $\psi = \langle t\rangle\psi' \in FL(\varphi)$, *if* $(\overline{w},\overline{w'}) \in \mathcal{R}_t^{FL(\varphi)}$ *and* $M,w' \models_\mathcal{A} \psi'$ *then* $M,w \models_\mathcal{A} \langle t\rangle\psi'$.

Proof. Assume that $\psi = \langle t\rangle\psi' \in FL(\varphi)$ then $\psi' \in FL(\varphi)$ by definition of the closure. Now, there are two cases which depend on whether $(\overline{w},\overline{w'}) \in \mathcal{R}_t^{FL(\varphi)}$ has been added to originary definition of filtration because an inclusion axiom of the form $[t]\alpha \supset [s_1]\ldots[s_m]\alpha \in \mathcal{A}$ or not.

Assume that it has not been added. Since by definition of $\mathcal{R}_t^{FL(\varphi)}$, there exist $w_1, w_1' \in W$ such that $(w_1,w_1') \in \mathcal{R}_t$, $w_1 \equiv w$, and $w_1' \equiv w'$. Since $M,w' \models_\mathcal{A} \psi'$, $M,w_1' \models_\mathcal{A} \psi'$ because $\psi' \in FL(\varphi)$ and $w' \equiv w_1'$. Hence, $M,w_1 \models_\mathcal{A} \langle t\rangle\psi'$ because $(w_1,w_1') \in \mathcal{R}_t$. Finally, $M,w \models_\mathcal{A} \langle t\rangle\psi'$ since $\langle t\rangle\psi' \in FL(\varphi)$ and $w \equiv w'$.

Assume that $(\overline{w},\overline{w'}) \in \mathcal{R}_t^{FL(\varphi)}$ but $(w,w') \notin \mathcal{R}_t$. The pair $(\overline{w},\overline{w'})$ has been added in $\mathcal{R}_t^{FL(\varphi)}$ by the closure operation in order to satisfy an inclusion property of an inclusion axiom of the form $[t]\alpha \supset [s_1]\ldots[s_m]\alpha \in \mathcal{A}$. Then, there exist $\overline{w_1}$, ..., $\overline{w_{m-1}}$ such that $(\overline{w_0},\overline{w_1}) \in \mathcal{R}_{s_1}^{FL(\varphi)}$, ..., $(\overline{w_{m-1}},\overline{w_m}) \in \mathcal{R}_{s_m}^{FL(\varphi)}$, where w_0 is w and w_m is w'. Now, in turn, for each $(\overline{w_{i-1}},\overline{w_i}) \in \mathcal{R}_{s_i}^{FL(\varphi)}$, for $i = 1,\ldots,n$, either the pair $(\overline{w_{i-1}},\overline{w_i})$, has been added by the closure operation or not. Going on this way, we have $(\overline{v_0},\overline{v_1}) \in \mathcal{R}_{t_1}^{FL(\varphi)}$, ..., $(\overline{v_{h-1}},\overline{v_h}) \in \mathcal{R}_{t_h}^{FL(\varphi)}$ such that the corresponding pairs belong to $\mathcal{R}_t$ and $t \Rightarrow_G^* t_1\cdots t_h$, v_0 is w_0 (that, in turn, is w), and v_h is w_m (that, in turn, is w'). By construction, there exist $v_{i-1}', v_i'' \in W$ such that $(v_{i-1}',v_i'') \in \mathcal{R}_t^{FL(\varphi)}$ and $v_{i-1} \equiv v_{i-1}'$ and $v_i \equiv v_i''$, for $i = 1,\ldots,h$.

Assume that $t \Rightarrow_G^* t_1\cdots t_h$ is the derivation $A_0 \Rightarrow_G \sigma_1 A_1 \Rightarrow_G \ldots \Rightarrow_G \sigma_1\cdots\sigma_n A_n \Rightarrow_G \sigma_1\cdots\sigma_n\sigma_{n+1}$, where A_0 is t and $A_n \to \sigma_{n+1}$ and $A_{i-1} \to \sigma_i A_i$, for $i = 1,\ldots,n$, are in P, and that σ_{n+1} is $d_1\cdots d_r$ ($= t_{h-r+1}\cdots t_h$). We know $M,v_h \models_\mathcal{A} \psi'$ and we have to prove that $M,v_{h-r+1} \models_\mathcal{A} \langle d_1\rangle\ldots\langle d_r\rangle\psi'$. Assuming that $\langle d_1\rangle\ldots\langle d_r\rangle\psi' \in FL(\varphi)$ then, we have that $M,v_h'' \models_\mathcal{A} \psi'$ since $v_h \equiv v_h''$ and $\psi' \in FL(\varphi)$. Since $(v_{h-1}',v_h'') \in \mathcal{R}_{t_h}$ and $M,v_h'' \models_\mathcal{A} \psi'$ then, $M,v_{h-1}' \models_\mathcal{A} \langle d_r\rangle\psi'$ and, since $\langle d_r\rangle\psi' \in FL(\varphi)$ and $v_{h-1}' \equiv v_{h-1}''$, we have that $M,v_{h-1}'' \models_\mathcal{A} \langle d_r\rangle\psi'$. We can proceed so on until we have $M,v_{h-r+1}'' \models_\mathcal{A} \langle d_1\rangle\ldots\langle d_r\rangle\psi'$ and $M,v_{h-r+1} \models_\mathcal{A} \langle d_1\rangle\ldots\langle d_r\rangle\psi'$ since $v_{h-r+1} \equiv v_{h-r+1}''$. Now, since the inclusion axiom $[A_n]\alpha \supset [d_1]\ldots[d_r]\alpha$ belongs to $\mathcal{A}$, $M,v_{h-r+1} \models_\mathcal{A} \langle A_n\rangle\psi'$. We can repeat the above argumentation for all derivation steps from A_0 obtaining $M,w \models_\mathcal{A} \langle A_0\rangle\psi'$.

We have now to prove that $\langle d_1\rangle\ldots\langle d_r\rangle\psi' \in FL(\varphi)$. By hypothesis $\langle A_0\rangle\psi' \in FL(\varphi)$ (A_0 is t) and $A_0 \Rightarrow_G^* \sigma_1\cdots\sigma_n\sigma_{n+1}$. Then, by Proposition 1, there exists a non-recursive derivation $A_0 \Rightarrow_G^* \sigma\sigma_{n+1}$, for some $\sigma \in T^*$. By definition of

Fischer-Ladner closure, since $\langle A_0 \rangle \psi' \in FL(\varphi)$, we have $\langle t'_1 \rangle \ldots \langle t'_{n'} \rangle \langle d_1 \rangle \ldots \langle d_r \rangle \psi' \in FL(\varphi)$, where σ is $t'_1 \cdots t'_{n'}$ and σ_{n+1} is $d_1 \cdots d_r$, and, hence, $\langle d_1 \rangle \ldots \langle d_r \rangle \psi' \in FL(\varphi)$.

Lemma 2 (Filtration Lemma). *For all $\psi \in FL(\varphi)$, $M, w \models_{\mathcal{A}} \psi$ if and only if $M^{FL(\varphi)}, \overline{w}, \models_{\mathcal{A}} \psi$.*

Proof. The proof is by induction on the structure of ψ. (*Base* step) For $\psi \in \text{VAR}$ the thesis holds trivially. (*Induction* step) The cases $\psi = \psi' \vee \psi''$ and $\psi = \neg \psi'$ are immediate from the definitions. Assume that $\psi = \langle t \rangle \psi'$. (*If*) If $M, w \models_{\mathcal{A}} \langle t \rangle \psi'$ then there exists w' such that $M, w' \models_{\mathcal{A}} \psi'$ and $(w, w') \in \mathcal{R}_t$. By definition, we have $(\overline{w}, \overline{w'}) \in \mathcal{R}_t^{FL(\varphi)}$ and, by induction hypothesis, $M^{FL(\varphi)}, \overline{w'} \models_{\mathcal{A}} \psi'$. Hence $M^{FL(\varphi)}, \overline{w} \models_{\mathcal{A}} \langle t \rangle \psi'$. (*Only if*) If $M^{FL(\varphi)}, \overline{w} \models_{\mathcal{A}} \langle t \rangle \psi'$ then, there exists $\overline{w'} \in W^{FL(\varphi)}$ such that $M^{FL(\varphi)}, \overline{w'} \models_{\mathcal{A}} \psi'$ and $(\overline{w}, \overline{w'}) \in \mathcal{R}_t^{FL(\varphi)}$. By inductive hypothesis, we have that $M, w' \models_{\mathcal{A}} \psi'$ and, by Lemma 1, since $(\overline{w}, \overline{w'}) \in \mathcal{R}_t^{FL(\varphi)}$, $M, w \models_{\mathcal{A}} \langle t \rangle \psi'$.

Theorem 4 (Small Model Theorem). *Let φ be a satisfiable formula of a grammar logic $\mathcal{I}_{\mathcal{L}}^{\mathcal{A}}$ based on a type-3 grammar G. Then, φ is satisfied in a Kripke $\mathcal{A}$-interpretation with no more that $2^{|FL(\varphi)|}$ states.*

Proof. If φ is satisfiable, then there is a Kripke $\mathcal{A}$-interpretation M and a state w in M such that $M, w \models_{\mathcal{A}} \varphi$. Let $FL(\varphi)$ be the Fischer-Ladner closure of φ. By Lemma 2, $M^{FL(\varphi)}, \overline{w} \models_{\mathcal{A}} \varphi$. Moreover, since, by Proposition 2, $|FL(\varphi)|$ is bounded, the filtration through $FL(\varphi)$ is a finite Kripke interpretation having at most $2^{|FL(\varphi)|}$ worlds, that being the maximum number of ways that worlds can disagree on sentences in $FL(\varphi)$.

Each right regular modal logics, by Theorem 4, is determined by a class of finite standard Kripke interpretations and, hence, it has the *finite model property* [22]. Then, we have the following corollary.

Corollary 3. *The validity problem for the class of right regular modal logics is decidable.*

6 Discussion and related work

In this paper we have established some undecidability results for multimodal logics, reducing well-known unsolvable problems of formal languages to satisfiability problems of multimodal systems by means of a tableau calculus based on prefixed formulas. Moreover, the decidability of the class of multimodal logics based on right regular grammars has been proved using the filtration method introduced by Fischer and Ladner in [12].

In order to have a general framework able to cope with any kind of grammar logics, we have chosen the simplest way of representing models: prefixes are worlds, and relations between them are built step by step by the rules of the calculus. In particular, axioms are used as rewrite rules which create new paths among worlds.

This approach is closely related to the approaches based on prefixes used by Fitting and other authors for classical modal systems (non-multimodal) [14, 25, 9]. There, prefixes are sequences of integers which represent a world as a path in the model that goes from the initial world to it. Thus, instead of representing a model as a graph, as in this paper, a model is represented as a set of paths, which can be considered as a spanning tree of the graph. Although this representation may be more efficient, it requires a specific ν-rule for each logic. Properties of accessibility relations are coded in these rules, and thus, depending on the logic, the ν-rules may express complex relations between prefixes, which instead in our case are explicitly available from the representation. Massacci [25] has proposed a "single step calculus", where ν-rules make use only of immediately accessible prefixes. His approach works for many logics, but it still requires the definition of specific ν-rules.

Besides the disadvantage of requiring specific ν-rules and the fact that they do not work with multimodal systems, we think that though the approach based on prefixes as sequences might be adapted for some subclasses of grammar logics it is difficult to extend it to the whole class. In particular, it can be shown that, for some grammar logic, a "generation lemma" like those used in [25, 17], does not hold, i.e. it is not true that, for any prefix occurring on a branch, all intermediate prefixes occur too. Let us consider, for instance, the derivation of Example 1. We can image to use the prefix $1.1_j.1_p$ to represent the world w_2. Now, by applying axiom A_1, the same world can also be represented with the sequence $1.1_p.1_j$, whose subprefix 1.1_p does not occur on the branch. On the other hand, this subprefix is needed in order to conclude with success the proof. Moreover, adding exsplicitly the subprefixes, as the one above, is not enough to solve the problem, since all prefixes representing the same world have to be identified. Similar consideration can be done for the proposals in [18, 32].

The proposals in [18, 32, 5] address the problem of an efficient implementation of the tableau calculi for a wide class of modal logics. They generalize the prefixes by allowing occurrences of variables and they use unification to show that two prefixes are names for the same world. While a straightforward implementation of our calculus is unlikely to be efficient, the generality of the approach makes it suitable to study the properties of different classes of logics.

Instead of developing specific proof techniques for modal logics, some authors have proposed the alternative approach of translating modal logics into classical first order logic [29]. The translation methods are based on the idea of making explicit reference to the worlds by adding to all predicates an argument representing the world where the predicate holds, so that the modal operators can be transformed into quantifiers of classical logic. In particular, the *functional translation* [30, 1] is based on the idea of representing paths in the possible worlds structure by means of compositions of functions which map worlds to accessible worlds. An advantage of this approach is that it keeps the structure of the original formula. However the approach is suitable mainly for *serial logics*, for which optimization technique have been studied [28, 15], and it requires a different equational unification algorithm for each logic. A way to avoid equational reasoning while retaining the advantages of the functional translation has been developed by Nonnengart [27]. Gasquet in [15] deals with the same class of multimodal logics we have presented, where, however, the seriality is assumed for each modal operator.

Though in this paper we have focused on a propositional language, the tableau calculus we have proposed can be naturally extended to the first order case by introducing the usual rules for quantifiers. Moreover, it can be extended to deal with a wider class of logics. In particular, in [2] a tableau calculus is developed for the class of multimodal logics characterized by "a, b, c, d-incestuality" axioms (defined by Catach in [7]) and, then, as a special case, also for the multimodal logics characterized by *serial*, *symmetric*, and *Euclidean* accessibility relations.

Acknowledgments. The authors would like to thank the referees for the precious advice.

References

1. Y. Auffray and P. Enjalbert. Modal Theorem Proving: An equational viewpoint. *Journal of Logic and Computation*, 2(3):247–297, 1992.
2. M. Baldoni. *Normal Multimodal Logics: Automatic Deduction and Logic Programming Extension*. PhD thesis, Dipartimento di Informatica, Università degli Studi di Torino, 1998.
3. M. Baldoni, L. Giordano, and A. Martelli. A Multimodal Logic to define Modules in Logic Programming. In *Proc. of ILPS'93*, pages 473–487. The MIT Press, 1993.
4. M. Baldoni, L. Giordano, and A. Martelli. A Framework for Modal Logic Programming. In *Proc. of the JICSLP'96*, pages 52–66. The MIT Press, 1996.
5. B. Beckert and R. Goré. Free Variable Tableaux for Propositional Modal Logics. In *Proc. of TABLEAUX'97*, volume 1227 of *LNAI*, pages 91–106. Springer-Verlag, 1997.
6. R. V. Book. Thue Systems as Rewriting Systems. *Journal of Symbolic Computation*, 3(1-2):39–68, 1987.
7. L. Catach. Normal Multimodal Logics. In *Proc. of the AAAI '88*, pages 491–495. Morgan Kaufmann, 1988.
8. L. Catach. TABLEAUX: A General Theorem Prover for Modal Logics. *Journal of Automated Reasoning*, 7(4):489–510, 1991.
9. G. De Giacomo and F. Massacci. Tableaux and Algorithms for Propositional Dynamic Logic with Converse. In *Proc. of CADE-15*, volume 1249 of *LNAI*, pages 613–627. Springer, 1996.
10. P. Enjalbert and L. Fariñas del Cerro. Modal Resolution in Clausal Form. *Theoretical Computer Science*, 65(1):1–33, 1989.
11. L. Fariñas del Cerro and M. Penttonen. Grammar Logics. *Logique et Analyse*, 121-122:123–134, 1988.
12. M. J. Fischer and R. E. Ladner. Propositional Dynamic Logic of Regular Programs. *Journal of Computer and System Sciences*, 18(2):194–211, 1979.

13. M. Fisher and R. Owens. An Introduction to Executable Modal and Temporal Logics. In *Proc. of the IJCAI'93 Workshop on Executable Modal and Temporal Logics*, volume 897 of *LNAI*, pages 1–20. Springer-Verlag, 1993.
14. M. Fitting. *Proof Methods for Modal and Intuitionistic Logics*, volume 169 of *Synthese library*. D. Reidel, Dordrecht, Holland, 1983.
15. O. Gasquet. Optimization of deduction for multi-modal logics. In *Applied Logic: How, What and Why?* Kluwer Academic Publishers, 1993.
16. M. Genesereth and N. Nilsson. *Logical Foundations of Artificial Intelligence*. Morgan Kaufmann, 1987.
17. R. A. Goré. Tableaux Methods for Modal and Temporal Logics. Technical Report TR-ARP-16-95, Automated Reasoning Project, Australian Nat. Univ., 1995.
18. G. Governatori. Labelled Tableaux for Multi-Modal Logics. In *Proc. of TABLEAUX '95*, volume 918 of *LNAI*, pages 79–94. Springer-Verlag, 1995.
19. J. Y. Halpern and Y. Moses. A Guide to Completeness and Complexity for Modal Logics of Knowledge and Belief. *Artificial Intelligence*, 54:319–379, 1992.
20. D. Harel, A. Pnueli, and J. Stavi. Propositional Dynamic Logic of Nonregular Programs. *Journal of Computer and System Sciences*, 26:222–243, 1983.
21. J. E. Hopcroft and J. D. Ullman. *Introduction to automata theory, languages, and computation*. Addison-Wesley Publishing Company, 1979.
22. G. E. Hughes and M. J. Cresswell. *A Companion to Modal Logic*. Meuthuen, 1984.
23. G. E. Hughes and M. J. Cresswell. *A New Introduciton to Modal Logic*. Routledge, 1996.
24. M. Kracth. Highway to the Danger Zone. *Journal of Logic and Computation*, 5(1):93–109, 1995.
25. F. Massacci. Strongly Analytic Tableaux for Normal Modal Logics. In *Proc. of the CADE'94*, volume 814 of *LNAI*, pages 723–737. Springer-Verlag, 1994.
26. A. Nerode. Some Lectures on Modal Logic. In F. L. Bauer, editor, *Logic, Algebra, and Computation*, volume 79 of *NATO ASI Series*. Springer-Verlag, 1989.
27. A. Nonnengart. First-Order Modal Logic Theorem Proving and Functional Simulation. In *Proc. of IJCAI'93*, pages 80–85, 1993.
28. H. J. Ohlbach. Optimized Translation of Multi Modal Logic into Predicate Logic. In *Proc. of the Logic Programming and Automated Reasoning*, volume 822 of *LNAI*, pages 253–264. Springer-Verlag, 1993.
29. H. J. Ohlbach. Translation methods for non-classical logics: An overview. *Bull. of the IGPL*, 1(1):69–89, 1993.
30. H.J. Ohlbach. Semantics-Based Translation Methods for Modal Logics. *Journal of Logic and Computation*, 1(5):691–746, 1991.
31. M.A. Orgun and W. Ma. An overview of temporal and modal logic programming. In *Proc. of the First International Conference on Temporal Logic*, volume 827 of *LNAI*, pages 445–479. Springer-Verlag, 1994.
32. J. Pitt and J. Cunningham. Distributed Modal Theorem Proving with KE. In *Proc. of the TABLEAUX'96*, volume 1071 of *LNAI*, pages 160–176. Springer-Verlag, 1996.
33. M. Wooldridge and N. R. Jennings. Agent Theories, Architectures, and Languages: A survey. In *Proc. of the ECAI-94 Workshop on Agent Theories*, volume 890 of *LNAI*, pages 1–39. Springer-Verlag, 1995.

Hyper Tableau — The Next Generation

Peter Baumgartner[*]

Universität Koblenz · Institut für Informatik
Rheinau 1 · D–56075 Koblenz · Germany
peter@informatik.uni-koblenz.de

Abstract. "Hyper tableau" is a sound and complete calculus for first-order clausal logic. The present paper introduces an improvement which removes the major weakness of the calculus, which is the need to (at least partially) blindly guess ground-instantiations for certain clauses. This guessing is now replaced by a unification-driven technique.
The calculus is presented in detail, which includes a completeness proof. Completeness is proven by using a novel approach to extract a model from an open branch. This enables semantical redundancy criteria which are not present in related approaches.

1 Introduction

In [BFN96] a clausal normal form tableau calculus called "hyper tableau" was introduced. This calculus was motivated by the possibility to keep many desirable features of analytic tableaux (such as a model construction for an open branch, branch-local and thus space-efficient clause generation and taking advantage of the rich structure of tableaux), while also taking advantage of the central idea from (positive) hyper resolution, namely to resolve away all negative literals of a clause in a single inference step. Unlike other tableau calculi, and similar to resolution calculi, hyper tableau permit a systematic branch saturation approach; a hyper tableau proof procedure thus does not have to start with a new tableau from scratch once the ressources are exhausted on the current tableau during itertive deepening.

Variants of (ground) hyper tableaux have been used for efficient minimal model reasoning [Nie96], for diagnosis applications [BFFN97] and to compute database updates [AB97]. Hyper tableau like calculi have been also applied in provers like SATCHMO [MB88,LRW95] and the MGTP system [FH91]. However, these calculi ground-instantiate all clauses during the tableau construction.

The hyper tableau calculus of [BFN96] improves on this by allowing branch-local universally quantified variables. Consider for instance, a disjunction $p(x,y) \lor q(x)$. When brought into the tableau, however, a ground instance for the variable x has to be guessed, say $f(a)$, because x appears in more than one positive literal. Hence extension with $p(f(a), y) \lor q(f(a))$ would be carried out. In general,

---

[*] Supported by the DFG within the research programme "Deduction" under grant Fu 263-2.

all ground instantiations of the clause have to be enumerated and brought into the tableau. But notice that this still improves on e.g. SATCHMO because y is universally quantified in $p(f(a), y)$, and e.g. the unit clause $p(f(a), b)$ would be subsumed and hence is redundant. The benefit of this use of universal variables was also demonstrated in [BFN96].

The purpose of the present paper is to describe an improvement of the hyper tableau calculus, such that its major weakness is eliminated, namely the guessing of ground instantiations for variables occurring in more than one positive literal. In order to achieve this, ideas from instance-generating calculi like Lee and Plaised's hyper-linking [LP92] and Billons disconnection calculus [Bil96] are adapted (differences to these calculi are discussed in Section 7 below).

The new hyper tableau calculus consists of an interplay between two inference rules: the Link rule generates branch-local instances of input clauses in a demand-driven way. The clauses generated in this way can be used by the other inference rule (the Ext rule) in hyper-resolution like extension steps. An important difference here is that the notion of branch closure is based on variant-ship of literals rather than syntactic equality (modulo negation).

Completeness is shown by constructing a model from an open branch which is closed under application of the inference rules. This construction is loosely related to the model-generation approaches for A-ordered tableaux [KH94] and for ordered resolution [BG94]. The model construction enables redundancy criteria which are not present in related approaches (cf. Section 7).

The rest of this paper is structured as follows: next we recall some preliminaries. Then we give a preview of the calculus by stating some examples. Then a more technical part comes which describes the calculus formally. Then the completeness of an improved version is proven. The last Section comments on related work and future improvements.

2 Preliminaries

We apply the usual notions of first-order logic, in a way consistent to [CL73]. For notions related to tableau calculi in general see [Fit90]; our primary interest however is in clausal tableaux similar to those in [LMG94].

A *clause* is a multiset of literals, written as a disjunction $A_1 \vee \cdots \vee A_m \vee \neg B_1 \vee \cdots \vee \neg B_n$ (where $m, n \geq 0$ and the A's and B's are atoms.), or in implication-style as $A_1, \ldots, A_m \leftarrow B_1, \ldots, B_n$ or $\mathcal{A} \leftarrow \mathcal{B}$, where $\mathcal{A} = \{A_1, \ldots, A_m\}$ and $\mathcal{B} = \{B_1, \ldots, B_n\}$. The literals $\mathcal{A}$ are called *head literals* and the literals $\mathcal{B}$ are called *body literals*. Clauses with $m \geq 1$ are also called *program clauses*.

A (Herbrand) interpretation $\mathcal{I}$ (for a given language) is represented as a (possibly infinite) set of atoms, such that atom A is true in $\mathcal{I}$ iff $A \in \mathcal{I}$. As usual, $\mathcal{I} \models X$ means X is true in $\mathcal{I}$ where X is a sentence or set of sentences (interpreted conjunctively). In particular, $\mathcal{I} \models \mathcal{A} \leftarrow \mathcal{B}$ iff $\mathcal{B}\gamma \subseteq \mathcal{I}$ implies $\mathcal{A}\gamma \cap \mathcal{I} \neq \emptyset$ for every ground substitution γ for $\mathcal{A} \leftarrow \mathcal{B}$.

We consider literal trees $\mathcal{T}$, i.e. finite, ordered trees, all nodes of which, except the root, are labelled with a literal. If L is a literal then $[L]$ ambiguously denotes some node N in $\mathcal{T}$ which is labelled with L. A branch of length n consisting of the nodes $N_0, N_1, \ldots, N_n$ with root N_0 and leaf N_n is usually written as $[L_1 \cdot \ldots \cdot L_n]$ where L_i is the label of N_i. The letters p and q are branch-valued variables, and if $p = [L_1 \cdot \ldots \cdot L_{n-1}]$ then $p \cdot [L_n]$ is the branch $[L_1 \cdot \ldots \cdot L_{n-1} \cdot L_n]$ (we assume that $[L_n]$ is a new node). Any (not necessarily strict) prefix $[L_1 \cdot \ldots \cdot L_m]$ of a branch $p = [L_1 \cdot \ldots \cdot L_m \cdot L_{m+1} \cdot \ldots \cdot L_n]$ is called a *partial branch (of p)*. By $[\,]$ we denote both the root node and the partial branch from the root node to the root node.

Branches may be labelled with a "$\star$" as *closed*; branches which are not closed are *open*. A tableau is *closed* if each of its branches is closed, otherwise it is *open*.

A literal tree is represented as the set of its branches; branch sets are denoted by the letters $\mathcal{P}, \mathcal{Q}$. We write $\mathcal{P}, \mathcal{Q}$ and mean $\mathcal{P} \cup \mathcal{Q}$. Similarly, $p, \mathcal{Q}$ means $\{p\}, \mathcal{Q}$. We write $X \in p$ iff X occurs in p, where X is a node or a literal label of some node in p.

The *extension of p with clause* $C = L_1 \vee \ldots \vee L_n$, written as $p \circ C$, is the branch set $p \cdot [L_1], \ldots, p \cdot [L_n]$. Equivalently, in tree view this operation extends the branch p by n new nodes $N_1, \ldots, N_n$ which are labelled with the respective literals from C. Here we say that C is the *tableau clause of N_i*, (for every i, $1 \leq i \leq n$). The tableau clause C of N_i is also denoted by $\mathsf{cl}(N_i)$.

For literals A and B we define $A \gtrsim B$, A *is more general than* B, iff there is a substitution σ_A such that $A\sigma_A = B$; A and B are *variants*, written as $A \sim B$, iff $A \gtrsim B$ and $B \gtrsim A$; A is *strictly more general* than B, $A > B$, iff $A \gtrsim B$ and not $A \sim B$. B is also said to be a *strict*, or *proper* instance of A then.

3 Informal Description of the Calculus

We want to preview our calculus of *hyper tableau*. For this we give two examples and show hyper tableau derivations. A hyper tableau derivation for a (possibly non-ground) clause set $\mathcal{C}$ is the construction of a closed clausal tableau (i.e. a tableau where every branch is labelled as closed), starting with the tableau which consists of the root node only. The tableaux are equipped with *branch selection*: for every open tableau exactly one open branch is *selected* (arbitrarily), and inferences may be carried out to this selected branch only.

The tableau construction must be *fair* to the application of the two inference rules Ext and Link modulo some redundancy. As usual, this means that every possible application of an inference rule must be carried out eventually unless shown to be redundant.

Besides an Init rule to set up the initial tableau, there are two inference rules: the Ext and the Link rule. The purpose of the Ext rule is to extend or close the selected branch. The Ext rule does not instantiate its "ressources" (i.e. branch literals). The purpose of the Link rule is to generate new instances of input clauses, so that Ext will be applicable again. Link is in a sense complementary to Ext in that at least one of its ressources *must* be properly instantiated.

We first consider the Ext rule; its application can be described as follows (cf. Def. 3): let p be the selected branch; take a clause $\mathcal{A} \leftarrow \mathcal{B}$ from the "current clause set" $\mathcal{C}^-$ (which is initialised with the given input clause set $\mathcal{C}$), and apply to p the β rule with $\mathcal{A} \leftarrow \mathcal{B}$, i.e. we split the clause below the leaf of p. *But this is done only if there is a most general substitution σ such that every element $B\sigma \in \mathcal{B}\sigma$ is equal to a variant of an literal L from p (see also Def. 3).* Then, all new branches with leaf $\neg B\sigma$ where $B\sigma \in \mathcal{B}\sigma$ are labelled as "closed"; the new branches (if any) with leafs from $\mathcal{A}\sigma$ are labelled as "open". If there is an open branch in the resulting tableau, select one.

Some terminology: this occurrence of the clause $\mathcal{A}\sigma \leftarrow \mathcal{B}\sigma$ is called a *tableau clause (of every branch passing through one of the literals of $\mathcal{A}\sigma \leftarrow \mathcal{B}\sigma$)*, and if the selected branch passes through a $A\sigma \in \mathcal{A}\sigma$ then we say that $A\sigma$ *is selected in $\mathcal{A}\sigma \leftarrow \mathcal{B}\sigma$*, which is denoted by $\mathsf{sel}(\mathcal{A}\sigma \leftarrow \mathcal{B}\sigma)$.

Example 1 (The Ext *Inference Rule).* Consider the following clause set[1] $\mathcal{C}_1$:

$$p(x,x) \leftarrow \qquad\qquad\qquad\qquad\qquad\qquad\qquad\qquad (R)$$
$$p(x,y), p(y,z) \leftarrow p(x,z) \qquad\qquad\qquad\qquad\qquad\qquad (T)$$

Figure 1 shows a hyper tableau derivation from $\mathcal{C}_1$. There, tableau $\boxed{A}$ is obtained from the tableau which consists of the root only (the body literal condition for (R) is trivially satisfied); tableau $\boxed{B}$ is obtained from $\boxed{A}$ by extension with (T); the substitution σ is $\{z \mapsto x\}$; similarly, $\boxed{C}$ is obtained from $\boxed{B}$ by extension with (T) and the empty substitution.

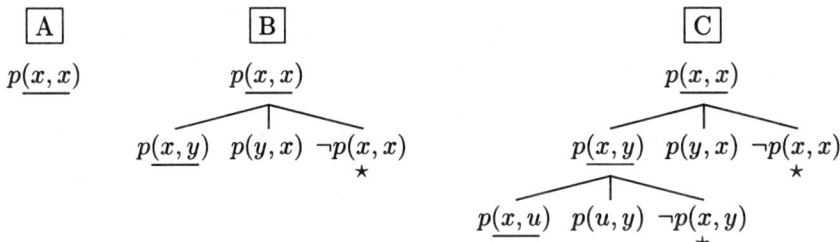

Fig. 1. A hyper tableau derivation from $\mathcal{C}_1$; underlining is used to indicate the selected branch and also to indicate the selected literals in the tableau clauses. For convenience only the same variable names are used for the tableau clauses; that is, each tableau clause is quantified individually. Closed branches are labelled with a $\star$.

It is fair not to apply the Ext rule any further, because extension with (R) (or (T)) would result in a new tableau clause (R) (or (T)) for which there is a variant as a tableau clause contained already. Consequently, the derivation stops here, because the other inference rule — Link — is not applicable.

Obviously, the Ext rule alone is not sufficient to achieve completeness, because the clause set $\{p(x) \leftarrow, \leftarrow p(a)\}$ would admit no refutation.

---

[1] The letters u, v, w, x, y, z denote variables.

The second inference rule of hyper tableau — the Link inference — can be described as follows (cf. Def. 3): let p be the selected branch; take a clause $A \leftarrow B$ from the current clause set C^-, and let σ be a most general multiset unifier $B\sigma = \{\mathsf{sel}(C_1), \ldots, \mathsf{sel}(C_n)\}\sigma$, where the C_i's are new variants of some tableau clauses of p. Furthermore, in order to avoid overlap with the Ext rule, we require that $C_i\sigma \not\sim C_i$, for some i, $1 \leq i \leq n$, i.e. at least one $C_i\sigma$ must be a proper instance of C_i.

If this holds, then consecutively add $C_1\sigma, \ldots, C_n\sigma$ to the current clause set C^-, except those $C_i\sigma$ for which a variant is present already.

Example 2 (The Link Inference Rule). Consider the following clause set[2] C_2:

$$p(x,y), q(x,y) \leftarrow \qquad (C)$$
$$\leftarrow p(y,a) \qquad (D)$$

Figure 2 shows a hyper tableau derivation from C_2. There, tableau $\boxed{A}$ is obtained by an Ext step with (C). Now, Ext is not applicable any more, in particular not with clause (D). However, $p(y,a)$ unifies with $p(x,y)$, the selected literal in the tableau clause $p(x,y), q(x,y) \leftarrow$ (more precisely, we have to take a new variant of $p(x,y), q(x,y) \leftarrow$ since variables are shared). Hence a Link step is applicable, resulting in the (proper) instance $(C') = p(x,a), q(x,a) \leftarrow$. This instance (C') is added to the current clause set C^-. Now, Ext becomes applicable again, resulting in the tableau $\boxed{B}$. Extension with (D) closes the branch, as indicated in tableau $\boxed{C}$. Now, to the selected branch p in $\boxed{C}$, Ext need no longer be applied, because both (C) and (C') are contained as a tableau clause of p already, and Ext with (D) cannot be applied. Further, a Link step with (D) applied to the tableau clause (C) p generates only a variant of clause which is already contained in C^-, namely (C'). Since it is fair never to add variants, the derivation stops now.

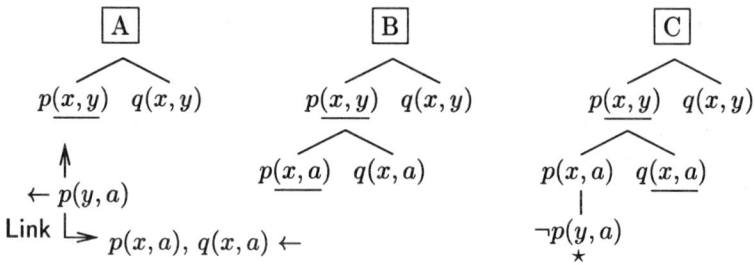

Fig. 2. A hyper tableau derivation from C_2.

In this example we left the impression that C^- is *global* to the tableau under construction. However, below we will define C^- to be *branch local* as $C^-(p)$

---
[2] The letters $a, b, c, \ldots$ denote constants.

such that $C^-(p)$ contains only those clauses which are in the input clause set C or added by Link inferences applied previously to the (not necessarily strict) prefixes of p. For instance $C^-(q)$, where q is the rightmost branch in $\boxed{C}$ in Figure 2, consists of the clauses (C) and (D) only, because (C') was generated on a different branch.

This branch-locality is expected to be important in practice in order to keep space consumption low. On the other side, since all generated clauses by Link are consequences of the input clause set, it is admissible (sound) to add them in any branch as well.

As a special case note that for the propositional logic, Link is never applicable, because Link can only be applied if some tableau clause is *properly* instantiated, which obviously is impossible for the ground case. Hence, Ext alone suffices then. Using propositional logic we can also indicate a weakness of this basic version. Consider the ground clause set $\{(A, B \leftarrow), (A, C \leftarrow)\}$. Using Ext two times a tableau can be constructed by first extending with $A, B \leftarrow$ and then below A with $A, C \leftarrow$. Clearly, there should be no need to apply the second Ext step as it duplicates the atom A on the branch. Every "serious" calculus prunes such steps. For instance, within connection calculi the "regularity check" accomplishes this [LMG94]. Within hyper tableau, the "semantical" reason for pruning the second step is that we are considering after the first step a model candidate which makes A true and hence renders $A, C \leftarrow$ true as well.

We solve this problem more generally by deriving from the selected branch an interpretation $\mathcal{I}(p)$ and (among other things) forbid extension with clauses which are true in $\mathcal{I}(p)$. This technique will be described in Section 5.

4 Hyper Tableau Calculus

In this section the inference rules of hyper tableau will be introduced. Then fairness of derivations is defined. In the next section fairness will then be refined to include a redundancy concept.

Some preliminaries: we consider literal trees equipped with a *branch selection function* which assigns to every open literal tree one of its open branches. We write $p, \mathcal{P}$ to indicate that p is selected in the branch set $p, \mathcal{P}$.

Further, every open branch p is labelled with a finite set of clauses, which is denoted by $C^-(p)$. Intentionally, $C^-(p)$ provides the "current clause set" whose members can be used for extension steps (cf. the informal presentation in Section 3). Alternatively, we will also write $\langle p, C^- \rangle$ and mean the branch p with $C^-(p) = C^-$.

The set $C^-(p)$ is complemented by the set $C^+(p)$ of tableaus clauses of p, i.e. those clauses which were used in extension steps to construct p. Since p is a "path" through $C^+(p)$ (in the connection method sense) it is natural that p determines a respective selection of head literals of the clauses in $C^+(p)$. More generally, a *clause with selection* is a program clause where one of its head literals L is labeled (in some distinguished way), and L is called the *selected literal*, which is denoted by $\text{sel}(C)$. A *clause set with selection* consists of clauses with selection only. In order to extract from a branch its clause set with selection we define:

$$\mathcal{C}^+([L_1 \cdot \ldots \cdot L_n]) = \{\mathsf{cl}([L_1]), \ldots, \mathsf{cl}([L_n])\}, \quad \text{where}$$
$$\mathsf{sel}(\mathsf{cl}([L_i])) = L_i, \quad \text{for } 1 \leq i \leq n.$$

We indicate selection by underlining, i.e. we write $A_1, \ldots A_{i-1}, \underline{A_i}, A_{i+1}, A_m \leftarrow B_1, \ldots, B_n$, where $i \in \{1, \ldots, m\}$, and mean $\mathsf{sel}(A_1, \ldots, A_m \leftarrow B_1, \ldots, B_n) = A_i$.

Two clauses with selection are considered as equal iff they they consist of the same literals and the same literals are selected. The qualification "disregarding selection" means to read a clause with selection as a clause without selection.

Two clauses with selection are variants iff they are variants disregarding selection; the same holds for the instance relation.

Definition 3 (Hyper Tableau Inference Rules). *The calculus of hyper tableau consists of the following inference rules:*

The Init *Inference Rule:*

$$\frac{\mathcal{C}}{\langle [], \mathcal{C}^- \rangle} \quad \text{Init}$$

for given finite clause set $\mathcal{C}$ without selection, where $\mathcal{C}^- = \mathcal{C}$.

The Ext *Inference Rule:*

$$\frac{\langle p, \mathcal{C}^- \rangle, \mathcal{P} \quad A \leftarrow B}{p \circ ((\mathcal{A} \leftarrow \mathcal{B})\sigma), \mathcal{P}} \quad \text{Ext}$$

where

1. *$p, \mathcal{P}$ is a branch set with selected branch p, and*
2. *$(A \leftarrow B) \in \mathcal{C}^-$, and*
3. *$C_1, \ldots, C_n$ are new and pairwise disjoint variants from clauses from $\mathcal{C}^+(p)$, with the same selected literals[3], and*
4. *σ is a most general multiset unifier $\mathcal{B}\sigma = \{\mathsf{sel}(C_1), \ldots, \mathsf{sel}(C_n)\}\sigma$, and*
5. *$C_i \sim C_i\sigma$, for every i, $1 \leq i \leq n$, and*
6. *every new branch $p \cdot [\neg B\sigma] \in p \circ ((\mathcal{A} \leftarrow \mathcal{B})\sigma)$, where $B \in \mathcal{B}$, is closed, and*
7. *every new branch $p \cdot [A\sigma] \in ((\mathcal{A} \leftarrow \mathcal{B})\sigma)$, where $A \in \mathcal{A}$, is open and $\mathcal{C}^-(p \cdot [A\sigma]) = \mathcal{C}^-$.*

The Link *Inference Rule:*

$$\frac{\langle p, \mathcal{C}^- \rangle, \mathcal{P} \quad A \leftarrow B}{\langle p, \mathcal{C}^- \cup \{C_1\sigma, \ldots, C_n\sigma\}\rangle, \mathcal{P}} \quad \text{Link}$$

where

---

[3] More precisely: the selected literal of the variant is such that when the variant is renamed back to the original clause, the selected literals will be the same.

1. $p, \mathcal{P}$ is a branch set with selected branch p, and
2. $(\mathcal{A} \leftarrow \mathcal{B}) \in \mathcal{C}^-$, and
3. $C_1, \ldots, C_n$ are new and pairwise disjoint variants from clauses from $\mathcal{C}^+(p)$, with the same selected literals, and
4. σ is a most general multiset unifier $\mathcal{B}\sigma = \{\text{sel}(C_1), \ldots, \text{sel}(C_n)\}\sigma$, and
5. $C_i \sigma \not\sim C_i$, for some i, $1 \leq i \leq n$.

The Init inference rule is used to setup an initial tableau consisting of the root only. For sample applications of the Ext and Link rules I refer back to Section 3.

In both Link and Ext, *all* the body literals $\mathcal{B}$ have to be "solved" simultaneously. This similarity to hyper resolution coined the name hyper tableau.

Notice that by the Conditions 5, the Link and Ext rule are exclusive wrt. the same $\mathcal{A} \leftarrow \mathcal{B}$, clauses $C_1, \ldots, C_n$ and σ.

With the Link inference rule it is not excluded that $\mathcal{C}^-(p)$ contains multiple variants of a clause. As a further improvement it is safe to keep only one variant for each clause in $\mathcal{C}^-$.

Definition 4 (Hyper Tableau Derivation). *A (hyper tableau) derivation from a set of clauses $\mathcal{C}$ is a (possibly infinite) sequence $\mathcal{P}_1, \ldots, \mathcal{P}_n, \ldots$ of branch sets (each element is also called a* hyper tableau (for $\mathcal{C}$)*), such that*

1. *$\mathcal{P}_1$ is obtained by an application of the* Init *inference rule to $\mathcal{C}$, and*
2. *for $i > 1$, $\mathcal{P}_i$ is obtained from $\mathcal{P}_{i-1}$ by one single application of either the* Ext *or* Link *inference rule.*

A derivation which contains (and thus ends in) a closed tableau is also called a (hyper tableau) refutation.

While completeness will be discussed separately below, we can comment on soundness here. In brief, the hyper tableau calculus is sound, because, first, the Link rule generates only instances of input clauses, hence logical consequences thereof. Second, in order to see that a closed hyper tableau $\mathcal{P}$ implies that the input clause set is unsatisfiable, think of replacing every variable in $\mathcal{P}$ by some constant (the same for every variable). Since variants of atoms become equal then, every branch contains a pair of complementary literals. Using the usual soundness result for tableau calculi we can conclude that the set of ground instances contained in the tableau is unsatisfiable. Since these all are instances of input clauses, the input clause set must be unsatisfiable itself.

As in resolution calculi, the calculus inference rules can be applied in a don't-care nondeterministic way, as long as no possible application of an inference rule is deferred infinitely long. In other words, a concept of fairness is needed.

Definition 5 (I-Paths, Finishedness, Fairness). *Let $\mathcal{D} = (p_1, \mathcal{P}_1), \ldots, (p_n, \mathcal{P}_n), \ldots$ be a derivation which is not a refutation, where p_i is the selected branch in $(p_i, \mathcal{P}_i)$ in the i-th hyper tableau. Any possibly infinite sequence*

$$P = \underbrace{(p_1 = N_0)}_{=: q_0}, \underbrace{(N_0 \cdot N_1)}_{=: q_1}, \ldots, \underbrace{(N_0 \cdot N_1 \cdot \ldots \cdot N_j)}_{=: q_j}, \ldots$$

such that $q_j = p_{s_j}$ for some selected branch p_{s_j} in $\mathcal{D}$ is called an i-path (of the derivation). An i-path P is finished iff for every of its elements the following two conditions hold:

1. If $\dfrac{\langle q_j, \mathcal{C}_j^- \rangle, \mathcal{P}_j \quad \mathcal{A} \leftarrow \mathcal{B}}{q_j \circ ((\mathcal{A} \leftarrow \mathcal{B})\sigma), \mathcal{P}_j}$ Ext then $(\mathcal{A} \leftarrow \mathcal{B})\sigma \in \mathcal{C}^+(q_m)$ modulo variants and disregarding selection for some selected branch q_m in P.

2. If $\dfrac{\langle q_j, \mathcal{C}_j^- \rangle, \mathcal{P}_j \quad \mathcal{A} \leftarrow \mathcal{B}}{\langle q_j, (\mathcal{C}_j^-)' \rangle, \mathcal{P}_j}$ Link then for every clause $C \in (\mathcal{C}_j^-)'$ there is a selected branch q_m in P with $C \in \mathcal{C}^-(q_m)$ modulo variants and disregarding selection. The relation "$C \in \mathcal{C}$ modulo variants" is defined as "$D \sim C$ for some $D \in \mathcal{C}$".

Derivation $\mathcal{D}$ is fair iff $\mathcal{D}$ is a refutation or some i-path of $\mathcal{D}$ is finished.

That is, by an i-path we trace the stepwise extension of some branch in $\mathcal{D}$. This need not necessarily be the sequence of selected branches in the derivation, as any derivation is free to temporarily shift the focus away from a branch and return to it later, or subtrees might be closed. For example, the selected branches of the derivation in Figure 2 from left to right are $[], [p(x,y)], [p(x,y), p(x,a)]$, $[p(x,y), q(x,a)]$, and (the only) i-path, which is also finished, is

$$P = [], [p(x,y)], [p(x,y), q(x,a)] \ .$$

Note 6 (Concrete Fair Strategy).
The finishedness conditions can easily be achieved by actually carrying out the inferences in the if-part eventually. One possible overall fair strategy consists of setting some resource bound, e.g. "maximal term depth", followed by alternate exhaustation in the following way: first exhaust on given selected branch p all Link inferences (it is straightforward to show that there are only finitely many of those, even without resource bound when function symbols are present). Then (finitely) exhaust all Ext inferences within the resource bound. If no refutation was found, then increase the resource bound by some value and continue on the exhausted branch.

5 Model Construction and Improvements

In this section we consider additional refinements which will allow to considerably restrict the application of the Link and Ext inference rules. These restrictions are based on semantical concepts which will be introduced next.

Definition 7 (Productive Clauses). *Let C be a program clause with selection, and let $\mathcal{C}_C^+$ be a possibly infinite set of clauses with selection of instances of C and let $\mathcal{C}_C^-$ be a set of clauses without selection of instances of C. We say that C produces[4] ground atom A wrt. $\langle \mathcal{C}_C^+, \mathcal{C}_C^- \rangle$, iff there is a ground substitution γ for C such that*

---
[4] This notion is borrowed from [BG94] and has a similar meaning for ordered resolution.

1. $A = \mathsf{sel}(C)\gamma$, and
2. there is no $D \in \mathcal{C}_C^+$ with $C > D$ and $D \gtrsim C\gamma$ and $\mathsf{sel}(D) \gtrsim A$, and
3. there is no $D \in \mathcal{C}_C^-$ with $C > D$ and $D \gtrsim C\gamma$.

Let $p, \mathcal{P}$ be a hyper tableau for $\mathcal{C}$ with selected branch p. Let $C \in \mathcal{C}$. Define

$$\mathcal{C}_C^+(p) = \{D \in \mathcal{C}^+(p) \mid C \gtrsim D\} \quad \text{and} \quad \mathcal{C}_C^-(p) = \{D \in \mathcal{C}^-(p) \mid C \gtrsim D\} \ .$$

We say that C produces *ground atom* A *wrt* p iff C produces A wrt $\langle \mathcal{C}_C^+(p), \mathcal{C}_C^-(p)\rangle$.

The intention of "producing clauses" is this: for given A we see if there is an instance of C in the given "positive" set $\mathcal{C}_C^+$ such that $\mathsf{sel}(C)$ can be instantiated to A. Condition 2 expresses that there is no more specific clause which achieves this. If condition 3 does not hold then there is a proper instance of C in the negative set $\mathcal{C}_C^-$ which "cancels" C.

For example, if $C = \underline{p(x,y)}, r(y,z) \leftarrow q(x)$ and $\mathcal{C}_C^+ = \{(p(x,x), \underline{r(x,z)} \leftarrow q(x))\}$ and $\mathcal{C}_C^- = \emptyset$ then C produces $p(a,b)$, but C does not produces $p(a,a)$ (neither does the clause in $\mathcal{C}_C^+$ produce $p(a,a)$). Now, if $\mathcal{C}_C^- = \{(p(x,c), r(c,z) \leftarrow q(x))\}$ instead, then C still produces $p(a,b)$ but no longer produces $p(a,c)$. This is because the clause in $\mathcal{C}_C^-$ "cancels" any appropriate ground substitution for C.

In order to take advantage of the just defined concepts (and to prove completeness below) we have to slightly generalise to i-paths:

Definition 8 (Semantics of i-paths). *Let $\mathcal{D}$ be a derivation from $\mathcal{C}$ and let $P = q_0, \ldots, q_j, \ldots$ be an i-path of $\mathcal{D}$. For any $C \in \mathcal{C}$ define*

$$\mathcal{C}_C^+(P) = \{D \in \mathcal{C}^+(P) \mid C \gtrsim D\}, \quad \text{where } \mathcal{C}^+(P) = \bigcup_{j \geq 0} \mathcal{C}^+(q_j) \ ,$$
$$\mathcal{C}_C^-(P) = \{D \in \mathcal{C}^-(P) \mid C \gtrsim D\}, \quad \text{where } \mathcal{C}^-(P) = \bigcup_{j \geq 0} \mathcal{C}^-(q_j) \ .$$

We say that C produces ground atom A wrt P iff C produces A wrt $\langle \mathcal{C}_C^+(P), \mathcal{C}_C^-(P)\rangle$. We assign an interpretation $\mathcal{I}(P)$ to P as follows:

$$\mathcal{I}(P) = \{A \mid C \text{ produces } A \text{ wrt } P \text{ for some } C \in \mathcal{C}^+(P)\}$$

That is, an atom A is true in this model construction iff it is produced by some clause coming up eventually as tableau clause in the derivation; the head literal producing A is determined by the open branch passing through it. In other words, only atoms can be true which are instances of atoms on the branch (this is "typical" for tableau model constructions), but the converse does not hold.

Note that the definition also gives the construction of an interpretation for given branch p, by taking the finite prefix ending in p of any i-path P which contains p.

For instance, in Figure 1, the selected branch in $\boxed{\mathrm{C}}$ renders all instances of $p(x,y)$ as true, and the interpretation associated to the selected branch in $\boxed{\mathrm{C}}$ in Figure 2 assigns true to $p(a,b)$, because $p(a,b)$ is an instance of the selected literal in $\underline{p(x,y)} \vee q(x,y)$; $q(a,b)$ is not an instance of the selected literal in $p(x,a) \vee \underline{q(x,a)}$ and is false. Similarly, $q(a,a)$ is true, but $p(a,a)$ is false.

A special case is propositional logic: whenever ground atom A is on a branch p, the clause C it stems from produces A wrt. p or any extension of p, because there are no strict instance of C which could prevent C from producing A. Hence, any Ext steps with a clause with head containing A are redundant and thus need not be carried out below p.

Note 9 (Chain Property and Compactness). Notice that as a property of the inference rules this *chain property* holds:

$$\mathcal{C}^+(q_0) \subseteq \mathcal{C}^+(q_1) \subseteq \cdots \subseteq \mathcal{C}^+(q_j) \subseteq \cdots \quad \text{and}$$
$$\mathcal{C}^-(q_0) \subseteq \mathcal{C}^-(q_1) \subseteq \cdots \subseteq \mathcal{C}^-(q_j) \subseteq \cdots$$

It is straightforward to show a compactness property, namely that $\mathcal{C} \subset \mathcal{C}^+(P)$ for finite $\mathcal{C}$ iff $\mathcal{C} \subseteq \mathcal{C}^+(q_j)$ for some q_j (and, of course, the same holds for $\mathcal{C} \subset \mathcal{C}^-(P)$).

We will use this property in the completeness proof below.

Definition 10 (Redundant Inferences, Restricted Fairness). *Let $\mathcal{D}$ be a derivation with i-path P.*

An application of the Ext *inference rule (cf. Def. 3) is redundant (in P) iff $\mathcal{I}(P) \models (\mathcal{A} \leftarrow \mathcal{B})\sigma$.*

An application of the Link *inference rule is redundant (in P) iff (a) some clause from $\{C_1, \ldots, C_n\}$ does not produce any ground instance of $B\sigma$ wrt. P, or (b) $\mathcal{I}(P) \models (\mathcal{A} \leftarrow \mathcal{B})\sigma$. In both cases non-redundant means "not redundant".*

We define the notion finished i-path wrt. non-redundant inferences *to be the same as* finished i-path *(cf. Def. 5) except that the qualification "or the inference is redundant in P" is added to the then-part in both conditions. A derivation $\mathcal{D}$ is fair wrt. non-redundant inferences iff $\mathcal{D}$ is a refutation or some i-path of $\mathcal{D}$ is finished wrt. non-redundant inferences.*

In practice these restrictions can be approximated by not carrying out an inference which is redundant in the current selected branch. This is possible because if an inference is redundant in all branches $p_j, p_{j+1}, \ldots$ starting from some j, then it is also redundant in P.

For instance, the Ext step leading to tableau $\boxed{C}$ in Figure 1 is redundant because (T) is true in the interpretation associated to the selected branch in $\boxed{B}$ (since all instances of $p(x,y)$ are true. If in Example 2 we assume the additional input clause $r(a) \leftarrow p(a,a)$ then a Link step with this clause in the tableau $\boxed{C}$ is redundant, because the tableau clause $p(x,y), q(x,y) \leftarrow$ used for that step does not produce $p(a,a)$.

6 Completeness

In this section we will prove the completeness of the improved version.

In the sequel let $\mathcal{D} = (p_1, \mathcal{P}_1), \ldots, (p_n, \mathcal{P}_n), \ldots$ be a fair derivation from $\mathcal{C}$ which is not a refutation, and let $P = q_0, \ldots, q_j, \ldots$ be a finished i-path of $\mathcal{D}$. Recall that according to Definition 5, for $q \geq 0$ we have $q_j = p_{s_j}$ for some selected branch p_{s_j} in $\mathcal{D}$.

Lemma 11 (Finite Production Property). *If $A \in \mathcal{I}(P)$ then for some j there is a clause $C \in \mathcal{C}^+(q_j)$ such that C produces A wrt. q_k for every $k \geq j$*

The relevance of this lemma is the property that every member A of $\mathcal{I}(P)$ will be produced by some C after finitely many steps (of course) and that C remains productive for C afterwards. In other words, A will be true at some point and remain true afterwards. We will need this for completeness below.

Proof. Let $A \in \mathcal{I}(P)$ be given. By Def. 8, there is a $C \in \mathcal{C}^+(P)$ such that C produces A wrt. P, i.e. C produces A wrt. $\langle \mathcal{C}_C^+(P), \mathcal{C}_C^-(P) \rangle$. By Def. 7, there is a ground substitution γ for C such that

1. $A = \mathsf{sel}(C)\gamma$, and
2. there is no $D \in \mathcal{C}_C^+(P)$ with $C > D$ and $D \gtrsim C\gamma$ and $\mathsf{sel}(D) \gtrsim A$, and
3. there is no $D \in \mathcal{C}_C^-(P)$ with $C > D$ and $D \gtrsim C\gamma$.

In order to have $C \in \mathcal{C}^+(P)$ it must be that $C \in \mathcal{C}^+(q_j)$ for some j.

To the contrary of the assumption that C produces A wrt. q_k for every $k \geq j$, suppose that (case 1) for some $k \geq j$ there is $D \in \mathcal{C}_C^+(q_k)$ with $C > D$ and $D \gtrsim C\gamma$ and $\mathsf{sel}(D) \gtrsim A$. But since also trivially $D \in \mathcal{C}_C^+(P)$ we immediately arrive at a contradiction to item 2.

Case 2, that for some $k \geq j$ there is a $D \in \mathcal{C}_C^-(q_k)$ with $C > D$ and $D \gtrsim C\gamma$ is similar: since trivially $D \in \mathcal{C}_C^-(P)$ we immediately arrive at a contradiction to item 3.

Consequently, C produces A wrt. q_k for every $k \geq j$. □

Theorem 12 (Model Existence of Open Hyper Tableaux). *Let $\mathcal{D}$ be a hyper tableau derivation from clause set $\mathcal{C}$ which is fair wrt. non-redundant inferences. If $\mathcal{D}$ is not a refutation then $\mathcal{C}$ is satisfiable. More specifically, there is at least one finished i-path P of $\mathcal{D}$, and $\mathcal{I}(P) \models \mathcal{C}$ for every finished i-path P of $\mathcal{D}$.*

Notice that this theorem immediately gives refutational completeness by taking the contrapositive.

Proof. Suppose that $\mathcal{D}$ is not a refutation. The existence of i-path P as claimed follows trivially from fairness. The non-trivial part is the model construction.

We need the following well-foundedness property: there is no infinite chain $(C = C_1) > C_2 > \ldots > C_n > \ldots$ of instances of clause C such that every $C_i \gtrsim C\gamma$ for given ground instance $C\gamma$. This is, because every C_i is a proper instance of C_{i-1}, and any infinite instantiation of C will necessarily result in a clause C_j with higher term depth than $C\gamma$, and hence $C_j \gtrsim C\gamma$ would not hold.

Hence for given (possibly infinitely many) C_i's and $C\gamma$ we can always find some $C_j \gtrsim C\gamma$ such that there is no C_k with $C_j > C_k \gtrsim C\gamma$. Let us refer to this situation by "C_j is the most specific generalisation of $C\gamma$ (wrt. the given C_i's)".

Assume, to the contrary of the claim, that $\mathcal{I}(P) \not\models \mathcal{C}$. Then, there is a ground instance $(\mathcal{A} \leftarrow \mathcal{B})\gamma$ of a clause $C = (\mathcal{A} \leftarrow \mathcal{B}) \in \mathcal{C}$ such that $\mathcal{B}\gamma \subseteq \mathcal{I}(P)$ and $\mathcal{A}\gamma \cap \mathcal{I} = \emptyset$. There is a most specific generalisation of $C\gamma$ in $\mathcal{C}^-(P)$ wrt. $\mathcal{C}^-(P)$. Without loss of generality let C itself be this clause.

Write $\mathcal{B}$ as the multiset $\mathcal{B} = \{B_1, \ldots, B_n\}$. By Lemma 11 for every $B_i\gamma$ there is a j_i and a clause $C_i \in \mathcal{C}^+(q_{j_i})$ such that C_i produces $B_i\gamma$ wrt. q_k for every $k \geq j_i$. By taking $m = \max\{j_1, \ldots, j_n\}$ we conclude that C_i produces $B_i\gamma$ wrt. q_l, for every $l \geq m$. Further, by the chain property (Note 9) we have $C_i \in \mathcal{C}^+(q_m)$.

By definition of productivity, there are ground substitutions $\gamma_1, \ldots, \gamma_n$ such that $\mathsf{sel}(C_i)\gamma_i = B_i\gamma$. Without loss of generality we can assume that the C_i's are pairwise variable disjoint and also disjoint from $\mathcal{A} \leftarrow \mathcal{B}$ (we can always find such variants and modify the γ_i's). But then it holds (possible after restricting the domains of the γ_i's to B_i) that $\mathsf{sel}(C_i)\gamma_1 \cdots \gamma_n\gamma = B_i\gamma_1 \cdots \gamma_n\gamma$. Then there is also a most general multiset unifier $\mathcal{B}\sigma = \{\mathsf{sel}(C_1), \ldots, \mathsf{sel}(C_n)\}\sigma$ and a substitution δ such that $\mathsf{sel}(C_i)\sigma\delta = \mathsf{sel}(C_i)\gamma_i = B_i\gamma = B_i\sigma\delta$.

Case 1. If for some i, $1 \leq i \leq n$, $C_i\sigma \not\sim C_i$ (*), then all the conditions to apply a Link step to q_m in the following way hold (for some $\mathcal{P}$):

$$\frac{\langle q_m, \mathcal{C}^-\rangle, \mathcal{P} \quad \mathcal{A} \leftarrow \mathcal{B}}{\langle q_m, \mathcal{C}^- \overset{.}{\cup} \{C_1\sigma, \ldots, C_n\sigma\}\rangle, \mathcal{P}} \text{ Link}$$

Since $(\mathcal{A} \leftarrow \mathcal{B})\gamma$ is false in $\mathcal{I}(P)$, it follows that $\mathcal{I}(P) \not\models (\mathcal{A} \leftarrow \mathcal{B})\sigma$. Furthermore, as shown, every C_i produces some ground instance of B_i, namely $B_i\gamma$. Hence, this inference is not redundant. Thus, since $\mathcal{D}$ is fair wrt. non-redundant inferences, every element from $\mathcal{C}^- \overset{.}{\cup} \{C_1\sigma, \ldots, C_n\sigma\}$ will be contained in $\mathcal{C}^-(P)$ eventually (either nothing has to be added, or by carrying out this Link step or any other Link step which add the $C_i\sigma$'s). In particular, that $C_i\sigma$ with $C_i\sigma \not\sim C_i$ must be generated eventually, say $C_i\sigma \in \mathcal{C}^-(q_g)$ modulo variants and disregarding selection for some selected branch q_g.

Notice that from $C_i\sigma \not\sim C_i$ it follows $C_i > C_i\sigma$. Now, if $g \leq m$ then also $C_i\sigma \in \mathcal{C}^-(q_m)$, which contradicts the fact that C_i produces $B_i\gamma$ wrt. q_m (by virtue of $C_i\sigma$, cf. Def. 7).

If $g > m$ then recall that C_i was shown above to produce $B_i\gamma$ wrt. q_l, for every $l \geq m$. But then C_i must produce $B_i\gamma$ wrt. q_g as well; again, this is impossible due to the existence of $C_i\sigma \in \mathcal{C}^-(q_g)$.

Hence, in both cases we get a contradiction, which renders case (1) impossible.

Case 2. (Complement to case 1). For every i, $1 \leq i \leq n$, $C_i\sigma \sim C_i$ (*), then all the conditions to apply a Ext step to q_m in the following way hold (for some $\mathcal{P}$):

$$\frac{\langle q_m, \mathcal{C}^-\rangle, \mathcal{P} \quad \mathcal{A} \leftarrow \mathcal{B}}{q_m \circ ((\mathcal{A} \leftarrow \mathcal{B})\sigma), \mathcal{P}} \text{ Ext}$$

Since $(\mathcal{A} \leftarrow \mathcal{B})\gamma$ is false in $\mathcal{I}(P)$, it follows that $\mathcal{I}(P) \not\models (\mathcal{A} \leftarrow \mathcal{B})\sigma$. Hence, this inference is not redundant. Thus, since $\mathcal{D}$ is fair wrt. non-redundant inferences, $(\mathcal{A} \leftarrow \mathcal{B})\sigma \in \mathcal{C}^+(q_g)$ modulo variants and disregarding selection for some selected branch q_g (this can be achieved by carrying out this Ext step or any other Ext step which adds a variant of $(\mathcal{A} \leftarrow \mathcal{B})\sigma$). Hence $(\mathcal{A} \leftarrow \mathcal{B})\sigma \in \mathcal{C}^+(P)$.

Notice that $\mathcal{A} = \emptyset$ is impossible because then q_g would be closed, contradicting the fact that q_g is contained in the considered i-path P.

By the well-foundedness property above we can find a most specific generalisation $C' \in \mathcal{C}^+(P)$ of $C\gamma$ wrt. $\mathcal{C}^+(P)$ (disregarding selection). This C' can either be $C\sigma$ itself or some (later or previously) added instance of $C\sigma$.

Now the case that C' produces some literal from $\mathcal{A}\gamma$ wrt. $\mathcal{C}^+(P)$ is impossible, because then $C\gamma$ would be true in $\mathcal{I}(P)$.

Hence (cf. Def. 7) for the ground instance $C'\gamma'$ of C' with $C'\gamma' = C\gamma$ there must be some $D' \in \mathcal{C}^-(P)$ with $C' > D'$ and $D' \gtrsim C'\gamma'$ (the other case, item 2 in Def. 7 is impossible because C' is a most specific generalisation of $C\gamma$ wrt. $\mathcal{C}^+(P)$).

But now from $C \gtrsim C\sigma \gtrsim C' > D' \gtrsim C'\gamma' = C\gamma$ we have a contradiction to the assumption that C is a most specific generalisation of $C\gamma$ (because D' is). Hence $\mathcal{I}(P) \models C\gamma$ and the claim follows. □

Corollary 13 (Decision Procedure for Bernays-Schönfinkel Class). *The hyper tableau calculus (both versions) is a decision procedure for the Bernays-Schönfinkel class, i.e. for formulas of the form $\exists^*\forall^*Q$ where Q is some quantifier-free formula.*

Proof. The clausal form of the formula contains no function symbols, but no other restrictions on the syntactic structure apply. It suffices to show that any fair derivation will neccessarily end after finitely many steps. The reason for this is simple: whenever Link adds a clause C to the current set $\mathcal{C}^-(p)$, C must be a *strict* instance of some clause in $\mathcal{C}^-(p)$. Clearly, in the absence of function symbols this cannot be done infinitely often. Further, application of the Ext rule is also finitely bounded, because extension is never carried out if a variant of the clause to be extended with is a tableau clause already. □

This property is remarkable, as there seems to be no resolution or free variable tableau variant developed until now which achieves this (at least without taking bounds on the maximal length of clauses into consideration, which depends on the number of constants of the clause set). No resolution variant in [Joy76] or in Leitsch's recent book [Lei97] accomplishes this. For instance, hyper resolution will loop on the clause set $\mathcal{C}_1$ in Example 1 because all resolvents of the form $p(x_1, x_2) \vee p(x_2, x_3) \vee \ldots \vee p(x_{n-1}, x_n) \vee p(x_n, x_1)$ will be generated, and subsumption is not powerful enough to prevent this.

Other calculi which decide this class are Billon's disconnection calculus [Bil96] or the hyper linking family of calculi [LP92]. This demonstrates that the hyper tableau and related calculi are very different from resolution calculi.

7 Conclusions

Related work. First of all, the hyper tableau calculus in [BFN96] can be seen as a predecessor of the calculus here. While the calculus in [BFN96] has to guess ground instantiation for variables occuring in clause heads in different literals, such as x in $p(x,y), q(x) \leftarrow$, the present version avoids this by a unification-based approach. This is the central contribution of the present paper.

In [Küh97] a variant of a hyper tableau calculus is described which treats the problematic variables (such as x) "rigidly". This calculus can be seen as an alternative to the method proposed here. Unfortunately, completeness is still an open problem.

Hyper tableau is different from hyper resolution [Rob65] in several aspects. Most importantly, hyper tableaux are analytical, i.e. no new clauses are added (short of instances). By this we achieve that hyper tableau is a decision procedure for the function-free case (cf. Corollary 13). Further, the generation of clause instances is branch-local (cf. Section 3), whereas the clauses in resolution are stored globally.

Another related calculus is analytic resolution [Bra76]. It can be setup to act similar to hyper resolution, but with the important difference that no clauses are merged when building resolvents; instead, the electron clauses and the resolvents would all be kept separately but obeying variable dependencies in a stack-like manner[5]. Analytic resolution then essentially handles positive disjunctions which are augmented by a list of conditions which are also positive disjunctions. An important difference of analytic resolution (in fact, all variants of resolution) to hyper tableau is that in resolution literals count as contradictory if they are complementary, whereas in hyper tableau literals count as contradictory if they are variants (with opposite sign). This is an essential difference, as it is needed to achieve that hyper tableau decide the Bernays-Schönfinkel class.

As indicated in Note 6, the fairness condition for hyper tableau can be implemented easily. This distinguishes hyper tableau from all implementations of free-variable tableau calculi developed so far[6]. Since free-variable tableaux are proof confluent, there is *in principle* a way to implement them without retracting a once derived tableau and without backtracking on a once applied substitution (provers like HARP [OS88] which use the γ-rule to substitute variables by *ground* terms do not count). However, no implementation takes advantage of this property (e.g. [BP95,BHOS96]).

The work most closely related to hyper tableau probably is the disconnection method (DCM) [Bil96]. Hyper tableau shares with the DCM the property of generating instances of clauses (this also holds for hyper-linking [LP92], but which does not employ the concept of a "path"). However, the closure condition is different. In DCM, $p(x,y)$ and $\neg p(u,u)$ constitute a link, because they are complementary when all variables are replace by the *same* constant; in hyper tableau a closure condition based on variants is used. In hyper tableau, clause instances are generated branch-local, whereas they are global in DCM. Importantly, as stated in [Bil96], DCM is not compatible to hyper-type of inferences. Hyper tableau can thus be seen as a way to bring in the hyper-type inferences to a DCM like calculi. As in resolution, hyper-type inferences restrict the possible inferences. For instance, if $p(a,a)$ and $p(b,b)$ both are on the current branch, then no inference with $\leftarrow p(x,a), p(x,b)$ is possible because the body cannot be

---

[5] It would be an interesting exercise to recast the very procedural formulation of analytic resolution in an analytic tableau framework.

[6] To my knowledge, admittedly.

solved simultaneously. However an inference in DCM is possible. Furthermore, DCM does not take advantage of a redundancy criterion as hyper tableau does. For instance, with $\underline{p(x,y)}$, $q(x)$ as single tableau clause and $p(y,x) \leftarrow p(x,y)$ as input clause, Ext is not applicable because $p(y,x)$ is true in the model given by $p(x,y)$, whereas DCM would have to build the connection.

Future work. The hyper tableau calculus of [BFN96] takes advantage of branch-local variables by treating them universally quantified. This technique should be made available to the current version of hyper tableau as well, and it would generalise the unit-clause improvements of DCM.

"Cutting out" clauses from a closed subtree which do not affect the closed-property of the subtree (introduced as *condensing* in [OS88]) is an important technique in the hyper tablaux of [BFN96]. For efficiency reasons I expect it to be mandatory here as well.

Alternatives for the model construction should be investigated to achieve a higher degree of redundant inferences. For instance, it seems possible in the model construction to forget the clause bodies and look at the heads only. Another way to achieve this is to alter the calculus and to combine it with ideas from analytic resolution [Bra76]; the combined calculus would essentially handle conditional positive disjunctions as in analytic resolution (see above) but keep the branch closure condition of hyper tableau. This calculus would be an even more natural generalization of hyper tableau (previous generation) for the Horn case.

Acknowledgements. Many thanks to Reiner Hähnle for numerous valuable comments.

References

[AB97]   Chandrabose Aravindan and Peter Baumgartner. A Rational and Efficient Algorithm for View Deletion in Databases. In Jan Maluszynski, editor, *Logic Programming - Proceedings of the 1997 International Symposium*, Port Jefferson, New York, 1997. The MIT Press.

[BFFN97]   Peter Baumgartner, Peter Fröhlich, Ulrich Furbach, and Wolfgang Nejdl. Semantically Guided Theorem Proving for Diagnosis Applications. In *15th International Joint Conference on Artificial Intelligence (IJCAI 97)*, pages 460–465, Nagoya, 1997. International Joint Conference on Artificial Intelligence.

[BFN96]   Peter Baumgartner, Ulrich Furbach, and Ilkka Niemelä. Hyper Tableaux. In *Proc. JELIA 96*, number 1126 in Lecture Notes in Aritificial Intelligence. European Workshop on Logic in AI, Springer, 1996.

[BG94]   Leo Bachmair and Harald Ganzinger. Rewrite-based equational theorem proving with selection and simplification. *Journal of Logic and Computation*, 4(3):217–247, 1994.

[BHOS96]   Bernhard Beckert, Reiner Hähnle, Peter Oel, and Martin Sulzmann. The tableau-based theorem prover $_3T^AP$, version 4.0. In *Proceedings, 13th International Conference on Automated Deduction (CADE), New Brunswick,*

[Bil96] Jean-Paul Billon. The Disconnection Method. In Miglioli et al. [MMMO96]. NJ, USA, volume 1104 of *Lecture Notes in Computer Science*, pages 303–307. Springer, 1996.

[BP95] Bernhard Beckert and Joachim Posegga. leanT^AP: Lean tableau-based deduction. *Journal of Automated Reasoning*, 15(3):339–358, 1995.

[Bra76] D. Brand. Analytic Resolution in Theorem Proving. *Artificial Intelligence*, 7:285–318, 1976.

[CL73] C. Chang and R. Lee. *Symbolic Logic and Mechanical Theorem Proving*. Academic Press, 1973.

[FH91] H. Fujita and R. Hasegawa. A Model Generation Theorem Prover in KL1 using a Ramified-Stack Algorithm. In *Proc. of the Eigth International Conference on Logic Programming*, pages 535–548, Paris, France, 1991.

[Fit90] M. Fitting. *First Order Logic and Automated Theorem Proving*. Texts and Monographs in Computer Science. Springer, 1990.

[Joy76] W.H. Joyner. Resolution Strategies as Decision Procedures. *Journal of the ACM*, 23(3):396–417, 1976.

[KH94] Stefan Klingenbeck and Reiner Hähnle. Semantic tableaux with ordering restrictions. In A. Bundy, editor, *Proc. CADE-12*, volume 814 of *LNAI*, pages 708–722. Springer, 1994.

[Küh97] Michael Kühn. Rigid Hypertableaux. In *Proc. of KI '97*, Lecture Notes in Aritificial Intelligence. Springer, 1997.

[Lei97] Alexander Leitsch. *The Resolution Calculus*. Springer, 1997.

[LMG94] R. Letz, K. Mayr, and C. Goller. Controlled Integrations of the Cut Rule into Connection Tableau Calculi. *Journal of Automated Reasoning*, 13, 1994.

[LP92] S.-J. Lee and D. Plaisted. Eliminating Duplicates with the Hyper-Linking Strategy. *Journal of Automated Reasoning*, 9:25–42, 1992.

[LRW95] D. Loveland, D. Reed, and D. Wilson. SATCHMORE: SATCHMO with RElevance. *Journal of Automated Reasoning*, 14:325–351, 1995.

[MB88] Rainer Manthey and François Bry. SATCHMO: a theorem prover implemented in Prolog. In Ewing Lusk and Ross Overbeek, editors, *Proceedings of the 9th Conference on Automated Deduction, Argonne, Illinois, May 1988*, volume 310 of *Lecture Notes in Computer Science*, pages 415–434. Springer, 1988.

[MMMO96] P. Miglioli, U. Moscato, D. Mundici, and M. Ornaghi, editors. *Theorem Proving with Analytic Tableaux and Related Methods*, number 1071 in Lecture Notes in Artificial Intelligence. Springer, 1996.

[Nie96] Ilkka Niemelä. A Tableau Calculus for Minimal Model Reasoning. In Miglioli et al. [MMMO96].

[OS88] F. Oppacher and E. Suen. HARP: A Tableau-Based Theorem Prover. *Journal of Automated Reasoning*, 4:69–100, 1988.

[Rob65] J. A. Robinson. Automated deduction with hyper-resolution. *Internat. J. Comput. Math.*, 1:227–234, 1965.

Fibring Semantic Tableaux

Bernhard Beckert[1,*] and Dov Gabbay[2]

[1] University of Karlsruhe, Institute for Logic, Complexity and Deduction Systems,
D-76128 Karlsruhe, Germany. E-mail: beckert@ira.uka.de
[2] Imperial College, Department of Computing, 180 Queen's Gate,
London SW7 2BZ, UK. E-mail: dg@ic.ac.uk

Abstract. The methodology of *fibring* is a successful framework for combining logical systems based on combining their semantics. In this paper, we extend the fibring approach to *calculi* for logical systems: we describe how to uniformly construct a sound and complete tableau calculus for the combined logic from calculi for the component logics.
We consider semantic tableau calculi that satisfy certain conditions and are therefore known to be "well-behaved"—such that fibring is possible. The identification and formulation of conditions that are neither too weak nor too strong is a main contribution of this paper.
As an example, we fibre tableau calculi for first order predicate logic and for the modal logic K.

1 Introduction

The methodology of *fibring* is a successful framework for combining logical systems based on combining their semantics [7, 6, 8]. The basic idea is to combine the structures defining the semantics of two logics $\mathbf{L}_1$ and $\mathbf{L}_2$ such that the result can be used to define semantics for expressions from the combined languages of $\mathbf{L}_1$ and $\mathbf{L}_2$. The general assumption is that these structures have components like, for example, the worlds in Kripke structures; to build fibred structures, *fibring functions* $F_{(1,2)}$ are defined assigning to each constituent w of an $\mathbf{L}_1$-model $\mathbf{m}_1$ an $\mathbf{L}_2$-model $\mathbf{m}_2$. An $\mathbf{L}_2$-expression is evaluated in w, where its value is undefined, by instead evaluating it in $\mathbf{m}_2 = F_{(1,2)}(w)$. The full power of the fibring method is revealed when this process it iterated to define a semantics for the logic $\mathbf{L}_{[1,2]}$, where the operators of the component logics can occur arbitrarily nested in formulae. Fibring has been successfully used in many areas of logic to combine systems and define their semantics; for an overview see [7].

In this paper, we extend the fibring approach to *calculi* for logical systems: we describe how to uniformly construct a sound and complete tableau calculus for the combined logic from calculi for the component logics. Since tableau calculi are known for most "basic" logics [5] (including classical logic, modal logic, intuitionistic logic, and temporal logic), calculi can be obtained for all "complex" logics that can be constructed by fibring basic logics, such as modal predicate logic, intuitionistic temporal logic, etc.

---

* This work was carried out during a visit at Imperial College, London, UK.

One cannot fibre just any proof procedures for two logics in a uniform way. First, "proving" can have different meanings in different logics: deciding (or semi-deciding) satisfiability or validity, computing a satisfying variable instantiation, etc. Second, it is not clear where to "plug in" the proof procedure for $\mathbf{L}_2$ into that for $\mathbf{L}_1$; a proof procedure may do something completely different from what (the definition of) the valuation function does that provides the truth value of a formula in a given model. For example, if the procedure P_1 is based on constructing a (counter) model, whereas the procedure P_2 uses a resolution calculus, they cannot be fibred (at least not uniformly).

Therefore, we consider semantic tableau calculi that satisfy certain conditions and are, thus, known to be "well-behaved"—such that fibring is possible (for some substructural logics, e.g. linear logic, no such "well-behaved" calculi exist). The identification and formulation of conditions that are neither too weak nor too strong is a main contribution of this paper.

If the components that are fibred satisfy these conditions, then the resulting calculus is automatically sound and complete. It may only be a semi-decision procedure, i.e., only terminate for unsatisfiable input formulae, even if its components are decision procedures; this, however, is not surprising because a fibred logic may be undecidable even if its components are decidable.

Related work includes [4], where a method for fibring tableau calculi for substructural implication logics has been presented. In [9], a method is described for fibring tableaux for modal logics to construct calculi for multi-modal logics; it can be seen as an instance of the general framework presented here.

We define the notion of a *logical system* in a very general way (Section 2); only indispensable properties of its syntax and semantics are part of the definition without which a useful tableau calculus for the logic cannot exist (or cannot be fibred with calculi for other logical systems).

Similarly, as few restrictions as possible are made regarding the type and form of tableau calculi. In particular, the calculus does not have to be analytical; and the tableau rules do not have to be given in form of rule schemata but can be described in an arbitrary way. The conditions that tableau calculi have to satisfy to be suitable for fibring are described in Section 3. We present two examples of calculi suitable for fibring in Sections 4 and 5: a calculus for first order predicate logic and a calculus for the modal logic K. In Section 6, the method of fibring logics is described in general and syntax and semantics of a fibred logic are defined, based on syntax and semantics of its component logics.

In Section 7, we present our uniform method for constructing a tableau calculus for a fibred logic from calculi for the component logics. The resulting calculus is shown to be sound and complete w.r.t. the semantics of the fibred logic and to be itself suitable for fibring with other calculi. The latter property makes it possible to iterate the fibring of tableau calculi and, thus, to construct a calculus for the fully fibred logic $\mathbf{L}_{[1,2]}$.

As an example, in Section 8, the calculi for first-order and for modal logic introduced in Sections 4 and 5 are fibred resulting in a calculus for modal predicate logic.

Finally, in Section 9 we draw conclusions from our work. Due to space restrictions, all proofs are omitted; they can be found in [2].

2 Logical Systems

In this section, we define the notion of a *logical system* in a very general way; only indispensable properties of its syntax and semantics are part of the definition without which a useful tableau calculus for the logic cannot exist (or cannot be fibred with calculi for other logical systems).

The logic has to have a model semantics that uses Kripke-style models, i.e., models consisting of *worlds* in which formulae are true or false; there are no restrictions on the relationship between these worlds. In fact, any kind of model can be considered to be a Kripke-style model with a *single* world (namely the model itself), including models of classical propositional and first-order logic. However, since the labels of tableau formulae are interpreted as worlds, if there is only one world in the models of a logic, then the interpretation of all labels is the same and they become useless for the calculus.

The restriction that only two-valued logics are considered is solely made for the sake of simplicity. All notions introduced in the following can easily be extended to many-valued logics (but no additional insight is gained).

Definition 1. *Associated with a logical system* **L** *(a logic for short) is a set Sig of signatures*[1] *of* **L**. *For each signature* $\Sigma \in Sig$, *syntax and semantics of the instance* $\mathbf{L}^\Sigma$ *of* **L** *are given by:*

Syntax: *A set* $Form^\Sigma$ *of formulae and a set* $Atom^\Sigma \subset Form^\Sigma$ *of atomic formulae (atoms), where the sets* $Atom^\Sigma$ *and* $Form^\Sigma$ *are decidable.*

Semantics: *A set* $\mathcal{M}^\Sigma$ *of models where each model* $\mathbf{m} \in \mathcal{M}^\Sigma$ *(at least) contains (a) a set* W *of worlds, (b) an initial world* $w^0 \in W$, *and (c) a binary relation* $\models$ *between* W *and* $Form^\Sigma$.

If $w \models \phi$ *for some world* $w \in W$ *and some formula* $\phi \in Form^\Sigma$, *then* ϕ *is said to be* true *in* w, *else it is* false *in* w. *A formula* $\phi \in Form^\Sigma$ *is* satisfied *by a model* $\mathbf{m} \in \mathcal{M}^\Sigma$ *if (and only if) it is true in the initial world* w^0 *of* $\mathbf{m}$. *A set* $G \subset Form^\Sigma$ *of formulae is satisfied by* $\mathbf{m}$ *iff all its elements are satisfied by* $\mathbf{m}$. *A formula* $\phi \in Form^\Sigma$ *(a set* $G \subset Form^\Sigma$ *of formulae) is* satisfiable *if there is a model* $\mathbf{m} \in \mathcal{M}$ *satisfying* ϕ *(resp. G).*

Although usually non-atomic formulae are constructed from atomic formulae, and their truth value is determined by the truth value of the atoms they consist of, this is *not* part of the above definition. However, the existence of a tableau calculus for a logic **L** that is suitable for fibring implies that the truth value of a formula ϕ is strongly related to the truth values of certain atoms (that may or may not be sub-formulae of ϕ).

---

[1] We do not further specify what a signature is; *Sig* can be seen as a set of indices for distinguishing different instances of **L** (which *usually* differ in the symbols they use).

Tableau calculi allow to check the *satisfiability* of a formula; we only consider this property. It may or may not be possible in a certain logic to check whether a formula is valid in some model (true in all worlds) or is a tautology (valid in all models) by reducing this problem to a satisfiability problem; in many logics—though not in all—a formula is a tautology if its negation is not satisfiable.

Often, formulae are used in tableau calculi that are not part of the original but of an extended signature (e.g., formulae containing Skolem symbols):

Definition 2. *Given a logic* L, *a signature* $\Sigma^* \in Sig$ *is an* extension *of a signature* $\Sigma \in Sig$ *(and* $\Sigma \in Sig$ *is a* restriction *of* $\Sigma^* \in Sig$*) if* $Form^\Sigma \subset Form^{\Sigma^*}$ *and* $Atom^\Sigma \subset Atom^{\Sigma^*}$.

In that case, a model $\mathbf{m} \in \mathcal{M}^\Sigma$ *is a* restriction *of a model* $\mathbf{m}^* \in \mathcal{M}^{\Sigma^*}$ *(to the signature* Σ*) if there is a function f that assigns to each world of* $\mathbf{m}$ *a world of* $\mathbf{m}^*$ *such that: (a) the initial world of* $\mathbf{m}^*$ *is assigned to the initial world of* $\mathbf{m}$*; and (b) for all formulae* $\phi \in Form^\Sigma$ *and worlds w of* $\mathbf{m}$*:* $w \models \phi$ *iff* $f(w) \models \phi$.

3 Tableau Calculi and the Conditions they Must Satisfy

As said above, only few restrictions are made regarding the type and form of tableau calculi. Any function that assigns to a tableau branch its (possible) extensions is regarded a tableau rule. Nevertheless, certain conditions have to be met, the first of which ensures that tableau rule applications do not transform the whole tableau in an arbitrary way:

Condition 1. Tableau rule applications have only *local* effects, in that they extend a single branch of a tableau, and do not alter or remove formulae already on the tableau.

The second assumption is that the applicability of a tableau rule to a branch and the result of its application are solely determined by the presence of certain formulae on the branch to which it is applied; no other pre-conditions are allowed such as, for example, the absence of certain formulae, the presence of formulae on different branches, or the order of formulae on a branch:

Condition 2. Whether a tableau branch B can be expanded in a certain way is solely determined by the presence of certain formulae on B (the *premiss* for that expansion).

Condition 2 implies that tableau branches are regarded as sets and that tableau rules are monotonic; thus, when formulae are added to the branch, previous tableau rule applications are not invalidated.

Conditions 1 and 2 intuitively prohibit "strange" behaviour of calculi. There are, however, useful calculi that violate these syntactical restrictions, including (a) calculi where variable substitutions are applied to the whole tableau, (b) calculi with resource restrictions that are not local to a branch (for example linear logic, where a formula can be "used up" globally), and (c) calculi using expansion rules that introduce *new* symbols, i.e., symbols that must not occur on the branch or even the whole tableau. At least the latter type of rules can often be replaced by similar rules satisfying Condition 2:

Example 1. In calculi for first-order predicate logic, often a tableau rule is used that allows to derive $\phi(c)$ from formulae of the form $(\exists x)(\phi(x))$, where c is a constant *new* to the tableau (or the branch); this rule violates Condition 2 because it demands the *absence* of formulae containing c.

If instead a special constant symbol c_ϕ is used, which does not have to be new, then the rule satisfies Condition 2 above. Soundness is preserved provided that c_ϕ is not introduced into the tableau in any other way than by skolemising $(\exists x)(\phi(x))$; in particular, the Skolem constant c_ϕ must not occur in the initial tableau (this is an adaptation of the rule for existential formulae presented in [3] to the ground case [1]).

As said before, we allow formulae from an extended signature Σ^* to be used in a tableau proof: Only the tableau formulae that are tested for satisfiability have to be taken from the non-extended signature Σ; they are put on the inital tableau. During the proof it is allowed, for example, to introduce Skolem symbols that are not elements of Σ. We proceed to formally define our (syntactical) notions of tableaux and tableau calculi:

Definition 3. *Given a logic* **L**, *a signature* $\Sigma \in Sig$, *and a set Lab of labels, a tableau formula* $\sigma{:}S\,\phi$ *consists of a label* $\sigma \in Lab$, *a truth value sign* $S \in \{T, F\}$, *and a formula* $\phi \in Form^\Sigma$; *it is called* atomic *if* $\phi \in Atom^\Sigma$. *The set of all tableau formulae is denoted with* $TabForm^\Sigma$. *A tableau is a finitely branching tree whose nodes are labelled with tableau formulae. A branch of a tableau* T *is a maximal path in* T. *The set of formulae on a branch* B *is denoted with* $Form(B)$.

A tableau calculus $\mathcal{C}$ for a logic **L** has (different) "instances" $\mathcal{C}^\Sigma$ for each signature $\Sigma \in Sig$:

Definition 4. *A* tableau calculus $\mathcal{C}$ *for a logic* **L** *is, for each signature* $\Sigma \in Sig$, *specified by: (a) an extension* $\Sigma^* \in Sig$ *of the signature* Σ; *(b) a set Lab of labels and an initial label* $\sigma^0 \in Lab$; *(c) a tableau (expansion and closure) rule* $\mathcal{R}^\Sigma$, *i.e., a function that assigns to each finite set* $\Pi \subset TabForm^{\Sigma^*}$ *of tableau formulae (each premiss)—and thus to each tableau branch* B *with* $\Pi \subset Form(B)$—*a set* $\mathcal{R}^\Sigma(\Pi)$ *of (possible) conclusions, where a conclusion is a finite set of branch extensions or the symbol* $\bot$ *(branch closure), and a branch extension is a finite set of tableau formulae from* $TabForm^{\Sigma^*}$. *The rule* $\mathcal{R}^\Sigma$ *must satisfy the following conditions: (i)* $\mathcal{R}^\Sigma(\Pi)$ *may be infinite but has to be enumerable; (ii)* $\mathcal{R}^\Sigma(\Pi) \subset \mathcal{R}^\Sigma(\Pi \cup \Pi')$ *for all* $\Pi, \Pi' \subset TabForm^{\Sigma^*}$ *(monotonicity).*

In practice, tableau rules are often described by means of rule schemata. This fits perfectly in our framework, with the exception that different rule schemata are usually considered to define different rules, whereas we consider them to define different sub-cases of one (single) rule.

We now have everything at hand to define what the tableaux for a set G of formulae is and when a tableau is closed. The construction of tableaux for G is in general a non-deterministic process, since there may be any—even an infinite—number of possible conclusions that can be derived from a given premiss.

Definition 5. *Given a tableau calculus C for a logic $\mathbf{L}$ and a signature $\Sigma \in \mathit{Sig}$, the set of all tableaux for a finite set $\Gamma \subset \mathit{TabForm}^\Sigma$ of tableau formulae is inductively defined as follows: (1) A linear tableau whose nodes are labelled with the formulae in Γ is a tableau for Γ (an initial tableau). (2) Let T be a tableau for Γ, B a branch of T, and $C \neq \bot$ a conclusion in $\mathcal{R}^\Sigma(\Pi)$ for a premiss $\Pi \subset \mathit{Form}(B)$. Then a new tableau for Γ can be constructed from T as follows: the branch B is extended by a new sub-branch for each extension E in C, where the nodes in that sub-branch are labelled with the tableau formulae in E.*

T is a tableau for a finite set $G \subset \mathit{Form}^\Sigma$ of formulae if it is a tableau for the set $\{\sigma^0\mathsf{:T}\,\phi \mid \phi \in G\}$ of tableau formulae.

Definition 6. *Given a tableau calculus C for a logic $\mathbf{L}$ and a signature $\Sigma \in \mathit{Sig}$, a tableau branch B is closed iff $\bot \in \mathcal{R}^\Sigma(\Pi)$ for a premiss $\Pi \subset \mathit{Form}(B)$. A tableau is closed if all its branches are closed.*

Conditions 1 and 2 above, which are purely syntactical, still allow calculi to behave "strangely". Formulae could be added to the tableau that syntactically encode knowledge derived from a premiss Π, but whose semantics (i.e., truth value) has nothing to do with that of Π. An extreme example for this is that two symbols of the signature are used to encode the formulae in Π in a binary representation, and tableau rules are employed that operate on that binary representation. Such calculi—though they may be sound and complete—cannot be fibred in a uniform way as an understanding of the encoding would be needed. To assure a more "conservative" behaviour one could impose additional syntactical restrictions, for example only allow tableau rules that are analytic. However, the property of tableau rules that has to be guaranteed is more of a semantic nature: the result of a rule application must be semantically related to its premiss. The first semantical condition (Cond. 3) is part of our definition of the semantics of tableau formulae and tableaux (Def. 7):

Condition 3. The labels that are part of tableau formulae represent worlds in models, and the truth value signs encode truth and falsehood of a formula; they do not contain other information.

Definition 7. *Given a tableau calculus C for a logic $\mathbf{L}$ and a signature $\Sigma \in \mathit{Sig}$, a tableau interpretation for C^Σ is a pair $\langle \mathbf{m}, I \rangle$ where $\mathbf{m} \in \mathcal{M}^{\Sigma^*}$ is a model for the extended signature Σ^* and I is a partial function that assigns to labels $\sigma \in \mathit{Lab}^\Sigma$ worlds of $\mathbf{m}$ such that $I(\sigma^0) = w^0$ (i.e., I assigns to the initial label σ^0 the initial world w^0 of $\mathbf{m}$). A tableau interpretation $\langle \mathbf{m}, I \rangle$ satisfies a tableau formula $\sigma\mathsf{:S}\,\phi \in \mathit{Form}^{\Sigma^*}$ iff $I(\sigma)$ is defined and (a) $\mathsf{S} = \mathsf{T}$ and ϕ is true in $I(\sigma)$ or (b) $\mathsf{S} = \mathsf{F}$ and ϕ is false in $I(\sigma)$. It satisfies a tableau branch B iff it satisfies all tableau formulae on B. It satisfies a tableau iff it satisfies at least one of its branches.*

Often, only a subset of all possible tableau interpretations is used to define the semantics of a tableaux. For example, to define the semantics of first-order tableaux, only tableau interpretations are used whose first part is an Herbrand model. In the following, the set of these tableau interpretations that are actually used to define the semantics of a calculus $\mathcal{C}^\Sigma$ is denoted with $TabInterp^\Sigma$.

The next four conditions we impose to make calculi "well-behaved", which are semantical, resemble the properties that a tableau calculus is shown to have in a classical soundness and completeness proof.

Condition 4. *Appropriateness of the set of tableau interpretations:* If a set $G \subset Form^\Sigma$ is satisfiable, then there is a tableau interpretation in $TabInterp^\Sigma$ that satisfies the initial tableau for G *(which is important for soundness)*; and, if $\langle \mathbf{m}^*, I \rangle$ is such a tableau interpretation, then $\mathbf{m}^*$ can be restricted to a model $\mathbf{m} \in \mathcal{M}^\Sigma$ that satisfies G *(which is important for completeness)*.

Condition 5. *Soundness of expansion (preliminary version):* If there is a tableau interpretation in $TabInterp^\Sigma$ satisfying a tableau T and T' is the result of applying the expansion rule to T then there is a tableau interpretation in $TabInterp^\Sigma$ satisfying T'.

Condition 6. *Soundness of Closure:* If a tableau branch is closed then it is *not* satisfied by any tableau interpretation in $TabInterp^\Sigma$.

Before Condition 7 can be formulated that establishes completeness of a calculus, the notion of a *fully expanded* tableau branch has to be defined. The definition relies on the fact that tableau rules are monotonic (Condition 2); without that property of tableau rules, it is difficult to define the notion of fully expanded branches in a uniform way. Intuitively a branch is fully expanded if no expansion rule application can add any new formulae to the branch.

Definition 8. *Given a tableau calculus $\mathcal{C}$ for a logic $\mathbf{L}$ and a signature $\Sigma \in Sig$, a tableau branch B is* fully expanded *if $E \subset Form(B)$ for all extensions E in all conclusions $C \in \mathcal{R}^\Sigma(\Pi)$ for all premisses $\Pi \subset Form(B)$.*

Condition 7. *Completeness:* If a tableau branch B is fully expanded and not closed then there is a tableau interpretation in $TabInterp^\Sigma$ satisfying B.

Conditions 4–7 ensure soundness and completeness of a tableau calculus:

Theorem 1. *If a tableau calculus $\mathcal{C}$ for a logic $\mathbf{L}$ satisfies Conditions 4–7 for all signatures $\Sigma \in Sig$ then the following holds for all finite sets $G \subset Form^\Sigma$: There is a closed tableau for G if and only if G is not satisfiable.*

To be suitable for fibring, a calculus has to satisfy two additional conditions. The first of these replaces Condition 5:

Condition 8. *Soundness of expansion:* If a tableau T is satisfied by a tableau interpretation in $TabInterp^\Sigma$ and T' is the result of applying the expansion rule to T, then T' is satisfied by *the same* tableau interpretation.

Intuitively, the reason why Condition 8 has to be used instead of Condition 5 is the following: Suppose T is a tableau for a fibred logic $\mathbf{L}_{(1,2)}$, the tableau interpretation $\langle \mathbf{m}_1, I_1 \rangle$ satisfies the $\mathbf{L}_1$-formulae on some branch B of T, the tableau interpretation $\langle \mathbf{m}_2, I_2 \rangle$ satisfies the $\mathbf{L}_2$-formulae on B, and together they form a tableau interpretation of the fibred logic $\mathbf{L}_{(1,2)}$ satisfying the whole branch B and, thus, the tableau T. Now, if the expansion rule for $\mathbf{L}_1$ only preserved satisfiability in *some* model, i.e., the $\mathbf{L}_1$-formulae on an extension B' of B were only satisfied by some different tableau interpretation $\langle \mathbf{m}'_1, I'_1 \rangle$, then a problem would arise if $\langle \mathbf{m}'_1, I'_1 \rangle$ and $\langle \mathbf{m}_2, I_2 \rangle$ are incompatible and do not form a fibred model.

Condition 9. If a tableau branch B is fully expanded then every tableau interpretation in $TabInterp^\Sigma$ satisfying the *atoms* on B satisfies *all* formulae on B.

This last condition ensures that the calculus is "analytical down to the atomic level". It is *not* a syntactical condition and it does *not* imply that the calculus is analytic in the classical sense. The condition is needed to ensure completeness when the calculus is used for fibring.

Example 2. In a tableau calculus for a modal logic that satisfies Condition 9, it must be possible to add the formula $\tau{:}\mathsf{T}\,p$ to a tableau branch containing $\sigma{:}\mathsf{T}\,\Box p$ for all labels τ representing a world reachable from the world represented by σ. In a tableau calculus for classical propositional logic it must be possible to expand a branch containing $\sigma{:}\mathsf{T}\,p \vee q$ by sub-branches containing $\sigma{:}\mathsf{T}\,p$ resp. $\sigma{:}\mathsf{T}\,q$, even if one of these atoms is *pure*, i.e., occurs only positively on the branch.

When the two calculi for propositional and for modal logic are fibred, then a propositional atom may indeed be a modal formula; even if it is pure (viewed as a propositional atom), it may be unsatisfiable as a modal formula. Thus, for example, a propositional calculus must expand the formula $\sigma{:}\mathsf{T}\,\Diamond(r \wedge \neg r) \vee q$ so that $\Diamond(r \wedge \neg r)$ can be passed on to the modal component of the fibred calculus, and its unsatisfiability can be detected.

Definition 9. *A tableau calculus $\mathcal{C}$ for a logic $\mathbf{L}$ is suitable for fibring if, for all signatures $\Sigma \in Sig$, there is a set $TabInterp^\Sigma$ of tableau interpretations such that Conditions 4–9 are satisfied (Condition 1–3 are part of the definition of tableau calculi resp. tableau interpretations).*

4 Example: First-order Predicate Logic

4.1 The Logical System of First-order Predicate Logic

To specify the logical system $\mathbf{L}_{\mathrm{PL1}}$ of first-order predicate logic, the set Sig_{PL1} of signatures and the syntax and semantics of $\mathbf{L}_{\mathrm{PL1}}$ have to be defined.

Signatures: The set Sig_{PL1} consists of all *first-order signatures* $\Sigma = \langle P_\Sigma, F_\Sigma \rangle$ where P_Σ is a set of predicate symbols and F_Σ is a set of function symbols. For skolemisation we do not use symbols from F_Σ but from a special infinite set F_Σ^{sko} of *Skolem function symbols* that is disjoint from F_Σ. The symbols in P_Σ, F_Σ and

F_Σ^{sko} may be used with any arity $n \geq 0$; in particular, function symbols can be used as constant symbols (arity 0).

Syntax: In addition to the predicate and function symbols there is an infinite set *Var* of *object variables*. The *logical operators* are $\vee$ (disjunction), $\wedge$ (conjunction), $\rightarrow$ (implication), and $\neg$ (negation), and the quantifiers $\forall$ and $\exists$. Terms, atoms, and formulae over a signature Σ are constructed as usual. As we use a calculus without free variables, $Form_{\text{PL1}}^\Sigma$ is the set of all formulae over Σ not containing free variables, and $Atom_{\text{PL1}}^\Sigma \subset Form_{\text{PL1}}^\Sigma$ is the set of all ground atoms.

Semantics: A first-order *structure* $\langle D, \mathcal{I} \rangle$ for a signature Σ consists of a domain D and an interpretation $\mathcal{I}$, which gives meaning to the function and predicate symbols of Σ. A *variable assignment* is a mapping $\mu : Var \rightarrow D$ from the set of variables to the domain D. The *evaluation function val* is defined as usual; that is, given a structure $\langle D, \mathcal{I} \rangle$ and a variable assignment μ, it assigns to each formula $\phi \in Form^\Sigma$ a truth value $val_{\mathcal{I},\mu}(\phi) \in \{true, false\}$. As all models must contain a set of worlds (Def. 1), we define $\mathcal{M}_{\text{PL1}}^\Sigma$ to consist of models where the initial and only world w^0 is a first-order structure. The relation $\models_{\text{PL1}}$ is defined by: $w^0 \models_{\text{PL1}} \phi$ iff, for all variable assignments μ, $val_{\mathcal{I},\mu}(\phi) = true$.

4.2 A Tableau Calculus for First-order Predicate Logic

To describe our calculus $\mathcal{C}_{\text{PL1}}$ for first-order predicate logic $\mathbf{L}_{\text{PL1}}$, we have to define, for each signature $\Sigma \in Sig_{\text{PL1}}$, the extension Σ^* to be used for constructing tableaux, the set of labels, the initial label, and the expansion and closure rule.

Extended signature: Since the function symbols in F_Σ^{sko} are used for skolemisation, the extended signature Σ^* is $\langle P_\Sigma, F_\Sigma \cup F_\Sigma^{sko} \rangle$.

Labels: The models of first-order logic consist of only one world; it is represented by the label $*$. Thus, $Lab^\Sigma = \{*\}$, and $*$ is the initial label.

Expansion and closure rule: The set of tableau formulae in $TabForm^\Sigma$ that are not literals is divided into four classes as shown on the right: α for formulae of conjunctive type, β for formulae of disjunctive type, γ for quantified formulae of universal type, and δ for quantified formulae of existential type (unifying notation). To comply with Condition 1, which does not allow the application of substitutions (to the whole tableau), we use the classical *ground* version of tableaux for first-order logic (universally quantified variables are replaced by *ground* terms when the γ-rule is applied.) To comply with Condition 2, we use a δ-rule that does not introduce a *new* Skolem function symbol. Rather, each class of δ-formulae identical up to variable renaming is assigned its own unique Skolem symbol:

α	$\alpha_1, \quad \alpha_2$
$*{:}\mathsf{T}\,(\phi \wedge \psi)$	$*{:}\mathsf{T}\,\phi, *{:}\mathsf{T}\,\psi$
$*{:}\mathsf{F}\,(\phi \vee \psi)$	$*{:}\mathsf{F}\,\phi, *{:}\mathsf{F}\,\psi$
$*{:}\mathsf{F}\,(\phi \rightarrow \psi)$	$*{:}\mathsf{T}\,\phi, *{:}\mathsf{F}\,\psi$
$*{:}\mathsf{T}\,\neg\phi$	$*{:}\mathsf{F}\,\phi, *{:}\mathsf{F}\,\phi$
$*{:}\mathsf{F}\,\neg\phi$	$*{:}\mathsf{T}\,\phi, *{:}\mathsf{T}\,\phi$

β	$\beta_1, \quad \beta_n$
$*{:}\mathsf{T}\,(\phi \vee \psi)$	$*{:}\mathsf{T}\,\phi, *{:}\mathsf{T}\,\psi$
$*{:}\mathsf{F}\,(\phi \wedge \psi)$	$*{:}\mathsf{F}\,\phi, *{:}\mathsf{F}\,\psi$
$*{:}\mathsf{F}\,(\phi \rightarrow \psi)$	$*{:}\mathsf{F}\,\phi, *{:}\mathsf{T}\,\psi$

$\gamma(x)$	$\gamma_1(x)$
$*{:}\mathsf{T}\,(\forall x)(\phi(x))$	$*{:}\mathsf{T}\,\phi(x)$
$*{:}\mathsf{F}\,(\exists x)(\phi(x))$	$*{:}\mathsf{F}\,\phi(x)$

$\delta(x)$	$\delta_1(x)$
$*{:}\mathsf{F}\,(\forall x)(\phi(x))$	$*{:}\mathsf{F}\,\phi(x)$
$*{:}\mathsf{T}\,(\exists x)(\phi(x))$	$*{:}\mathsf{T}\,\phi(x)$

Definition 10. *Given a signature $\Sigma \in Sig_{\mathrm{PL1}}$, the function sko assigns to each δ-formula $\phi \in TabForm^{\Sigma^*}$ a symbol $sko(\phi) \in F_\Sigma^{sko}$ such that (a) $sko(\phi) > f$ for all $f \in F_\Sigma^{sko}$ occurring in ϕ, where $>$ is an arbitrary but fixed ordering on F_Σ^{sko}, and (b) for all δ-formulae $\phi, \phi' \in TabForm^{\Sigma^*}$ the symbols $sko(\phi)$ and $sko(\phi')$ are identical if and only if ϕ and ϕ' are identical up to renaming of quantified variables.*

The purpose of condition (a) in the above definition of *sko* is to avoid cycles like: $sko(\phi) = f$, f occurs in ϕ', $sko(\phi') = g$, and g occurs in ϕ.

The expansion and closure rule $\mathcal{R}_{\mathrm{PL1}}$ of our calculus $\mathcal{C}_{\mathrm{PL1}}$ is formally defined as follows: For all premises $\Pi \subset TabForm_{\mathrm{PL1}}^{\Sigma^*}$, the set $\mathcal{R}_{\mathrm{PL1}}^\Sigma(\Pi)$ of possible conclusions is the smallest set containing the following conclusions (where $\alpha, \beta, \gamma, \delta$ denote tableau formulae of the corresponding type): (a) $\{\{\alpha_1, \alpha_2\}\}$ for all $\alpha \in \Pi$, (b) $\{\{\beta_1\}, \{\beta_2\}\}$ for all $\beta \in \Pi$, (c) $\{\{\gamma_1(t)\}\}$ for all $\gamma \in \Pi$ and all ground terms t over Σ^*, (d) $\{\{\delta_1(c)\}\}$ for all $\delta \in \Pi$ where $c = sko(\delta)$ (Def. 10), (e) $\bot$ if $*\!:\!\mathrm{T}\,\phi, *\!:\!\mathrm{F}\,\phi \in \Pi$ for any $\phi \in Form_{\mathrm{PL1}}^{\Sigma^*}$.

Semantics: We define the semantics of $\mathcal{C}_{\mathrm{PL1}}$-tableaux using tableau interpretations that are *canonical* in the following sense:

Definition 11. *A tableau interpretation for $\mathcal{C}_{\mathrm{PL1}}$ is canonical if its first-order structure $\langle D, \mathcal{I} \rangle$ satisfies the following conditions: (a) D is the set of all ground terms over Σ^*; (b) for all δ-formulae $\delta(x) \in TabForm^{\Sigma^*}$ and all variable assignments μ: if $val_{I,\mu}(\delta(x)) = \mathrm{true}$ then $val_{I,\mu}(\delta_1(c)) = \mathrm{true}$ where $c = sko(\delta)$.*

Using the set $TabInterp_{\mathrm{PL1}}^\Sigma$ of canonical tableau interpretations, the calculus $\mathcal{C}_{\mathrm{PL1}}$ satisfies Conditions 4–9. In particular, if a tableau T is satisfied by a canonical tableau interpretation, then all tableaux constructed from T are satisfied by the same interpretation; and every fully expanded tableau branch that is not closed is satisfied by a canonical interpretation.

Theorem 2. *The tableau calculus $\mathcal{C}_{\mathrm{PL1}}$ for $\mathbf{L}_{\mathrm{PL1}}$ is suitable for fibring.*

5 Example: The Logic $\mathbf{L}_{\mathrm{K}}$ of Modalities

5.1 The Logical System $\mathbf{L}_{\mathrm{K}}$

As a second example, we use the modal logic **K** without binary logical connectives; that is, all formulae are of the form $\circ_1 \cdots \circ_n p$ ($n \geq 0$), where p is a propositional variable and $\circ_i$ is one of the modalities $\Box, \Diamond$ or the negation symbol $-$ (which is used to avoid confusion with first-oder negation $\neg$). We call this logic $\mathbf{L}_{\mathrm{K}}$. The missing connectives are not needed, since $\mathbf{L}_{\mathrm{K}}$ is later fibred with first-oder logic where they are available (Sect. 8).

Signatures: A signature Σ in Sig_{K} is an enumerable non-empty set of primitive propositions.

Syntax: The formulae in $Form_{\mathrm{K}}^\Sigma$ consist of a single element of Σ prefixed by a sequence of the logical operators $\Box, \Diamond, -$. The set $Atom_{\mathrm{K}}^\Sigma$ is identical to Σ.

Semantics: The semantics of $\mathbf{L_K}$ is defined in the usual way using Kripke structures: A model $\mathbf{m}$ in $\mathcal{M}_K^\Sigma$ consists of (a) a non-empty set W of worlds, one of which is the initial world w^0, (b) a binary reachability relation on W, and (c) a valuation V, which is a mapping from Σ to subsets of W. Thus, $V(p)$ is the set of worlds at which p is "true". For primitive propositions p, the relation $\models_K$ is defined by: $w \models_K p$ iff $w \in V(p)$; for complex formulae it is recursively defined by: (a) $w \models_K \neg\phi$ iff not $w \models_K \phi$, (b) $w \models_K \Box\phi$ iff $w' \models_K \phi$ for *all* w' reachable from w, and (c) $w \models_K \Diamond\phi$ iff $w' \models_K \phi$ for *some* w' reachable from w.

5.2 A Tableau Calculus for the Logic $\mathbf{L_K}$

We define a calculus $\mathcal{C}_K$ for $\mathbf{L_K}$ that uses sequences of natural numbers as labels; the world named by $\sigma.n$ is reachable from the world named by σ.

Extended signature: No extension of the signature is needed, thus $\Sigma = \Sigma^*$.

Labels: The set Lab_K of labels is for all Σ inductively defined by: the initial label 1 is a label, and if σ is a label then so is $\sigma.n$ for all natural numbers n.

Expansion and closure rule: To comply with Condition 2, we use a π-rule that does not introduce a *new* label but—similar to the δ-rule in Section 4.2—uses a label that is uniquely assigned to the formula to which the rule is applied.

The expansion and closure rule of our calculus $\mathcal{R}_K^\Sigma$ for the logic $\mathbf{L_K}$ is formally defined as follows: For all premisses $\Pi \subset TabForm_K^\Sigma$, the set $\mathcal{R}_K(\Pi)$ of possible conclusions is the smallest set containing the following conclusions (where *goedel* is any bijection from $Form_K^\Sigma$ to the set of natural numbers): (a) $\{\{\sigma.n:\mathsf{T}\,\phi\}\}$ for all $\sigma:\mathsf{T}\,\Box\phi \in \Pi$ and all labels of the form $\sigma.n$ occurring in Π, (b) $\{\{\sigma.n:\mathsf{F}\,\phi\}\}$ for all $\sigma:\mathsf{F}\,\Diamond\phi \in \Pi$ and all labels of the form $\sigma.n$ occurring in Π, (c) $\{\{\sigma.n:\mathsf{F}\,\phi\}\}$ for all $\sigma:\mathsf{F}\,\Box\phi \in \Pi$ where $n = goedel(\phi)$, (d) $\{\{\sigma.n:\mathsf{T}\,\phi\}\}$ for all $\sigma:\mathsf{T}\,\Diamond\phi \in \Pi$ where $n = goedel(\phi)$, (e) $\{\{\sigma:\mathsf{F}\,\phi\}\}$ for all $\sigma:\mathsf{T}\,\neg\phi \in \Pi$, (f) $\{\{\sigma:\mathsf{T}\,\phi\}\}$ for all $\sigma:\mathsf{F}\,\neg\phi \in \Pi$, (g) $\bot$ if $\sigma:\mathsf{T}\,\phi, \sigma:\mathsf{F}\,\phi \in \Pi$ for any $\phi \in Form_K^\Sigma$.

Semantics: The set $TabInterp_K^\Sigma$ contains *canonical* tableau interpretation satisfying the following condition:

Definition 12. *A tableau interpretation $\langle \mathbf{m}, I \rangle$ for $\mathbf{L_K}$ is canonical if: (a) if $I(\sigma)$ is defined and satisfies $\sigma:\mathsf{T}\,\Diamond\phi$, then $I(\sigma.n)$ is defined and satisfies $\sigma:\mathsf{T}\,\phi$ where $n = goedel(\phi)$; and (b) for all numers n, if $w = I(\sigma)$ and $w' = I(\sigma.n)$ are defined, then the world w' is reachable from w.*

Theorem 3. *The tableau calculus $\mathcal{C}_K$ for $\mathbf{L_K}$ is suitable for fibring.*

6 Fibring Logical Systems

To fibre two logics $\mathbf{L_1}$ and $\mathbf{L_2}$ means to consider a logic whose formulae are constructed from symbols and operators from both logics [7,8]. In a first step we consider a logic $\mathbf{L_{(1,2)}}$ where $\mathbf{L_2}$-formulae can occur inside $\mathbf{L_1}$-formulae but not vice versa.

Example 3. If $\mathbf{L_1} = \mathbf{L_{PL1}}$ and $\mathbf{L_2} = \mathbf{L_K}$, then $(\forall x)(p(x))$, $\Box q$, $(\forall x)(\Box p(x))$, and $(\forall x)(\Box p(x)) \to (\exists x)(\Diamond q(x))$ are formulae of $\mathbf{L_{(1,2)}}$, but $\Box(\forall x)(p(x))$ is not.

The logic $\mathbf{L}_{[1,2]} \equiv \mathbf{L}_{[2,1]}$ that is the full combination of $\mathbf{L}_1$ and $\mathbf{L}_2$, where expressions from the two logics can be nested arbitrarily, can be handled by inductively repeating the construction presented in this section. Similarly, it is possible to combine three or more logics.

We consider $\mathbf{L}_{(1,2)}$ to be a special case of $\mathbf{L}_1$: it contains the formulae of $\mathbf{L}_2$ as (additional) atoms. And, in each world w of an $\mathbf{L}_1$-model, the truth value of the additional atoms, which are $\mathbf{L}_2$-formulae, is the same as that in the initial world of an $\mathbf{L}_2$-model assigned to w. Thus, an $\mathbf{L}_{(1,2)}$-model consists of an $\mathbf{L}_1$-model $\mathbf{m}_1$ and a *fibring function* F that assigns to each world w in $\mathbf{m}_1$ an $\mathbf{L}_2$-model. Intuitively, when an $\mathbf{L}_2$-formula is to be evaluated in w, where its value is undefined, it is evaluated in $\mathbf{m}_2 = F(w)$ instead. In most cases, certain restrictions have to be imposed on F to make sure that the fibred models define the desired semantics. These restrictions are given in form of a relation $\mathcal{P}$ between $\mathbf{L}_1$-models, $\mathbf{L}_1$-worlds, and $\mathbf{L}_2$-models; a fibring function can be used for an $\mathbf{L}_1$-model $\mathbf{m}_1$ if $\mathcal{P}(\mathbf{m}_1, w, F_{(1,2)}(w))$ holds for all worlds w of $\mathbf{m}_1$.

Example 4. A proposition may be represented by different atoms p_1 and p_2 in $\mathbf{L}_1$ and $\mathbf{L}_2$. Then, for the semantics defined by the fibred models to be useful, one imposes the restriction that, if p_1 is true in a world w of $\mathbf{m}_1$ then p_2 is true in the initial world of $F(w)$.

Definition 13. *Logics $\mathbf{L}_1, \mathbf{L}_2$ are suitable for fibring iff, for all $\Sigma_1 \in Sig_1$ and $\Sigma_2 \in Sig_2$, there is a signature $\Sigma_{(1,2)} \in Sig_1$ such that $Form_2^{\Sigma_2} \subset Atom_1^{\Sigma_{(1,2)}}$.*

Let $\mathcal{P}$ be a restricting relation between $\mathbf{L}_1$-models, $\mathbf{L}_1$-worlds, and $\mathbf{L}_2$-models. Then, the fibred logic $\mathbf{L}_{(1,2)}$ is given by:

Signatures: $Sig_{(1,2)} = \{\Sigma_{(1,2)} \mid \Sigma_1 \in Sig_1, \Sigma_2 \in Sig_2\}$.

Syntax: *For all $\Sigma_{(1,2)} \in Sig_{(1,2)}$, $Form_{(1,2)}^{\Sigma_{(1,2)}}$ is identical to $Form_1^{\Sigma_{(1,2)}}$ and $Atom_{(1,2)}^{\Sigma_{(1,2)}}$ is identical to $Atom_1^{\Sigma_{(1,2)}}$.*

Semantics: *A model $\mathbf{m}_{(1,2)} \in \mathcal{M}_{(1,2)}^{\Sigma_{(1,2)}}$ consists of an $\mathbf{L}_1$-model $\mathbf{m}_1 \in \mathcal{M}_1^{\Sigma_{(1,2)}}$ and a fibring function F that assigns to each world w in $\mathbf{m}_1$ an $\mathbf{L}_2$-model $\mathbf{m}_2$ in $\mathcal{M}_2^{\Sigma_2}$ such that (a) $\mathcal{P}(\mathbf{m}_1, w, \mathbf{m}_2)$, and (b) $w \models_1 \phi$ iff $F(w) \models_2 \phi$ for all $\phi \in Form_2^{\Sigma_2}$. We define $\models_{(1,2)} = \models_1$, $W_{(1,2)} = W_1$, and $w_{(1,2)}^0 = w_1^0$.*

Example 5. To fibre $\mathbf{L}_{\text{PL1}}$ and $\mathbf{L}_K$, we assume that there is an $\mathbf{L}_K$-signature Σ_K for every $\mathbf{L}_{\text{PL1}}$-signature Σ_{PL1} such that the atoms over Σ_{PL1} are the primitive propositions in Σ_K. Then, $\Sigma_{(\text{PL1},K)}$ is an $\mathbf{L}_{\text{PL1}}$-signature such that the predicate symbols are of the form $\circ_1 \cdots \circ_n p$ ($n \geq 0$) where $\circ_i \in \{\square, \diamond, -\}$ and p is a predicate symbol in Σ_{PL1}.

The fibred logic $\mathbf{L}_{(\text{PL1},K)}$ is a first-order modal logic, where the modal operators can only occur on the atomic level. If, however, the fibring process is iterated, then the result is full modal predicate logic, because then the logical connectives $\vee, \wedge, \neg$ of $\mathbf{L}_{\text{PL1}}$ can be used inside modal formulae.

7 Fibring Tableau Calculi

In this section, we describe how to construct—in a uniform way—a calculus for a fibred logic $\mathbf{L}_{(1,2)}$ from two calculi $\mathcal{C}_1$ and $\mathcal{C}_2$ for $\mathbf{L}_1$ and $\mathbf{L}_2$.

Expanding a tableau can be seen as an attempt to construct a model for the formula in the root node. If the tableau is closed, then there is no model and the formula in the root node is unsatisfiable. A tableau formula $\sigma{:}\mathsf{T}\,\phi$ represents the fact that, in the constructed model, ϕ is true in the world corresponding to σ.

Now, we have to construct a fibred model and, thus, to represent knowledge about a fibred model by tableau formulae. Therefore, labels now are either of the form $\sigma_1 \in Lab_1$ denoting a world in the $\mathbf{L}_1$-model or of the form $(\sigma_1;\sigma_2)$ (where $\sigma_1 \in Lab_1$ and $\sigma_2 \in Lab_2$) denoting a world in the $\mathbf{L}_2$-model that is assigned by the fibring function to the world represented by σ_1 in the $\mathbf{L}_1$-model. A tableau formula $\sigma_1{:}\mathsf{T}\,\phi$ still means that ϕ is true in $I_1(\sigma_1)$; a tableau formula $(\sigma_1;\sigma_2){:}\mathsf{T}\,\phi$ means that ϕ is true in the world $I_2(\sigma_2)$ of the model assigned to $I_1(\sigma_1)$.

The combined calculus does not construct separate tableaux for $\mathbf{L}_1$- and $\mathbf{L}_2$-formulae but a single tableau, using a unified (set of) tableau rule(s).

The only additional assumption we have to make is that the extension of the restricting relation $\mathcal{P}$ (Def. 13) to tableau interpretations can be characterised using *finite* sets of tableau formulae:

Definition 14. *Let $\mathbf{L}_1$ and $\mathbf{L}_2$ be logics suitable for fibring, let $\mathcal{C}_1$ and $\mathcal{C}_2$ be calculi for $\mathbf{L}_1, \mathbf{L}_2$, let $\mathcal{P}$ be a restricting relation (Def. 13), and let $\Sigma_1 \in Sig_1$ and $\Sigma_2 \in Sig_2$. A function $\mathcal{P}^T$ that assigns to a finite subset Π of $TabForm_1^{\Sigma_1^*}$ and a label $\sigma_1 \in Lab_1$ a finite set $\mathcal{P}^T(\Pi, \sigma_1)$ of $\mathbf{L}_2$-tableau formulae over the non-extended signature Σ_2 characterises $\mathcal{P}$ if the following holds for all finite or infinite sets $\tilde{\Pi} \subset TabForm_1^{\Sigma_1^*}$, all labels $\sigma_1 \in Lab_1$, and all tableau interpretations $\langle \mathbf{m}_1, I_1 \rangle \in TabInterp_1^{\Sigma_1}$ and $\langle \mathbf{m}_2, I_2 \rangle \in TabInterp_2^{\Sigma_2}$:*

$\mathcal{P}(\mathbf{m}_1, I_1(\sigma_1), \mathbf{m}_2)$ holds if and only if (a) $I_1(\sigma_1)$ is defined, (b) $I_1(\sigma_1) \models_1 \tilde{\Pi}$, and (c) $\langle \mathbf{m}_1, I_1 \rangle$ satisfies $\mathcal{P}^T(\Pi, \sigma_1)$ for all finite subsets Π of $\tilde{\Pi}$.

Of course, the fibred calculus can only be implemented if the function $\mathcal{P}^T$ is computable; for a semi-decision procedure, it is sufficient if $\mathcal{P}^T(\Pi, \sigma_1)$ is enumerable for all Π and σ_1.

Example 6. The following function can be used to characterise the (simple) restriction from Example 4: $\mathcal{P}^T(\Pi, \sigma_1) = \{\sigma_2^0{:}\mathsf{S}\,p_2 \mid \sigma_1{:}\mathsf{S}\,p_1 \in \Pi\}$ where σ_2^0 is the initial label of $\mathcal{C}_2$.

The expansion and closure rule of the fibred calculus $\mathcal{C}_{(1,2)}$ constructed from $\mathcal{C}_1$ and $\mathcal{C}_2$ has four components: (1) the expansion rule of $\mathcal{C}_1$, which can be applied to $\mathbf{L}_1$-tableau formulae; (2) the expansion rule of $\mathcal{C}_2$, which can be applied to $\mathbf{L}_2$-tableau formulae with a label of the form $(\sigma_1;\sigma_2)$; (3) a transition rule that allows to derive $(\sigma_1;\sigma_2^0){:}\mathsf{S}\,\phi_2$ from $\sigma_1{:}\mathsf{S}\,\phi_2$ if ϕ_2 is an $\mathbf{L}_2$-formula (in that case ϕ_2 has to be expanded by the $\mathcal{C}_2$-rule), i.e., if an $\mathbf{L}_2$-formula ϕ_2 is true in an $\mathbf{L}_1$-world $w = I_1(\sigma_1)$ then it is true in the initial world of the $\mathbf{L}_2$-model assigned to w; (4) a rule implementing the restriction relation, i.e., if the formulae in Π occur on a branch and $\sigma_2{:}\mathsf{S}\,\phi_2 \in \mathcal{P}^T(\Pi, \sigma_1)$ then $(\sigma_1;\sigma_2){:}\mathsf{S}\,\phi_2$ may be added.

Definition 15. Let $\mathbf{L}_1, \mathbf{L}_2$ be logics suitable for fibring; let $\mathcal{C}_1, \mathcal{C}_2$ be calculi for logics $\mathbf{L}_1, \mathbf{L}_2$, and let these calculi be suitable for fibring; let $\mathcal{P}$ be a restricting relation characterised by the function $\mathcal{P}^T$ (Def. 14). Then, the fibred calculus $\mathcal{C}_{(1,2)}$ is, for all $\Sigma_1 \in Sig_1, \Sigma_2 \in Sig_2$, defined by:

Extended Signature: The extension of $\Sigma_{(1,2)}$ is the signature $\Sigma^*_{(1,2)}$ that is associated with Σ_1^* and Σ_2^* according to Definition 13.

Labels: $Lab_{(1,2)}^{\Sigma_{(1,2)}} = Lab_1 \cup \{(\sigma_1; \sigma_2) \mid \sigma_1 \in Lab_1^{\Sigma_1}, \sigma_2 \in Lab_2^{\Sigma_2}\}$; the initial label $\sigma^0_{(1,2)}$ is the initial label σ^0_1 if $\mathcal{C}_1$.

Expansion and closure rule: For all premises $\Pi \subset TabForm_{(1,2)}^{\Sigma^*_{(1,2)}}$, the set $\mathcal{R}_{(1,2)}(\Pi)$ is the smallest set containing:

1. the conclusions in $\mathcal{R}_1(\Pi_1)$ where Π_1 consists of all tableau formulae of the form $\sigma_1 \text{:S} \phi$ in Π such that $\phi \in Form_1^{\Sigma^*_1}$ (expansion rule of $\mathcal{C}_1$),
2. for all $\sigma_1 \in Lab_1^{\Sigma_{(1,2)}}$, the conclusions that can be constructed from the conclusions in $\mathcal{R}_2(\Pi_{2,\sigma_1})$ replacing σ_2 by $(\sigma_1; \sigma_2)$; the set Π_{2,σ_1} consists of all tableau formulae of the form $\sigma_2 \text{:S} \phi$ such that $(\sigma_1; \sigma_2) \text{:S} \phi$ is in Π and $\phi \in Form_2^{\Sigma^*_2}$ (expansion rule of $\mathcal{C}_2$),
3. the conclusion $\{\{(\sigma_1; \sigma^0_2) \text{:S} \phi\}\}$ for all tableau formulae of the form $\sigma_1 \text{:S} \phi$ in Π such that $\phi \in Form_2^{\Sigma^*_2}$ (transition rule),
4. for all $\sigma_1 \in Lab_1^{\Sigma_{(1,2)}}$ and all subsets Π_1 of Π (see point 1 above), the conclusion $\{\mathcal{P}^T(\Pi_1, \sigma_1)\}$ (restriction relation).

Theorem 4. *The fibred calculus $\mathcal{C}_{(1,2)}$ that is constructed according to Definition 15 is suitable for fibring, i.e., it satisfies Conditions 4–9 in Section 3.*

Corollary 1. *The fibred calculus $\mathcal{C}_{(1,2)}$ that is constructed according to Definition 15 is a sound and complete calculus for $\mathbf{L}_{(1,2)}$, i.e., there is a closed tableau for $G \in Form^{\Sigma_{(1,2)}}$ if and only if G is not satisfiable.*

8 Fibring Calculi for Predicate and Modal Logic

As an example, we fibre the calculi $\mathcal{C}_{\text{PL1}}$ for first-order predicate logic $\mathbf{L}_{\text{PL1}}$ introduced in Section 4.2 and the calculus $\mathcal{C}_{\text{K}}$ for the logic $\mathbf{L}_{\text{K}}$ of modalities defined in Section 5.2. The result is a calculus $\mathcal{C}_{(1,2)}$ for first-order modal logic where the modal operators can only occur on the literal level (Example 5). Since, in this case, there is no additional restriction on which $\mathbf{L}_{\text{K}}$-models may be assigned to worlds in $\mathbf{L}_{\text{PL1}}$-models, the function $\mathcal{P}^T(\Pi, \sigma)$ characterising the fibring restriction (Def. 14) is empty for all formula sets Π and labels σ; therefore, the tableau expansion rule that implements the restriction relation is never applied.

Due to space restrictions, we cannot list the tableau expansion and closure rules of the fibred calculus, which can easily be constructed by instantiating the calculi $\mathcal{C}_1$ and $\mathcal{C}_2$ in Definition 15 with $\mathcal{C}_{\text{PL1}}$ resp. $\mathcal{C}_{\text{K}}$. Instead, we prove the formula

$$G = (\forall x)(\Box p(x)) \to [\neg(\exists y)(\Diamond - p(y)) \land \neg(\exists z)(\Diamond - p(z))]$$

to be valid in all models of the logic $\mathbf{L}_{(1,2)} = \mathbf{L}_{(PL1,K)}$, using the fibred calculus $\mathcal{C}_{(1,2)} = \mathcal{C}_{(PL1,K)}$ to construct a closed tableau for $\neg G$.

The closed tableau shown on the right is constructed as follows: Tableau formula 1 is put on the tableau initially; then formulae 2–7 are added using the α- and β-rules of $\mathcal{C}_{PL1}$. The δ-rule of $\mathcal{C}_{PL1}$ is applied to derive 8 from 7, using the Skolem constant $c_1 = sko((\exists y)(\Diamond - p(y)))$. Since 8 is an $\mathbf{L}_K$-formula, the transition rule is applied to add 9 to the branch, which then allows to apply the $\mathbf{L}_K$-expansion rule to derive 10 from 9 (we assume that $goedel(\Diamond - p(c_1)) = 1$) and to derive 11

1 *:T $\neg((\forall x)(\Box p(x)) \rightarrow [\neg(\exists y)(\Diamond-p(y)) \wedge \neg(\exists z)(\Diamond-p(z))])$	
2 *:F $(\forall x)(\Box p(x)) \rightarrow [\neg(\exists y)(\Diamond-p(y)) \wedge \neg(\exists z)(\Diamond-p(z))]$	
3 *:T $(\forall x)(\Box p(x))$	
4 *:F $\neg(\exists y)(\Diamond-p(y)) \wedge \neg(\exists z)(\Diamond-p(z))$	
5 *:F $\neg(\exists y)(\Diamond-p(y))$	6 *:F $\neg(\exists y)(\Diamond-p(y))$
7 *:T $(\exists y)(\Diamond-p(y))$	15 *:T $(\exists y)(\Diamond-p(y))$
8 *:T $\Diamond-p(c_1)$	16 *:T $\Diamond-p(c_1)$
9 $(*;1)$:T $\Diamond-p(c_1)$	17 $(*;1)$:T $\Diamond-p(c_1)$
10 $(*;1.1)$:T $-p(c_1)$	18 $(*;1.1)$:T $-p(c_1)$
11 $(*;1.1)$:F $p(c_1)$	19 $(*;1.1)$:F $p(c_1)$
12 *:T $\Box p(c_1)$	20 *:T $\Box p(c_1)$
13 $(*;1)$:T $\Box p(c_1)$	21 $(*;1)$:T $\Box p(c_1)$
14 $(*;1.1)$:T $p(c_1)$	22 $(*;1.1)$:T $p(c_1)$
$\bot$	$\bot$

from 10. At this point, the γ-rule of $\mathbf{L}_{PL1}$ is applied to 3 to derive 12, replacing the universally quantified variable x with the ground term c_1 (which shows that $\mathbf{L}_1$- and $\mathbf{L}_2$-rules can be applied in an arbitrary order). Finally, the transition rule is applied to 12 to derive 13, and the $\mathbf{L}_K$-rule for $\Box$-formulae is applied to derive 14. At this point, the left branch of the tableau is closed by the $\mathbf{L}_K$-closure rule, because it contains the complementary atoms 11 and 14. The right branch is expanded and closed in the same way.

The full power of the fibring method is revealed when the fibring process is iterated to construct a calculus $\mathcal{C}_{[PL1,K]}$ for the full modal predicate logic $\mathbf{L}_{[PL1,K]}$; this is possible because the calculi $\mathcal{C}_{(1,2)}, \mathcal{C}_{(1,(2,1))}, \ldots$ are all suitable for fibring. As an example, we use $\mathcal{C}_{[PL1,K]}$ to prove that the formula is valid in all models of $\mathbf{L}_{[PL1,K]}$ that is constructed from G replacing the literal $p(x)$ by $r(x) \wedge s(x)$ and replacing the literals $-p(y)$ and $-p(z)$ by

10' $(*;1.1)$:T $-r(c_1) \vee -s(c_1)$	
14' $(*;1.1)$:T $r(c_1) \wedge s(c_1)$	
23 $(*;1.1;*)$:T $r(c_1) \wedge s(c_1)$	
24 $(*;1.1;*)$:T $r(c_1)$	
25 $(*;1.1;*)$:T $s(c_1)$	
26 $(*;1.1;*)$:T $-r(c_1) \vee -s(c_1)$	
27 $(*;1.1;*)$:T $-r(c_1)$	31 $(*;1.1;*)$:T $-s(c_1)$
28 $(*;1.1;*;1)$:T $-r(c_1)$	32 $(*;1.1;*;1)$:T $-s(c_1)$
29 $(*;1.1;*;1)$:F $r(c_1)$	33 $(*;1.1;*;1)$:F $s(c_1)$
30 $(*;1.1;*;1)$:T $r(c_1)$	34 $(*;1.1;*;1)$:T $s(c_1)$
$\bot$	$\bot$

$-r(y) \vee -s(y)$ resp. $-r(z) \vee -s(z)$. The construction of the tableau starts as above for G. We only consider the left branch (the right branch can be closed in the same way). Instead of the literals 10 and 14, the branch now contains $10' = (*;1.1)$:T $-r(c_1) \vee -s(c_1)$ and $14' = (*;1.1)$:T $r(c_1) \wedge s(c_1)$. The expansion of the branch continues as shown above (to simplify notation, we write $(*;1;*)$ instead of $(*;(1;*))$, etc.). The tableau formula $14'$ contains an $\mathbf{L}_{PL1}$-formula. Therefore, the transition rule is applied, and 23 is derived from $14'$; this is the transition rule of the calculus $\mathcal{C}_{(K,PL1)}$ that, during the iteration process, has been fibred with $\mathcal{C}_{PL1}$ to construct $\mathcal{C}_{(PL1,(K,PL1))}$. The α-rule of $\mathbf{L}_{PL1}$ is used to derive 24 and 25 from 23; then, 26 is derived from $10'$ by again applying the transition rule, and the β-rule is applied to derive 27 and 31 from 26. The literal $-p(c_1)$ in 27 contains the modal and not the first-oder negation sign. Thus, the transition rule has to be applied again to derive 28, which then allows to derive 29 by applying the rule for modal negation. The atomic tableau formulae 24 and 29 cannot be used to close the branch, because their labels are different. Thus, the transition rule is applied a last time to derive 30 from 24. Then, the branch is closed by 29 and 30.

9 Conclusion

We have presented a uniform method for constructing a sound and complete tableau calculus for a fibred logic from calculi for its component logics. Conditions have been identified that tableau calculi have to satisfy to be suitable for fibring; the conditions are neither too weak nor too strong. Since tableau calculi are already known for most "basic" logics, it is possible to construct calculi for all "complex" logics that can be constructed by fibring basic logics. The main advantages of a uniform framework for fibring calculi are:

To construct a calculus for the combination $\mathbf{L}_{[1,2]}$ of two particular logics, no knowledge is needed about the interaction between calculi for $\mathbf{L}_1$ and $\mathbf{L}_2$. Thus, a calculus for the combination $\mathbf{L}_{[1,2]}$ can be obtained quickly and easily.

Soundness and completeness of the fibred calculus does not have to be proven; it follows from Theorem 4 if the fibred calculi are suitable for fibring.

A calculus $\mathcal{C}_1$ for a logic $\mathbf{L}_1$ can be fibred with a calculus $\mathcal{C}_2$ for a "sublogic" $\mathbf{L}_2$ of $\mathbf{L}_1$ (for example, propositional logic is a sub-logic if predicate logic); although $\mathcal{C}_1$ can handle the whole logic $\mathbf{L}_1$, the calculus $\mathcal{C}_2$ may be more efficient for formulae from $\mathbf{L}_2$ such that the fibred calculus $\mathcal{C}_{(1,2)}$ is more efficient than $\mathcal{C}_1$. This can be seen as a generalisation of the theory reasoning method.

Acknowledgement. We thank Guido Governatori and two anonymous referees for useful comments on an earlier version of this paper.

References

1. W. Ahrendt and B. Beckert. An improved δ-rule for ground first-order tableaux. Unpublished draft available from the authors, 1997.
2. B. Beckert and D. Gabbay. A general framework for fibring semantic tableaux. Unpublished draft available from the authors, 1997.
3. B. Beckert, R. Hähnle, and P. H. Schmitt. The even more liberalized δ-rule in free variable semantic tableaux. In *Proceedings of KGC*, LNCS 713. Springer, 1993.
4. M. D'Agostino and D. Gabbay. Fibred tableaux for multi-implication logics. In *Proceedings of TABLEAUX*, LNCS 1071. Springer, 1996.
5. M. D'Agostino, D. Gabbay, R. Hähnle, and J. Posegga, editors. *Handbook of Tableau Methods*. Kluwer, Dordrecht, 1998. To appear.
6. D. Gabbay. Fibred semantics and the weaving of logics. Part 1: Modal and intuitionistic logics. *Journal of Symbolic Logic*, 61:1057–1120, 1996.
7. D. Gabbay. An overview of fibred semantics and the combination of logics. In *Proceedings of FroCoS*. Kluwer, Dordrecht, 1996.
8. D. Gabbay. *Fibring Logic*. Oxford University Press, 1998. Forthcoming.
9. D. Gabbay and G. Governatori. Fibred modal tableaux. Draft, 1997.

A Tableau Calculus for Quantifier-Free Set Theoretic Formulae

Bernhard Beckert[1] and Ulrike Hartmer[2,*]

[1] University of Karlsruhe, Institute for Logic, Complexity and Deduction Systems,
D-76128 Karlsruhe, Germany. E-mail: `beckert@ira.uka.de`
[2] Deutsche Telekom AG, Technologiezentrum Darmstadt,
D-64307 Darmstadt, Germany. E-mail: `hartmer@tzd.telekom.de`

Abstract. Set theory is the common language of mathematics. Therefore, set theory plays an important rôle in many important applications of automated deduction. In this paper, we present an improved tableau calculus for the decidable fragment of set theory called *multi-level syllogistic with singleton* (MLSS). Furthermore, we describe an extension of our calculus for the bigger fragment consisting of MLSS enriched with free (uninterpreted) function symbols (MLSSF).

1 Introduction

Set theory is the common language of mathematics. Therefore, set theory plays an important rôle in many important applications of automated deduction. For example, some of the most widely used specification languages, namely the Z and B specification languages, are completely based on set theory. For other languages, sets are at least a very important construct, frequently used in specifications either on the meta-level or as a data structure of the specified programs. Set theoretic proof obligations occur both as part of proving an implementation to be sound w.r.t. a specification and as part of immanent reasoning (such as consistency checks, proving invariants, pre- and post-conditions).

Set theoretic reasoning, i.e., employing special purpose techniques instead of using the axioms of set theory, is indispensable for automated deduction in real world domains. Automated deduction tools can, for example, be integrated into interactive software verification systems and relieve the user from the need to interactively handle simple set theoretic problems that do not require his or her intuition but merely a combinatorial search.

In this paper, we present an improved tableau calculus for the decidable fragment of set theory called *Multi-level Syllogistic with Singleton* (MLSS). Furthermore, we describe an extension of our calculus for the bigger fragment consisting of MLSS enriched with free (uninterpreted) function symbols (MLSSF).

Multi-level Syllogistic (MLS) consists of quantifier-free formulae built using the set theoretic predicates *membership, equality, set inclusion*, the binary functions *union, intersection, set difference*, and a constant representing the empty

---
[*] This work was carried out while the author stayed at the University of Karlsruhe.

set. In the extension MLSS of MLS, n-ary functions $\{\cdot\}_n$ can be used to construct singletons, pairs, etc.

The expressiveness of MLSS is sufficient for many applications. MLSS formulae can contain variables, which are implicitly universally quantified. The main restriction is that there is no existential quantification; thus, sentences such as "there is an infinite set" cannot be formalised within MLSS.

Our calculus for MLSS, which is a sound and complete decision procedure, is an extension of the tableau-based calculus for MLSS that Cantone described in [4]. It does not require formulae to be in normal form, whereas Cantone's calculus only contains rules for normalised MLSS literals (which are not allowed to contain complex terms) and relies on a pre-processing transformation for normalising formulae. The handling of free function symbols in the extended calculus for MLSSF employs E-unification techniques for reducing the search space by finding term pairs that, when shown to be equal, close a tableau branch.

Several methods for handling set theory in tableaux or the sequent calculus (without the restriction to a certain fragment) have been proposed: In [2], Brown presents a first-order sequent calculus that contains special rules for many set theoretic symbols. De Nivelle [10] and Pastre [14] introduce sequent calculi for set theory. Shults [15] describes a tableau calculus with special set theoretic rules. All these calculi, however, are incomplete (no semi-decision procedures).

Decision and semi-decision procedures for various extensions of MLS have been described in the literature; these, however, are not based on tableaux but are highly non-deterministic search procedures and are not suitable for implementation; an overview can be found in [5, 6]. Extensions of MLS that are known to be decidable include: MLS with powerset and singleton [3, 7], with relational constructs [9], with unary union [8], and with a choice operator [11].

This paper is structured as follows: In Sect. 2, we define the syntax and semantics of the fragments MLSS and MLSSF of set theory. In Sect. 3, we introduce those parts of our calculus that are not specific for set theoretic formulae. In Sect. 4, we describe our calculus for the fragment MLSS; and in Sect. 5, we extend the calculus for handling the fragment MLSSF including free function symbols. As an example, we present a proof for an MLSSF formula in Sect. 6. Finally, in Sect. 7, we draw conclusions and discuss future work. Due to space restrictions, proofs are not included in this paper; they can be found in [12].

2 Syntax and Semantics

2.1 Syntax of MLSS and MLSSF

We handle two classes of set theoretic formulae: The first are formulae in the fragment *multi-level syllogistic with singletons* (MLSS); this is the quantifier free fragment of set theoretic formulae using (a) the set theoretic predicate symbols $\in$ (membership), $\approx$ (equality), $\sqsubseteq$ (set inclusion), (b) the set theoretic function symbols $\sqcap$ (intersection), $\sqcup$ (union), $\setminus$ (set difference), and $\{\cdot\}_n$ with arity $n \geq 1$ (singleton, pair, etc.), and (c) the set theoretic constant $\emptyset$ (the empty set). As usual, the binary function and predicate symbols are written in infix notation, and $\{\cdot\}_n$ is written in circumfix notation.[1] The second fragment, called MLSSF, is the extension of MLSS by free function symbols that have no special set theoretic interpretation.

In the following, we assume that a fixed signature is given consisting of a set *Var* of variables, a set *Const* of constants, and a set *Func* of function symbols.

Definition 1. *The set of* pure set terms *is inductively defined by: (1) All variables $x \in Var$, all constants $c \in Const$, and $\emptyset$ are pure set terms. (2) If t_1, t_2 are pure set terms, then $t_1 \sqcap t_2$, $t_1 \sqcup t_2$, and $t_1 \setminus t_2$ are pure set terms. (3) For all $n \geq 1$, if $t_1, \ldots, t_n$ are pure set terms, then $\{t_1, \ldots, t_n\}_n$ is a pure set term.*

The set of set terms *is inductively defined by: (1) All pure set terms are set terms. (2) If $f \in Func$ is a function symbol of arity $n \geq 1$ and $t_1, \ldots, t_n$ are set terms, then $f(t_1, \ldots, t_n)$ is a set term.*

A set term is called functional *if it is of the form $f(t_1, \ldots, t_n)$.*

Note that functional set terms can contain non-functional set terms (which are not necessarily pure) and vice versa.

MLSS and MLSSF are built using the logical connectives $\lor$ (disjunction), $\land$ (conjunction), $\neg$ (negation), and $\to$ (implication). Formulae that are identical up to associativity of $\lor$ and $\land$ are identified.

Definition 2. *If t_1, t_2 are pure set terms (resp. set terms), then $t_1 \mathrel{\mathsf{E}} t_2$, $t_1 \approx t_2$, and $t_1 \sqsubseteq t_2$ are MLSS (resp. MLSSF) atoms. If p is an MLSS (MLSSF) atom, then p and $\neg p$ are MLSS (MLSSF) literals.*

The sets of MLSS and MLSSF formulae *are inductively defined by: (1) All MLSS (MLSSF) literals are MLSS (MLSSF) formulae. (2) If ϕ, ψ are MLSS (MLSSF) formulae, then $\neg\phi$ and $\phi \to \psi$ are MLSS (MLSSF) formulae. (3) If $\phi_1, \ldots, \phi_n$ are MLSS (MLSSF) formulae, then $\phi_1 \land \cdots \land \phi_n$ and $\phi_1 \lor \cdots \lor \phi_n$ are MLSS (MLSSF) formulae $(n \geq 2)$.*

To simplify notation, we use the negative versions $\not\mathrel{\mathsf{E}}, \not\approx$, and $\not\sqsubseteq$ of the predicate symbols $\mathrel{\mathsf{E}}, \approx$, and $\sqsubseteq$, where $s \not\mathrel{\mathsf{E}} t$ is an abbreviation for $\neg(s \mathrel{\mathsf{E}} t)$, etc.

2.2 Semantics

We use the semantics of set theory (and thus its fragments MLSS and MLSSF) as it is defined by the ZF axiom system or, equivalently, by the von Neumann hierarchy (cumulative hierarchy) of sets (see for example [13] for a detailed discussion of the semantics of set theory).

Definition 3. *Let Ord denote the class of all ordinals. The* von Neumann hierarchy *of sets is defined by $\mathcal{V} = \bigcup_{\alpha \in Ord} \mathcal{V}_\alpha$ where (1) $\mathcal{V}_0 = \emptyset$, (2) $\mathcal{V}_\alpha = \bigcup_{\beta < \alpha} \mathcal{V}_\beta$ for each limit ordinal α, and (3) $\mathcal{V}_{\alpha+1}$ is the powerset of $\mathcal{V}_\alpha$ for each ordinal α.*

---

[1] To avoid confusion we use the non-standard symbols $\mathrel{\mathsf{E}}, \approx, \sqsubseteq, \sqcap, \sqcup, \emptyset$ on the object level and the standard symbols $\in, =, \subset, \cap, \cup, \emptyset$ on the meta level.

We only define the semantics of the fragment MLSSF; the semantics of MLSS is the same as that of MLSSF for the case of an empty set of function symbols.

Definition 4. *A set structure $M = \langle D, I \rangle$ consists of a domain D and an interpretation I with the following properties: The elements of D are sets in the von Neumann hierarchy; D is closed under the set operations $\cap$, $\cup$, $\setminus$, and $\{\cdot\}_n$ ($n \geq 1$), and it contains the empty set; I interprets (1) each constant symbol $c \in \mathit{Const}$ by an element of D, (2) each function symbol $f \in \mathit{Func}$ of arity n by a function $D^n \to D$, (3) the constant $\emptyset$ by the empty set, (4) the predicate symbols by their canonical interpretations, i.e., E by $\in$, $\approx$ by the identity relation, and $\sqsubseteq$ by $\subseteq$, (5) the set theoretic function symbols by their canonical interpretations, i.e., $\sqcup$ by $\cup$, $\sqcap$ by $\cap$, $\setminus$ by $\setminus$, and $\{\cdot\}_n$ by $\{\cdot\}_n$ ($n \geq 1$).*

Definition 5. *Given a set structure $M = \langle D, I \rangle$, a variable assignment is a mapping $\mu : \mathit{Var} \to D$ from the set of variables to the domain D. The combination of M and a variable assignment μ associates (by structural recursion) with each set term t an element $\mathit{val}_{M,\mu}(t)$ of D; and it associates with each MLSSF formula ϕ either true or false. A formula ϕ is true in M (and M is a model of ϕ) if, for all variable assignments μ, $\mathit{val}_{M,\mu}(\phi) = \mathit{true}$; else ϕ is false in M.*

Definition 6. *An MLSSF formula ϕ is* satisfiable *if there is a set structure M such that ϕ is true in M; ϕ is* valid *if it is true in all set structures.*

3 Tableaux for Quantifier-free Formulae

In this section, we introduce those parts of our calculus that are not specific for set theory. In particular, we define the expansion rules for logical connectives.

The non-literal MLSS and MLSSF formulae are divided into two classes: α for formulae of conjunctive type and β for formulae of disjunctive type. In the left part of Table 1, the expansion rules for α- and β-formulae are given schematically. Premises and conclusions are separated by a horizontal bar, while vertical bars in the conclusion denote different *extensions*. The tableau expansion rule corresponding to a formula ϕ is obtained by looking up the formula type of ϕ in the right part of Table 1 and instantiating the matching rule schema. The formulae in an extension are implicitly conjunctively connected, and different extensions are implicitly disjunctively connected. We use n-ary α- and β-rules, i.e., when the β-rule is applied to a formula $\psi = \phi_1 \vee \ldots \vee \phi_n$, then ψ is broken up into n subformulae (instead of splitting it into two formulae $\phi_1 \vee \ldots \vee \phi_r$ and $\phi_{r+1} \vee \ldots \vee \phi_n$).

Below, tableaux and tableau proofs are defined in general; which expansion rules (besides those for the logical connectives) and which closure rules are to be used is described in the following sections.

Definition 7. *An MLSS tableau (an MLSSF tableau) is a finitely branching tree whose nodes are MLSS formulae (MLSSF formulae). A branch in a tableau $\mathcal{T}$ is a maximal path in $\mathcal{T}$ (where no confusion can arise, a branch is often identified with the set of formulae it contains).*

Table 1. Rule schemata for α- and β-formulae, and correspondence between non-literal formulae and rule types.

$$\frac{\alpha}{\begin{array}{c}\alpha_1\\ \vdots\\ \alpha_n\end{array}} \qquad \frac{\beta}{\beta_1 \mid \cdots \mid \beta_n}$$

α	$\alpha_1, \ldots, \alpha_n$	β	$\beta_1, \ldots, \beta_n$
$\phi_1 \wedge \ldots \wedge \phi_n$	$\phi_1, \ldots, \phi_n$	$\phi_1 \vee \ldots \vee \phi_n$	$\phi_1, \ldots, \phi_n$
$\neg(\phi_1 \vee \ldots \vee \phi_n)$	$\neg\phi_1, \ldots, \neg\phi_n$	$\neg(\phi_1 \wedge \ldots \wedge \phi_n)$	$\neg\phi_1, \ldots, \neg\phi_n$
$\neg(\phi \rightarrow \psi)$	$\phi, \neg\psi$	$\phi \rightarrow \psi$	$\neg\phi, \psi$
$\neg\neg\phi$	ϕ		

Given an MLSS (MLSSF) formula ϕ and a set of tableau expansion rules, the tableaux for ϕ are (recursively) defined by: (1) The tree consisting of a single node labelled with ϕ is a tableau for ϕ (initialisation rule). (2) Let $\mathcal{T}$ be a tableau for ϕ, B a branch of $\mathcal{T}$, and let the premiss of one of the expansion rules occur on B. If the tree $\mathcal{T}'$ is constructed by extending B by as many new linear subtrees as the tableau expansion rule has extensions, where the nodes of the new subtrees are labelled with the formulae in the extensions, then $\mathcal{T}'$ is a tableau for ϕ (expansion rule).

Since the free variables in quantifier-free formulae are implicitly universally quantified, a formula $\phi(x)$ is valid if and only if a *Skolemisation* $\neg\phi(c)$ of its negation is unsatisfiable. Thus, free variables can be eliminated, and a tableau calculus for formulae without free variables is sufficient for checking the validity of a given formula ϕ.

Definition 8. *Given an MLSSF formula $\phi(x_1, \ldots, x_n)$, where $x_1, \ldots, x_n$ are the (free) variables in ϕ $(n \geq 0)$, a formula $\phi(c_1, \ldots, c_n)$ is a Skolemisation of ϕ if $c_1, \ldots, c_n$ are constants that do not occur in $\phi(x_1, \ldots, x_n)$.*

Definition 9. *A tableau $\mathcal{T}$ is a tableau proof for an MLSS/MLSSF formula ϕ, if (1) $\mathcal{T}$ is a tableau for a Skolemisation of $\neg\phi$ (Def. 8), and (2) all branches of $\mathcal{T}$ are closed (Def. 10).*

4 A Tableau Calculus for MLSS

In this section, we present tableau expansion rules that—in combination with the expansion rules for the logical connectives—represent a sound and complete calculus for MLSS, i.e., for formulae built using only *pure* set literals. It can easily be turned into a decision procedure (see Sect. 4.5).

Since the (negation of) the formula to be proven is first Skolemised and is then split into literals using the rules for logical connectives, it is sufficient to define expansion rules for handling pure, variable free set literals.

4.1 Expansion Rules for Splitting Complex Set Terms

The first group of expansion rules applies simple set theoretic lemmata such as "if $s \in t_1 \cup t_2$ then $s \in t_1$ or $s \in t_2$" to (a) eliminate literals containing the set inclusion predicate $\sqsubseteq$ and replace them with (in-)equalities, and to (b) split complex terms on the right side of the membership predicate $\in$ into their constituents. These rules can be described using the α- and β-rule schemata (left part of Table 1); the formula and rule types are listed in Table 2.

Table 2. Rule types for splitting complex set terms.

Name	α	$\alpha_1, \ldots, \alpha_n$	Name	β	$\beta_1, \ldots, \beta_n$
(R1)	$s \sqsubseteq t$	$s \approx s \sqcap t$	(R7)	$s \in t_1 \sqcup t_2$	$s \in t_1,\ s \in t_2$
(R2)	$s \not\sqsubseteq t$	$s \not\approx s \sqcap t$	(R8)	$s \notin t_1 \sqcap t_2$	$s \notin t_1,\ s \notin t_2$
(R3)	$s \in t_1 \sqcap t_2$	$s \in t_1,\ s \in t_2$	(R9)	$s \notin t_1 \setminus t_2$	$s \notin t_1,\ s \in t_2$
(R4)	$s \in t_1 \setminus t_2$	$s \in t_1,\ s \notin t_2$	(R10)	$s \in \{t_1, \ldots, t_n\}_n$	$s \approx t_1, \ldots, s \approx t_n$
(R5)	$s \notin t_1 \sqcup t_2$	$s \notin t_1,\ s \notin t_2$			
(R6)	$s \notin \{t_1, \ldots, t_n\}_n$	$s \not\approx t_1, \ldots, s \not\approx t_n$			

4.2 Expansion Rules for Handling Equality and Inequality

There are three types of special rules for handling the equality and inequality of sets. First, there are two rules ((EQ1) and (EQ2) in Table 3) that allow to "apply" an equality $t_1 \approx t_2$ to other literals in a very restricted way: an equality can only be applied at the top level and only to the right side of an atom whose predicate symbol is $\in$. That is, an equality can only be applied to derive one of the atoms $s \in t_1$ and $s \in t_2$ from the other one. This restriction is important, because the possibility to apply equalities arbitrarily to other literals would lead to a much larger search space.

Second, it is possible to derive $s_1 \not\approx s_2$ from $s_1 \in t$ and $s_2 \notin t$ (rule (R11) in Table 3). This rule is based on the fact that two objects are different if one of them is an element of some set and the other is not.

Third, the opposite of the above holds as well: if two sets t_1 and t_2 are different, then one of them contains an element c that is not element of the other set. Unfortunately, this leads to a branching rule (rule (R12) in Table 3), because c can be an element of t_1 (and not of t_2) or of t_2 (and not of t_1). A new constant has to be introduced representing the unknown element c.

Table 3. Rules for handling equality and inequality, and the restricted cut rule.

$$\frac{t_1 \approx t_2 \quad s \in t_1}{s \in t_2} \quad \frac{t_1 \approx t_2 \quad s \in t_2}{s \in t_1} \quad \frac{s_1 \in t \quad s_2 \notin t}{s_1 \not\approx s_2} \quad \frac{t_1 \not\approx t_2}{c \in t_1 \mid c \notin t_1} \quad \frac{}{s \in t \mid s \notin t}$$
$$\qquad\qquad\qquad\qquad\qquad\qquad\qquad\qquad\qquad\qquad c \notin t_2 \mid c \in t_2$$

(EQ1) (EQ2) (R11) where c is a constant new to the tableau (R12) where s resp. $\{\ldots, s, \ldots\}$ and t resp. $\{\ldots, t, \ldots\}$ are top-level terms on the branch (Cut)

4.3 The Cut Rule

The cut rule (Table 3) may be applied to extend a tableau branch B using as cut formula atoms $s \mathrel{\mathsf{E}} t$ where the set terms s and t occur (a) as top-level arguments of a literal on B, or (b) as arguments on the second level if the top-level function symbol is $\{\cdot\}_n$. In practice the cut rule is rarely needed to find a proof; it is, for example, needed to detect implicit membership cycles on a branch; see Sect. 4.4.

Example 1. If $t_1 \mathrel{\mathsf{E}} \{t_2, (t_3 \sqcap t_4)\}$ and $t_5 \sqcap t_6 \approx t_7$ are literals on the branch, then $t_1, t_2, (t_3 \sqcap t_4), (t_5 \sqcap t_6), t_7$ may be used in a cut rule application and t_3, t_4, t_5, t_6 may not be used.

4.4 The Closure Rules

Tableau expansion rules add formulae to a tableau branch being true in all set structures that are models of the expanded branch; the purpose of closure rules is to detect inconsistencies, i.e., formulae on a branch that are false in all set structures. There are four types of inconsistencies that have to be considered: (1) In no set structure both a formula ϕ and its complement $\neg \phi$ are true; thus, a pair $\phi, \neg \phi$ is inconsistent (for completeness it is sufficient to only consider complementary *literals*). (2) No object is an element of the empty set; therefore, a literal of the form $t \mathrel{\mathsf{E}} \emptyset$ is inconsistent. (3) As no object is different from itself, literals of the form $t \not\approx t$ are inconsistent. (4) The existence of a membership cycle, i.e., of sets $u_1, \ldots, u_k$ such that $u_i \in u_{i+1}$ ($1 \leq i < k$) and $u_k \in u_1$, would contradict the Axiom of Foundation. In fact, there are by construction no sets in the von Neumann hierarchy that form a membership cycle. Thus, literals defining a membership cycle are inconsistent; in particular, $t \mathrel{\mathsf{E}} t$ is inconsistent.

Definition 10. *A tableau branch B is closed if it contains (1) a complementary pair ϕ and $\neg \phi$ of literals, (2) a literal of the form $t \mathrel{\mathsf{E}} \emptyset$, (3) a literal of the form $t \not\approx t$, or (4) for some $k \geq 1$, literals $t_i \mathrel{\mathsf{E}} t_{i+1}$ ($1 \leq i < k$) and $t_k \mathrel{\mathsf{E}} t_1$.*

4.5 Soundness, Completeness, Termination

The calculus for MLSS described in the previous sections is sound and complete:

Theorem 1. *An MLSS formula ϕ is valid iff there is a tableau proof for ϕ using the expansion rules from Tables 1–3 and the closure rule from Def. 10.*

Without further restrictions, the calculus is not a decision procedure. The rule for inequalities ((R12) in Table 3) introduces new constants, and the cut rule can—in connection with rule (R11)—construct new inequalities from the new constants; the interaction of these rules can lead to infinite branches.

Fortunately, the calculus can easily be turned into a decision procedure, observing the fact that chains $c_1, c_2, \ldots$ where c_i is derived applying the inequality rule (R12) to an inequality that contains the constant c_{i-1} cannot be infinite; their length is bounded by the number of (sub-)terms in the initial tableau:

Definition 11. *The rank $rank(s)$ of a set term s in a tableau for an MLSS formula ϕ that has been constructed using the expansion rules from Tables 1–3 and the closure rule from Def. 10 is defined as follows: If s occurs in ϕ or has been generated by an application of rules (R1) and (R2), then $rank(s) = 0$; otherwise, i.e., if s is a constant that has been introduced by applying rule (R12) to an inequality $t_1 \not\equiv t_2$, then its rank is $rank(s) = 1 + \max\{rank(t_1), rank(t_2)\}$.*

Definition 12. *A tableau $\mathcal{T}$ for an MLSS formula ϕ is* exhausted, *if no tableau expansion rule can be applied to $\mathcal{T}$ without either adding a constant whose rank is greater than the number of (sub-)terms in the root node of $\mathcal{T}$ or adding only formulae to a branch B that already occur on B.*

Theorem 2. *There is an exhausted tableau for an MLSS formula ϕ if and only if ϕ is satisfiable.*

Thus, if a tableau for the Skolemisation of the negation of an MLSS formula ϕ is constructed in a *fair* way (i.e., all possible rule applications are executed sooner or later), then the construction will terminate after a finite number of steps with a tableau that is (a) closed, in which case ϕ is valid, or (b) exhausted, in which case ϕ is not valid.

4.6 Restricting the Search Space

Although it is finite, the search space for a tableau proof is large because of the indeterminism of the cut rule, and because the number of new constants that can be introduced is exponential in the size of the formula to be proven.

Fortunately, it is possible to impose a strong restriction on cut rule applications, which at the same time restricts the number of new constants that are introduced, because a constant c_k of rank k can only be deduced from a constant c_{k-1} of rank $k-1$ after the cut rule has been applied to a literal containing c_{k-1}. The idea is to apply all rules except the cut rule until no further applications are possible, and then to construct a *realisation* of open branches. The realisation of a branch B approximates a model for B (if the branch is satisfiable); it satisfies at least all literals of the form $t_1 \in t_2$ on B. If the realisation does not satisfy all the other literals on B as well, it can be used to find cut rule applications that are (at least potentially) useful.

The switching between the expansion of tableau branches and the construction of possible models, and the way in which we construct models are similar to the method Cantone describes in [4].

Definition 13. *Let $\mathcal{T}$ be a tableau for an MLSS formula ϕ, and let B be a branch of $\mathcal{T}$. Then, G is the set of all (sub-)terms occurring in ϕ; V is the set of all terms $t \in G$ such that $t \in s$ occurs on B and of all constants in ϕ; T is the set of all constants on B that are not in V; $\sim$ is the equivalence relation on $G \cup T$ induced by the equalities on B; T' is the set of all $c \in T$ such that $c \not\sim s$ for all $s \in G$; V' is the set $(V \cup T) \setminus T'$; u_c is, for each $c \in T'$, an element of V different from all $u_{c'}$ for $c \neq c'$.*

Note, that T' contains the new constants that have been introduced by applying the inequality rule (R12) and that are not equal to other terms (w.r.t. the equalities on the branch). The interpretation of these constants has to be different from the interpretation of all other terms, whereas different terms in V' may have the same interpretation.

Definition 14. *Let B be a branch of a tableau for an MLSS formula ϕ, and let t be a set term on B. Then the set $P(t)$ of implicit predecessors of t is defined by: (1) $P(\emptyset) = \emptyset$; (2) $P(c) = \{s \in V \cup T \mid s \in c \text{ on } B\}$ if $c \in \text{Const}$; (3) $P(t_1 \sqcup t_2) = P(t_1) \cup P(t_2)$; (4) $P(t_1 \sqcap t_2) = P(t_1) \cap P(t_2)$; (5) $P(t_1 \setminus t_2) = P(t_1) \setminus P(t_2)$; and (6) $P(\{t_1, \ldots, t_n\}_n) = \{s \in V \cup T \mid s \in \{t_1, \ldots, t_n\}_n \text{ on } B\} \cup \{t_1, \ldots, t_n\}$.*

The sets of implicit predecessors can be used to detect implicit membership cycles. If, for example, $s \in P(t), t \in P(s)$ for some terms s, t, then the branch can be closed, and it is not necessary to apply expansion rules (especially the cut rule) to make the cycle explicit. Thus, using the predecessor sets we can add another closure rule:

Definition 15. *A tableau branch B is closed if it is (a) closed according to Def. 10 or (b) its sets of implicit predecessors contain a cycle, i.e., there are (sub-)terms $t_1, \ldots, t_n$ on B such that $t_1 \in P(t_2), \ldots, t_{n-1} \in P(t_n), t_n \in P(t_1)$.*

The set $P(t)$ of implicit predecessors contains those terms denoting elements of t whose membership can be deduced from literals on B of the form $s \in a$ (where $a \in \text{Const}$) and applying the definition of the set operators. The *realisation* of a branch goes beyond that: it is a partial definition of a set interpretation (different terms may be interpreted by the same set).

Definition 16. *Let B be a branch of a tableau for an MLSS formula ϕ, and let t be a set term on B. If B is not closed (Def. 15), then the realisation $\mathcal{R}$ of B is defined by:[2] (1) $\mathcal{R}(t) = \emptyset$ if $t = \emptyset$, (2) $\mathcal{R}(t) = \{\mathcal{R}(s) \mid s \in P(t)\} \cup \{u_t\}$ if $t \in T'$, and (3) $\mathcal{R}(t) = \{\mathcal{R}(s) \mid s \in P(t)\}$ otherwise.*

The realisation can be effectively computed and can be used to restrict the application of the cut rule: provided B is exhausted w.r.t. all other expansion rules, the cut rule has only to be applied to terms occurring in literals which are not satisfied by the realisation of B (if there is no such literal, then B is satisfiable and we are done). If, for example, $t_1 \not\in t_2$ occurs on B but $\mathcal{R}(t_1) \in \mathcal{R}(t_2)$, then there has to be a term s such that (a) $\mathcal{R}(s) = \mathcal{R}(t_1)$, i.e., the realisation of s is the same as that of t_1, and (b) s is an implicit member of t_2, i.e., $s \in P(t_2)$—but that membership is not (yet) made explicit on the branch (there is no literal $s \in t_2$ on B). In that case, the cut rule is applied to the literal $s \in t_2$.

Now everything is at hand to define the restricted version of the cut rule:

---

[2] One has to make sure that the u_c's are different from $\mathcal{R}(t)$ for all terms t; it is always possible to choose such u_c's.

Definition 17. *The restricted cut rule (Cut') is identical to rule (Cut) in Table 3 with the exception that (1) it may only be applied to extend a tableau branch B that is not closed (Def. 15) and is exhausted w.r.t. all other expansion rules; and (2) it may only be applied to a cut formula $s \mathrel{\mathsf{E}} t$ satisfying one of the following conditions*

- $t \approx t'$ is on B, $\mathcal{R}(t) \neq \mathcal{R}(t')$, and (a) $s \in P(t)$, $s \notin P(t')$, $s \not\mathrel{\mathsf{E}} t$ is not on B, or (b) $s \in P(t')$, $s \notin P(t)$, and $s \mathrel{\mathsf{E}} t'$ is not on B;
- $t \not\approx t'$, $c \not\mathrel{\mathsf{E}} t$, and $c \mathrel{\mathsf{E}} t'$ are on B (for some constant c), $\mathcal{R}(t) = \mathcal{R}(t')$, $\mathcal{R}(s) = \mathcal{R}(c)$, $s \in P(t)$, $s \notin P(t')$, and $s \mathrel{\mathsf{E}} t$ is not on B;
- $t' \not\mathrel{\mathsf{E}} t$ is on B, $\mathcal{R}(t') \in \mathcal{R}(t)$, $\mathcal{R}(s) = \mathcal{R}(t')$, $s \in P(t)$, and $s \mathrel{\mathsf{E}} t$ is not on B.

Using the restricted version of the cut rule preserves completeness:

Theorem 3. *An MLSS formula ϕ is valid if and only if there is a tableau proof for ϕ using the expansion rules from Tables 1–3 with the restriction of the cut rule according to Def. 17, and the closure rule from Def. 15.*

4.7 A Comparison with Cantone's Calculus

The calculus for MLSS described in the previous sections is similar to that presented by Cantone in [4]. The main difference is that Cantone's calculus is restricted to *normalised* literals, i.e., literals not containing complex set terms:

Definition 18. *A set literal ϕ is normalised iff it is of the form $a \mathrel{\mathsf{E}} b$, $a \not\mathrel{\mathsf{E}} b$, $a \approx b$, $a \not\approx b$, $a \approx b \sqcup c$, $a \approx b \sqcap c$, $a \approx b \setminus c$, or $a \approx \{b_1, \ldots, b_n\}_n$ ($n \geq 1$), where a, b, c and $b_1, \ldots, b_n$ are constants.*

There is a satisfiability preserving transformation of any finite set Γ of set literals into a set of normalised set literals by introducing new constants for complex set terms. For example, $a \mathrel{\mathsf{E}} (b \sqcap b')$ is replaced by $c \approx (b \sqcap b')$ and $a \mathrel{\mathsf{E}} c$ where c is a new constant. The overhead for computing the transformation is negligible, because its complexity is polynomial in the size of the set to be transformed. However, the introduction of new constants leads to a much bigger search space, even more so as all these new constants occur in equalities.

Our rules (R7), (R3), (R4), and (R10) are—in combination with rules (EQ1) and (EQ2) extensions for handling literals with *complex* set terms of the corresponding rules in Cantone's calculus. For example, our rule (R3), that allows to derive $a \mathrel{\mathsf{E}} b$ and $a \mathrel{\mathsf{E}} b'$ from $a \mathrel{\mathsf{E}} (b \sqcap b')$, corresponds to Cantone's rule that allows to derive $a \mathrel{\mathsf{E}} b$ and $a \mathrel{\mathsf{E}} b'$ from $c \approx (b \sqcap b')$ and $a \mathrel{\mathsf{E}} c$ (for all a, b, c).

There are no rules in Cantone's calculus corresponding to our rules (R5), (R8), and (R9) for literals expressing negated membership. Consider the three literals $\phi = c \not\mathrel{\mathsf{E}} (b_1 \sqcup b_2) \setminus b_3$, $\psi_1 = c \mathrel{\mathsf{E}} b_1$, and $\psi_2 = c \not\mathrel{\mathsf{E}} b_3$, whose conjunction is inconsistent. To close a branch containing these literals, our rules (R9) and (R5) are applied to split the literal ϕ and derive that one of $\neg \psi_1$ and $\neg \psi_2$ holds, thus closing the two resulting sub-branches. Since no rules for splitting ϕ exist in Cantone's calculus, instead rules for positive membership literals have to be used to derive $\neg\phi$ from ψ_1 and ψ_2: first, ϕ has to be normalised, the result are the literals $c \not\in d_1$, $d_1 \approx d_2 \setminus b_3$, and $d_2 \approx b_1 \sqcup b_2$ where d_1 abbreviates $(b_1 \sqcup b_2) \setminus b_3$ and d_2 abbreviates $b_1 \sqcup b_2$. Then, with two rule applications, $c \in d_2$ and $c \in d_1$ are derived. The latter literal can be used to close the branch; it corresponds to the non-normalised literal $\neg\phi$.

The need (and possibility) to derive more complex terms from simpler ones leads to a larger search space. Our rules, that split complex terms into simpler ones, are more goal directed.

5 A Tableau Calculus for MLSSF

5.1 A Simple Extension of MLSS

To extend the calculus described in the previous sections from MLSS to MLSSF, it suffices to (a) relax the restrictions on the equality rules ((EQ1') and (EQ2') in Table 4), and (b) add a cut rule that uses equalities as cut formulae ((Cut') in Table 4). The new rules only need to be applied to functional set terms. Non-functional terms, even if they are not pure, can be handled by the MLSS rules. The result of using these additional rules is a sound and complete calculus for MLSSF; it is, however, not a decision procedure.

Table 4. Additional expansion rules for MLSSF.

$$\frac{s \approx t \quad \phi[s]}{\phi[t]} \qquad \frac{t \approx s \quad \phi[s]}{\phi[t]}$$

where the occurrence of s in ϕ is inside a functional term

(EQ1')   (EQ2')

$$\frac{}{t_1 \approx t_2 \mid t_1 \not\approx t_2}$$

where t_1, t_2 occur on the branch and at least one is a functional term
(Cut')

Theorem 4. *An MLSSF formula ϕ is valid iff there is a tableau proof for ϕ using the expansion rules from Tables 1–4, and the closure rule from Def. 10.*

5.2 Using Rigid E-Unification to Restrict the Equality Cut Rule

The additional rules for MLSSF introduced in the previous section are highly non-deterministic. In this section, we describe an expansion rule for MLSSF that is much more goal-directed and leads to a smaller search space. It is based on the concept of *rigid E-unification*.

Definition 19. *A rigid E-unification problem $\langle E, s, t\rangle$ consists of a finite set E of equalities and terms s and t; the equalities in E and the terms s and t may contain free variables (and may have variables in common). A substitution σ is a solution to the problem iff $E\sigma \models (s\sigma \approx t\sigma)$ where the free variables in $E\sigma$ are "held rigid", i.e., treated as constants.*

The problem of deciding whether a given rigid E-unification problem has a solution is decidable (it is NP-complete). In general, the number of solutions is infinite. An overview of methods for rigid E-unification can be found in [1].

The basic idea is to use rigid E-unification for handling the functional part of formulae on a branch and to use the tableau rules for handling the non-functional (i.e. set theoretic) part. The additional tableau rule we describe in the following forms the connecting link between the two parts.

Consider, for example, a branch B containing the two literals $f(a) \approx b$ and $g(f(a \sqcap (b \sqcup a))) \sqsubseteq g(b)$. They are inconsistent, because $a \sqcap (b \sqcup a) = a$ and, thus, $g(f(a \sqcap (b \sqcup a))) = g(f(a)) = g(b)$; this implies $g(b) \in g(b)$, which is a membership cycle. To close the branch, one first has to find out what the important set theoretic identities are that have to be proven[3], in this case $a \sqcap (b \sqcup a) = a$. It is impossible to do this using only heuristics; here, for example, it is futile to try to show that $a \sqcap (b \sqcup a) = b$.

The question of which set theoretic identities have to be proven to close the branch is transformed into rigid E-unification problems as follows: for each pair s, t of terms that, if they were identical would allow to close the branch (e.g. if $s \not\approx t$ is on B), one rigid E-unification problem is generated. In s and t all maximal non-functional sub-terms are replaced by (new) variables; the resulting terms t^x and s^x and the equalities on the branch form a rigid E-unification problem. Each solution to the problem corresponds to identities between non-functional sub-terms that, when proven, allow to close the branch. The corresponding inequalities are (disjunctively connected) added to the branch.

Definition 20. *Given a set L of set literals, the set L^x is constructed by replacing all non-functional (sub-)terms t in L by a new variable x_t. Let the substitution τ_L be defined by: $\tau(x_t) = t$ for all terms t in L that have been replaced (i.e., τ_L is the inverse of the transformation that turns L into L^x: $\tau(L^x) = L$).*

Example 2. If $L = \{(a \sqcap c) \sqcup b \approx c, f(c) \sqsubseteq g(a \sqcap c, f(d \setminus e))\}$, then the result of the transformation is $L^x = \{x_1 \approx x_2, f(x_2) \sqsubseteq g(x_3, f(x_4))\}$.

Definition 21. *The* rigid E-unification expansion rule *(EU) is defined as follows: Let B be a branch in a tableau for an MLSSF formula, and let L_B be the set of all literals on B of the form $t_1 \approx t_2$, $t_1 \sqsubseteq t_2$, or $t_1 \not\sqsubseteq t_2$. Let E_B^x be the set of all equalities in L_B^x. Further let $\mu = \{x_1 \leftarrow r_1, \ldots, x_n \leftarrow r_n\}$ ($n \geq 1$) be a solution to (1) a rigid E-unification problem $\langle E_B^x, \langle s_1, t_1 \rangle, \langle s_2, t_2 \rangle \rangle$ such that $s_1 \sqsubseteq t_1$ and $s_2 \not\sqsubseteq t_2$ are in L_B^x or (2) a rigid E-unification problem $\langle E_B^x, \langle t_1, \ldots, t_n \rangle, \langle t_1', \ldots, t_n' \rangle \rangle$ such that literals $t_1 \sqsubseteq t_2', \ldots, t_{n-1} \sqsubseteq t_n'$, and $t_n \sqsubseteq t_1'$ in L_B^x form a potential membership cycle. Then B may be extended by n new linear subtrees where the nodes of the new subtrees are labelled with the literals $\tau_{L_B}(x_1) \not\approx \tau_{L_B}(r_1), \ldots, \tau_{L_B}(x_n) \not\approx \tau_{L_B}(r_n)$.*

Example 3. We continue the example from the beginning of this section and apply the rule (EU) to show that a branch containing the literals $f(a) \approx b$ and

---

[3] An identity is proven by using it as a cut formula; after the branch that contains its negation has been closed, it is available on the remaining open branch.

$g(f(a \sqcap (b \sqcup a))) \in g(b)$ is inconsistent. The only rigid E-unification problem that can be extracted from these literals is $\langle \{f(x_a) \approx x_b\}, g(f(x_{a \sqcap (b \sqcup a)})), g(x_b) \rangle$. Its simplest solution is the substitution $\{x_a \leftarrow x_{a \sqcap (b \sqcup a)}\}$. Thus, the rule (EU) allows to add $a \not\approx a \sqcap (b \sqcup a)$ to the branch. The complete proof is shown in Fig. 1.

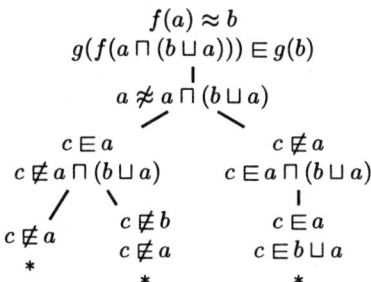

Fig. 1. A tableau proof using the rule (EU) (Example 3).

It is not necessary to consider rigid E-unification problems constructed from inequalities $s \not\approx t$ because, when rule (R11) has been applied, L_B^x contains literals $x_c \in s, x_c \not\in t$ or $x_c \not\in s, x_c \in t$.

The (EU) expansion rule partly overlaps with other expansion rules. It allows, for example, to derive $s_1 \not\approx s_2$ from $s_1 \in t$ and $s_2 \not\in t$ if s_1 and s_2 are non-functional set terms. This is also possible applying the rule (R11).

Theorem 5. *An MLSSF formula ϕ is valid if and only if there is a tableau proof for ϕ using the expansion rules from Tables 1–4, the rule (EU) (Def. 21), and the closure rule from Def. 10.*

The rule (EU) is sound and helps to reduce the search space; we conjecture that completeness is preserved if the rules (EQ1'), (EQ2'), and (Cut') are replaced by (EU), but have not proven this yet.

6 An Example

As an example, we proof that the MLSSF formula

$$\phi = [x \in [(f(x) \setminus f(x \sqcup (y \sqcap x))) \sqcup z \sqcup w] \wedge w \sqcup y \in x] \to x \in z$$

is valid; it contains the free function symbol f. Intuitively, the reason for the validity of ϕ is the following: We assume that x is an element of (at least) one of the three sets $u = f(x) \setminus f(x \cup (y \cap x))$, z, and w, and that $w \cup y$ is an element of x. Now, the set u cannot contain x, because $x = (x \cup (y \cap x))$ and therefore u is empty for all interpretations of f; the set w cannot contain x, otherwise there would be a membership cycle $x \in (w \cup y) \in x$. Therefore, z contains x.

Figure 2 shows a tableau proof for ϕ. Its root is labelled with the Skolemisation $\neg [a \in [(f(a) \setminus f(a \sqcup (b \sqcap a))) \sqcup d \sqcup e] \wedge e \sqcup b \sqsubseteq a] \to a \sqsubseteq d$ of $\neg \phi$. The i-th formula in the tableau is labelled with $[i; j; R]$, which indicates that it has been derived from the j-th formula applying the expansion rule R.

Formula 9 is derived from formulae 7 and 8 applying the E-unification rule. A solution to the E-unification problem $\langle \emptyset, \langle x_a, x_a \rangle, \langle f(x_a), f(x_{a \sqcup (b \sqcap a)}) \rangle \rangle$, which is constructed from 7 and 8, is the substitution $\{x_a \leftarrow x_{a \sqcup (b \sqcap a)}\}$. Accordingly, the inequality $a \not\approx a \sqcup (b \sqcap a)$ is added to the branch.

The branch ending in formula 21 is closed by the membership cycle $e \sqcup b \sqsubseteq a$ and $a \sqsubseteq e \sqcup b$ (formulae 2 and 21). All other branches are closed by complementary literals; their leaves are labelled with the numbers of the closing literals.

If the closure rule that uses the sets of implicit predecessors to detect implicit membership cycles is used (Def. 15), the cut rule application that generates formulae 22 and 23 is not needed. Instead, the branch ending in the literal 21 is already closed; it contains an implicit cycle because $a \sqsubseteq e$ implies $a \sqsubseteq e \sqcup b$ (this cycle is made explicit by the cut rule application).

Implicit cycles can be detected by calculating the predecessor relation for the branch. The set of possible predecessors for the branch ending in formula 21 is $\{a, b, d, e, f(a), f(a \sqcup (b \sqcap a)), (e \sqcup b)\}$. The predecessor sets of the constants are $P(a) = \{e \sqcup b\}$, $P(b) = \emptyset$, $P(d) = \emptyset$, and $P(e) = \{a\}$. The predecessor set of $e \sqcup b$ is $P(e \sqcup b) = P(e) \cup P(b) = \{a\}$. Thus, we have $a \in P(e \sqcup b)$ and $e \sqcup b \in P(a)$, which indicates the presence of an implicit membership cycle.

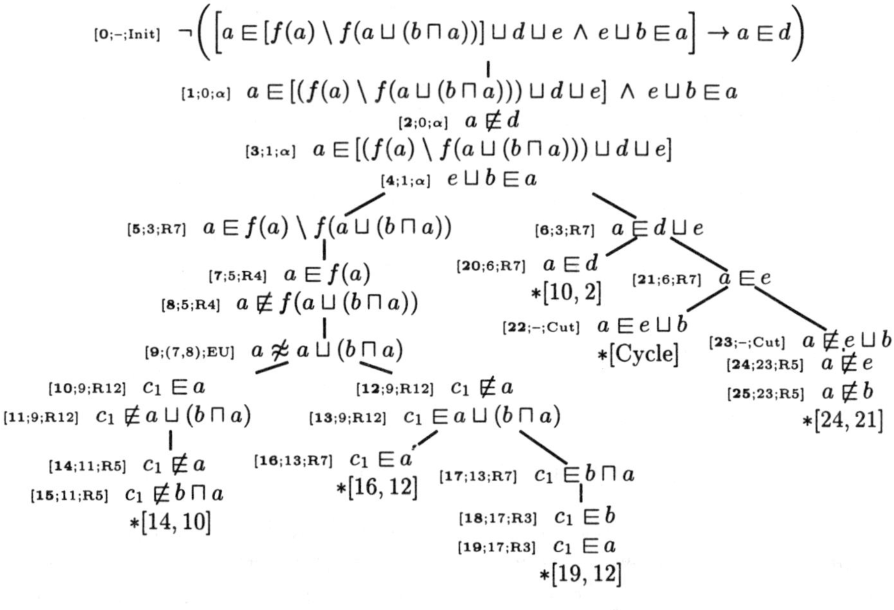

Fig. 2. Tableau proof for the formula ϕ from Sect. 6.

7 Conclusion

We have presented an improved tableau calculus for the fragment MLSS of set theory that extends the calculus described in [4]. Our tableau expansion rules are more goal-directed; this leads to a smaller search space, which is important for the efficiency of an implementation. Our calculus is a sound and complete decision procedure for MLSS. In addition, we have described a version of the calculus for the larger fragment MLSSF (MLSS with free function symbols); and we have shown how to use a special tableau rule based on *rigid E-unification* to reduce the search space in the case of MLSSF.

Future work includes, besides an implementation and practical evaluation of our calculus, its extension to larger (and undecidable) fragments that (a) contain additional set theoretic operators such as, for example, the power set operator, and that (b) allow existential quantification of variables.

Acknowledgement. We thank Domenico Cantone, Sebastiano Battiato, and three anonymous referees for useful comments on an earlier version of this paper.

References

1. B. Beckert. Rigid E-unification. In W. Bibel and P. H. Schmitt, editors, *Automated Deduction – A Basis for Applications*, volume I. Kluwer, 1998. To appear.
2. F. M. Brown. Towards the automation of set theory. *J. of AI*, 10:218–316, 1978.
3. D. Cantone. Decision procedures for elementary sublanguages of set theory. X. *Journal of Automated Reasoning*, 7:193–230, 1991.
4. D. Cantone. A fast saturation strategy for set-theoretic tableaux. In *Proceedings, TABLEAUX, Pont-a-Mousson, France*, LNCS 1227, pages 122–137. Springer, 1997.
5. D. Cantone and A. Ferro. Techniques of computable set theory with applications to proof verification. *Comm. on Pure and Applied Mathematics*, 48:901–946, 1995.
6. D. Cantone, A. Ferro, and E. Omodeo. *Computable Set Theory*. Oxford University Press, 1989.
7. D. Cantone, A. Ferro, and T. J. Schwartz. Decision procedures for elementary sublanguages of set theory. VI. *Comm. on Pure and Applied Mathematics*, 38:549–571, 1985.
8. D. Cantone, A. Ferro, and T. J. Schwartz. Decision procedures for elementary sublanguages of set theory. V. *J. of Computer and Syst. Sciences*, 34:1–18, 1987.
9. D. Cantone and T. J. Schwartz. Decision procedures for elementary sublanguages of set theory. XI. *Journal of Automated Reasoning*, 7:231–256, 1991.
10. H. de Nivelle. Implementation of sequent calculus and set theory. Draft, Feb. 1997.
11. A. Ferro and E. Omodeo. Decision procedures for elementary sublanguages of set theory. VII. *Communications on Pure and Applied Mathematics*, 40:265–280, 1987.
12. U. Hartmer. Erweiterung des Tableaukalküls mit freien Variablen um die Behandlung von Mengentheorie. Diplomarbeit, Universität Karlsruhe, 1997.
13. T. Jech. *Set Theory*. Academic Press, New York, 1978.
14. D. Pastre. Automatic theorem proving in set theory. *J. of AI*, 10:1–27, 1978.
15. B. Shults. Comprehension and description in tableaux. Draft, May 1997.

A Tableau Method for Interval Temporal Logic with Projection

Howard Bowman and Simon Thompson

Computing Laboratory,
University of Kent at Canterbury,
Canterbury, Kent, CT2 7NF, United Kingdom
{H.Bowman,S.J.Thompson}@ukc.ac.uk

Abstract. This paper introduces a tableau method for propositional interval temporal logic (ITL) [14]. Beyond the usual operators of linear temporal logic, ITL contains sequencing and iterative operators, ';' and **proj** akin to programming combinators. Central to our approach is a normal form for the formulas of ITL, particularly ';' and **proj** , in terms of the '○' operator of the logic.

1 Introduction

Interval Temporal Logic (ITL) is an important class of temporal logic. Early work on the topic was performed by Moszkowski [14] with a number of researchers progressing the topic since then, e.g. Hale [9], Kono [10], Duan [7], Cau *et. al.* [6], Bowman *et. al.* [3] and Thompson [17].

Standard temporal logics are defined over infinite state models, for example, the models LTL, the linear time temporal logic developed by Manna and Pnueli [12] are infinite state sequences. However, in interval temporal logic the model theory is restricted to finite state sequences, called *intervals*, though it is possible to extend the interpretatin to infinite sequences, and thus to see ITL as an extension of LTL.

There are a number of reasons for being interested in such logics. One reason is that interval temporal logic lends itself to execution. This is apparent from Moszkowski's work [14]. In addition, a number of interesting and powerful operators arise naturally from ITL. In fact, it is straightforward to derive operators very like the constructs of imperative programming (e.g. assignment, conditionals, iteration etc). This then yields the possibility that abstract specifications and concrete implementations can be realised in the same notation, with refinement mappings between. This we have used in our work in the field of multimedia document description, [3].

An additional aspect of interval temporal logic is that it provides a very simple real-time model in which one unit of time is past when moving from state to state. Consequently, timings can be obtained by measuring interval lengths. The ITL operator **len**(n) is used for this purpose. This operator is satisfied by any interval with $n+1$ states (transitions between states are counted rather than numbers of states).

Two operators which are characteristic of interval temporal logic are the Chop operator ; and the projection operator **proj** . The former of these implements a form of sequential composition; an interval will satisfy P ; Q if it can be divided into two contiguous sub-intervals such that P holds over the first sub-interval and Q holds over the second. The operator is illustrated in figure 1, where line segments depict intervals.

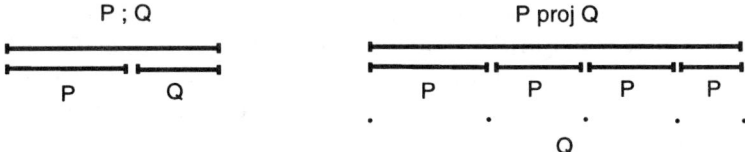

Fig. 1. Chop and Projection

In contrast, the projection operator yields repetitive behaviour; an interval satisfies P **proj** Q if it can be sub-divided into a series of sub-intervals, each of which satisfies P - we call P *the projection formula* - and the new interval formed from the end points of these sub-intervals satisfies Q, which we call *the projected formula*. The operator is illustrated in figure 1, notice the interval that Q holds over is not depicted as a line segment, rather it is the concatenation of the depicted sequence of points (each of which represents a state).

The value of the Chop operator should be self evident, however, the motivation behind projection will perhaps be clarified by some examples. An important use of projection is in deriving iteration constructs. For example, *for* loops and *while* loops can be derived using projection. The *for* loop is defined as:

$$\textbf{for } n \textbf{ times do } P \equiv P \textbf{ proj } \textbf{len}(n)$$

Notice how **len**(n) is used to count the number of iterations, by counting endpoints of sub-intervals. In contrast, the *while* loop is defined as:

$$\textbf{while } P \textbf{ do } Q \equiv (P \wedge Q)^* \wedge \Box(\textbf{len}(0) \Rightarrow \neg P)$$

$\Box$ is the ITL operator henceforth, $\Box P$ holds over an interval in which P holds over all suffixes of the interval, and R^* gives arbitrary repetition of R; it is defined as:

$$R^* \equiv R \textbf{ proj } \textbf{True}$$

In the *while* loop P will typically be a *point formula*, that is a formula whose interpretation depends only on the first point of an interpreting interval, rather than the whole interval. For example, an atomic proposition, p say, when lifted to the interval level, will hold over any interval in which p is true at the first state of the interval. Point formula are called *local formula* in some work [10].

It should also be pointed out that projection has proved a valuable operator in the real-time setting where it can be used to realise temporal abstraction [13] and hence, for example, it can describe speeding up or slowing down real-time presentations [3].

Tableau Methods have been extensively investigated in the standard (infinite) temporal logic setting, e.g. [11]. In addition, there has been some tableau work in the interval temporal logic setting, e.g. Kono [10], but this work is far less mature than that found in the (infinite) temporal logic setting. The reason for this disparity is that ITL operators are in many senses more difficult to deal with. For example, inductive definitions of the until operator of standard temporal logic are straightforward, e.g.

$$P \mathcal{U} Q \equiv Q \vee (P \wedge \bigcirc(P \mathcal{U} Q))$$

In contrast, inductive definitions of chop and projection are inherently more complex. In particular, with chop an inductive definition must cope with its first argument evolving when the operator is unfolded. This issue will become clear when we present our normal forms for chop.

Research work that is closest to ours is that by Rosner and Pnueli [16]. The logic considered in their work is a version of standard (infinite) temporal logic which includes the chop operator and they do give a tableau algorithm. However, although this work gives a number of relevant insights it is rather complex and does not handle the projection operator.

In this paper we do present a complete tableau method for interval temporal logic and we include the projection operator. Central to our strategy is the identification of normal forms for all the operators of our logic. In effect, these normal forms give inductive definitions of the ITL operators. Then, in the style of Wolper [18], we define a tableau decision procedure to check satisfiability of our logic.

Structure of the Paper. Section 2 presents background on interval temporal logic while section 3 presents our normal form. Finally, section 4 describes our tableau algorithm.

Acknowledgement. We would like to acknowledge the support – in the form of travel assistance – given to us by the British Council.

2 Interval Temporal Logic

The interval temporal logic that we use is defined in the following subsections. In the context of this paper we will call this logic PITL, for Propositional Interval Temporal Logic.

We begin by defining intervals, which will give the semantic models for our logic.

2.1 Intervals

PITL is defined over finite state sequences. Each sequence is called an interval and $\mathcal{I}$ denotes the set of all possible intervals; $\sigma \in \mathcal{I}$ has the form: $\sigma_0, \sigma_1, ..., \sigma_{|\sigma|}$, where $|\sigma|$ denotes the length of an interval and σ_i denotes the ith state in an interval. By convention the length of an interval is the number of states minus one and all intervals must have at least one state. We use $[\sigma]^i$ to denote the ith prefix of an interval and $(\sigma)^i$ to denote the ith suffix of an interval; formally, $[\sigma]^i = \sigma_0, ..., \sigma_i$ and $(\sigma)^i = \sigma_i, ..., \sigma_{|\sigma|}$. Each state is a set containing all the atomic propositions that are true at that state.

2.2 The Logic

The set of formulas of propositional logic is denoted $\mathcal{P}$ and $P \in \mathcal{P}$ is constructed as follows, where $n \in \mathcal{N}$:

$$P ::= p \mid \textbf{False} \mid \neg P \mid P \vee P \mid \bigcirc P \mid \textbf{empty} \mid P\,;P \mid P \textbf{ proj } P$$

Much of this logic will be well known to a reader familiar with interval temporal logic.

- **False**, $\neg$ and $\vee$ are the familiar connectives of classical propositional logic. A full set of logical operators can be derived in the usual way.
- p is chosen from a set of atomic propositions.
- $\bigcirc$ is the (strong) next operator. Thus, $\bigcirc P$ holds if and only if P holds over an interval of length one less than the current interval, resulting from moving one state into the future. In particular, $\bigcirc P$ is **False** on an **empty** interval.
- **empty** holds over any interval of length zero, i.e. which has one state.
- ; and **proj** are as described in the introduction.

The reader who requires a more detailed discussion of these operators is referred to [14] and also [3] where we also argue that the **proj** operator is not definable from the other operations.

Also, we have the following standard derived operators. A wealth of other operators can be defined, see for example [4].

$$\circledcirc P \equiv \bigcirc P \vee \textbf{empty}\quad,\quad \Diamond P \equiv \textbf{True}\,;P\quad,\quad \Box P \equiv \neg \Diamond \neg P$$
$$\textbf{len}(0) \equiv \textbf{empty} \quad \text{and} \quad \textbf{len}(n+1) \equiv \bigcirc \textbf{len}(n).$$

Thus, in addition, to the standard logical derivations, we have an operator to measure the length of an interval, **len**; a weak next operator, $\circledcirc$; eventually, $\Diamond$, and henceforth, $\Box$.

$$\sigma \models \textbf{empty} \quad \text{iff} \quad |\sigma| = 0$$

$$\sigma \models \bigcirc P \quad \text{iff} \quad (\sigma)^1 \models P$$

$$\sigma \models P_1; P_2 \quad \text{iff} \quad \exists k \in \mathcal{N} \ (k \leq |\sigma| \text{ and } [\sigma]^k \models P_1 \text{ and } (\sigma)^k \models P_2)$$

$$\sigma \models P_1 \ \textbf{proj} \ P_2 \quad \text{iff} \quad \exists m \in \mathcal{N} \text{ and } \exists \tau_0, \tau_1, ..., \tau_m \in \mathcal{N}$$
$$(0 = \tau_0 < \tau_1 < ... < \tau_m = |\sigma| \text{ and}$$
$$\forall j < m \ ([\sigma]^{\tau_{j+1}})^{\tau_j} \models P_1 \text{ and}$$
$$\sigma_{\tau_0} \sigma_{\tau_1} ... \sigma_{\tau_m} \models P_2)$$

Fig. 2. Satisfaction for PITL

Interpreting PITL Our satisfaction relation, $\models$, interprets PITL formulae over intervals. The notation $\sigma \models P$ denotes that the interval σ satisfies (or models) the formula P.

We interpret PITL propositions in the usual way, by induction over the structure of propositions. The satisfaction relation for the main operators is shown in figure 2; others which are standard can be found in [5]. In particular, chop subdivides the given interval into two sub-intervals the first of which satisfies P_1 and the second of which satisfies P_2. The two sub-intervals arising from chop have one common state, the kth state in the above definition. Thus, the last state of the sub-interval over which P_1 holds is the first state of the sub-interval over which P_2 holds.

The semantics of projection also require some explanation. The definition states that for an interval to satisfy P_1 **proj** P_2 there must exist a sequence of m increasing points (or states) in the interval, $\tau_0, \tau_1, ..., \tau_m$, such that the first and last points bound the interval. (It is common only to constrain this sequence to be non-decreasing, e.g. [14], however, this generates a number of pathological cases that we manage to avoid in our definition.) This sequence of points effectively divides the interval into a series of m sub-intervals, each of which comprises the states between τ_j and τ_{j+1}. We require that P_1 holds over each of these sub-intervals and, in addition, that P_2 holds when $\sigma_{\tau_0}, \sigma_{\tau_1}, ..., \sigma_{\tau_m}$ is viewed as an interval.

Satisfaction and Validity. We define that an interval σ *satisfies* a proposition P if and only if $\sigma \models P$. In addition, in the usual way, we state that P is *valid* if and only if for all σ in $\mathcal{I}$, $\sigma \models P$. If P is valid we write $\models P$.

3 Normal Forms in Interval Temporal Logic

3.1 A Normal Form for PITL

The use of normal forms in temporal logic is now relatively common, e.g. [1], [8]. Furthermore, a number of authors have proposed normal forms for interval temporal logic, e.g. [14], [10], [7].

Our normal form is based on that of previous workers; it has the general format:

$$(\textbf{empty} \land P_e) \lor \bigvee_i (P_i \land \bigcirc P_i')$$

where P_e and P_i are point formulas and P_i' is a general PITL formula. The left disjunct characterises under what circumstances a formula can be satisfied over an empty interval, while the second disjunct characterises the possible ways in which a formula can be satisfied over an interval of length greater than zero, i.e. a point property must hold at the initial state and then an arbitrary property must hold over the remainder of the interval.

You should also note that this normal form embodies a recipe for evaluating PITL formula. The first disjunct embodies a base case, i.e. what must hold of a one state interval, while the second disjunct embodies an inductive step, i.e. one of the P_i's must hold now and the associated P_i' must hold from the next state onwards.

We claim that any arbitrary PITL formula can be mapped into this normal form. The next subsection illustrates this claim and the validity of the mapping is proved in [5], for space reasons we have had to exclude this proof from this paper.

3.2 Inductive Definition of the Normal Form

We proceed through PITL highlighting how each construct of the logic can be expressed in our normal form. We thereby give an inductive definition of the normal form. So, assume Q is an arbitrary formula of PITL and assume that P_1, P_2 and P are already in normal form, as follows:

$$P_1 \equiv (\textbf{empty} \land P_{e,1}) \lor \bigvee_i (P_{i,1} \land \bigcirc P_{i,1}')$$

$$P_2 \equiv (\textbf{empty} \land P_{e,2}) \lor \bigvee_j (P_{j,2} \land \bigcirc P_{j,2}')$$

$$P \equiv (\textbf{empty} \land P_e) \lor \bigvee_i (P_i \land \bigcirc P_i')$$

Non-temporal Operators Apart from negation, which will be discussed in the next subsection, all the non-temporal operators can be mapped into the normal form in a relatively straightforward fashion. Full details can be found in [5], but by way of illustration, disjunction is handled as follows:

$$Q \equiv P_1 \lor P_2 \equiv (\textbf{empty} \land (P_{e,1} \lor P_{e,2})) \lor \bigvee_i (P_{i,1} \land \bigcirc P_{i,1}') \lor \bigvee_j (P_{j,2} \land \bigcirc P_{j,2}')$$

Partitioned normal forms Before we cover the normal form for negations of propositions, we take a detour into elementary propositional calculus. We say that a collection of formulas $\langle P_i \rangle_i$ *partitions* **True** when

- $\bigvee_i P_1 \equiv$ **True** – the partition is exhaustive; and
- $\bigvee_{i \neq j}(P_i \wedge P_j) \equiv$ **False** – the partition is exclusive.

It is an exercise to show that in the case that $\langle P_i \rangle_i$ partitions **True** then,

$$\neg \bigvee_i (P_i \wedge Q_i) \equiv \bigvee_i (P_i \wedge \neg Q_i)$$

We call a normal form

$$(\textbf{empty} \wedge P_e) \vee \bigvee_i (P_i \wedge \bigcirc P_i')$$

partitioned when $\langle P_i \rangle_i$ partitions **True**.

We can turn an arbitrary normal form into partitioned form – the details of this procedure can be found in [5]. Some logical operations (including conjunction and projection) preserve the property of being partitioned; others, notably disjunction and chop, do not. Using the technique outlined in this section we can put normal forms into partitioned form should that be necessary for what follows.

Negation Using the material in the previous section, if P is in partitioned normal form then

$$Q \equiv \neg P \equiv (\textbf{empty} \wedge \neg P_e) \vee \bigvee_i (P_i \wedge \bigcirc \neg P_i')$$

Any normal form can be transformed into partitioned normal form, and so can be negated by this method.

Temporal Operators Empty. If Q is **empty** then,

$$Q \equiv \textbf{empty} \equiv (\textbf{empty} \wedge \textbf{True}) \vee (\textbf{False} \wedge \bigcirc \textbf{True})$$

Thus, **empty** holds over any arbitrary "one state" interval, but fails to hold over any "larger" interval.

Next. If Q is $\bigcirc P$ then,

$$Q \equiv \bigcirc P \equiv (\textbf{empty} \wedge \textbf{False}) \vee (\textbf{True} \wedge \bigcirc P)$$

Thus, $\bigcirc P$ cannot hold over an empty interval (this is because it is a strong next), while it puts no constraints on the first state of a non-empty interval and requires that P holds over the remainder of the interval.

Chop. If Q is $P_1 ; P_2$ then,

$$Q \equiv P_1 \,;\, P_2 \equiv (\textbf{empty} \wedge (P_{e,1} \wedge P_{e,2})) \vee$$
$$\bigvee_i (P_{i,1} \wedge \bigcirc(P'_{i,1} \,;\, P_2)) \vee$$
$$\bigvee_j ((P_{e,1} \wedge P_{j,2}) \wedge \bigcirc P'_{j,2})$$

The normal form embodies three cases.

- The first disjunct gives the condition under which the chop formula holds over an empty interval, namely when the two constituent formulas hold over the empty interval;
- the second disjunct is the condition in which P_1 holds over a non-empty interval, and P_2 holds over the remainder; and,
- the third disjunct embodies the case where P_1 holds over an empty interval and P_2 holds over the entire (non-empty) interval.

The negation of chop is given by transforming the normal form given here into the partitioned normal form, as explained above.

Projection. If Q is $(P_1 \textbf{ proj } P_2)$ then,

$$Q \equiv P_1 \textbf{ proj } P_2 \equiv (\textbf{empty} \wedge P_{e,2}) \vee$$
$$\bigvee_i \bigvee_j ((P_{i,1} \wedge P_{j,2}) \wedge \bigcirc(P'_{i,1} \,;\, P_1 \textbf{ proj } P'_{j,2}))$$

The normal form embodies two cases.

- The first disjunct gives the condition under which the projection holds over an empty interval, namely when P_2 holds over the interval.
- The other disjunct gives the condition where $(P_1 \textbf{ proj } P_2)$ holds over a non-empty interval. A choice of initial-state conditions of P_1 and P_2, such as $P_{i,1}$ and $P_{j,2}$, must hold at the first state; over the remaining part of the interval a chop has to hold. First the remainder of the P_1 condition – $P'_{i,1}$ – must hold, then the derivative of the projection – $(P_1 \textbf{ proj } P'_{j,2})$ – must be valid. This is illustrated in Figure 3

Note that if P_1 and P_2 are in partitioned normal form then this normal form will also be partitioned.

3.3 Example

Assume that the formula P is in normal form and consider the derived formula eventually P, $\Diamond P$. The standard derivation of eventually is from chop,

$$\Diamond P \equiv \textbf{True} \,;\, P$$

Now, using the normal form rule for chop, $\Diamond P$ can be placed in normal form as follows:

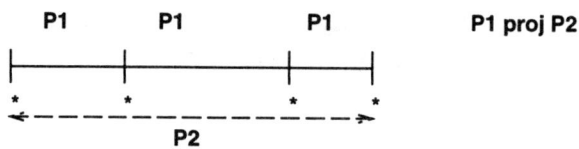

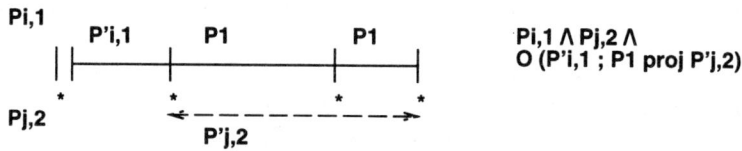

Fig. 3. The Normal Form for Projection

$$\Diamond P \equiv (\text{empty} \wedge P_e)$$
$$\vee\ (\text{True} \wedge \bigcirc(\text{True}\ ; P))$$
$$\vee \bigvee_i (P_i \wedge \bigcirc P_i')$$

Notice that using associativity and symmetry of disjunction and **True** as unit of conjunction we can reduce this to:

$$\Diamond P \equiv P \vee \bigcirc \Diamond P$$

which is a more common "inductive" interpretation of $\Diamond$.

4 The Tableau Method

In this section we present a tableau decision procedure for checking the satisfiability of PITL formulae. The decision procedure is influenced by the tableau procedures developed for standard temporal logic, e.g. [2] [15] and [18]. Our work particularly builds from the tableau method defined by P. Wolper [18].

The tableau decision algorithm is a graph construction (and then reduction) algorithm. Nodes of the graph are sets of PITL formula; these are said to *label the node*. Edges in the graph represent steps to satisfaction of the formulas of a node.

These steps are made according to a set of *tableau (decomposition) rules*. Branches in the tableau reflect disjunctive choices.

An important element of the tableau rules is that they subdivide the requirements imposed by temporal formulas into requirements on the present (i.e. the first state of an interval) and requirements on the remainder of the interval, the latter being embodied in a formula of the form $\bigcirc P$. This is in fact exactly the subdivision in our normal form rules. Accordingly our normal forms will play a central role in these tableau rules.

A set of *graph construction* rules build the graph using the tableau rules. In particular, the graph construction rules generate time passing steps in the tableau that unravel next formulas.

The final stage of the tableau algorithm is a *graph reduction* algorithm which systematically removes unsatisfiable nodes in the graph. A formula can then be deduced to be satisfiable if it appears in the reduced graph.

We will work through the different stages of the tableau algoorithm in turn, but first some notation.

Notation. The following notation is related to that used in [18].

- The formula being evaluated is called the *initial formula*.
- The *labelling of node n* is denoted T_n.
- A formula P is called *elementary* if
  1. it is an atomic proposition or the negation of an atomic proposition;
  2. it is **empty** or ¬**empty**; or
  3. it has $\bigcirc$ as the main connective.
- A node containing only elementary formulae is called a *state*.
- A *pre-state* is a node that is either initial or the immediate son of a state.

Tableau Rules. The *tableau rules* (also called *decomposition rules*) drive the reduction of formulae into their components. They map a formula into a set of sets of formulas, i.e. they have the form:

$$P \longrightarrow \{ S_1, S_2, ..., S_n \}$$

where S_i is a set of PITL formulas. The interpretation that such rules embody is that P can be satisfied if there exists a j $(1 \leq j \leq n)$ such that all formulas in S_j can together be satisfied.

Our first set of rules are standard and are unchanged from those in [18]:

$$\neg\neg P \longrightarrow \{\{ P \}\}$$
$$P_1 \wedge P_2 \longrightarrow \{\{ P_1 , P_2 \}\}$$
$$\neg(P_1 \wedge P_2) \longrightarrow \{\{ \neg P_1 \},\{ \neg P_2 \}\}$$
$$P_1 \vee P_2 \longrightarrow \{\{ P_1 \},\{ P_2 \}\}$$
$$\neg(P_1 \vee P_2) \longrightarrow \{\{ \neg P_1 , \neg P_2 \}\}$$

We also need the following rule which enables the next operator to be moved to the top level through negation. Notice that since we are using a strong next operator, $\bigcirc P$ will fail over any **empty** interval.

$$\neg \bigcirc P \longrightarrow \{\{ \textbf{empty} \},\{ \bigcirc \neg P \}\}$$

The remaining rules are direct extrapolations from our normal form rules. They all assume that P_1 and P_2 are in the normal forms we highlighted earlier. The next two rules decompose chop and projection:

$$P_1 \; ; P_2 \longrightarrow \{\{ \, (\textbf{empty} \wedge (P_{e,1} \wedge P_{e,2})) \vee$$
$$\bigvee_i (P_{i,1} \wedge \bigcirc(P'_{i,1} \; ; P_2)) \vee$$
$$\bigvee_j ((P_{e,1} \wedge P_{j,2}) \wedge \bigcirc P'_{j,2}) \, \}\}$$

$$P_1 \; \textbf{proj} \; P_2 \longrightarrow \{\{ \, (\textbf{empty} \wedge P_{e,2}) \vee$$
$$\bigvee_i \bigvee_j ((P_{i,1} \wedge P_{j,2}) \wedge \bigcirc(P'_{i,1} \; ; P_1 \; \textbf{proj} \; P'_{j,2})) \, \}\}$$

$\neg(P_1 \; ; P_2)$ and $\neg(P_1 \; \textbf{proj} \; P_2)$ can be obtained by the negation of normal forms previously highlighted. In fact, negation of all the operators could be obtained from negation of the normal forms, but in the case of the propositional operators and strong next it is more straightforward to handle negation outside the normal form expansion.

The following two rules encapsulate the required normal form negation. Both rules assume that $\langle P_{i,1} \rangle_i$ and $\langle P_{i,2} \rangle_i$ are in partition form.

$$\neg(P_1 \; ; P_2) \longrightarrow \{\{ \, (\textbf{empty} \wedge (\neg P_{e,1} \vee \neg P_{e,2})) \vee$$
$$\bigvee_i ((\neg P_{e,1} \wedge P_{i,1}) \wedge \bigcirc \neg(P'_{i,1} \; ; P_2)) \vee$$
$$\bigvee_i \bigvee_j ((P_{e,1} \wedge P_{i,1} \wedge P_{j,2}) \wedge \bigcirc(\neg P'_{j,2} \wedge \neg(P'_{i,1} \; ; P_2))) \, \}\}$$

$$\neg(P_1 \; \textbf{proj} \; P_2) \longrightarrow \{\{ \, (\textbf{empty} \wedge \neg P_{e,2}) \vee$$
$$\bigvee_i \bigvee_j ((P_{i,1} \wedge P_{j,2}) \wedge \bigcirc \neg(P'_{i,1} \; ; P_1 \; \textbf{proj} \; P'_{j,2})) \, \}\}$$

We have the following lemma.

Lemma 1. *All the decomposition rules preserve satisfaction. Thus, for a rule of the form,*

$$P \longrightarrow \{ \, S_i \, \}$$

the following holds:

$$\sigma \models P \;\; \textit{iff} \;\; \exists j \, . \, \sigma \models \bigwedge S_j \;\;\; \textit{where} \;\; \bigwedge \{P_1, ..., P_n\} = P_1 \wedge ... \wedge P_n$$

Proof
The propositional rules and the rule for $\neg \bigcirc P$ are straightforward. The rules for ;, **proj**, $\neg$; and $\neg$ **proj** follow from validity of the normal forms.

The Graph Construction. The tableau graph is constructed using the following rules:

1. Where P is the formula to test, label the *initial node* with { P, **True** ; **empty** } and apply steps (2) and (3) repeatedly until neither can be applied to any node in the graph.

   Note that when we create a son node in rules (2) and (3) with labelling T, we only add a new node if a node labelled T does not already exist, otherwise we add an edge back to the original node.
2. If a node n contains a non-elementary formula P and the tableau rule for P is $P \longrightarrow \{S_i\}$, then, for all i create a son of n labelled: $(T_n - \{P\}) \cup S_i$.
3. If a node n contains only: elementary formulas and does not contain **empty**, then create a son of n labelled by the next formulas of T_n with outermost nexts removed.

We call rule 2 *the decomposition rule* and rule 3 is called *the step rule*.

So, we start with the formula under test and we add ◊**empty** to the initial node. This is needed in order to enforce the requirement that a successful path through the tableau terminates. This termination arises through the final node in the path containing **empty**. Another way of looking at this is that it ensures that a marker for termination of a path is included in every state (note ◊**empty** is either satisfied immediately and a path is terminated or it is preserved in the next state, i.e. ◊**empty** arises in all descendent states, lemma 2 encapsulates this fact).

Graph Reduction. The next stage in the tableau algorithm is to eliminate unsatisfiable nodes by repeatedly applying the following rules:

- (E1) Eliminate nodes containing formula P and $\neg P$.
- (E2) If a node has no successors and does not contain **empty**, eliminate it.
- (E3) If a node contains **empty** and $\bigcirc P$, eliminate it.
- (E4) If a pre-state contains an unaccepted formula of the form $P_1 \; ; \; P_2$, $\neg(P_1 \; ; \; P_2)$, P_1 **proj** P_2 or $\neg(P_1$ **proj** $P_2)$ then eliminate the pre-state.

This procedure terminates when all unsatisfiable nodes have been eliminated. The first three of these rules embody straightforward cases of an unsatisfiable node. The fourth rule hinges on the notion of a chop or projection formula failing to be accepted. The following rule defines what it means for a formula to be accepted.

- (F) A formula P is *accepted* in a pre-state if there is a path in the tableau leading from that pre-state to a node containing **empty**.

We claim that the initial formula is satisfiable if we can generate a tableau according to the graph construction rules, reduce it according to rules (E1) - (E4) and the initial formula is not eliminated.

Discussion. The tableau algorithm that we present here is simpler in a number of respects than that defined in [18]. For example, a central concept in Wolper's work is that of a *fulfilling path*. For any temporal formula, other than ○, to be satisfied a path which witnesses the satisfaction of the formula must exist.

Our accepting paths play a similar role. However, many of the complexities associated with fulfilling paths do not arise with accepting paths since we are searching for *finite* models/satisfying paths and thus, we can use **empty** as a marker for successful completion of such a satisfying model. In addition, Wolper includes a marking mechanism, whereby formulas are marked when they are being considered for fulfillment. This means that formulas that have been fulfilled are carried through the tableau. In contrast, we simply discard formulas once we unfold them with the decomposition rule. This simpler approach is also justified on the grounds that we are working with finite models.

Example. The tableau expansion for $(\mathbf{len}(1)\,;\,\bigcirc q) \wedge \mathbf{len}(1)$ is shown in figure 4. In order to simplify the presentation the decomposition arrows shown reflect multiple applications of the decomposition rules. These multiple applications will simplify the normal form expansions, yielding the following relationships:

$$\bigcirc\mathbf{empty}\,;\,\bigcirc q \equiv \bigcirc(\mathbf{empty}\,;\,\bigcirc q)$$
$$\mathbf{True}\,;\,\mathbf{empty} \equiv \mathbf{empty} \vee \bigcirc(\mathbf{True}\,;\,\mathbf{empty})$$
$$\mathbf{empty}\,;\,\bigcirc q \equiv \bigcirc q$$

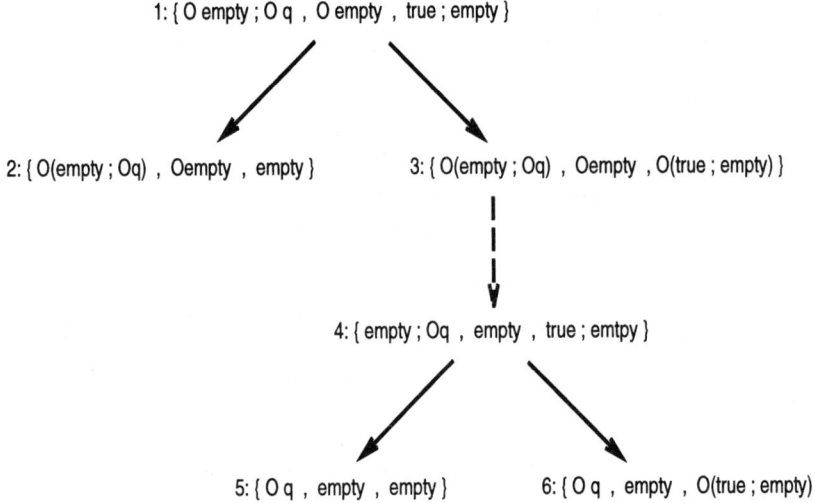

Fig. 4. Tableau expansion of $\mathbf{len}(1)\,;\,\bigcirc q \wedge \mathbf{len}(1)$

It turns out that all the "terminal" nodes here, i.e. those containing **empty**, will be eliminated because they contain unsatisfied next state formulae. The elimination of all leaf nodes will cause all new leaf nodes (which neither contain **empty** or have any successors) to be eliminated which in turn will have a knock on effect on the new set of leaf nodes and so on, until the initial node has been eliminated. Thus, showing that the formula,

$$(\mathbf{len}(1) \mathbin{;} \bigcirc q) \wedge \mathbf{len}(1)$$

is unsatisfiable.

5 Correctness of the Tableau

This section investigates correctness of the tableau algorithm. First we need a small lemma which clarifies how the $\Diamond\mathbf{empty}$ behaves in the initial formula. We give sketches of the proofs in this section; full proofs can be found in [5].

Lemma 2. *All states in the tableau contain either* **empty** *or* $\bigcirc(\mathbf{True} \mathbin{;} \mathbf{empty})$.

Proof sketch:
We proceed by induction over the structure of the graph, noting that the normal form for **True ; empty** is

$$\mathbf{True} \mathbin{;} \mathbf{empty} \equiv (\mathbf{empty} \wedge \mathbf{True}) \vee$$
$$(\mathbf{True} \wedge \bigcirc(\mathbf{True} \mathbin{;} \mathbf{empty}))$$

and that the rules of the tableau expand the formula according to its normal form. □

The following theorem shows that our tableau algorithm does indeed check satisfiability.

Theorem 1. *P is satisfiable if and only if the initial node of the graph generated by the tableau method just presented is not eliminated.*

Proof sketch:
($\Longrightarrow$) For this we prove the contrapositive:

The initial node is eliminated implies that P is unsatisfiable.

First we prove the general result that a node, n say, labelled $\{P_1, .., P_m\}$ is eliminated implies $P_1 \wedge P_2 \wedge ... \wedge P_m$ is unsatisfiable. We prove this by induction on the structure of the graph. Thus, we assume the result for all successors of node n. We consider in turn the possible reasons for node n being eliminated; cases (E1) and (E3) are the base cases; (E2) and (E4) give the induction steps. The cases (E1)–(E3) are standard.

In the case of (E4), it is sufficient to show that if there is a model σ for a formula P then there is an accepting path for P. The proof is by induction over the length of σ and with reference to the normal forms of the formulas involved.

This completes our proof that a node n is eliminated implies that the conjunction of its constituent formulas is unsatisfiable. The required property concerning the initial node is a special case of this result.

($\Longleftarrow$) We need to verify the property:

The initial node of the graph is not eliminated implies P is satisfiable

First note that if the initial node of the graph is not eliminated then an accepting path for P must exist in the (reduced) tableau. Our strategy will be to derive a model for P from this accepting path.

Assume P has an accepting path. Now from that path extract the *subaccepting* path which contains just the states in the accepting path, i.e. all pre-states and intermediate states are removed. Note that an accepting path must contain at least one state since the terminating node of the path will be a state. This is because if the terminating node contains a non-elementary formula then the node would be eliminated by rule (E4) since an accepting path for the formula would not exist. We will argue by induction on the size of this subaccepting path that an interval σ exists. In fact, the induction states that σ corresponds exactly to this subaccepting path.

Thic completes the second half of the proof and thus the proof of the result itself. □

References

1. H. Barringer, M. Fisher, D. Gabbay, G. Gough, and R. Owens. METATEM : A framework for programming in temporal logic. In *Lecture Notes in Artificial Intelligence, vol. 430*. Springer–Verlag, 1989.
2. M. Ben-Ari, Z. Manna, and A. Pnueli. The temporal logic of branching time. In *8th ACM Symposium on Principles of Programming Languages*, pages 164–176. ACM, 1981.
3. Howard Bowman, Helen Cameron, Peter King, and Simon Thompson. Specification and prototyping of structured multimedia documents using interval temporal logic. In *ICTL'97, International Conference on Temporal Logic, Manchester*. Kluwer, Applied Logic Series, 1997.
4. Howard Bowman, Helen Cameron, Peter King, and Simon Thompson. Specification and prototyping of structured multimedia documents using interval temporal logic. Technical Report 3-97 (Kent), Computing Laboratory, University of Kent, 1997.
5. Howard Bowman and Simon Thompson. A tableau method for interval temporal logic. Technical Report 12-97, Computing Laboratory, University of Kent, 1997.
6. A. Cau and H. Zedan. Refining interval temporal logic specifications. In *Fourth AMAST Workshop on Real-Time Systems, Concurrent, and Distributed Software (ARTS'97), LNCS 1231*, May 1997.
7. Z. H. Duan. *An Extended Interval Temporal Logic and A Framing Technique for Temporal Logic Programming*. PhD thesis, University of Newcastle Upon Tyne, May 1996.
8. Dov Gabbay. The declarative past and imperative future. In B. Banieqbal, H. Barringer, and A. Pnueli, editors, *Temporal Logic in Specification*. Lecture Notes In Computer Science 389, Springer-Verlag, 1989.
9. R. Hale. Using temporal logic for prototyping: the design of a lift controller. In *Lecture Notes in Computer Science, vol. 379*, pages 375–408. Springer–Verlag, 1989.
10. S. Kono. A combination of clausal and non-clausal temporal logic programs. In *Lecture Notes in Artificial Intelligence, vol. 897*, pages 40–57. Springer–Verlag, 1993.

11. O. Lichtensteim, A. Pnueli, and L. Zuck. The glory of the past. In *Proceedings of Conference on Logics of Programs, LNCS 193*. Springer-Verlag, 1985.
12. Z. Manna and A. Pnueli. *The Temporal Logic of Reactive and Concurrent Systems*. Springer-Verlag, 1992.
13. T. Melham. *Higher Order Logic and Hardware Verification*. Cambridge Tracts in Theoretical Computer Science (31), 1993.
14. B. Moszkowski. *Executing Temporal Logic*. Cambridge University Press, 1986.
15. N. Rescher and A. Urquart. *Temporal Logic*. Springer-Verlag, 1971.
16. R. Rosner and A. Pnueli. A choppy logic. In *Proceedings of 1st IEEE Symposium on Logic in Computer Science*, pages 306–314. IEEE, 1986.
17. Simon Thompson. Constructive interval temporal logic in alf. In *ICTL'97, International Conference on Temporal Logic, Manchester*. Kluwer, Applied Logic Series, 1997.
18. P. Wolper. Temporal logic can be more expressive. *Information and Computation*, 56:72–99, 1983.

Bounded Model Search in Linear Temporal Logic and Its Application to Planning

Serenella Cerrito[1] and Marta Cialdea Mayer[2]*

[1] Université de Paris-Sud, LRI, Bât 490,
F-91405 Orsay Cedex, France.
e-mail: serena@lri.fr

[2] Dipartimento di Informatica e Automazione, Università di Roma Tre,
via Vasca Navale 79, 00146 Roma, Italia.
e-mail: cialdea@dia.uniroma3.it

Abstract. In this work a tableau calculus is proposed, that checks whether a finite set of formulae in propositional linear temporal logic (LTL) has a finite model whose cardinality is bounded by a constant given in input, and constructs such a model, if any. From a theoretical standpoint, the method can also be used to check finite satisfiability *tout court*. The following properties of the proposed calculus are proved: termination, soundness and completeness w.r.t. finite model construction. The motivation behind this work is the design of a logical language to model planning problems and an associated calculus for plan construction, integrating the declarativity, expressiveness and flexibility typical of the logical languages with the capability of embedding search-based techniques well established in the planning community.

1 Introduction

This work investigates the issue of the construction of *finite* models of a given set of linear temporal formulae and, in particular, its application to model and solve planning problems.

In [12] the view of a planning activity is proposed, as the search for a finite model of the specification of the planning problem. In such a perspective, planning amounts to prove that *preconditions* ∧ *goals* has a finite model and such a model represents, in fact, a plan achieving the desired goals. Except for this and few other proposals, the logical approach to planning is usually based on deduction, and plan generation is carried out by constructively *proving* plan specification formulae, having the form *preconditions* → *goals*. Within this approach, different logics have been used: classical logic [16], linear logic [14,15], temporal logics [21,13]. Works using modal temporal logics rely on the branching model of time of interval-based temporal logics.

---

* This work has been partially supported by MURST, ASI (Agenzia Spaziale Italiana) and CNR (SCIxSIA Project).

The other main approach to planning, deriving from classical STRIPS, is based on *ad hoc*, more or less powerful formalisms, and solves planning problems by means of search based techniques (see, for example, [8, 19]). Some of the planning languages in this category have a definitely unclear semantics. The problem specification sometimes codes procedural information, used to guide the search, that leads the planner to prune part of the potential search space, thus gaining in efficiency but risking to lose completeness. Other planning formalisms have been given a clean semantics, but they generally have a limited expressive power, corresponding to significant restrictions of first order logic. The problem of designing more expressive and *easy-to-use* planning languages, with a clear and natural semantics, is felt as an important one in the planning community. On the other hand, the use of specialized algorithms and techniques, such as partial order, regression planning, least commitment (see Section 4), as well as control strategies to guide the search, may overperform the use of general theorem provers as tools for plan construction. In fact, in many planning problems, methods based on regression and plan space search lead to a significant reduction of the search space.

The motivation of this work is to lay the grounds for the design of a logical language and an associated calculus for plan construction having, on one side, the declarativity, expressiveness and flexibility typical of the logical approach, together with a clean semantics based on a natural model of time and action, and, on the other, the capability of embedding search-based techniques, typically partial order, regression and least commitment. Viewing a plan as a finite model of the problem specification, this work proposes the use of linear temporal logic (LTL) as the specification language (differently, [12] uses propositional classical logic). The main advantage of the use of LTL as a planning language derives from its rich expressive power[1] and the underlying simple model of time.

The search for finite LTL models is carried out by means of tableau construction. Semantical tableaux bear in fact a significant advantage with respect to other proof systems strongly based on normal forms (resolution oriented, for instance) when proofs are expected to convey understandable information to the user. This is in fact the case in the planning domain, since the design of mixed-initiative systems can be a good choice in order to control the complexity of real world planning problems. Moreover, tableau methods have proved to be efficient reasoning methods for practical purposes, when the proof search is guided by control knowledge that help keep the search space to a manageable size. The tableau calculus described in this work carries strong similarities with [18]. However, while temporal structures are alway infinite in that work, we focus on finite temporal models, that are obviously more suitable to represent plans.

Planning is a notoriusly hard problem: plan existence is PSPACE-complete for propositional STRIPS-planning [5]. However, the existence of polynomial-length plans is NP-complete [12], hence practical solutions are often required not to exceed a given (polynomial) length known in advance. For this reason,

---

[1] Mainly, with respect to classical propositional logic, the "since and "until" operators add the strength of a sort of bounded quantification.

this work focuses on *bounded* planning problems, i.e., in the temporal logic approach, the search of finite models whose size is not greater than a given limit. Whenever such a limit is not known in advance, the shortest plan can be found by iterative deepening search. Moreover, from a theoretical standpoint, the calculus can also be used to check finite satisfiability *tout court*, just by fixing a suitable (exponential) limit to the plan length (see Section 3.4).

As a byproduct of the focus on finite models, the "since" and "until" operators can be given an intuitive and simple treatment, without hindering termination. As already remarked, the naturalness of the expansion rules is an important feature for the intended application.

We stress that in this work we focus on the theoretical grounds of the integration of search-based techniques in an LTL tableau calculus, without specifically addressing efficiency issues. Practical systems for model construction can be built on this basis, by studying meaningful subclasses of the language, allowing the encoding of planning problems. These two phases are typical of the development of logical approaches to AI problems.

2   The Tableau System for Linear Temporal Logic

The language of linear temporal logic we consider extends classical propositional logic, with the logical operators $\neg, \wedge, \vee, \bot$ (always false) and $\top$ (always true), by means of the unary modal operators $\Box$ (always in the future), $\boxminus$ (always in the past), $\Diamond$ (sometime in the future), $\Diamondminus$ (sometime in the past), and the binary ones $\mathcal{S}$ (since) and $\mathcal{U}$ (until).

The semantics of the language is defined as follows. A *temporal structure* T is a finite initial segment of the natural numbers: $\langle 0, ..., k \rangle$; its elements are called *time points*. If $\mathcal{L}$ is an LTL temporal language and P the set of propositional letters in $\mathcal{L}$, an $\mathcal{L}$-*interpretation* $\mathcal{M}$ is a pair $\langle T, \sigma \rangle$, where:

- T is a temporal structure;
- $\sigma : T \to \mathcal{P}(P)$ is a function on time points, providing an interpretation to the propositional letters in P for any point in T. I.e. if $i \in T$, then $\sigma(i) \subseteq P$ is the set of propositional letters true at i.

The satisfiability relation $\mathcal{M}_i \models A$, for $i \in T$, is inductively defined by addition of the following clauses to the usual ones for the classical connectives:

1. $\mathcal{M}_i \models \Box A$ iff for all $j > i$, $\mathcal{M}_j \models A$.
2. $\mathcal{M}_i \models \Diamond A$ iff there exists $j \in T$ such that $j > i$ and $\mathcal{M}_j \models A$.
3. $\mathcal{M}_i \models A\mathcal{U}B$ iff $\exists j \in T$ such that $j > i$ and $\mathcal{M}_j \models B$ and for any k with $i < k < j$ $\mathcal{M}_k \models A$.
4. $\mathcal{M}_i \models \boxminus A$ iff for all $j < i$, $\mathcal{M}_j \models A$.
5. $\mathcal{M}_i \models \Diamondminus A$ iff there exists $j \in T$ such that $j < i$ and $\mathcal{M}_j \models A$.
6. $\mathcal{M}_i \models A\mathcal{S}B$ iff $\exists j \in T$ such that $j < i$ and $\mathcal{M}_j \models B$ and for any k with $j < k < i$ $\mathcal{M}_k \models A$.

Truth is satisfiability in the initial state: a formula A is true in $\mathcal{M}$ (and $\mathcal{M}$ is a model of A) iff $\mathcal{M}_0 \models A$. Truth of sets of formulae is defined as usual.

Note that, due to the "strong" interpretation of the modal operators (excluding the present time point also in the case of future time operators), the weak and strong "Next" and "Last" operators are definable. Although also the $\Diamond$ and $\overleftarrow{\Diamond}$ operators are definable, we prefer to give them a separate treatment in the tableau system. The semantics of such "existential" operators is again the strong one, so that, for example, $\Box \Diamond A$ can be true only in temporal structures consisting of a single time point (it is in fact always true in such cases).

A formula is in *negation normal form* (nnf) iff no logical operator is in the scope of a negation. Two formulae A and B are equivalent iff for all $\mathcal{M}$ and i, $\mathcal{M}_i \models A$ iff $\mathcal{M}_i \models B$. It can easily be shown that, under this strong notion of equivalence, every formula can be transformed into an equivalent formula in nnf, by applying the usual equivalences for $\rightarrow, \wedge, \vee, \Box, \overleftarrow{\Box}, \Diamond, \overleftarrow{\Diamond}$, and:

$$\neg(A\mathcal{U}B) \equiv (\neg B \mathcal{U}(\neg A \wedge \neg B)) \vee \Box \neg B \qquad \neg(A\mathcal{S}B) \equiv (\neg B \mathcal{S}(\neg A \wedge \neg B)) \vee \overleftarrow{\Box} \neg B$$

In the rest of this section a tableau system is defined, that allows one to test whether a set LTL formulae in nnf admits models whose underlying temporal structure does not exceed a given size. In case of a positive answer, a model can be easily extracted from the tableau construction. The restriction to nnf formulae is introduced only to simplify the presentation of the rules.

Let $C = \{start, finish, d_1, d_2, d_3, ...\}$ be a set of *constants* (intuitively denoting time points). A *state* is any expression of the form $c+n$, for $c \in C$ and $n \in \mathbb{Z}$. The set of states is denoted by Σ. It is intended that $C \subset \Sigma$ (c can be rewritten as $c+0$). If $s, t \in \Sigma$, then $s \leq t$ is a *temporal constraint*. A *labelled formula* is an expression of the form $[s,t]A$, where $s, t \in \Sigma$ and A is an LTL formula in nnf. $[s,s]A$ will be abbreviated by $[s]A$.

Tableau nodes are labelled either by temporal constraints or labelled formulae (that are called *logical nodes*). If S is a finite set of formulae in nnf and $K = \{finish \leq start + k\}$ for some integer $k \geq 0$ (representing the maximal size of the searched models), then tableaux for $S \cup K$ are initialized with the set $\{[start]A \mid A \in S\} \cup K$ and expanded by application of the rules in Table 1, where c denotes an element of C, and $s, t, s', t', ..$ elements of Σ. The set of nodes occurring above the line of a rule is called the *premise* of the rule, while the sets of nodes occurring below are the *expansions* of the premise.

Note that a sort of contraction is implicit in the β-rule: the rightmost expansion of the rule contains a node with the same formula already occurring in the premise, even though the labels (intervals) of the nodes are different. The intuition behind the β-rule is the following: either A is true in the whole interval (leftmost branch), or there exists a smallest time point c in the interval where A is false, hence B is true; since c is chosen to be the first of such points, A is true in the (possibly empty) subinterval before it. The rule could also be formulated in a symmetric way, distinguishing two cases according to whether A is true at s or B is true at s. However, this would force the rule to be reapplied once for each point of the interval; on the contrary, with the proposed asymmetric formulation the interval may be cut into larger pieces at each application (obviously, the behaviour is the same in the worst case, where A and B are interleaved). The β-rule is a delicate point: note that it is indirectly charged to expand also $\mathcal{U}$- and $\mathcal{S}$-formulae.

Logical rules		
Propositional	*Future time*	*Past time*
α-**rule**	$\Box$-**rule**	$\overleftarrow{\Box}$-**rule**
$\dfrac{[s,t]\,A \wedge B \quad s \leq t}{[s,t]\,A \quad [s,t]\,B}$	$\dfrac{[s,t]\,\Box A \quad s \leq t}{[s+1, finish]\,A}$	$\dfrac{[s,t]\,\overleftarrow{\Box} A \quad s \leq t}{[start, t-1]\,A}$
β-**rule**	$\mathcal{U}$-**rule**	$\mathcal{S}$-**rule**
$\dfrac{[s,t]\,A \vee B \quad s \leq t}{[s,t]\,A \quad \begin{array}{c} s \leq c \\ c \leq t \\ [s,c-1]\,A \\ [c]\,B \\ [c+1,t]\,A \vee B \\ c \in C\ fresh \end{array}}$	$\dfrac{[s,t]\,A\mathcal{U}B \quad s \leq t}{\begin{array}{c}[c]\,B \\ t+1 \leq c \\ [t+1, c-1]\,A \\ [s+1,t]\,A \vee B \\ c \in C\ fresh\end{array}}$	$\dfrac{[s,t]\,A\mathcal{S}B \quad s \leq t}{\begin{array}{c}[c]\,B \\ c \leq s-1 \\ [c+1, s-1]\,A \\ [s,t-1]\,A \vee B \\ c \in C\ fresh\end{array}}$
	$\Diamond$-**rule**	$\overleftarrow{\Diamond}$-**rule**
	$\dfrac{[s,t]\,\Diamond A \quad s \leq t}{\begin{array}{c}[c]\,A \\ t+1 \leq c \\ c \in C\ fresh\end{array}}$	$\dfrac{[s,t]\,\overleftarrow{\Diamond} A \quad s \leq t}{\begin{array}{c}[c]\,A \\ c \leq s-1 \\ c \in C\ fresh\end{array}}$
Interval rule	**Conflict resolution rules**	
$\dfrac{[s,t]\,A}{t \leq s-1 \quad s \leq t}$	$\dfrac{\begin{array}{c}s \leq t \\ s' \leq t' \\ [s,t]\,p \\ [s',t']\,\neg p\end{array}}{t \leq s'-1 \quad t' \leq s-1}$	$\dfrac{[s,t]\,\bot}{t \leq s-1}$

Table 1. Tableau expansion rules

When the interval rule is applied to expand $[s,t]A$, we say that it is applied to the interval $[s,t]$, independently of the formula A. This rule distinguishes the cases where an interval is empty or not. Its role is to provide the preconditions for the application of the logical and resolution rules. Intuitively, it is useless – and sometimes even incorrect – to expand a node $[s,t]A$ when the interval is empty and, given two nodes $[s,t]p$, $[s',t']\neg p$, there is no conflict to be solved if either $[s,t]$ or $[s',t']$ (or both) are empty. Note that such rule could be dispensed with, and a corresponding branching added to most of the other rules, handling the case where the considered interval is empty. Its distinction as a separate rule makes the formulation of the calculus more compact and clearer: obviously, a test on the "emptyness" of an interval $[s,t]$ needs to be done just once in a branch, independently on the number of logical nodes labelled by $[s,t]$.

When the leftmost conflict resolution rule is applied, we say that it is applied to the nodes $[s,t]\,p$ and $[s',t']\,\neg p$.

In the following, if $\mathcal{B}$ is a tableau branch, $const(\mathcal{B})$ denotes the set of constants occurring in $\mathcal{B}$ and including $start$ and $finish$.

Definition 1. *Let C be a set of constants (including $start$ and $finish$) and $\mathcal{I}$ a mapping from C to the integers. The notation $\mathcal{I}^*$ is used to denote the extension of $\mathcal{I}$ from states to the integers such that $\mathcal{I}^*(c+n) = \mathcal{I}(c)+n$ for every $c \in C, n \in Z$.*

1. *Let $T = \langle 0,...,k\rangle$ be a finite sequence of integers starting at 0. $\mathcal{I}$ is a temporal mapping for C with range T if $min\{\mathcal{I}(c) \mid c \in C\} = \mathcal{I}(start) = 0$ and $max\{\mathcal{I}(c) \mid c \in C\} = \mathcal{I}(finish) = k$. Hence, in particular, the range of a temporal mapping is always finite.*
2. *If K is a set of temporal constraints over C, then $\mathcal{I}$ is a solution to K iff:*
   (a) $\mathcal{I}$ is a temporal mapping for C;
   (b) if $s \leq t \in K$, then $\mathcal{I}^*(s) \leq \mathcal{I}^*(t)$.
3. *Let $\mathcal{B}$ be a tableau branch, $C = const(\mathcal{B})$ and $\mathcal{M}$ a temporal interpretation with domain T.*
   (a) If $\mathcal{I}$ is a temporal mapping for C with range T, then $\langle \mathcal{M}, \mathcal{I}\rangle$ satisfies $\mathcal{B}$ ($\langle \mathcal{M}, \mathcal{I}\rangle \models \mathcal{B}$) iff:
      i. $\mathcal{I}$ is a solution to the set of temporal constraints occurring in $\mathcal{B}$;
      ii. if $[s,t]A$ occurs in $\mathcal{B}$, then for every integer i, if $i \in T$ and $I^*(s) \leq i \leq I^*(t)$, then $\mathcal{M}_i \models A$.
   (b) $\mathcal{B}$ is satisfiable in $\mathcal{M}$ iff there exists a temporal mapping $\mathcal{I}$ for C such that $\langle \mathcal{M}, \mathcal{I}\rangle \models \mathcal{B}$.

Definition 2. *Let $\mathcal{B}$ be a tableau branch and K the set of temporal constraints occurring in $\mathcal{B}$. $\mathcal{B}$ is* open *iff there exists a solution to K. Otherwise it is* closed.

Later on (Lemma 2) we show that every (non redundant) infinite branch is closed. Hence, we are only concerned with checking closure for finite branches, that amounts to checking satisfiability of a finite set of integer constraints. This can be done by means of well known graph algoritms (see any standard textbook on algorithms, e.g. [7]).

The following definition captures the intuitive idea of tableaux where no wasteful expansions are ever performed. In particular, closed branches are never expanded.

Definition 3. *A tableau branch $\mathcal{B}$ is canonical iff:*

- *The interval rule is applied no more than once to each interval.*

- *Every labelled formula is expanded no more than once by means of a logical rule.*
- *The conflict resolution rule is applied no more than once to each node or pair of nodes.*
- *No proper initial subsegment of $\mathcal{B}$ is closed.*

A tableau is canonical iff all its branches are canonical.

Definition 4. *If $\mathcal{B}$ is a tableau branch, then $\mathcal{B}$ is* complete *iff there exists no canonical expansion of $\mathcal{B}$. A tableau is* complete *if all its branches are complete.*

Here follows an example, showing the partial development of a tableau for $A = \Diamond p \wedge \Box(\neg p \vee q)$. Below, the application of the interval rule to intervals of the form $[c]$ is not shown, and neither are obvious premises of the form $c \leq c$. The nodes are numbered in order to comment the tableau. We assume that the tableau is initialized with some limit k, for any $k \geq 1$.

1. $finish \leq start + k$
2. $[start] A$
3. $[start] \Diamond p$
4. $[start] \Box(\neg p \vee q)$
5. $[d_1] p$
6. $start + 1 \leq d_1$
7. $[start + 1, finish] \neg p \vee q$

8. $finish \leq start$		9. $start + 1 \leq finish$
	10. $[start + 1, finish] \neg p$	13. $start + 1 \leq d_2$
11. $d_1 \leq start$	12. $finish \leq d_1 - 1$	14. $d_2 \leq finish$
		15. $[start + 1, d_2 - 1] \neg p$
		16. $[d_2] q$
		17. $[d_2 + 1, finish] \neg p \vee q$
	18. $d_2 - 1 \leq start$	19. $start + 1 \leq d_2 - 1$
20. $finish \leq d_2$	21. $d_2 + 1 \leq finish$	

Nodes 3 and 4 are the expansion of 2, and their expansions are 5, 6 and 7, respectively. Nodes 8 and 9 result from the application of the interval rule to 7; the branch with node 8 is closed, since, implicitly, $d_1 \leq finish$. Nodes 10 and 13–17 are the branches obtained from 7 and 9. The conflict resolution rule is applied to 5, 9 and 10, giving 11 and 12, that both close. Nodes 18–19 and 21–21 derive from 15 and 17, respectively, by the interval rule. The branches passing through 19 and 21 are not further developed. The open and complete branch ending at 20 represents the smallest (partial) model, with points $0 = \mathcal{I}(start) < 1 = \mathcal{I}(d_1) = \mathcal{I}(d_2) = \mathcal{I}(finish)$, with both p and q holding at 1. One of the further expansions of the branch passing through 21 would yield a more general description, with $0 = \mathcal{I}(start) < 1 = \mathcal{I}(d_2) \leq \mathcal{I}(d_1)$, where p holds at $\mathcal{I}(d_1)$ and q holds in the whole interval $[1, \mathcal{I}(finish)]$. The development of 19 explores models where $d_1, d_2 > 1$.

3 Termination, Completeness and Soundness

In this section we show that the construction of any canonical tableau terminates and that the calculus is complete and sound, i.e., if a given set of formulae S has a model $\mathcal{M}$ whose size is not greater than k, then any tableau for $S \cup \{finish \leq start + k\}$ has an open branch that is satisfiable in $\mathcal{M}$ (completeness) and any complete and open branch, in any tableau for $S \cup \{finish \leq start + k\}$, is satisfiable in some model whose size is bounded by k (soundness). It is also shown that, from a theoretical standpoint, the calculus can be used to check satisfiability in models of any finite size. Complete proofs can be found in [6].

3.1 Termination

Here we restrict our attention to canonical tableaux, thus ensuring, in particular, that the construction of a branch is abandoned as soon as it is recognized to be closed. As already remarked, the existence of a solution for a given set of constraints can be tested by means of shortest-path graph algorithms.

The following definition captures the idea of a sequence of applications of the β-rule, each of which expands an expansion of the previous one.

Definition 5. *Let $\mathcal{B}$ be a tableau branch. A β-node in $\mathcal{B}$ is a node of the form $[s,t]A \vee B$. A β-chain in $\mathcal{B}$ is a sequence of β-nodes $X_0, X_1, ...$ such that, for every $i \geq 1$, X_{i-1} is expanded in $\mathcal{B}$ by application of the β-rule and X_i is the β-node in the corresponding rightmost expansion. A k-length-β-chain is a finite β-chain $X_0, X_1, ..., X_{k+1}$, constituted by $k + 2$ nodes.*

A node X in $\mathcal{B}$ is the root node of a β-chain if it is the first node in a β-chain and it is not itself obtained by an application of the β-rule, i.e. there exists a maximal length β-chain in $\mathcal{B}$ having X as its first node.

The proof that the construction of canonical tableaux terminates uses the following lemmas.

Lemma 1. *Let $\mathcal{B}$ be a branch in a tableau and $\mathcal{I}$ a temporal mapping satisfying the constraints in $\mathcal{B}$. If $[s,t]A$ occurs in $\mathcal{B}$ and $\mathcal{I}^*(s) \leq \mathcal{I}^*(t)$, then $0 \leq \mathcal{I}^*(s) \leq \mathcal{I}^*(t) \leq \mathcal{I}(finish)$.*

Lemma 2.
1. *If $\mathcal{B}$ is a canonical open branch in a tableau for $S \cup \{finish \leq start + n\}$, then $\mathcal{B}$ contains no k-length-β-chain with $k > n$.*
2. *Any canonical infinite branch contains an infinite β-chain.*
3. *Every open branch is finite.*

Proof. We give here only a proof of the first item. Let us assume that $\mathcal{B}$ does contain a k-length-β-chain with $k > n$. Then $\mathcal{B}$ contains a sequence of nodes having the form: $[s,t]\,A \vee B, [a_1+1,t]\,A \vee B, [a_2+1,t]\,A \vee B, ..., [a_{n+2}+1,t]\,A \vee B$ and the sequence of constraints: $s \leq a_1, a_1 + 1 \leq a_2, ..., a_{n+1} + 1 \leq a_{n+2}$, and $a_{n+2} \leq t$. If $\mathcal{I}$ is is a solution for the constraints in $\mathcal{B}$, then, by Lemma 1, $0 \leq \mathcal{I}^*(s) \leq \mathcal{I}^*(t) \leq \mathcal{I}^*(finish)$ (in fact $s \leq t$ is in $\mathcal{B}$), so that $\mathcal{I}^*(a_{n+2}) \geq n+1$. However, $\mathcal{I}^*(a_{n+2}) \leq \mathcal{I}^*(t) \leq \mathcal{I}^*(finish) \leq n$ holds too, that is absurd.

Theorem 1. *Any canonical tableau for $S \cup \{finish \leq start+n\}$, for any $n \geq 0$, is finite.*

3.2 Completeness with Respect to Model Construction

The proof that the tableau calculus is complete w.r.t. model construction uses the following lemmas.

Lemma 3. *Let $\mathcal{B}$ be a branch in a tableau $\mathcal{T}$, $\mathcal{M}$ a temporal interpretation and $\mathcal{I}$ a temporal mapping for the constants occurring in $\mathcal{B}$. If $\langle \mathcal{M}, \mathcal{I} \rangle \models \mathcal{B}$ and $\mathcal{B}$ can be expanded, $\mathcal{B}_1$ (and $\mathcal{B}_2$) being its expansion(s), then for some $i = 1, 2$ there exists an extension $\mathcal{I}'$ of $\mathcal{I}$ to the fresh constants of $\mathcal{B}_i$ (if any) such that $\langle \mathcal{M}, \mathcal{I}' \rangle \models \mathcal{B}_i$*

Proof. Different cases must be considered according to the applied expansion rule. Let T be the temporal structure underlying $\mathcal{M}$ (and the range of $\mathcal{I}$). Here we show the treatment of the β-rule and the $\mathcal{S}$-rule.

1. In the case of the β-rule, by hypothesis $\langle \mathcal{M}, \mathcal{I} \rangle \models [s,t]A \vee B$ and $\mathcal{I}^*(s) \leq \mathcal{I}^*(t)$, so that $0 \leq \mathcal{I}^*(s) \leq \mathcal{I}^*(t) \leq \mathcal{I}^*(finish)$, by Lemma 1. It follows that for every $i \in [\mathcal{I}^*(s), \mathcal{I}^*(t)]$, $\mathcal{M}_i \models A \vee B$. If for every such i, $\mathcal{M}_i \models A$, then the leftmost branch is clearly satisfied by $\mathcal{M}$ and $\mathcal{I}$. Otherwise, let k be the smallest element in the interval $[\mathcal{I}^*(s), \mathcal{I}^*(t)]$ such that $\mathcal{M}_k \not\models A$, hence $\mathcal{M}_k \models B$. The mapping $\mathcal{I}$ is extended to $\mathcal{I}'$ such that $\mathcal{I}'(a) = k$. By the choice of k, clearly $\langle \mathcal{M}, \mathcal{I}' \rangle \models [s, a-1]A$ and $\langle \mathcal{M}, \mathcal{I}' \rangle \models [a]B$. If $k = \mathcal{I}^*(t)$, $\langle \mathcal{M}, \mathcal{I}' \rangle \models [a+1,t]A \vee B$ is voidly satisfied. Otherwise, the interval $[\mathcal{I}^*(a+1), \mathcal{I}^*(t)]$ is a subinterval of $[\mathcal{I}^*(s), \mathcal{I}^*(t)]$, hence again $\langle \mathcal{M}, \mathcal{I}' \rangle \models [a+1,t]A \vee B$.

2. Let the $\mathcal{S}$-rule be applied to $[s,t]A\mathcal{S}B$. By hypothesis, $\langle \mathcal{M}, \mathcal{I} \rangle \models A\mathcal{S}B$ and $\mathcal{I}(s) \leq \mathcal{I}^*(t)$. By Lemma 1, $0 \leq \mathcal{I}^*(s) \leq \mathcal{I}^*(t) \leq \mathcal{I}(finish)$. It follows that $\mathcal{I}^*(s) \in T$ and that there exists $i \in T$ such that $i < \mathcal{I}^*(s)$, $\mathcal{M}_i \models B$, and for any $j \in T$ such that $i < j < \mathcal{I}^*(s)$, $\mathcal{M}_j \models A$. Hence if we set $\mathcal{I}'(a) = i$, where a is the new constant introduced by the rule, we get that $\langle \mathcal{M}, \mathcal{I}' \rangle$ satisfies $[a]B$, $a \leq s-1$ and $[a+1, s-1]A$. Let now consider any time point $j \in T$ such that $\mathcal{I}^*(s) \leq j \leq \mathcal{I}^*(t-1)$ and assume that $\mathcal{M}_j \not\models A \vee B$. Let k be the smallest of such points; then $\mathcal{I}^*(s) \leq k+1 \leq \mathcal{I}^*(t)$ but $\mathcal{M}_{k+1} \not\models A\mathcal{S}B$, a contradiction.

As a consequence of the above lemma and Theorem 1 we have the following:

Theorem 2. *Let $\mathcal{M} = \langle T, \sigma \rangle$ be a temporal interpretation, with $T = \langle 0, ..., n \rangle$, S a finite set of LTL formulae in nnf, and $K = \{finish \leq start + k\}$, for some integer $k \geq n$. If $\mathcal{M}$ is a model of S then any canonical tableau for $S \cup K$ has a branch that is finite, open and satisfiable in $\mathcal{M}$.*

3.3 Soundness with Respect to Model Construction

Conversely, any complete open branch describes some model of its initial formula. Such a soundness result follows from the invertibility of the logical rules, with respect to their logical nodes, established by the following lemma.

Lemma 4. *Let $\mathcal{M} = \langle T, \sigma \rangle$ be a temporal interpretation and $\mathcal{I}$ a temporal mapping with range T. For every logical rule*

$$\frac{[s,t]\,F \quad s \leq t}{\mathcal{B}_1 \quad \mathcal{B}_2} \quad or \quad \frac{[s,t]\,F \quad s \leq t}{\mathcal{B}_1}$$

if $\langle \mathcal{M}, \mathcal{I} \rangle \models \mathcal{B}_i$ (for some $i = 1, 2$), then $\langle \mathcal{M}, \mathcal{I} \rangle \models [s,t]F$.

Proof. If $\mathcal{I}^*(t) < \mathcal{I}^*(s)$ then $\langle \mathcal{M}, \mathcal{I} \rangle \models [s,t]F$ for any F. Hence we assume $\mathcal{I}^*(s) \leq \mathcal{I}^*(t)$. Here, we prove the lemma only for the case of the $\mathcal{U}$-rule.

If F is $A\mathcal{U}B$, then $\langle \mathcal{M}, \mathcal{I} \rangle$ satisfies $[a]B$, $[t+1, a-1]A$, $[s+1,t]A \vee B$, where a is a fresh constant such that $\mathcal{I}^*(t+1) \leq \mathcal{I}(a) = n$. Since $\langle \mathcal{M}, \mathcal{I} \rangle \models [t+1, a-1]A$, clearly $\mathcal{M}_{\mathcal{I}^*(t)} \models A\mathcal{U}B$. Let now $i \in [\mathcal{I}^*(s), \mathcal{I}^*(t-1)]$. Since for all $j \in [\mathcal{I}^*(s+1), \mathcal{I}^*(t)]$, $\mathcal{M}_j \models A \vee B$, also for all $j \in [i+1, \mathcal{I}^*(t)]$, $\mathcal{M}_j \models A \vee B$. We distinguish two cases:

- If for all $j \in [i+1, \mathcal{I}^*(t)]$, $\mathcal{M}_j \models A$, then n itself is the "witness" for i: $n > j$, $\mathcal{M}_n \models B$ and for all j, $i < j < n$, $\mathcal{M}_j \models A$. Hence in this case $\mathcal{M}_i \models A\mathcal{U}B$.
- Otherwise, let k be the smallest $j \in [i+1, \mathcal{I}^*(t)]$ such that $\mathcal{M}_k \not\models A$. Since $\mathcal{M}_k \models A \vee B$, $\mathcal{M}_k \models B$. Moreover $k > i$ and, by construction, for all j, $i < j < k$, $\mathcal{M}_j \models A$. Hence again $\mathcal{M}_i \models A\mathcal{U}B$.

Theorem 3. *If $\mathcal{B}$ is a complete and open tableau branch and $\mathcal{I}$ is a solution of the set of the temporal constraints occurring in $\mathcal{B}$, then there is an interpretation $\mathcal{M}$ such that $\langle \mathcal{M}, \mathcal{I} \rangle \models \mathcal{B}$.*

Proof. First of all, observe that by Lemma 1 the branch $\mathcal{B}$ is finite. Let T be the (finite) range of $\mathcal{I}$ and $\mathcal{M} = \langle T, \sigma \rangle$, where σ is such that:

for each atom p and $i \in T$, $p \in \sigma(i)$ iff there is a node $[s,t]p$ in $\mathcal{B}$ such that $\mathcal{I}^*(s) \leq i \leq \mathcal{I}^*(t)$.

The interpretation function σ is well defined, since, if $[s,t]\bot$ is in $\mathcal{B}$, then K contains $t \leq s - 1$, because $\mathcal{B}$ is complete, so that there are no elements $i \in T$ with $\mathcal{I}^*(s) \leq i \leq \mathcal{I}^*(t)$.

By hypothesis, $\mathcal{I}$ satisfies all the temporal constraints in $\mathcal{B}$, so it rests to be shown that for any logical node $[s,t]F$ in $\mathcal{B}$, $\langle \mathcal{M}, \mathcal{I} \rangle \models [s,t]F$. This is done by induction on the number of application of logical rules in $\mathcal{B}$. Note that if the constraint $t \leq s - 1$ occurs in $\mathcal{B}$, then trivially $\langle \mathcal{M}, \mathcal{I} \rangle \models [s,t]F$. Hence, the only interesting cases arise when the constraint $s \leq t$ occurs in $\mathcal{B}$ (since $\mathcal{B}$ is complete, either $t \leq s-1$ or $s \leq t$ occurs in $\mathcal{B}$, for any $[s,t]F$ in $\mathcal{B}$). In the following, K denotes the set of constraints occurring in $\mathcal{B}$.

1. The base case is when there is no application of logical rules in $\mathcal{B}$, hence, since $\mathcal{B}$ is complete, the only logical nodes in $\mathcal{B}$ have the form $[s,t]F$, where F is a literal.
   - If F is an atom, the result holds by definition of σ.
   - If F is $\neg p$, assume that $\langle \mathcal{M}, \mathcal{I} \rangle \not\models [s,t]\neg p$, i.e. there exists $i \in T$ such that $\mathcal{I}^*(s) \le i \le \mathcal{I}^*(t)$ and $p \in \sigma(i)$. Hence, by definition of σ, there is a node $[s',t']p$ in $\mathcal{B}$ where $\mathcal{I}^*(s') \le i \le \mathcal{I}^*(t')$. Since $\mathcal{B}$ is complete, both $s \le t$ and $s' \le t'$ are in K (otherwise, $t \le s - 1$ or $t' \le s' - 1$ would be in K, contradicting the fact that neither $[\mathcal{I}^*(s), \mathcal{I}^*(t)]$ nor $[\mathcal{I}^*(s'), \mathcal{I}^*(t')]$ are empty), and the conflict resolution rule is applied to $s \le t$, $s' \le t'$, $[s,t]\neg p$ and $[s',t']p$. Hence, either $t \le s' - 1$ or $t' \le s - 1$ is a node of $\mathcal{B}$. In both cases, the intersection of $[\mathcal{I}^*(s), \mathcal{I}^*(t)]$ and $[\mathcal{I}^*(s'), \mathcal{I}^*(t')]$ should be empty.
2. The inductive step follows from Lemma 4.

3.4 Complexity and the Search for Models of Any Finite Size

Since the rules of the calculus are invertible, the search for a model can be done by construction of a single (canonical) tableau, following any strategy to choose the nodes to be expanded. It can be shown that the length of a branch in a tableau for $S \cup \{finish \le start + k\}$ is polynomial in $n \times k$, where n is the size of S. Since testing whether a branch is open or not can be done in polynomial time, a non deterministic algorithm implementing the proposed tableau system is $O(n \times k)$.[2] A deterministic algorithm is $O(2^{n \times k})$, since the number of branches must also be taken into account. However, the space required by depth first search is still polynomial in $n \times k$.

Suitable versions of Theorems 2 and 3 also hold when tableaux are initialized without any constraint of the form $finish \le start + k$: if $\mathcal{M}$ is a model of S, then any canonical tableau for S has a finite open branch that is satisfiable in $\mathcal{M}$, and, conversely, any complete and open tableau branch is satisfiable. This gives a semi-decision procedure for testing satisfiability in models of finite size.

However, the calculus also induces a decision procedure for checking LTL finite satisfiability. It is already known that finite satisfiability in LTL is decidable; for instance, this can be proved by use of Lemma 4.5 in [20] (that can be easily adapted to our finite semantics). Roughly, the lemma says that, given any model $\mathcal{M}$ of an LTL formula F, if two time points satisfy exactly the same subformulae of F then they may collapse (and the interval between them disappear), so that a smaller model $\mathcal{M}'$ may be obtained. As a consequence of such a result, if F has a model, then it has a model whose size does not exceed the cardinality of the powerset of the set of its subformulae, i.e. $2^{|sub(F)|}$. In our framework, this implies that by setting $K = \{finish \le start + 2^{|sub(F)|}\}$ we immediately get a decision procedure for finite satisfiability in LTL.

---

[2] This corresponds to the fact that, although the general plan-existence problem for STRIPS-like operators (see Section 4) is PSPACE-complete, planning is NP-complete when only polynomial length plans are considered.

This may be interesting from a theoretical standpoint, but clearly unpractical: if the bound k on the model size is exponential in n, a doubly exponential upper bound to the deterministic time complexity is obtained, and the space complexity is exponential in n.

4 Planning in LTL

In this section, we give a brief overview of the main features of some important search based planning system, in order to illustrate the advantages induced by the use of LTL as a specification language for planning problems and the use of the proposed tableau calculus as the basis for a plan search engine.

Most of the planning languages in the non logical approach, whose semantics has been been given a formal characterisation, are extensions of the language of STRIPS, the first major planning system [10]. In such languages, the description of an action a consists of the specification of the preconditions for its executability, a set of formulae describing, for some relations R, the set of elements that will newly enjoy of R in the situation that results from the execution of a (the "add list") and formulae describing the set of elements that are going to lose some property R after executing a (the "delete list"). Syntactic restrictions are imposed on the formulae that are allowed to occur in the precondition, add and delete list, although different languages may vary in their expressive power.

Planners are planning algorithms that use the representation of the problem, given by action descriptions together with a description of the initial situation and goals, in order to synthesise a plan, i.e. a sequence of actions that, if executed, would lead from the initial situation to the desired goals. The search for the plan can be performed in different ways. Mainly, the following features characterise different strategies:

- The search can either proceed in a data driven manner, starting from the initial situation (progression planners) or backward, taking the goals as the starting point (regression planners), considering those actions that achieve the goal and, in turn, their preconditions, until the initial situation is reached.
- The search space can be constituted either by the set of situations themselves, actions transforming a situation into another one, and a plan is therefore a path leading from the initial situation to a situation where the goals are satisfied (linear planners), or by a set of *partial plans* (partial order planning). A partial plan is essentially a set of actions related by a (partial) ordering relation. Partial order planning starts with the empty plan and successively refines it, either by addition of new actions or new temporal constraints between them. The result of the search (the final plan) is itself a node in the search space, fulfilling some requirement that, roughly, ensures that every goal has been reached. This approach reflects the *least commitment principle*: postpone constraining decisions till they are actually needed.

Although the preferability of one or the other of such approaches may strongly depend on the problem structure, consensus suggests that regression and partial order planning are generally more adequate to handle real world problems.

In [19], a partial order planning algorithm (UCPOP), working by regression, is described. Succinctly, UCPOP starts with a plan consisting of two "dummy" actions: *start*, whose effects are the initial conditions, and *goal*, whose preconditions are the goals of the problem. Then, new actions and constraints are successively added to the plan, until all preconditions are satisfied. At each step, an "open" precondition P of an action a_i is chosen (that is not satisfied yet) and either a new action or an existing action, having such precondition as effect, is nondeterministically chosen and, if new, added to the plan. The plan is also added a *causal link* from the chosen action a_j to a_i, labelled by P, in order to record the fact that a_j is present in the plan exactly to produce P. After that, UCPOP resolves possible *threats*: there may be actions a_{threat} in the plan whose effect would destroy P. The threat is then solved either by *promotion*, adding the time constraint $a_{threat} < a_j$, or *demotion*, by addition of the constraint $a_i < a_{threat}$.

The use of LTL as a specification language allows one to encode planning problems in a very flexible way. In general, the encoding of a planning problem in LTL consists of the following sets of formulae:

- a set S of *initial state formulae*, describing the initial situation, where no temporal operator occurs;
- a set F of *goal formulae*, describing what is expected hold in the final situation, each of which having the form $\Diamond^-(A \wedge \Box \bot)$, where $\Diamond^- B \equiv_{def} (B \vee \Diamond B)$;
- a set G of *global assumptions*, describing what holds in every situation (description of action preconditions and effects, general truths, etc.), having the form $A \wedge \Box A$.

Moreover, a specification can include a set T of *task* formulae, describing intermediate tasks that have to be accomplished before reaching the goal, in the form $\Diamond^- A$. In some problems, in fact, the activity is more significant than the goal and the specification of the goal state can even be equal to the initial state one, but intermediate tasks must be performed: "do this and that, then go back home". Intermediate tasks cannot be directly modeled in STRIPS-like languages, which lack the capability to refer to what happens between the initial state and the goal. They are taken into account by the *Hierarchical Task Networks* approach (see, for example [9, 23]), as well as in [1].[3]

The richer expressive power of LTL as a planning language, with respect to STRIPS-like formalisms, can be exploited in several other directions. For example, as a consequence of the fact that a plan consists of a partially ordered set of *actions* – i.e. an action is identified with the state where it is performed – STRIPS-like formalisms cannot cope in a natural manner with problems requiring two or more actions to be performed contemporarily in order to achieve a desired effect. In LTL, on the contrary, situations are not identified with actions (which are represented by means of ordinary propositions), so that nothing prevents a model from containing a situation where two or more actions are performed. Furthermore, parallel actions can be modeled.

---

[3] Obviously, such tasks could be represented by means of goals of the form "having this and that done", but this is an unnecessary complication.

The same advantages are shared by the "planning as satisfiability" paradigm, proposed by Kautz and Selman [12], the logical language used to encode planning problems being propositional logic. However, with respect to the use of propositional logic, in an LTL specification time is implicitly represented and the presence of the binary temporal operators "since" and "until" gives the language a very rich expressive power, resulting in the possibility to assert what must hold in a whole interval.

When model search for LTL formulae is performed by means of the proposed tableau system, different plan search strategies may be simulated, by use of different ways of writing global assumptions. For instance, the regression strategy followed by UCPOP may be simulated by meand of "regression encodings", where effect and frame axioms have the form $\Box(p \rightarrow (\neg B \mathcal{S} A) \vee \boxminus(p \wedge \neg B))$, where A (resp. B) encodes all the conditions that may lead p to become true (resp. false): if p is true sometime, then it must be the case that either some action was performed before (A), having p as effect, and such an effect has not been destroyed (by B) since then, or p has always been true and no action destroyng p has ever been performed. The search in the tableau resulting from the expansion of such regression axioms is goal driven. The partial ordering of time points in tableau construction reflects a form of plan-space search planning, while the conflict resolution rules are also used for solving UCPOP "threats". Data driven plan construction may be expressed as well, by means of global assumptions ("progression encodings") roughly stating that if an action is performed at point i, then its effects hold at point $i+1$ (or even in a whole interval, until (possibly) such effects are destroyed), and classical frame conditions for all literals. A detailed descritption of such encodings, and the induced plan search mechanisms, would override space limits and will be given elsewhere.

The application of LTL to planning has been considered in other works. For example, [3] applies the executable temporal language METATEM [2] to planning and scheduling. F. Bacchus and F. Kabanza [1] use a version of temporal logic to specify temporally extended goals as sets of acceptable sequences of states, i.e. temporal models, and define correct plans as those whose execution results in one of such sequences.[4] In both cited works, only a form of linear, data driven planning is obtained.

5 Conclusions and Related Work

In this work, the possibility of using linear temporal logic as a planning language is investigated, in the view that a plan is a finite model of the specification of the problem. With respect to formalisms used in the search based approach to planning, the use of LTL shares the advantages of any logical approach: a formal semantics, generality and expressivity. Furthermore, using a well studied logic

---

[4] Temporal logic is not used to encode the planning problem entirely: actions are described in an ADL format, and an *Expand* operation is used, that, applied to a state s, generates all the successors of s that are produced by performing any allowed action.

allows the exploitation of meta-theoretical results. In this case, we know that completeness (with respect to plan construction of any length) is not lost if the search is restricted to consider plans whose maximal length is $2^{|sub(S)|}$, where S is the problem specification.

On the other hand, the tableau calculus by means of which model search is performed allows the embedding of different strategies and techniques mutuated from the search based approach to planning, depending on the encoding of the problem. In particular, plan space search can be simulated, since the calculus uses labelled formulae, labels corresponding to time intervals where the corresponding formulae are true, and time points are only partially ordered.

Different proof systems for linear temporal logic can be found in the literature, that can be turned into model search methods (the underlying model of time being however always infinite). Most of them (for example, [22, 4, 2]) are essentially based on the following equivalences

$$\Diamond A = A \vee \bigcirc \Diamond A \qquad A\mathcal{U}B = B \vee (A \wedge \bigcirc(A\mathcal{U}B))$$

where $\bigcirc$ is the "next" operator and $\Diamond$ and $\mathcal{U}$ have the weak semantics, including the present time point. Rewriting a formula of the form $\Diamond A$ as $A \vee \bigcirc \Diamond A$ leads to choose whether "executing" A in the present time point or postponing its execution, the same problem passing on to the next time point. Consequently, in the model description that results from the construction, the ordering of time points is always total. Moreover, such systems must be equipped with some mechanism to check whether all "eventualities" (formulae in the scope of a $\Diamond$ or $\overleftarrow{\Diamond}$, or in the right scope of a $\mathcal{U}$ or $\mathcal{S}$ operator) are sooner or later satisfied.

The systems reported in [18, 17], carrying on the work in [11], are significantly different from the above ones. Several important points of such works are resumed in the present paper. In [18], where only unary temporal operators are considered, the construction of tableau branches terminates and branch closure is reduced to the satisfiability of the set of integer constraints in the branch. The introduction of the "since" and "until" operators – that is however essential in coding planning problems – raises new difficulties. In [17] the tableau expansion rules for such operators are mainly based on the equivalences $A\mathcal{U}B = B \vee (A \wedge \bigcirc(A\mathcal{U}B))$ and the symmetric one for $\mathcal{S}$, but an ingenious rewriting of the "contracted" formula into a propositional letter guarantees termination. After the terminating tableau expansion, branch closure is checked by reduction to a model checking problem in CTL with fairness constraints, in the style of [4]. Because of the special form of the set of active formulae in the branch, the model checking problem is somewhat simplified. Surely, focusing on finite models could lead to simpler solutions. In this case, in fact, some hard problems disappear, such as the problem of fairness conditions (in connection with formulae of the form $\Box \Diamond A$).

This work lays the theoretical grounds for the use of LTL, with the proposed tableau system, to solve planning problems. A prototype implementation exists, TabPlan. It is written in Standard ML and uses a modification of Bellman-Ford shortest path algorithm to check branch closure, allowing for incremental tests, so that as soon as a new constraint is produced, the set is tested for consistency considering only what has been newly modified. Moreover, some obvious shortcuts are added to the rule set. In order to have a practical planning system and collect experimental results on significant planning problem examples, however, some theoretical work still has to be done.

One of the first issues to be addressed is the characterization of the class of formulae involved in different encodings of planning problems. In fact, one does not expect that a general purpose system for full temporal logic over finite domains can compete with special purpose planners. One of the main reasons is that the encoding of a planning problem never exploits the full expressive power of LTL. For example, the nesting of temporal operators is always limited; regression encodings mainly use past time operators and progression ones future time operators; the occurrences of β-subformulae can be recognized to have a regular structure, that allows one to solve the general problem raised by β-chains. Hence, refinements of the calculus for planning formulae can be defined, in order to improve TabPlan performance.

A further issue concerns the extension to a limited first order language. In fact, although from a theoretical standpoint propositional logic suffices to represent planning problems over finite domains, needless to say that treating an existential quantified formula like the finite disjunction of its instances is not an excellent solution. It would be a gross violation of the *least commitment* principle.

Finally, the possibility to apply TabPlan to other domains, such as, for example, the management of dynamical integrity constraints in data bases, is to be investigated.

Acknowledgements. The authors are strongly indebted with Jean Goubault-Larrecq for reading a first version of this paper and making helpful remarks. We also thank Wolfgang Gehrke, for implementing TabPlan first prototype.

References

1. Fahiem Bacchus and Froduald Kabanza. Planning for temporally extended goals. In *Proceedings of the Thirteenth National Conference on Artificial Intelligence (AAAI-96)*, pages 1215–1222. AAAI Press / The MIT Press, 1996.
2. H. Barringer, M. Fisher, D. Gabbay, G. Gough, and R. Owens. METATEM: a framework for programming in temporal logic. In *Proc. of REX Workshop on Stepwise Refinement of Distributed Systems: Models, Formulisms, Correctness*, volume 430 of *LNCS*. Springer, 1989.
3. H. Barringer, M. Fisher, D. Gabbay, and A. Hunter. Meta-reasoning in executable temporal logic. In *Proc. of the Second Int. Conf. on Principles of Knowledge Representation and Reasoning*, 1991.
4. J. R. Burch, E. M. Clarke, K. L. McMillan, D. L. Dill, and L. J. Hwang. Symbolic model checking: 10^{20} states and beyond. *Information and Computation*, 98(2):142–170, 1992.
5. T. Bylander. Complexity results for planning. In *Proc. of the 12th Int. Joint Conf. on Artificial Intelligence (IJCAI-91)*, pages 274–279, 1991.

6. S. Cerrito and M. Cialdea Mayer. TabPlan: Planning in linear temporal logic. Technical Report 1141, Université de Paris Sud, Laboratoire de Recherche en Informatique, 1997.
7. T. H. Cormen, C. E. Leiserson, and R. L. Rivest. *Introduction to Algorithms*. The MIT press, 1990.
8. K. Currie and A. Tate. O-Plan: the open planning architecture. *Journal of Artificial Intelligence*, 52:49–86, 1991.
9. K. Erol, J. Hendler, and D. S. Nau. Complexity results for HTN planning. In *Proceedings of AAAI-94*, 1994.
10. R. E. Fikes and N.J. Nilsson. STRIPS: a new approach to the application of theorem proving to problem solving. *Artificial Intelligence*, 2(3–4):189–208, 1971.
11. R. Hähnle and O. Ibens. Improving temporal logic tableaux using integer constraints. In D. M. Gabbay and H. J. Ohlbach, editors, *Proceedings of the First International Conference on Temporal Logic (ICTL 94)*, volume 827 of *LNCS*, pages 535–539. Springer, 1994.
12. H. Kautz and B. Selman. Planning as satisfiability. In B. Neumann, editor, *10th European Conference on Artificial Intelligence (ECAI)*, pages 360–363. Wiley & Sons, 1992.
13. J. Koehler and R. Treinen. Constraint deduction in an interval-based temporal logic. In M. Fisher and R. Owens, editors, *Executable Modal and Temporal Logics, (Proc. of the IJCAI'93 Workshop)*, volume 897 of *LNAI*, pages 103–117. Springer, 1995.
14. M. Masseron. Generating plans in linear logic: II. A geometry of conjunctive actions. *Theoretical Computer Science*, 113:371–375, 1993.
15. M. Masseron, C. Tollu, and J. Vauzeilles. Generating plans in linear logic: I. Actions as proofs. *Theoretical Computer Science*, 113:349–370, 1993.
16. R. Reiter. The frame problem in the situation calculus: A simple solution (sometimes) and a completeness result for goal regression. In V. Lifschitz, editor, *Artificial Intelligence and mathematical theory of computation: Papers in honor of John McCarthy*, pages 359–380. Academic Press, 1991.
17. P.H. Schmitt and J. Goubault-Larrecq. A tableau system for full linear temporal logic. Draft.
18. P.H. Schmitt and J. Goubault-Larrecq. A tableau system for linear-time temporal logic. In E. Brinksma, editor, *3rd Workshop on Tools and Algorithms for the Construction and Analysis of Systems (TACAS'97)*, LNCS. Springer Verlag, 1997.
19. J. Scott Penberthy and D.S Weld. UCPOP: A sound, complete, partial order planner for ADL. In *Proc. of the Third Int. Conf. on Principles of Knowledge Representation and Reasoning (KR'92)*, pages 103–114. Morgan Kauffman Publ., 1992.
20. A. P. Sistla and E. M. Clarke. The complexity of propositional linear temporal logics. *Journal of the ACM*, 32(3):733–749, 1985.
21. B. Stephan and S. Biundo. Deduction based refinement planning. In B. Drabble, editor, *Proceedings of the 3rd International Conference on Artificial Intelligence Planning Systems (AIPS-96)*, pages 213–220. AAAI Press, 1996.
22. P. Wolper. The tableau method for temporal logic: an overview. *Logique et Analyse*, 28:119–152, 1985.
23. Q. Yang. Formalizing planning knowledge for hierarchical planning. *Computational Intelligence*, 6:12–24, 1990.

On Proof Complexity of Circumscription

Uwe Egly and Hans Tompits

Technische Universität Wien
Abt. Wissensbasierte Systeme 184/3
Treitlstraße 3, A–1040 Wien, Austria
e-mail: [uwe,tompits]@kr.tuwien.ac.at

Abstract. Circumscription is a non-monotonic formalism based on the idea that objects satisfying a certain predicate expression are considered as the *only* objects satisfying it. Theoretical complexity results imply that circumscription is (in the worst case) computationally harder than classical logic. This somehow contradicts our intuition about commonsense reasoning: non-monotonic rules should help to *speed up* the reasoning process, and not to slow it down.
In this paper, we consider a first-order sequent calculus for circumscription and show that the presence of circumscription rules can tremendously simplify the search for proofs. In particular, we show that certain sequents have only long "classical" proofs, but short proofs can be obtained by using circumscription.

1 Introduction

One motivation for the introduction of non-monotonic reasoning principles was the hope to *speed up* the reasoning process. Instead of specifying an exhaustive list of procedures, "rules of thumb" should enable an automated reasoning machine (like, e.g., a robot) to draw inferences in a more efficient and timesaving way. Unfortunately, these expectations were somewhat shattered as soon as complexity results appeared for these logics (see, e.g., [12, 14, 15, 18, 26], an overview is given in [8]). Basically, it turned out that almost all non-monotonic formalisms are "harder" than classical logic (for propositional systems, this holds under the proviso that the polynomial hierarchy does not collapse). However, these results are just one side of the coin. They only show how non-monotonic reasoning behaves *in the worst case*, but they give no indication on *how we can profit from non-monotonic rules*. One of the few investigations emphasizing this point are the works by Cadoli, Donini and Schaerf [6, 7]. Roughly speaking, they show that, unless the polynomial hierarchy collapses, propositional non-monotonic systems allow a "super-compact" representation of knowledge as compared with (monotonic) classical logic.

Recently, tableau and sequent-style calculi for various forms of non-monotonic reasoning have been introduced [1, 3–5, 21–23]. In this paper, we consider a generalization, CIRC*, of the cut-free sequent calculus for propositional circumscription, introduced by Bonatti and Olivetti in [5]. CIRC* consists of three parts, namely a classical (first-order) LK-calculus, a "complementary" sequent calculus for propositional logic, and certain inference rules formalizing circumscription. In the complementary calculus, "anti-sequents" of the form $\Gamma \not\vdash \Theta$ state the non-derivability of Θ from Γ.

The basic idea of our approach is the following. We compare in the calculus CIRC* the minimal proof length of "purely classical" proofs, i.e., of proofs *without* applications of circumscription rules, with proofs where circumscription is applied. More precisely, we show that there are infinite sequences $(C_k)_{k \in \mathbb{N}}$, $(S_k)_{k \in \mathbb{N}}$ of sequents with the following properties:

1. The minimal proof length of C_k in CIRC* is *non-elementary* in k, i.e., the proof length of C_k is of the order $\mathsf{s}(k)$, where $\mathsf{s}(0) := 1$ and $\mathsf{s}(n+1) := 2^{\mathsf{s}(n)}$ for all $n \geq 0$.
2. The minimal proof length of S_k in CIRC* is linear in k.

C_k represents the fact that a certain formula H_k is "classically" derivable from a given theory T, whereas S_k represents the fact that H_k is proved with the help of circumscription. Although the derivation of H_k with the non-monotonic rules involves both a classical derivation *and* a derivation in the complementary calculus, for sufficiently large k, the length of this proof is much shorter than the proof of H_k without circumscription. The reason is that the length of any cut-free proof of H_k is non-elementary in the size of the input formula, but short proofs can be obtained by using the cut rule — and circumscription provides additional information such that "one part" of the short classical proof with cut is sufficient. Moreover, since for first-order cut-free sequent calculi, the size of the search space is elementarily related to the minimal proof length, a non-elementary decrease of the search space is also achieved.

A motivation of our method can be given as follows. Usually, non-monotonic techniques are applied in case a classical proof cannot be found. Although this is a reasonable procedure in decidable systems, it is not appropriate for undecidable systems like first-order logic. Indeed, if we integrate non-monotonic rules into first-order theorem provers, we have to invoke non-monotonic mechanisms *after a certain amount of time*, whenever the goal formula has not been proven classically up to this point. Accordingly, it may happen that a formula is provable both classically and with the help of non-monotonic rules. Our result shows therefore that, in certain cases, the theorem prover may easier find a proof because the presence of circumscription rules yields a much smaller search space.

The paper is organized as follows. In Section 2 we introduce basic definitions and notations. Moreover, the sequent calculus CIRC* is described. In Section 3 we prove our main result, and in Section 4 we conclude with some general remarks.

2 Preliminaries

Throughout this paper we use a first-order language consisting of *variables, function symbols, predicate symbols, logical connectives, quantifiers* and *punctuation symbols*. *Terms* and *formulae* are defined according to the usual formation rules. We will identify 0-ary predicate symbols with *propositional atoms*, and 0-ary function symbols with *(object) constants*.

Let t be the function with $t(x,0) := 2^x$ and $t(x, n+1) := 2^{t(x,n)}$, for all $n \in \mathbb{N}$. A function $f : \mathbb{N} \to \mathbb{N}$ is called *elementary* iff there is a Turing machine M and a constant $c \in \mathbb{N}$ such that M computes f and the computing time $T_M(x)$ of M with input x obeys the relation $T_M(x) \leq t(x,c)$, for all $x \in \mathbb{N}$.

Note that, for each fixed $n \in \mathbb{N}$, the function $t(\cdot, n)$ itself is elementary. On the other hand, the function s, defined by $s(n) := t(0, n)$, is *non-elementary*.

2.1 Circumscription

Reasoning under circumscription is based on the idea that objects satisfying a certain predicate expression P are considered as the *only* objects satisfying it. Roughly speaking, the positive information about P is treated as a *sufficient part* of a definition of P, and the circumscription of P "completes" this definition by *minimizing the extension of* P.[1] In this process, one can determine which predicate symbols shall be minimized, which predicate symbols shall retain their meaning, and which ones can be varied. In contrast to classical reasoning, under circumscription, logical consequence is evaluated in terms of models which are *minimal* in a certain sense.

Definition 1. *Let S be a finite set of closed formulae, let M, N be two models of S, and let P, R be finite sets of predicate symbols such that P and R are disjoint. We call M a $(P; R)$-submodel of N, symbolically $M \preceq_{P;R} N$, iff the following conditions are satisfied:*

1. *M and N have the same domain;*
2. *M and N have the same interpretation for each predicate symbol in R;*
3. *the interpretation of each predicate symbol $F \in P$ in M is a subset of the interpretation of F in N.* □

Note that in condition 3, if F is a propositional atom, then F must be true in N whenever it is true in M.

Clearly, the relation $\preceq_{P;R}$ is a pre-order. The minimal objects with respect to this ordering are called $(P; R)$-*minimal*. Observe that in the relation $\preceq_{P;R}$, all function symbols are allowed to vary.

Definition 2. *Under the circumstances of Definition 1, we say that a formula A is entailed by S (with circumscribed predicate symbols P and fixed predicate symbols R), written $S \models_{P;R} A$, iff A is true in all $(P; R)$-minimal models of S.* □

Historically, circumscription was proposed by McCarthy in [19, 20], where the minimization principle is encoded in the form of a certain *second-order formula*. Subsequently, circumscription has been advanced by many AI researchers resulting in a whole family of different circumscription techniques. Our version of circumscription is a special form of that given by Lifschitz [16].

---
[1] Circumscription is closely related to *predicate completion* in clause logic [9], and to the well-known *closed world assumption* [25].

2.2 Classical Proof Machinery

A (*classical*) *sequent* S is an ordered tuple of the form $\Gamma \vdash \Sigma$, where Γ, Σ are finite sequences of first-order formulae. Γ is the *antecedent* of S, and Σ is the *succedent* of S. The informal meaning of a sequent $A_1, \ldots, A_n \vdash B_1, \ldots, B_m$ is the same as the informal meaning of the formula $(\bigwedge_{i=1}^{n} A_i) \rightarrow (\bigvee_{i=1}^{m} B_i)$.

As proof system we use the (cut-free) sequent calculus **LK**. Axioms (or *initial sequents*) are sequents of the form $A \vdash A$, where A is any first-order formula. The inference rules of **LK** are given below, consisting of the *logical rules*, the *quantifier rules* and the *structural rules*.

SYSTEM **LK**: LOGICAL RULES

$$\frac{\Gamma_1, A, \Gamma_2 \vdash \Sigma}{\Gamma_1, (A \wedge B), \Gamma_2 \vdash \Sigma} \wedge l_1 \qquad \frac{\Gamma_1, A, \Gamma_2 \vdash \Sigma}{\Gamma_1, (B \wedge A), \Gamma_2 \vdash \Sigma} \wedge l_2$$

$$\frac{\Gamma \vdash \Sigma_1, A, \Sigma_2 \quad \Lambda \vdash \Pi_1, B, \Pi_2}{\Gamma, \Lambda \vdash \Sigma_1, \Pi_1, (A \wedge B), \Sigma_2, \Pi_2} \wedge r$$

$$\frac{\Gamma_1, A, \Gamma_2 \vdash \Sigma_1 \quad \Pi_1, B, \Pi_2 \vdash \Sigma_2}{\Gamma_1, \Pi_1, (A \vee B), \Gamma_2, \Pi_2 \vdash \Sigma_1, \Sigma_2} \vee l$$

$$\frac{\Gamma \vdash \Sigma_1, A, \Sigma_2}{\Gamma \vdash \Sigma_1, (A \vee B), \Sigma_2} \vee r_1 \qquad \frac{\Gamma \vdash \Sigma_1, A, \Sigma_2}{\Gamma \vdash \Sigma_1, (B \vee A), \Sigma_2} \vee r_2$$

$$\frac{\Gamma \vdash \Sigma_1, A, \Sigma_2 \quad \Gamma_1, B, \Gamma_2 \vdash \Pi}{\Gamma, (A \rightarrow B), \Gamma_1, \Gamma_2 \vdash \Sigma_1, \Sigma_2, \Pi} \rightarrow l$$

$$\frac{\Gamma_1, A, \Gamma_2 \vdash \Sigma_1, B, \Sigma_2}{\Gamma_1, \Gamma_2 \vdash \Sigma_1, (A \rightarrow B), \Sigma_2} \rightarrow r$$

$$\frac{\Gamma_1, \Gamma_2 \vdash \Sigma_1, A, \Sigma_2}{\Gamma_1, \neg A, \Gamma_2 \vdash \Sigma_1, \Sigma_2} \neg l \qquad \frac{\Gamma_1, A, \Gamma_2 \vdash \Sigma_1, \Sigma_2}{\Gamma_1, \Gamma_2 \vdash \Sigma_1, \neg A, \Sigma_2} \neg r$$

SYSTEM **LK**: QUANTIFIER RULES

$$\frac{\Sigma_1, A(t), \Sigma_2 \vdash \Gamma}{\Sigma_1, \forall x A(x), \Sigma_2 \vdash \Gamma} \forall l \qquad \frac{\Gamma \vdash \Sigma_1, A(y), \Sigma_2}{\Gamma \vdash \Sigma_1, \forall x A(x), \Sigma_2} \forall r$$

$$\frac{\Sigma_1, A(y), \Sigma_2 \vdash \Gamma}{\Sigma_1, \exists x A(x), \Sigma_2 \vdash \Gamma} \exists l \qquad \frac{\Gamma \vdash \Sigma_1, A(t), \Sigma_2}{\Gamma \vdash \Sigma_1, \exists x A(x), \Sigma_2} \exists r$$

$\forall r$ and $\exists l$ must fulfill the *eigenvariable condition*: the (free) variable y must not occur in the conclusion of the rule. For $\forall l$ and $\exists r$, the term t must be free for x in A.

SYSTEM LK: STRUCTURAL RULES

WEAKENING

$$\frac{\Gamma_1, \Gamma_2 \vdash \Sigma}{\Gamma_1, A, \Gamma_2 \vdash \Sigma} \; wl \qquad\qquad \frac{\Gamma \vdash \Sigma_1, \Sigma_2}{\Gamma \vdash \Sigma_1, A, \Sigma_2} \; wr$$

CONTRACTION

$$\frac{\Gamma_1, A, \Gamma_2, A, \Gamma_3 \vdash \Sigma}{\Gamma_1, A, \Gamma_2, \Gamma_3 \vdash \Sigma} \; cl \qquad\qquad \frac{\Gamma \vdash \Sigma_1, A, \Sigma_2, A, \Sigma_3}{\Gamma \vdash \Sigma_1, A, \Sigma_2, \Sigma_3} \; cr$$

We also use the following two systems: $\mathsf{LK_0}$ is the propositional version of LK (i.e., without quantifier rules and restricted to sequents which contain only propositional formulae), and $\mathsf{LK_{cut}}$ is LK together with the *cut rule*:

$$\frac{\Gamma_1 \vdash \Sigma_1, A \qquad A, \Gamma_2 \vdash \Sigma_2}{\Gamma_1, \Gamma_2 \vdash \Sigma_1, \Sigma_2} \; cut$$

Let α be a proof in LK, $\mathsf{LK_0}$ or $\mathsf{LK_{cut}}$. The *length* of α is the number of sequents occurring in α. We denote the length of α by $|\alpha|$.

2.3 Proof Machinery for Circumscription

Recently, Bonatti and Olivetti [5] introduced a sequent calculus for propositional circumscription, following closely their earlier contributions of presenting similar systems for default logic and autoepistemic logic [3,4]. Like its predecessors, the newly proposed calculus consists of three parts, namely a classical propositional sequent calculus, a propositional anti-sequent calculus (the *complementary* system), and certain inference rules involving circumscription. The ingenious part in their approach is the use of the complementary system *formalizing invalid statements*, i.e., an *anti-sequent* $\Gamma \not\vdash \Theta$ is provable in the complementary sequent calculus iff the corresponding classical sequent $\Gamma \vdash \Theta$ is invalid. In general, two logical systems are *complementary* iff objects derivable in one system are *not* derivable in the other system and vice versa.[2]

The study of logical systems describing invalid statements is a less known branch of logic tracing back as early as Aristotle's theory of syllogisms. The first modern author who introduced a logical calculus for rejected statements was Łukasiewicz in his attempt to translate Aristotle's system of syllogisms into present-day logic [17]. However, his system is of a somewhat hybrid nature:

---

[2] The term "complementary proof system" is due to Varzi [27].

it defines rejected statements in terms of valid and invalid ones. Preparatory for his default logic calculus, Bonatti introduced in [2] a sequent calculus for invalid formulae whose description *relies purely on the notion of unprovability*. Independently, Goranko [13] presented essentially the same calculus as part of several complementary sequent calculi for certain modal logics.

Since we use first-order formulae for our result, we must slightly generalize the system of Bonatti and Olivetti. Due to the undecidability of first-order logic, we cannot have a sound and complete formalization of first-order non-theorems. If one wants to construct such a sound and complete axiomatization of invalid statements, only a *decidable* subclass of first-order formulae can be used. In fact, for our purpose, it suffices to generalize only the "classical part" of their system, but the "complementary part" remains propositional. It is straightforward to check that the soundness and completeness results given in [5] hold for our version of the calculus as well.

We will introduce now this slightly generalized calculus for circumscription, called CIRC*.

In the remainder of this paper, we use upper-case Greek letters (possibly with subscripts) in the following way: Γ and Θ shall denote finite sequences of propositional formulae, Σ shall denote finite sequences of propositional atoms, and Δ shall denote finite sequences of closed first-order formulae. Furthermore, P and R shall always represent finite sets of propositional atoms.

For a sequence s of formulae, the expression $\hat{s}$ stands for the set of elements of s; the empty sequence will be denoted by ϵ.

Definition 3.

1. An anti-sequent *is an ordered pair of the form* $\Gamma \not\vdash \Theta$.
2. A circumscription sequent *is an ordered 5-tuple of the form* $\Sigma; \Gamma \vdash_{P;R} \Delta$, *provided that* $(P \cup \hat{\Sigma}) \cap R = \emptyset$. □

The sequence Σ occurring in a circumscription sequent $\Sigma; \Gamma \vdash_{P;R} \Delta$ is introduced for technical reasons only; its elements are called *constraints*.

Definition 4.

1. An anti-sequent $\Gamma \not\vdash \Theta$ *is true iff there is a first-order interpretation such that the classical sequent* $\Gamma \vdash \Theta$ *is false.*
2. A circumscription sequent $\Sigma; \Gamma \vdash_{P;R} \Delta$ *is true iff for any* $(P \cup \hat{\Sigma}; R)$-*minimal model* M *of* $\hat{\Gamma}$ *satisfying* $\hat{\Sigma}$, *at least one element of* $\hat{\Delta}$ *is true in* M. □

Obviously, $\Gamma \not\vdash \Theta$ is true iff $\Gamma \vdash \Theta$ is invalid, and $\epsilon; \Gamma \vdash_{P;R} A_1, \ldots, A_n$ is the syntactical counterpart of $\hat{\Gamma} \models_{P;R} A_1 \vee \ldots \vee A_n$.

The *complementary sequent calculus* LK_0^r is defined as follows. The axioms of LK_0^r are anti-sequents of the form $\Phi \not\vdash \Psi$, where Φ and Ψ are finite sequences of propositional atoms such that $\hat{\Phi} \cap \hat{\Psi} = \emptyset$. The inference rules of LK_0^r comprise the logical rules and the structural rules, which are given below.

System LK_0^r: Logical Rules

$$\frac{\Gamma_1, A, \Gamma_2, B, \Gamma_3 \not\vdash \Theta}{\Gamma_1, (A \wedge B), \Gamma_2, \Gamma_3 \not\vdash \Theta} \wedge l^r$$

$$\frac{\Gamma \not\vdash \Theta_1, A, \Theta_2}{\Gamma \not\vdash \Theta_1, (A \wedge B), \Theta_2} \wedge r_1^r \qquad \frac{\Gamma \not\vdash \Theta_1, A, \Theta_2}{\Gamma \not\vdash \Theta_1, (B \wedge A), \Theta_2} \wedge r_2^r$$

$$\frac{\Gamma_1, A, \Gamma_2 \not\vdash \Theta}{\Gamma_1, (A \vee B), \Gamma_2 \not\vdash \Theta} \vee l_1^r \qquad \frac{\Gamma_1, A, \Gamma_2 \not\vdash \Theta}{\Gamma_1, (B \vee A), \Gamma_2 \not\vdash \Theta} \vee l_2^r$$

$$\frac{\Gamma \not\vdash \Theta_1, A, \Theta_2, B, \Theta_3}{\Gamma \not\vdash \Theta_1, (A \vee B), \Theta_2, \Theta_3} \vee r^r$$

$$\frac{\Gamma_1, \Gamma_2 \not\vdash \Theta_1, A, \Theta_2}{\Gamma_1, (A \to B), \Gamma_2 \not\vdash \Theta_1, \Theta_2} \to l_1^r \qquad \frac{\Gamma_1, B, \Gamma_2 \not\vdash \Theta}{\Gamma_1, (A \to B), \Gamma_2 \not\vdash \Theta} \to l_2^r$$

$$\frac{\Gamma_1, A, \Gamma_2 \not\vdash \Theta_1, B, \Theta_2}{\Gamma_1, \Gamma_2 \not\vdash \Theta_1, (A \to B), \Theta_2} \to r^r$$

$$\frac{\Gamma_1, \Gamma_2 \not\vdash \Theta_1, A, \Theta_2}{\Gamma_1, \neg A, \Gamma_2 \not\vdash \Theta_1, \Theta_2} \neg l^r \qquad \frac{\Gamma_1, A, \Gamma_2 \not\vdash \Theta_1, \Theta_2}{\Gamma_1, \Gamma_2 \not\vdash \Theta_1, \neg A, \Theta_2} \neg r^r$$

System LK_0^r: Structural Rules
Contraction

$$\frac{\Gamma_1, A, \Gamma_2, A, \Gamma_3 \not\vdash \Theta}{\Gamma_1, A, \Gamma_2, \Gamma_3 \not\vdash \Theta} cl^r \qquad \frac{\Gamma \not\vdash \Theta_1, A, \Theta_2, A, \Theta_3}{\Gamma \not\vdash \Theta_1, A, \Theta_2, \Theta_3} cr^r$$

Observe that these rules bear a close resemblance to their classical counterparts, except that each binary rule of LK gives rise to *two* rules in LK_0^r. Intuitively, we can say that what is exhaustive search in the classical calculus becomes non-determinism in the complementary calculus. If read from bottom to top, an LK_0^r-proof corresponds to the (successful) construction of a counterexample, given by its (unique) axiom.

Theorem 1 ([2]). *The anti-sequent $\Gamma \not\vdash \Theta$ is provable in LK_0^r iff it is true.*

Corollary 1 ([2]). *The anti-sequent $\Gamma \not\vdash \Theta$ is provable in LK_0^r iff the classical sequent $\Gamma \vdash \Theta$ is not provable in LK_0.*

The length of an LK_0^r-proof is defined analogously to the classical case, i.e., as the number of anti-sequents occurring in it.

Next we define the *circumscription sequent calculus* CIRC*. It consists of classical sequents, anti-sequents and circumscription sequents. Furthermore, it incorporates the systems LK for classical sequents and LK$_0^r$ for anti-sequents. The additional inference rules for circumscription sequents are given below.

SYSTEM CIRC*: LOGICAL RULES

$$\frac{\Gamma_1, \neg p_1, \ldots, \neg p_n, \Gamma_2 \not\vdash q}{\Sigma_1, q, \Sigma_2; \Gamma_1, \Gamma_2 \vdash_{\{p_1,\ldots,p_n\};\emptyset} \Delta} C_1 \qquad \frac{\Sigma, \Gamma \vdash \Delta}{\Sigma; \Gamma \vdash_{P;R} \Delta} C_2$$

$$\frac{\Sigma_1, q, \Sigma_2; \Gamma_1, \Gamma_2 \vdash_{P;R} \Delta_1 \qquad \Sigma_1, \Sigma_2; \Gamma_1, \neg q, \Gamma_2 \vdash_{P;R} \Delta_2}{\Sigma_1, \Sigma_2; \Gamma_1, \Gamma_2 \vdash_{P \cup \{q\};R} \Delta_1, \Delta_2} C_3$$

$$\frac{\Sigma; \Gamma_1, r, \Gamma_2 \vdash_{P;R} \Delta_1 \qquad \Sigma; \Gamma_1, \neg r, \Gamma_2 \vdash_{P;R} \Delta_2}{\Sigma; \Gamma_1, \Gamma_2 \vdash_{P;R \cup \{r\}} \Delta_1, \Delta_2} C_4$$

For the rule C_1, the propositional atom q must be present.

SYSTEM CIRC*: STRUCTURAL RULES
CONTRACTION

$$\frac{\Sigma_1, q, \Sigma_2, q, \Sigma_3; \Gamma \vdash_{P;R} \Delta}{\Sigma_1, q, \Sigma_2, \Sigma_3; \Gamma \vdash_{P;R} \Delta} cl_1^c \qquad \frac{\Sigma; \Gamma_1, A, \Gamma_2, A, \Gamma_3 \vdash_{P;R} \Delta}{\Sigma; \Gamma_1, A, \Gamma_2, \Gamma_3 \vdash_{P;R} \Delta} cl_2^c$$

$$\frac{\Sigma; \Gamma \vdash_{P;R} \Delta_1, A, \Delta_2, A, \Delta_3}{\Sigma; \Gamma \vdash_{P;R} \Delta_1, A, \Delta_2, \Delta_3} cr^c$$

Let us give a brief explanation of these rules. The rules C_1 and C_2 represent two opposing situations how circumscription sequents can be introduced: for rule C_1 it holds that if its premise is true then its conclusion is *vacuously* true, while rule C_2 states that classical entailment implies minimal entailment. Rule C_3 distinguishes the case when a minimized atom q is either true or false in a minimal model; similar considerations apply for rule C_4 and a fixed atom r. Incidentally, the latter rule implements a variant of the atomic cut rule, restricted to propositional variables from the fixed atoms R as cut formulae.

Theorem 2 ([5]). *The circumscription sequent* $\Sigma; \Gamma \vdash_{P;R} \Delta$ *is derivable in* CIRC* *iff it is true.*

As ususal, the length of a proof in CIRC* is the number of sequents occurring in it (and this includes *a fortiori* the length of the LK$_0^r$-proofs for anti-sequents occurring as premises in applications of rule C_1, and the length of LK-proofs of classical sequents occurring as premises in applications of rule C_2).

$$C_1(\alpha, \beta, \gamma) := \exists z \, (p(\alpha, \beta, z) \wedge p(z, \beta, \gamma))$$
$$C_2(\alpha, \beta, \gamma) := \exists y \, (p(y, b_0, \alpha) \wedge C_1(\beta, y, \gamma))$$
$$C := \forall u \forall v \forall w \, (C_2(u, v, w) \to p(v, u, w))$$
$$B_0(\alpha) := \exists v_0 \, p(b_0, \alpha, v_0)$$
$$B_{i+1}(\alpha) := \exists v_{i+1} \, (p(b_0, \alpha, v_{i+1}) \wedge B_i(v_{i+1}))$$
$$A_0(\alpha) := \forall w_0 \exists v_0 \, p(w_0, \alpha, v_0)$$
$$A_{i+1}(\alpha) := \forall w_{i+1} \, (A_i(w_{i+1}) \to \overline{A}_{i+1}(w_{i+1}, \alpha))$$
$$\overline{A}_0(\alpha, \delta) := \exists v_0 \, p(\alpha, \delta, v_0)$$
$$\overline{A}_{i+1}(\alpha, \delta) := \exists v_{i+1} \, (A_i(v_{i+1}) \wedge p(\alpha, \delta, v_{i+1}))$$

Fig. 1. Abbreviations used in the following.

3 Main Result

In this section, we show how circumscription can speed up proofs. We use a sequence of formulae for which Orevkov [24] showed a non-elementary lower bound on proof length in (cut-free) LK, but which possess short LK_{cut}-proofs. We show that these short LK_{cut}-proofs yield short CIRC^*-proofs if the cut formulae can be derived by applying circumscription rules, but any such CIRC^*-proof *without* circumscription has a non-elementary lower bound on proof length.

Definition 5. *Let F_k occur in the infinite sequence of formulae $(F_k)_{k \in \mathbb{N}}$ where*

$$F_k := \forall b \, ((\forall w_0 \exists v_0 \, p(w_0, b, v_0) \wedge$$
$$\forall uvw \, (\exists y \, (p(y, b, u) \wedge \exists z \, (p(v, y, z) \wedge p(z, y, w))) \to p(v, u, w)))$$
$$\to \exists v_k \, (p(b, b, v_k) \wedge \exists v_{k-1} \, (p(b, v_k, v_{k-1}) \wedge \ldots \wedge \exists v_0 \, p(b, v_1, v_0) \ldots))). \quad \square$$

Intuitively, $p(x, y, z)$ represents the relation $x + 2^y = z$, and F_k "computes" certain numbers using a recursive definition of this relation.

Abbreviations shown in Figure 1 are used in the following in order to simplify the notation. Using these abbreviations, F_k looks as follows:

$$F_k = \forall b \, ((A_0(b) \wedge C) \to B_k(b)).$$

The formulae F_k ($k \in \mathbb{N}$) have the following properties with respect to proof length.

Proposition 1 ([24]). *Let $(F_k)_{k \in \mathbb{N}}$ be the infinite sequence of formulae defined above.*

(a) *There is an LK_{cut}-proof ψ_k of $\vdash F_k$ such that $|\psi_k| \leq c_1 \cdot k + c_2$, for some constants c_1, c_2.*
(b) *For any (cut-free) LK-proof α of $\vdash F_k$ it holds that $|\alpha| \geq 2 \cdot s(k) + 1$.*

Thus, eliminating the cut yields a non-elementary increase of proof length.

In the following, we need the short LK_{cut}-proof ψ_k of $\vdash F_k$. However, we will not present ψ_k in all details but sketch the proof stressing the relevant details.

There are two kinds of cut-free LK-derivations, namely β_k and $\delta_k(t)$, which are relevant in the following. The cut-free LK-derivations β_k and $\delta_k(t)$ have end sequents $A_0(b_0), C \vdash A_k(b_0)$ and $A_0(b_0), C, A_k(t) \vdash B_k(t)$, respectively. Then, the LK_{cut}-proof ψ_k is as follows:

$$\dfrac{\dfrac{\beta_k \qquad \delta_k(b_0)}{A_0(b_0), C \vdash B_k(b_0)}\; cut, cl, cl}{\vdash \forall b\, ((A_0(b) \wedge C) \to B_k(b))}\; \wedge l_1, \wedge l_2, cl, \to r, \forall r$$

The derivation ψ_k of $\vdash F_k$ in LK_{cut} discussed so far has one application of the cut rule where the cut formula $A_k(b_0)$ has a free variable.

Definition 6. Let H_k ($k \in \mathbb{N}$) be formulae of the form

$$H_k := (\forall x\, (A_k(x) \to q)) \to F_k,$$

where q is a propositional atom which does not occur elsewhere in $A_i(x)$ or F_i ($0 \leq i \leq k$). □

An LK_{cut}-derivation of $\vdash H_k$ is obtained from the derivation of $\vdash F_k$ in LK_{cut} presented above by simply adding a wl inference with weakening formula $\forall x\, (A_k(x) \to q)$, followed by an $\to r$ inference.

$$\dfrac{\dfrac{\dfrac{\beta_k \qquad \delta_k(b_0)}{A_0(b_0), C \vdash B_k(b_0)}\; cut, cl, cl}{\vdash \forall b\, ((A_0(b) \wedge C) \to B_k(b))}\; \wedge l_1, \wedge l_2, cl, \to r, \forall r}{\vdash (\forall x\, (A_k(x) \to q)) \to \forall b\, ((A_0(b) \wedge C) \to B_k(b))}\; wl, \to r$$

Clearly, the length of this proof is also linear in k.

In contrast to this short LK_{cut}-derivation, any derivation of the same end sequent in LK has length which is non-elementary in k.

Lemma 1. Let α be an LK-proof of $\vdash H_k$ ($k \in \mathbb{N}$). Then the length of α is greater than $c \cdot s(k)$, for some constant c.

Proof. There are two possible inferences by which the formula $Q := (A_k(t) \to q)$ can be introduced into α (where t is some term), namely wl and $\to l$. Also, there are only two possible inferences by which the formula $Q' := \forall x\, (A_k(x) \to q)$ can be introduced, namely wl and $\forall l$. Without loss of generality we assume that α contains no weakenings of the latter kind, because such inferences can be simulated by wl with weakening formula Q and an application of $\forall l$, resulting in at most doubling the proof length.

We first eliminate all occurrences of Q introduced by $\to l$. Then, all occurrences of Q introduced by wl are eliminated, together with all inferences introducing Q'. The resulting derivation, β, is an LK-derivation of $\vdash F_k$ and, by Proposition 1(b), the length of β is greater than $2 \cdot \mathsf{s}(k) + 1$. Moreover, since $|\alpha| \geq |\beta|$, the lemma will be proved. The details follow.

STEP 1. Q is introduced by $\to l$. Select the first $\to l$ inference (with respect to some tree ordering) such that α_1 and α_2 do not have an $\to l$ inference introducing the formula Q. If there is no such inference, then go to Step 2. Otherwise, this first inference has the following form (I_1 and I_2 are LK-inferences).

$$\cfrac{\cfrac{\alpha_1}{\Gamma \vdash \Delta_1, A_k(t), \Delta_2} I_1 \quad \cfrac{\alpha_2}{\Pi_1, q, \Pi_2 \vdash \Delta_3} I_2}{\underset{\gamma}{\Gamma, (A_k(t) \to q), \Pi_1, \Pi_2 \vdash \Delta_1, \Delta_2, \Delta_3}} \to l$$

Construct an LK-derivation of $\vdash H_k$ of the following form:

$$\cfrac{\cfrac{\alpha'_2}{\Pi'_1, \Pi'_2 \vdash \Delta'_3} I'_2}{\underset{\gamma}{\cfrac{\vdots \; wl, wr \; (*)}{\cfrac{\Pi_1, \Pi_2 \vdash \Delta_3}{\cfrac{\vdots \; wl, wr \; (**)}{\cfrac{\Gamma, \Pi_1, \Pi_2 \vdash \Delta_1, \Delta_2, \Delta_3}{\Gamma, (A_k(t) \to q), \Pi_1, \Pi_2 \vdash \Delta_1, \Delta_2, \Delta_3}}}}} wl$$

The derivation α'_2 is obtained from α_2 by omitting all weakenings introducing the formula q, and by omitting contractions upon q. I'_2 is either I_2 or an inference occurring in α_2. The length of the resulting derivation is not greater than $|\alpha|$, because occurrences of q in Π_1, Π_2, or Δ_3 are placed down to $(*)$, and $|\alpha_1|$ is not less than the number of wl and wr in $(**)$.

Replace all such $\to l$ inferences without increasing the number of sequents, resulting in an LK-derivation where all occurrences of Q are introduced by wl.

STEP 2. Omit all wl introducing Q, and adjust the derivation by omitting all contractions upon formulae Q or Q' and all inferences using auxiliary formulae which contain the subformula Q. Since all occurrences of Q are introduced by wl, $\vdash F_k$ is derived. ∎

The circumscription sequent we are interested in is

$$S_k := \epsilon; r \to q \vdash_{q;\emptyset} H_k,$$

where r is a propositional atom different from q.

In the following, we present a short CIRC*-proof of this sequent. We start with the classical derivation ϑ_k. This derivation includes the short LK-proof β_k described above. The LK-proof ϑ_k is as follows:

$$\dfrac{\dfrac{\dfrac{\dfrac{\dfrac{A_0(b_0), C \vdash A_k(b_0)}{A_0(b_0) \wedge C \vdash A_k(b_0)} \wedge l_1, \wedge l_2, cl}{A_0(b_0) \wedge C \vdash B_k(b_0), r, A_k(b_0)} wr, wr \qquad \dfrac{q \vdash q}{q, \neg q \vdash} \neg l}{A_k(b_0) \to q, \neg q, A_0(b_0) \wedge C \vdash B_k(b_0), r} \to l}{\dfrac{\forall x\, (A_k(x) \to q), \neg q \vdash F_k, r}{\neg q \vdash H_k, r} \forall l, \to r, \forall r}}{r \to q, \neg q \vdash H_k} \to r \qquad \dfrac{q \vdash q}{q, \neg q \vdash} \neg l}{\to l, cl}$$

with β_k over the top inference, and the right branch derivation continuing to $\to l, cl$.

Next, a CIRC*-derivation of S_k is obtained.

$$\dfrac{\dfrac{\dfrac{\not\vdash r, q}{r \to q \not\vdash q} \to l^r}{q; r \to q \vdash_{\emptyset;\emptyset} H_k} C_1 \qquad \dfrac{\vartheta_k}{\epsilon; r \to q, \neg q \vdash_{\emptyset;\emptyset} H_k} C_2}{\epsilon; r \to q \vdash_{q;\emptyset} H_k} C_3, cr^c$$

According to Proposition 1(a), $|\beta_k|$ is linear in k, hence $|\vartheta_k|$ is linear in k. Consequently, the following result holds:

Lemma 2. *Let ϕ_k be the CIRC*-proof of $S_k = \epsilon; r \to q \vdash_{q;\emptyset} H_k$ described above. Then $|\phi_k| \leq d_1 \cdot k + d_2$, for some constants d_1, d_2.*

On the other hand, deriving the formula H_k "classically" from the theory $r \to q$ in CIRC* is tantamount to deriving the circumscription sequent

$$C_k := \epsilon; r \to q \vdash_{\emptyset;\emptyset} H_k$$

in CIRC*. But the length of any CIRC*-proof of this sequent is non-elementary in k for the following reasons. The sequent C_k can only be derived in CIRC* by an application of the rule C_2, using the classical sequent $W_k := r \to q \vdash H_k$ as premise. However, it is easy to see that the additional implication $r \to q$ in the antecedent of W_k does not reduce proof length mainly because q is a *pure atom* in W_k, i.e., q occurs negatively only. To put it another way, by an argument similar to the proof of Lemma 1, any cut-free LK-derivation of W_k can be transformed into an LK-proof of H_k without increasing the proof length, hence any LK-proof of W_k must be non-elementary in k.

Lemma 3. *Any CIRC*-proof of $C_k := \epsilon; r \to q \vdash_{\emptyset;\emptyset} H_k$ has length $\geq c \cdot \mathsf{s}(k)$, for some constant c.*

Let us examine the above short CIRC*-proof ϕ_k of S_k in more detail. The important consequence of circumscribing q is the propagation of the literal $\neg q$ into the right upper sequent of the inference C_3. This newly introduced literal $\neg q$ is used in the classical deduction as an "axiom partner" for the pure atom q in W_k. As a consequence, there is a sequent of the form

$$U_k := A_0(b_0) \wedge C \vdash B_k(b_0), r, A_k(b_0)$$

occurring in the classical proof ϑ_k. Since $A_k(b_0)$ is the only cut formula in the short LK_{cut}-deduction ψ_k of $\vdash F_k$, the sequent U_k has a short *cut-free* derivation. Consequently, the proof above is also short. Of course, this argumentation fails for the "classical" sequent C_k, because q occurs in one polarity only.

Theorem 3. *There is an infinite sequence* $(H_k)_{k \in \mathbb{N}}$ *of first-order formulae for which the following holds.*

(a) *There exists a* CIRC^*-*proof of* $S_k = \epsilon; r \to q \vdash_{q;\emptyset} H_k$ *whose length is linear in* k.
(b) *The minimal proof length of* $C_k = \epsilon; r \to q \vdash_{\emptyset;\emptyset} H_k$ *in* CIRC^* *is greater than* $c \cdot \mathsf{s}(k)$, *for some constant* c.

Not only does the proof length decrease non-elementarily, but also the size of the search space. The reason is the elementary relation between proof length and search-space size for cut-free sequent calculi. Hence, if circumscription is possible, the overall effort remains elementary, whereas the overall effort in the classical (monotonic) case is non-elementary.

So far, we have considered S_k and C_k which are *both* derivable. Let us slightly modify these sequents by defining $S'_k := \epsilon; r \to q \vdash_{q;\emptyset} H'_k$ and $C'_k := \epsilon; r \to q \vdash_{\emptyset;\emptyset} H'_k$, where

$$H'_k := \forall x \, (A_k(x) \to q) \to F'_k,$$
$$F'_k := \forall b \, ((A_0(b) \land C) \to r).$$

The sequent $r \to q \vdash H'_k$ is *not* classically derivable mainly because $A_0(b) \land C$ is a logic program without a goal and, therefore, is satisfiable but not valid. The sequent S'_k, however, remains derivable. This is easily checked by considering the LK-derivation ϑ_k. If we replace in this derivation the formula H_k by H'_k, we obtain an LK-derivation ϑ'_k containing a sequent of the form $A_0(b_0) \land C \vdash r, r, A_k(b_0)$. However, we know that this sequent has a short classical proof, because $A_0(b_0), C \vdash A_k(b_0)$ has one. Hence, the length of the whole deduction remains short.

4 Conclusion and Discussion

In most works, non-monotonic reasoning is studied with no relation to a particular calculus. However, if non-monotonic techniques are embedded into automated deduction systems, the relative efficiency of the calculus becomes a crucial property. We used a cut-free sequent calculus for our analysis and showed that the presence of circumscription can tremendously simplify not only the proofs themselves but also the *search* for proofs. Although circumscription is in the worst case harder than classical logic, our result demonstrates that the other way around is also possible. After all, making things easier and not more complicated is a desired property when it comes to formalizations of common-sense reasoning.

Let us compare our result here with the result obtained in [10]. There, we used circumscription as a "preprocessing" activity to complete the precondition of the main connective in H_k resulting in the first-order formula H'_k, where

$$H_k := (\forall b\ (A_k(b) \rightarrow q(b))) \rightarrow F_k;$$
$$H'_k := (\forall b\ (A_k(b) \leftrightarrow q(b))) \rightarrow F_k.$$

Then we estimated for both formulae the minimal proof length in analytic calculi,[3] with the result that H'_k has short proofs whereas H_k has non-elementary proofs only. The reason for the tremendous speed-up is the simulation of instances of the cut rule with cut formula $A_k(b_0)$ by the newly introduced definition. In the circumscription process, the predicate $q(\cdot)$ is minimized while all other predicates and functions remain *fixed*. In contrast, the chosen circumscription in our sequent S_k minimizes the (propositional) predicate q but allows all other predicates and function symbols to *vary*. This is the main reason why we do *not* get the speed-up result for the classical counterpart of S_k in the "old" circumscription variant. Obviously, different variants of circumscription yield different behaviours with respect to proof length.

Although the class of formulae used to establish our result is constructed in regard to show the best case for the speed-up, one should observe that even simpler and more natural examples may exist, which become easier to prove by considering additional (relevant) knowledge.

In a related paper [11], we show that a generalization of Bonatti's sequent calculus for default logic [3] allows a similar non-elementary speed-up of proof length.

References

1. G. Amati, L. C. Aiello, D. Gabbay and F. Pirri. A Proof Theoretical Approach to Default Reasoning I: Tableaux for Default Logic. *Journal of Logic and Computation*, 6(2):205–231, 1996.
2. P. A. Bonatti. *A Gentzen System for Non-Theorems.* Technical Report CD-TR 93/52, Christian Doppler Labor für Expertensysteme, Technische Universität Wien, Paniglgasse 16, A–1040 Wien, 1993.
3. P. A. Bonatti. Sequent Calculi for Default and Autoepistemic Logics. *Proceedings TABLEAUX'96*, Springer LNAI 1071, pp. 127–142, 1996.
4. P. A. Bonatti and N. Olivetti. A Sequent Calculus for Skeptical Default Logic. *Proceedings TABLEAUX'97*, Springer LNAI 1227, pp. 107–121, 1997.
5. P.A. Bonatti and N. Olivetti. A Sequent Calculus for Circumscription. *Preliminary Proceedings CSL'97*, BRICS Notes Series NS-97-1, pp. 95–107, 1997.
6. M. Cadoli, F. M. Donini, and M. Schaerf. Is Intractability of Non-Monotonic Reasoning a Real Drawback? In *Proceedings of the AAAI National Conference on Artificial Intelligence*, pages 946–951. MIT Press, 1994.

---

[3] We used Herbrand complexity for our analysis, resulting in an estimation for essentially all analytic cut-free calculi.

7. M. Cadoli, F. M. Donini, and M. Schaerf. On Compact Representation of Propositional Circumscription. In *Proceedings STACS '95*, Springer LNCS 900, pp. 205–216, 1995.
8. M. Cadoli and M. Schaerf. A Survey of Complexity Results for Non-Monotonic Logics. *Journal of Logic Programming*, 17:127–160, 1993.
9. K. L. Clark. Negation as Failure. In *Logic and Databases*, Plenum, New York, pp. 293–322, 1978.
10. U. Egly and H. Tompits. Is Nonmonotonic Reasoning Always Harder? *Proceedings LPNMR'97*, Springer LNAI 1265, pp. 60–75, 1997.
11. U. Egly and H. Tompits. Non-Elementary Speed-Ups in Default Reasoning. *Proceedings ECSQARU-FAPR'97*, Springer LNAI 1244, pp. 237–251, 1997.
12. T. Eiter and G. Gottlob. Propositional Circumscription and Extended Closed World Reasoning are Π_2^P-complete. *Journal of Theoretical Computer Science*, 114(2):231–245, 1993. Addendum: vol. 118, p. 315, 1993.
13. V. Goranko. Refutation Systems in Modal Logic. *Studia Logica*, 53(2):299–324, 1994.
14. G. Gottlob. Complexity Results for Nonmonotonic Logics. *Journal of Logic and Computation*, 2:397–425, 1992.
15. H. A. Kautz and B. Selman. Hard Problems for Simple Default Logics. *Artificial Intelligence*, 49:243–379, 1990.
16. V. Lifschitz. Computing Circumscription. In *Proceedings of IJCAI'85*, Morgan Kaufmann, Los Altos, CA, pp. 121–127, 1985.
17. J. Lukasiewicz. *Aristotle's Syllogistic from the Standpoint of Modern Formal Logic*. Clarendon Press, Oxford, 1951.
18. W. Marek, A. Nerode and J. Remmel. A Theory of Nonmonotonic Rule Systems II. *Annals of Mathematics and Artificial Intelligence*, 5:229–264, 1992.
19. J. McCarthy. Epistemological Problems of Artificial Intelligence. In *Proceedings of IJCAI'77*, Cambridge, MA, pp. 1038–1044, 1977.
20. J. McCarthy. Circumscription – A Form of Non-Monotonic Reasoning. *Artificial Intelligence*, 13:27–39, 1980.
21. I. Niemelä. A Tableau Calculus for Minimal Model Reasoning. *Proceedings TABLEAUX'96*, Springer LNAI 1071, pp. 278–294, 1996.
22. I. Niemelä. Implementing Circumscription Using a Tableau Method. *Proceedings ECAI'96*, John Wiley & Sons Ltd., Chichester, pp. 80–84, 1996.
23. N. Olivetti. Tableaux and Sequent Calculus for Minimal Entailment. *Journal of Automated Reasoning*, 9:99–139, 1992.
24. V. P. Orevkov. Lower Bounds for Increasing Complexity of Derivations after Cut Elimination. *Zapiski Nauchnykh Seminarov Leningradskogo Otdeleniya Matematicheskogo Instituta im V. A. Steklova AN SSSR*, 88:137–161, 1979. English translation in *Journal of Soviet Mathematics*, 2337–2350, 1982.
25. R. Reiter. On Closed-World Data Bases. In *Logic and Databases*, pp. 55–76, Plenum, New York, 1978.
26. J. Schlipf. Decidability and Definability with Circumscription. *Annals of Pure and Applied Logic*, 35:173–191, 1987.
27. A. Varzi. Complementary Sentential Logics. *Bulletin of the Section of Logic*, 19:112–116, 1990.

Tableaux for Finite-Valued Logics with Arbitrary Distribution Modalities[*]

Christian G. Fermüller and Herbert Langsteiner

Institut für Computersprachen
Technische Universität
Karlsplatz 13/E185.2
A-1040 Wien, Austria

Abstract. We generalize finite-valued modal logics by introducing the concept of distribution modalities in analogy to distribution quantifiers. Sound and complete proof search procedures are provided using prefixed signed tableaux. Examples indicate that our generalized concept of modalities is indeed needed to formalize different types of statements in contexts of "graded truth" and inconsistent or incomplete databases.

1 Introduction

Typical applications of logic in computer science – and Artificial Intelligence in particular – differ from mathematical applications (in the narrow sense) mainly in two aspects. First, one is urged to study not only classical or intuitionistic logic, but a very wide range of non-classical logics. Various applications trigger the invention and investigation of ever new logics, where 'classic' logics appear to narrow a basis for adequate formalization. Second, efficient proof search is a central pre-requisite for many applications. Usually it is not sufficient to find a logical formalism that allows to express salient features of the phenomena to be modeled. Rather one aims at the construction of programs that find proofs of corresponding statements. Whenever possible, also concrete decision procedures (not just proofs of decidability) should be provided. In this paper we want to contribute to both aspects of logic in computer science. We investigate a very broad family of many-valued modal logics and provide computationally adequate tableau-based proof systems for them.

The idea to generalize possible world semantics to a many-valued context is not new. Modal logics based on three-valued logics have been studied in [24], [16], and [23]. Broader families of many-valued modal logics are investigated, e.g., in [25], [15], [19], [20]. Undoubtedly, the most advanced treatment of the topic consists in a series of papers by M.C. Fitting [7–9]. All authors consider the generalization of classical (i.e., two-valued) evaluation of formulas in possible worlds to many-valued evaluations. In addition, Fitting introduced a second class of many-valued modal logics by making also the accessibility relation many-valued. Although we consider this second approach to be well motivated

---
[*] Partly supported by COST-Action No. 15.

and attractive, too, we aim here at a generalization and proof search theoretic consolidation of the first approach.[1]

The central motivation of our contribution is the following. We are convinced that various applications call for a broader understanding of "modality" that has not yet been fully captured to our best knowledge. All cited papers only consider rather straightforward counterparts of the classic modal operator □ ("necessarily") and its dual ◊ ("possibly"). However, our examples in Section 7 intend to show that modal operators that do not correspond to such modalities arise naturally in different many-valued contexts. For this purpose we introduce the concept of "distribution modalities" (in Section 2.3). Moreover, we aim at a very general, uniform and modular representation. Prefixed signed tableaux as presented by Fitting in [6] turned out to be an almost perfect tool for this purpose.

The paper can also be read as another exercise in the very topical subject of "combining logics", most inspiringly propagated, e.g., by D. Gabbay (see also [2]). Indeed we like to view the introduced class of logics as the space of all possible combinations between the following three building blocks:

- an arbitrary finite-valued "base logic",
- any Kripke semantics with standard accessibility relation, and
- most importantly: an arbitrary (finite) collection of distribution modalities.

Once the particular choice for these three parameters is made, a sound, complete and even optimized tableau based calculus for the corresponding logic can (in principle) be generated automatically using procedures like those implemented in the system MULTLOG [3]. The many-valued modal logics described in [25], [15], [19], and elsewhere appear as simple instances. Thus one can also see this work as a contribution to the exciting field of "logic engineering" (see, [17, 18]).

For sake of a concise and clear presentation we only describe propositional logics here. The generalization to the first-order level does not present any difficulties beyond those for the underlying many-valued and (standard) modal logics themselves.

2 Logical building blocks

2.1 Finite-valued logics

Literature on many-valued logics abounds (see, eg., [11, 21]). Here, we consider the class of all finite-valued propositional logics. A **language** $\mathcal{L}$ for a finite-valued logic consists of countably many propositional variables and a finite number of propositional connectives, each of which has some fixed arity. (The arity may be 0; in that case the connectives are called truth constants.) $\mathcal{L}$-formulas are

---

[1] We think that the work presented here can rather straightforwardly be extended to many-valued accessibility relations as well as whole collections of different accessibility relations for single logics. However, to keep things reasonably simple we restrict ourselves here to the propositional case of the first approach.

defined as usual. A corresponding **matrix** $\mathbf{M}_{\mathcal{L}}$ consists in a non-empty set of **truth values** $N = \{a_1, \ldots, a_n\}$ and an abstract algebra with domain N of appropriate type. For every n-place connective $\circ$ of $\mathcal{L}$ there is an associated **truth table** $\tilde{\circ}: N^m \to N$. An **interpretation** $I_{\mathbf{M}}$ is a matrix $\mathbf{M}$ together with an assignment of truth values to the propositional variables. The corresponding evaluation function $v_{I_{\mathbf{M}}}$ is defined as usual. Thus any pair $\langle \mathcal{L}, \mathbf{M}_{\mathcal{L}} \rangle$ determines an m-**valued logic L**. Usually one also distinguishes a subset D of N as "designated truth values" and defines formulas X as valid if $v_{I_{\mathbf{M}}}(X) \in D$ for all interpretations $I_{\mathbf{M}}$. However, we consider the more general setting, where one is interested in proofs of statements of the form $\forall I_{\mathbf{M}}: v_{I_{\mathbf{M}}}(X) \in A$ (or $\notin A$) for arbitrary formulas X and $A \subset N$.

2.2 Normal modal logics

The literature on modal logics is even more immense than that on many-valued logics. Our starting point are **normal modal logics** characterized by simple defining conditions on the accessibility relations of their Kripke semantics.

Definition 1. *A* **(Kripke-)frame** *is a pair* $\langle \mathcal{G}, \mathcal{R} \rangle$ *where* $\mathcal{G}$ *is a non-empty set of possible worlds and* $\mathcal{R}$ *is a binary relation on* $\mathcal{G}$. *Members of* $\mathcal{G}$ *will be referred to as* **possible worlds***. A world* Δ *is* **accessible** *from* Γ *if* $\Gamma \mathcal{R} \Delta$.

A **valuation** *in a frame* $\langle \mathcal{G}, \mathcal{R} \rangle$ *is a mapping* $v : \mathcal{G} \times \text{VAR} \to \{\text{true}, \text{false}\}$, *where* VAR *is the set of propositional variables.*

A **(Kripke-)model** $\mathcal{M}$ *is a triple* $\langle \mathcal{G}, \mathcal{R}, v \rangle$ *where* $\langle \mathcal{G}, \mathcal{R} \rangle$ *is a frame and* v *is a valuation in it.*

The usual language of classical propositional logic is enriched by a unary connective (modal operator) "□". Let FORM denote the set of all formulas over this language.

Definition 2. *The evaluation function corresponding to a model* $\mathcal{M}$ *is defined as an extension of the valuation* v *to a mapping* $\bar{v}_{\mathcal{M}} : \mathcal{G} \times \text{FORM} \to \{\text{true}, \text{false}\}$:

1. $\bar{v}_{\mathcal{M}}(\Gamma, P) = v(\Gamma, P)$ *for every variable* $P \in \text{VAR}$ *and world* $\Gamma \in \mathcal{G}$.
2. $\bar{v}_{\mathcal{M}}(\Gamma, \circ(X_1, \ldots, X_m)) = \tilde{\circ}(\bar{v}(\Gamma, X_1), \ldots, \bar{v}(\Gamma, X_m))$ *for every classical connective* $\circ$ *and its corresponding truth function* $\tilde{\circ}$.
3. $\bar{v}_{\mathcal{M}}(\Gamma, \Box X) = \textbf{true}$ *iff* $(\forall \Delta \in \mathcal{G}) \Gamma \mathcal{R} \Delta$ *implies* $\bar{v}_{\mathcal{M}}(\Delta, X) = \textbf{true}$.

Particular logics are characterized by certain properties of the accessibility relation. For sake of a concise presentation, which nevertheless allows to recognize the generality of our approach we consider the following concrete logics as reference points of our generalizations:

L	L-property	serial
K	—	no
T	reflexive	yes
K4	transitive	no
KB	symmetric	no
B	reflexive and symmetric	yes
S4	reflexive and transitive	yes
S5	reflexive, symmetric and transitive	yes
D	—	yes
D4	transitive	yes
DB	symmetric	yes

A model is called **L-model** if its accessibility relation satisfies the L-property and is serial[2] in the above table.

2.3 Distribution modalities

As mentioned above Ostermann and others already have considered generalizations of standard modal logics to a many-valued context. In the Definition 1 of a Kripke-model one only has to replace {**true**, **false**} by an arbitrary set of truth values N. An ordering on N, with **true** as maximal element and **false** as minimal element is assumed to reflect "grades ot truth". The relevant step consists in generalizing the interpretation of the standard "necessity" operator $\Box$ from

$$\bar{v}_{\mathcal{M}}(\Gamma, \Box X) = \textbf{true} \text{ iff } \forall \Delta : \Gamma \mathcal{R} \Delta \text{ implies } \bar{v}_{\mathcal{M}}(\Delta, X) = \textbf{true}$$

to

$$\bar{v}_{\mathcal{M}}(\Gamma, \Box X) = a \text{ iff } \inf\{\bar{v}_{\mathcal{M}}(\Delta, X) \mid \Delta \in \mathcal{G}, \Gamma \mathcal{R} \Delta\} = a,$$

for every truth value $a \in N$. In the corresponding definition for the dual "possibility" operator $\Diamond$ the supremum is taken in place of the infimum.

The semantics of the non-modal connectives is defined exactly as in Definition 2.

We want to use many-valued Kripke-models in a more general way. Instead of considering only "inf" and "sup" as basis for the definition of finite-valued modalities we suggest to investigate all functions from $2^N \to N$ as candidates for determining the semantics of modalities. The examples, below and in Section 7, intend to demonstrate that modalities that do not arise as direct generalizations of $\Box$ ("necessarily") or $\Diamond$ ("possibly") may be needed in different contexts.

Definition 3. *Syntactically, a* **distribution modality** μ *is a unary connective. Given a many-valued (Kripke-)model $\mathcal{M}$ its semantics is determined by*

$$\bar{v}_{\mathcal{M}}(\Gamma, \mu X) = \tilde{\mu}(\{\bar{v}(\Delta, X) \mid (\forall \Delta \in \mathcal{G}) \ \Gamma \mathcal{R} \Delta\})$$

---
[2] A binary relation $\mathcal{R}$ is serial if $\forall x \exists y : x \mathcal{R} y$.

where $\tilde{\mu}$ is a function of type $2^N \to N$.

The name "distribution modalities" is suggested by the close relation of $\tilde{\mu}$ to the truth functions of distribution quantifiers (see, e.g., [5,13]). The truth value that is assigned to the formula μX in a world Γ of a Kripke model $\mathcal{M}$ is determined by the distribution of truth values for X in the worlds that are accessible from Γ in $\mathcal{M}$. Similarly, the truth value of $QxA(x)$ for a distribution quantifier Q is determined by the distribution of truth values over all instances of $A(x)$. However, the analogy between quantifiers and modalities is not perfect. Whereas the domain of an interpretation is required to be non-empty, the set of accessible worlds may be empty (for non-serial logics).

Example 1. Many, if not most, important applications of many-valued logics are induced by their interpretation as logics of graded truth or fuzzy logics. One singles out truth values, say t and f, as denoting "absolute truth" and "absolute falsehood", respectively. The other truth values are intended to refer to intermediate grades of truth. A central notion in this context is that of "crispness". Propositions are called crisp if they always evaluate either to t or to f.

One can usually define a formula $F[X]$ such that F evaluates to t iff its subformula X evaluates either to t or to f. However, observe that this fact is not sufficient to be able to claim that crispness can be expressed within the logic. After all, we only want to call a proposition crisp if its truth value is "absolute" with respect to **any** interpretation.[3] To express crispness within the logic we propose to view it as a distribution modality which is added to the fuzzy logic of your choice. The following definition seems a reasonable choice for the interpretation of **C** as a modality, intended to express the property "crisp" of propositions:

$$\tilde{\mathbf{C}}(A) = \begin{cases} t & \text{if } A = \emptyset, \{t\}, \{f\} \text{ or } \{t,f\} \\ f & \text{otherwise} \end{cases}$$

If the underlying logic allows to define a unary propositional connective df with $\widetilde{df}(a) = t$ iff $a = t$ or $a = f$ then we can alternatively use the standard necessity operator $\Box$ (in the sense of Ostermann) to express the crispness of a proposition X by $\Box df(X)$.

However, one might prefer to interpret the statement "X is crisp" itself as a non-crisp statement. For example, let some subset N of the real interval $[0,1]$ containing 0 and 1 be the set of truth values, where $0 = f$ and $1 = t$. Suppose N is closed under $\dot{-}$ and *mean*, where $mean(A)$ denotes a suitable type of mean value of $A \subseteq N$. Moreover, let $A_f := \{a \mid a \leq 0.5, a \in A\}$ and $A_t := \{a \mid a \geq 0.5, a \in A\}$. Then

$$\tilde{\mathbf{C}}(A) = mean(A_t) - mean(A_f)$$

---

[3] In a fuzzy context, we hardly want to call the proposition "this is a fine day" crisp only because it was absolutely true for one of the authors of this paper on September 6th, 1983.

could be a plausible candidate for the interpretation of $\mathbf{C}X$ as "X is crisp". ($\widetilde{\mathbf{C}}$ measures how "close" the distribution of truth values in the accessible worlds gets to t or f, respectively.)

We do not claim that these examples contribute deeply to the logical foundation of fuzzy logic. But we hope that they enable the reader to see that the concept of distribution modalities opens a wide space of possible formalizations of natural properties of propositions at the object level that are usually only expressible at the meta-level.

Definition 4. *Given a finite set of truth values N, an N-valued logic with (normal) distribution modalities is given by an arbitrary combination of three components:*

- *a non-empty, finite set of connectives $\{\circ_1, \ldots, \circ_c\}$ and corresponding truth functions $\{\tilde{\circ}_1, \ldots, \tilde{\circ}_c\}$*
- *a standard property of the accessibility relation in Kripke-frames*
- *a non-empty, finite set of distribution modalities $\{\mu_1, \ldots, \mu_d\}$ and corresponding truth functions $\{\tilde{\mu}_1, \ldots, \tilde{\mu}_d\}$*

Again, one may want to round off this definition by designating at least one truth value. For our context only the following observation is of importance: Any truth functional concept of "tautology" or "entailment" can be reduced to questions about the status of prefixed signed formulas as defined in the next section.

3 Prefixed signed tableaux

In recent years, **signed tableaux** – i.e., tableaux where formulas are labeled by (sets of) truth values – became recognized as an almost ideal frame for proof search for finite-valued logics (see, e.g., [12, 13]). Similarly, **prefixed tableaux**, as introduced by Fitting [6] referring to ideas of Kripke, are a very flexible tool for proof search in modal logics; in particular if one aims at generality and conceptional clarity. Considering finite-valued modal logics, what would be more natural than to combine these two versions of tableaux?

Definition 5. *A **prefix** is a finite non-empty sequence of positive integers. The concatenation of two prefixes σ, τ will be denoted by $\sigma \cdot \tau$. Let Π be the set of all prefixes.*

*A prefix τ is a **simple extension** of σ if $\tau = \sigma \cdot \langle n \rangle$ for some integer n. The **K-accessibility relation** $\mathcal{R}_{\mathbf{K}}$ on prefixes is defined by:*

$$(\forall \sigma, \tau \in \Pi) \; \sigma \mathcal{R}_{\mathbf{K}} \tau \Leftrightarrow \tau \text{ is a simple extension of } \sigma.$$

For every $\mathbf{L}$ let $\mathcal{R}_{\mathbf{L}}$ be the $\mathbf{L}$-property-closure of $\mathcal{R}_{\mathbf{K}}$.

Prefixes are intended to name worlds of a model. This is made precise in the following definition.

Definition 6. *Let $\Sigma \subseteq \Pi$ be a set of prefixes and let $\langle \mathcal{G}, \mathcal{R}, v \rangle$ be an **L**-model. An **L**-interpretation in the **L**-model $\mathcal{M} = \langle \mathcal{G}, \mathcal{R}, v \rangle$ is a mapping $I_\mathcal{M} : \Sigma \to \mathcal{G}$ from prefixes in Σ to worlds in $\mathcal{G}$ such that*

$$\sigma \mathcal{R}_\mathbf{L} \tau \Rightarrow I_\mathcal{M}(\sigma) \mathcal{R} I_\mathcal{M}(\tau)$$

for all $\sigma, \tau \in \Sigma$.

Definition 7. *A **prefixed signed formula**[4] is an expression of the form $\sigma:[a]:X$ or $\sigma^*:A:X$, where σ is a prefix, a a truth value and A a finite set of truth values. Let PSF be the set of all prefixed signed formula.*

The intended interpretation of $\sigma:[a]:X$ is that X evaluates to a in the world denoted by σ. The intended interpretation of $\sigma^*:A:X$ is that X evaluates to some truth value in A in every world that is accessible from the one denoted by σ.

Definition 8. *The relation $\models_\mathbf{L}$ between an **L**-interpretation $I_\mathcal{M}$ into an **L**-model $\mathcal{M} = \langle \mathcal{G}, \mathcal{R}, v \rangle$ and prefixed signed formulas is defined by:*

$$I_\mathcal{M} \models_\mathbf{L} \sigma:[a]:X \iff \sigma \in \Sigma \text{ and } v(I_\mathcal{M}(\sigma), X) = a,$$
$$I_\mathcal{M} \models_\mathbf{L} \sigma^*:A:X \iff \sigma \in \Sigma \text{ and } \{v(\Delta, X) \mid I_\mathcal{M}(\sigma) \mathcal{R} \Delta\} \subseteq A.$$

$\models_\mathbf{L}$ *is extended to sets S of prefixed signed formulas by:*

$$I_\mathcal{M} \models_\mathbf{L} S \iff (\forall Z \in S) \, I_\mathcal{M} \models_\mathbf{L} Z.$$

*A set S of prefixed signed formulas is **L**-satisfiable iff there exists an **L**-interpretation $I_\mathcal{M}$ with $I_\mathcal{M} \models_\mathbf{L} S$.*

Definition 9. *For a finite non-empty set E, $\mathbf{T}(E)$ denotes any labeled linear tree with $|E|$ nodes whose labels are exactly the members of E.*

*Let $\mathbf{T}$ be a labeled tree, $\mathbf{B}$ a branch of $\mathbf{T}$ and suppose $\mathcal{C} = \{E_1, \ldots, E_m\}$ is a finite collection of finite non-empty sets. To **extend B with** $\mathcal{C}$ means to construct a new labeled tree by adjoining the trees $\mathbf{T}(E_1), \ldots, \mathbf{T}(E_m)$ at the leave node of $\mathbf{B}$.*

Whenever there is no misunderstanding to be expected we identify a branch with the sets of prefixed signed formulas that label its nodes.

Definition 10. *A prefix σ occurs in S iff $\sigma:[a]:X \in S$ or $\sigma^*:A:X \in S$ for some formula X, $a \in N$, $A \subseteq N$. Let $\pi(S)$ be the set of prefixes occurring in S.*

We prefer to specify the tableau extension rules in a more general and abstract way than usual. For this purpose we first single out certain sets of sets of prefixed signed formulas that directly correspond to the truth functions of connectives and modal operators, respectively, if interpreted as disjunctive normal forms (on the meta-level).

---

[4] We follow Fitting [6] in using this arguably oxymoronic expression.

Definition 11. *Let $\mathcal{C} = \{E_1, \ldots, E_n\}$ be a finite collection of finite non-empty sets of prefixed signed formulas.*

*$\mathcal{C}$ is an **analysis** for $Z = \sigma\colon[a]\colon \circ(X_1, \ldots, X_m)$ if the members of E_i ($1 \le i \le n$) are of form $\sigma\colon[b]\colon X_j$ ($1 \le j \le m$) and for all **L**-interpretations $I_\mathcal{M}$:*

$$I_\mathcal{M} \models_\mathbf{L} Z \iff (\exists E \in \mathcal{C})\, I_\mathcal{M} \models_\mathbf{L} E.$$

*$\mathcal{C}$ is an **analysis** for $Z = \sigma\colon[a]\colon \mu X$ if the members of the E_i ($1 \le i \le n$) are of form $\tau\colon[b]\colon X$ or $\sigma^*\colon A\colon X$ and for all **L**-interpretations $I_\mathcal{M}$ there exists an **L**-interpretation $I'_\mathcal{M}$ with $I'_\mathcal{M}(\sigma) = I_\mathcal{M}(\sigma)$ such that:*

$$I_\mathcal{M} \models_\mathbf{L} Z \iff (\exists E \in \mathcal{C})\, I'_\mathcal{M} \models_\mathbf{L} E.$$

The set of all analysis of Z is denoted by $\mathcal{A}(Z)$.

As is well known from the literature (see, eg.,[5, 13, 1]) one can easily compute an analysis for signed formulas lead by any truth functional connective. An example of a general schema for an analysis of a prefixed signed formula $\sigma\colon[a]\colon\circ(X_1, \ldots, X_m)$ is

$$\{\{\sigma\colon[a_1]\colon X_1, \ldots, \sigma\colon[a_m]\colon X_m\} \mid \alpha = \langle a_1, \ldots, a_m\rangle \in \tilde{\circ}^{-1}(a)\}.$$

For a prefixed signed formula of form $\sigma\colon[a]\colon\mu X$ it is straightforward to check that

$$\{\{\sigma_{a_1}\colon[a_1]\colon X, \ldots, \sigma_{a_m}\colon[a_m]\colon X, \sigma^*\colon A\colon X\} \mid A = \{a_1, \ldots, a_m\} \in \tilde{\mu}^{-1}(a)\},$$

is an analysis. (Here, the σ_{a_i} are pairwise distinct new prefixes.)

A complete set of analysis for all types of signed formulas essentially constitutes a tableau calculus.

Definition 12. *Let S be a finite non-empty set of prefixed signed formulas. The set $\mathcal{T}_\mathbf{L}(S)$ of all **L**-tableau for S is defined as the set of PSF-labeled trees that can be constructed by finitely many applications of the following rules:*

Initial Tableau Rule: *A finite linear tree whose node labels are the prefixed signed formulas of S is an **L**-tableau for S.*

L-Tableau Extension Rules: *If $\mathbf{T}$ is an **L**-tableau for S and Z occurs on a branch $\mathbf{B}$ of $\mathbf{T}$ then construct a new **L**-tableau for S by extending $\mathbf{B}$ with some $\mathcal{C}$ satisfying the following conditions depending on Z.*

Z	condition on $\mathcal{C}$
$\sigma\colon[a]\colon\circ(X_1,\ldots,X_m)$	$\mathcal{C} \in \mathcal{A}(Z)$
$\sigma\colon[a]\colon\mu X$	$\mathcal{C} \in \mathcal{A}(Z)$, $(\forall \tau \in \pi(\mathcal{C}))\, \tau = \sigma$ or τ is an unrestricted simple extension of σ
$\sigma^*\colon A\colon X$	$\mathcal{C} = \{\{\tau\colon[a]\colon X\} \mid a \in A\}$, $\tau \in \pi(\mathbf{B})\colon \sigma\mathcal{R}_\mathbf{L}\tau$

*Here, τ is called **unrestricted** if it is not an initial segment of any prefix occurring on $\mathbf{B}$.*

Definition 13. *A set $S \subseteq \text{PSF}$ of tableau formulas is* **closed** *iff one of the following conditions holds:*

	S contains	and
C_1	$\sigma\colon [a]\colon X$, $\sigma\colon [b]\colon X$	$a \neq b$
C_2	$\sigma\colon [a]\colon \circ$	$\circ$ is a constant so that $\tilde{\circ} \neq a$
C_3	$\sigma\colon [a]\colon \circ(X_1,\ldots,X_m)$	$\tilde{\circ}^{-1}(a) = \emptyset$
C_4	$\sigma\colon [a]\colon \mu X$	$\tilde{\mu}^{-1}(a) = \emptyset$
C_5	$\sigma^*\colon \emptyset\colon X$, $\tau\colon [a]\colon Y$	τ is **L**-accessible from σ

S *is* **atomically closed** *iff C_1 holds for a propositional variable X or C_2 holds. An* **L**-*tableau branch* **B** *is* **closed** *if the set of tableau formulas occurring on* **B** *is closed. A branch is called* **open** *if it is not closed. An* **L**-*tableau is closed if every branch is closed.*

Definition 14. *A tableau* **T** *is called* **L**-*satisfiable if it has at least one* **L**-*satisfiable branch.*

4 Soundness

The soundness proof closely follows Fitting's corresponding proof for classical prefixed tableaux (§8,♯3 in [6]).

Lemma 1. *Let* **T** *be* **L**-*satisfiable and suppose* **T'** *is created from* **T** *by an application of an* **L**-*tableau extension rule. Then* **T'** *is also* **L**-*satisfiable.*

Proof. **T** contains at least one **L**-satisfiable branch **B**. Let $I_\mathcal{M}$ be some **L**-interpretation with $I_\mathcal{M} \models_\mathbf{L} \mathbf{B}$. If any branch different from **B** is extended to construct **T'** then **B** is unchanged and hence is still **L**-satisfiable.

Now suppose branch **B** is extended by applying a **L**-tableau extension rule to some prefixed signed formula $Z \in \mathbf{B}$. We have to consider three cases.

$Z = \sigma\colon [a]\colon \circ(X_1,\ldots,X_m)$. Then **B** is extended by $\mathcal{C} \in \mathcal{A}(Z)$. Obviously $I_\mathcal{M} \models_\mathbf{L} Z$. Hence, by definition of an analysis, there is an $E \in \mathcal{C}$ such that $I_\mathcal{M} \models_\mathbf{L} E$. Therefore $\mathbf{B}' := \mathbf{B} \cup E$ is a branch of **T'** with $I_\mathcal{M} \models_\mathbf{L} \mathbf{B}'$.

$Z = \sigma\colon [a]\colon \mu X$. Then **B** is extended by $\mathcal{C} \in \mathcal{A}(Z)$, where all prefixes occurring in $\mathcal{C}$ are unrestricted simple extensions of σ. Obviously $I_\mathcal{M} \models_\mathbf{L} Z$. By definition there is an **L**-interpretation $I'_\mathcal{M}$ and an $E \in \mathcal{C}$ such that $I'_\mathcal{M} \models_\mathbf{L} E$. Let $\mathbf{B}' := \mathbf{B} \cup E$. Hence $\pi(\mathbf{B}') = \pi(\mathbf{B}) \cup \pi(E)$. All prefixes occurring in E are unrestricted for **B** or equal σ. Therefore $\pi(\mathbf{B}) \cap \pi(E) \subseteq \{\sigma\}$. Now we define a mapping $I''_\mathcal{M}\colon \pi(\mathbf{B}') \to \Pi$ by:

$$I''_\mathcal{M}(\tau) := \begin{cases} I_\mathcal{M}(\tau) \text{ if } \tau \in \pi(\mathbf{B}) \\ I'_\mathcal{M}(\tau) \text{ if } \tau \in \pi(E) \end{cases}$$

$I''_\mathcal{M}$ is well-defined since $\pi(\mathbf{B}) \cap \pi(E) \subseteq \{\sigma\}$ and $I_\mathcal{M}(\sigma) = I'_\mathcal{M}(\sigma)$. Finally it is an easy matter to check that $I''_\mathcal{M}$ is an **L**-interpretation with $I''_\mathcal{M} \models_\mathbf{L} \mathbf{B}'$.

$Z = \sigma^*\colon A\colon X$. τ is a prefix occurring in **B** with $\sigma\mathcal{R}_\mathbf{L}\tau$. **B** is extended by $\mathcal{C} = \{\{\tau\colon [a]\colon X\} \mid a \in A\}$. $\sigma\mathcal{R}_\mathbf{L}\tau$ implies $I_\mathcal{M}(\sigma)\mathcal{R}I_\mathcal{M}(\tau)$. From this and $I_\mathcal{M} \models_\mathbf{L} Z$, that is $\{v(\Delta, X) \mid \Delta \in \mathcal{G},\ I_\mathcal{M}(\sigma)\mathcal{R}\Delta\} \subseteq A$, we conclude that there must be an $a \in A$ with $v(I_\mathcal{M}(\tau), X) = a$, i.e. $I_\mathcal{M} \models_\mathbf{L} \tau\colon [a]\colon X$. But then $I_\mathcal{M} \models_\mathbf{L} \mathbf{B}' := \mathbf{B} \cup \{\tau\colon [a]\colon X\}$. Therefore $\mathbf{B}'$ is an **L**-satisfiable branch of $\mathbf{T}'$.

Theorem 1 (Soundness). *Let S be a finite non-empty set of prefixed signed formulas. If there is a closed **L**-tableau for S then S is not **L**-satisfiable.*

Proof. Let **T** be a closed **L**-tableau for S and suppose S is **L**-satisfiable. Then the initial tableau $\mathbf{T}_0$ of **T** is **L**-satisfiable. By Lemma 1 every extension of $\mathbf{T}_0$ is **L**-satisfiable. Therefore **T** is **L**-satisfiable. But it is easy to see that a closed **L**-tableau can never be **L**-satisfiable.

5 Completeness

Again, the proof follows ideas of [6]. I.e., a model is extracted from a branch that is left open by a systematic proof search procedure.

Definition 15. *A set $S \subseteq \mathrm{PSF}$ of prefixed signed formulas is **L-downward saturated** iff the following conditions hold:*

1. *S is not atomically closed.*
2. *$Z = \sigma\colon [a]\colon \circ(X_1, \ldots, X_m) \in S \Rightarrow (\exists \mathcal{C} \in \mathcal{A}(Z))\ (\exists E \in \mathcal{C})\ E \subseteq S$.*
3. *$Z = \sigma\colon [a]\colon \mu X \in S \Rightarrow (\exists \mathcal{C} \in \mathcal{A}(Z))\ (\exists E \in \mathcal{C})\ E \subseteq S$.*
4. *$\sigma^*\colon A\colon X \in S \Rightarrow (\forall \tau \in \pi(S)\colon \sigma\mathcal{R}_\mathbf{L}\tau)\ (\exists a \in A)\ \tau\colon [a]\colon X \in S$*

Lemma 2. *If S is **L**-downward saturated then S is **L**-satisfiable in an **L**-model whose possible worlds are simply the prefixes occurring in S.*

Proof. Suppose S is **L**-downward saturated. Construct an **L**-model as follows. Let $\mathcal{G} := \pi(S)$ be the set of prefixes that occur in S and let $\mathcal{R}$ be the restriction of $\mathcal{R}_\mathbf{L}$ to $\mathcal{G}$. For all logics **L**– except possibly those where the accessibility relation is of type **D**, **D4** or **DB** – $\langle \mathcal{G}, \mathcal{R} \rangle$ is an **L**-frame. $\mathcal{G}$ might contain worlds σ, for which there is no τ with $\sigma\mathcal{R}\tau$. This means that $\langle \mathcal{G}, \mathcal{R} \rangle$ need not be an **L**-frame if **L** is based on a serial, but not necessarily reflexive accessibility relation. However, it is easy to check that we obtain a proper frame if we augment $\mathcal{R}$ by $\{(\sigma, \sigma) \mid \neg(\exists \tau \in \mathcal{G})\sigma\mathcal{R}\tau\}$.[5] Let

$$v(\sigma, P) := \begin{cases} a & \text{if } \sigma\colon [a]\colon P \in S \\ a_0 & \text{otherwise} \end{cases}$$

where a_0 is an arbitrary element of N. v is well-defined since S is not atomically closed. Hence $\mathcal{M} = \langle \mathcal{G}, \mathcal{R}, v \rangle$ is an **L**-model. Obviously, the identity map

---

[5] This corrects a minor error in [6].

$I_\mathcal{M}: \pi(S) \to \mathcal{G}$, $I_\mathcal{M}(\sigma) := \sigma$, is an L-interpretation in $\mathcal{M}$. Finally, it is easy to show by induction on the degree of the formula in Z:

$$Z \in S \Rightarrow I_\mathcal{M} \models Z$$

for every prefixed signed formula Z. Therefore $I_\mathcal{M} \models S$.

Theorem 2 (Completeness). *Let S be a finite non-empty set of prefixed signed formulas. If S is not L-satisfiable then there is a closed L-tableau for S.*

Proof. Analogous to Fitting's proof of Theorem 6.2 in §8 of [6].

6 Optimized rules

We have presented not just particular tableau calculi for particular logics, but have described a broad class of such calculi in a uniform and abstract manner. Depending on the choice of the analysis – in the sense of Definition 11 – for a given connective and truth value – we obtain different rules, that may be of very different complexity. By complexity we mean the number of new branches (and, secondarily, of formulas on this branches) that are introduced by an application of the rule. For actual proof search one obviously wants to keep this "branching degree" as small as possible. For the case of many-valued connectives and quantifiers this optimization problem is well investigated (see, e.g., [22], [1], [14]). R. Hähnle [13] discovered, that – for standard connectives – drastic reductions of the branching factors can be gained by taking sets of truth values instead of single truth values as signs of formulas. In fact, we made use of this fact in the case of prefixed signed formulas of type $\sigma^*: A: X$.

The important point here is that, after fixing the set of connectives and the set of possible signs, optimal rules can be computed by suitable programs.[6]

The strong analogy between distribution modalities and distribution quantifiers allows us to apply the same optimization techniques. (The case of non-serial logics where the empty distribution is allowed – in contrast to quantifier distributions, were this is excluded – does not pose principal problems.)

There are still many interesting open questions concerning the "right" choice of sets of signs for given (families of) logics. Fortunately, these optimization problems are largely independent from our subject. Here it suffices to remark that for implementations of our proof method, one should certainly employ the sets-as-signs paradigm and the various optimization algorithms cited above.

7 Examples

7.1 Expressing crispness

We already outlined in Section 2.3 how distribution modalities can be used to express "crispness" within a logic of graded truth. Having presented a generic

---

[6] In the current version MULTLOG only computes optimal *conjunctive* normal forms. The *disjunctive* normal forms needed for tableaux can be read of from MULTLOG's output if one specifies the "complementary" logic as input.

tableau calculus for such logics in the previous sections we are now in the position to demonstrate how a concrete statement about "crispness" can be proved. For this purpose we choose the three-valued Gödel logic [10] as a base logic that allows to evaluate statements to *true* (t), *false* (f), or *intermediate* (u), respectively. The truth tables for the standard connectives $\neg$, $\wedge$, $\vee$ and $\supset$ are as follows.

$\neg$	
f	t
u	f
t	f

$\wedge$	f	u	t
f	f	f	f
u	f	u	u
t	f	u	t

$\vee$	f	u	t
f	f	u	t
u	u	u	t
t	t	t	t

$\supset$	f	u	t
f	t	t	t
u	f	t	t
t	f	u	t

We augment this logic by the distribution modality **C**, where **C**X is intended to denote the statement "X is crisp"; or, more exactly, the statement "In all possible worlds X evaluates either to t or to f (but not to the intermediate value)". The truth function $\widetilde{\mathbf{C}}$ is defined by:

$$\widetilde{\mathbf{C}}(A) := \begin{cases} t & \text{if } A = \emptyset, \{t\}, \{f\} \text{ or } \{t,f\} \\ f & \text{otherwise} \end{cases}$$

Optimized tableau rules for the distribution modality **C** are given by:

$$\frac{\sigma\colon [f]\colon \mathbf{C}X}{\tau\colon [u]\colon X} \quad \text{where } \tau \text{ is an unrestricted simple extension of } \sigma$$

$$\frac{\sigma\colon [u]\colon \mathbf{C}X}{\text{closure}} \quad \text{closure rule } C_4 \ (\widetilde{\mathbf{C}}^{-1}(u) = \emptyset, \text{ see Definition 13})$$

$$\frac{\sigma\colon [t]\colon \mathbf{C}X}{\sigma^*\colon \{t,f\}\colon X}$$

Tableau rules for the connectives $\wedge$ and $\supset$ used in the sample proof below are given by:

$$\frac{\sigma\colon [u]\colon (X \supset Y)}{\sigma\colon [t]\colon X \quad \sigma\colon [u]\colon Y} \qquad \frac{\sigma\colon [u]\colon (X \wedge Y)}{\sigma\colon [t]\colon X \ \ \sigma\colon [u]\colon X \ \ \sigma\colon [u]\colon X}{\sigma\colon [u]\colon Y \ \ \sigma\colon [u]\colon Y \ \ \sigma\colon [t]\colon Y}$$

Remark. We used MULTLOG [3] to compute these rules.

To fix the appropriate conditions on the prefixes in applying the rules we have to choose defining properties for the accessibility relation. We choose a **K**-type relation; i.e., no restriction is imposed on accessibility.

Let us prove that the statement that "the disjunction of any two statements implies that both statements are crisp" is itself a crisp statement. Our modal extension of Gödel's logic allows us to express this sentence by the rather simple formula

$$\mathbf{C}((X \vee Y) \supset (\mathbf{C}X \wedge \mathbf{C}Y)).$$

To prove that this formula is valid – i.e. t in all models – we have to find closed tableaux for the prefixed signed formulas

$$\langle 1 \rangle \colon [f] \colon \mathbf{C}((X \vee Y) \supset (\mathbf{C}X \wedge \mathbf{C}Y))$$

and

$$\langle 1 \rangle \colon [u] \colon \mathbf{C}((X \vee Y) \supset (\mathbf{C}X \wedge \mathbf{C}Y)).$$

In the second case, the rule for $\mathbf{C}$ and u leads to immediate closure. For the first case we get the following:

```
(1)  ⟨1⟩: [f]: C((X ∨ Y) ⊃ (CX ∧ CY))
(2)  ⟨1,1⟩: [u]: ((X ∨ Y) ⊃ (CX ∧ CY))
(3)  ⟨1,1⟩: [t]: (X ∨ Y)
(4)  ⟨1,1⟩: [u]: (CX ∧ CY)
(5) ⟨1,1⟩:[t]:CX    (7) ⟨1,1⟩:[u]:CX    (9) ⟨1,1⟩:[u]:CX
(6) ⟨1,1⟩:[u]:CY    (8) ⟨1,1⟩:[u]:CY   (10) ⟨1,1⟩:[t]:CY
      closure            closure            closure
```

Comment. Line (2) is obtained from line (1) by the rule for $\mathbf{C}$ and f. Lines (3) and (4) are obtained from line (2) by the rule for $\supset$ and u. Lines (5) to (10) are obtained from line (4) by the rule for $\wedge$ and u. The left branch closes by applying the rule for $\mathbf{C}$ and u to line (6). The right branch closes by applying the rule for $\mathbf{C}$ and u to line (9). The branch branch in the middle closes by applying the same rule to either line (7) or line (8).

Remember that we did not impose any restriction on the accessibility relation. Therefore we have shown that $\mathbf{C}((X \vee Y) \supset (\mathbf{C}X \wedge \mathbf{C}Y))$ is valid in all (normal) extensions of Gödel's logic with modality $\mathbf{C}$.

7.2 Making Belnap's logic reflective

Belnap's four-valued logic [4] has repeatedly been suggested as a tool for reasoning about (possibly) inconsistent and incomplete information. The main intuition in this context is that a database may not only contain information that a certain statement is *false* or *true*, but such information may either be absent or over-determined in the sense that both, *true and false*, is assigned to some statement. The four possible states of knowledge are represented by the four truth values f (*false*), u (*undetermined*), $\bot$ (*inconsistent*), and t (*true*), respectively.

The truth functions for connectives $\neg$, $\wedge$, and $\vee$ are defined by:

¬		∧	f	u	⊥	t		∨	f	u	⊥	t
f	t	f	f	f	f	f		f	f	u	⊥	t
u	u	u	f	u	f	u		u	u	u	t	t
⊥	⊥	⊥	f	f	⊥	⊥		⊥	⊥	t	⊥	t
t	f	t	f	u	⊥	t		t	t	t	t	t

If, for every truth value a, we add unary connectives J_a such that $J_a X$ evaluates to t iff X evaluates to a, and to f otherwise, we can express inconsistency and under-determinedness within the language. However, observe that we still cannot "reason" within the logic about inconsistent or incomplete databases in a strong sense. For this it should be possible to express statements like

(a) the status of the database entry X remains stable under all possible updates, or
(b) X can never get over-determined or under-determined, or
(c) possible updates only add but never remove information about the truth of X.

Statements of this type become expressible if we "modalize" Belnap's logic. More exactly, we identify a database with a possible world in a Kripke-model. Accessible worlds represent possible updates.

To express "stability" in the sense of the statement (a) we introduce the distribution modality STABLE. We want to have: $\bar{v}(\Gamma, \text{STABLE}(X)) = t$ if for all Δ, Δ', such that $\Gamma \mathcal{R} \Delta$ and $\Gamma \mathcal{R} \Delta$, we have $\bar{v}(\Delta, X) = \bar{v}(\Delta', X)$, and $\bar{v}(\Gamma, X) = f$ in all other cases. Hence, the corresponding truth function is defined as follows:

$$\widetilde{\text{STABLE}}(A) = \begin{cases} t & A = \emptyset, \{f\}, \{u\}, \{\bot\} \text{ or } \{t\} \\ f & \text{otherwise} \end{cases}$$

Since STABLEX is intended to reflect a meta-linguistic, and therefore classical, statement about formulas of Belnap's logic with the object-language itself, it is not surprising that we specified it to evaluate either to t or f. However, one may want to express similar "reflective" statements that can be under- or over-determined, themselves. Consider the following modality that is intended to capture certain intuitions about the "conservativity" of statements with respect to updates.[7]

$A \subseteq N$	$\widetilde{\text{CON}}(A)$
$\emptyset, \{f\}, \{t\}$	t
$\{u\}, \{u, f\}, \{u, t\}$	u
$\{f, t\}, \{f, u, t\}, \{f, \bot, t\}, \{f, u, \bot, t\}$	f
$\{\bot\}, \{f, \bot\}, \{\bot, t\}, \{u, \bot\}, \{f, u, \bot\}, \{u, \bot\ t\}$	$\bot$

---

[7] We intend to clarify the semantical adequacy of this and many other types of distribution modalities for different many-valued base logics at another place. In particular, we plan to specify modalities with respect to appropriate mathematical models of databases and not just intuitions about them.

The defining properties of the accessibility relation $\mathcal{R}$ determine the structure of possible updates. E.g., symmetry of $\mathcal{R}$ means that updates can always be revoked. If one wants to call the result of any sequence of updates also an update of the original database then $\mathcal{R}$ has to be transitive. If at least one update is possible for every database than $\mathcal{R}$ is serial, etc.

Let us use our logical framework to investigate the stability of the claim that a statement is not conservative but stable (in the sense of Belnap's logic enriched by the distribution modality STABLE (stable) and CON (conservative) as defined above.) The corresponding formula is $F = \text{STABLE}(\neg\text{CON}X \wedge \text{STABLE}X)$. It is easy to find a model that evaluates F to false if the accessibility relation is not symmetric:

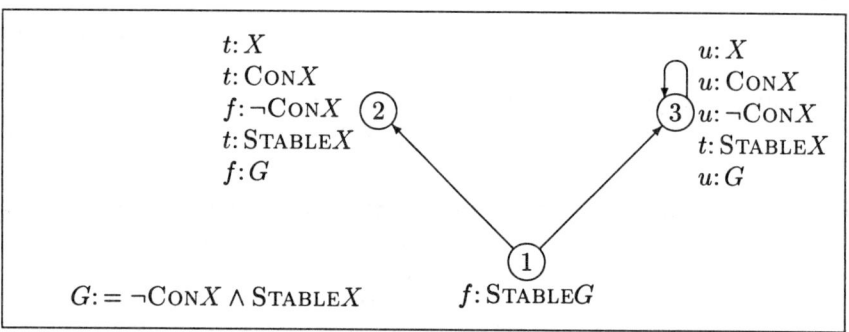

In contrast, a closed tableau for $\langle 1 \rangle: [f]: F$ can be constructed if we require the accessibility relation to satisfy the **S5**-property. Such a tableau is easy to generate using rules that can be read off from the truth tables as indicated in Section 3. However, it is quite large. Its explicit construction is left as an exercise for the industrious reader or — even better: to the future implementation of our proof search procedure.

References

1. Matthias Baaz and Christian G. Fermüller. Resolution-based theorem proving for many-valued logics. *Journal of Symbolic Computation*, 19:353–391, 1995.
2. Matthias Baaz and Christian G. Fermüller. Combining many-valued and intuitionistic tableaux. In *Theorem Proving with Analytic Tableaux and Related Methods (TABLEAUX'96)*, volume 1071 of *Lecture Notes in Artificial Intelligence*, pages 65–79. Springer, 1996.
3. Matthias Baaz, Gernot Salzer, Christian G. Fermüller, and Richard Zach. MUltlog 1.0: Towards an expert system for many-valued logics. In *13th Int. Conf. on Automated Deduction (CADE'96)*, volume 1104 of *Lecture Notes in Artificial Intelligence*, pages 226–230. Springer, 1996.
4. N. D. Belnap. A useful four-valued logic. In J.M. Dunn and G. Epstein, editors, *Modern Uses of Multiple-Valued Logic*, pages 8–37. D. Reidel Publishing Co., Boston, 1977.
5. Walter A. Carnielli. Systematization of finite many-valued logics through the method of tableaux. *Journal of Symbolic logic*, 52(2):437–493, 1987.

6. Melvin C. Fitting. *Proof Methods for Modal and Intuitionistic Logics*. D. Reidel Publishing Co., Dordrecht, 1983.
7. Melvin C. Fitting. Many-valued modal logics. *Fundamenta Informaticae*, 15:235–254, 1992.
8. Melvin C. Fitting. Many-valued modal logics II. *Fundamenta Informaticae*, 17:55–73, 1992.
9. Melvin C. Fitting. Tableaus for many-valued modal logic. *Studia Logica*, 15:63–87, 1995.
10. K. Gödel. Zum intuitionistischen Aussagenkalkül. *Anzeiger der Akademie der Wissenschaften in Wien*, 69:65–66, 1932.
11. Siegfried Gottwald. *Mehrwertige Logik*. Akademie-Verlag, Berlin, 1989.
12. Reiner Hähnle. Towards an efficient tableau proof procedure for multiple-valued logics. In *Computer Science Logic (CSL'90)*, volume 533 of *Lecture Notes in Computer Science*, pages 248–260. Springer, 1991.
13. Reiner Hähnle. *Automated Deduction in Multiple-valued Logics*. Clarendon Press, Oxford, 1993.
14. Reiner Hähnle. Commodious axiomatization of quantifiers in multiple-valued logic. In *Proc. 26th International Symposium on Multiple-valued Logic*, pages 118–123, Los Alamitos, 1996. IEEE Press.
15. Charles G. Morgan. Local and global operators and many-valued modal logics. *Notre Dame Journal of Formal Logic*, 20:401–411, 1979.
16. Osamu Morikawa. Some modal logics based on a three-valued logic. *Notre Dame Journal of Formal Logic*, 30:130–137, 1989.
17. Hans Jürgen Ohlbach. Translation methods for non-classical logics – an overview. *Journal of the IGPL*, 1(1):69–90, 1993.
18. Hans Jürgen Ohlbach. Computer support for the development and investigation of logics. Technical Report MPI-I-94-228, Max-Planck-Institut für Informatik, Saarbrücken, Germany, 1994.
19. Pascal Ostermann. Many-valued modal propositional calculi. *Zeitschrift für mathematische Logic und Grundlagen der Mathematik*, 34:343–354, 1988.
20. Pascal Ostermann. Many-valued modal logics: Uses and predicate calculus. *Zeitschrift für mathematische Logic und Grundlagen der Mathematik*, 36:367–376, 1990.
21. Nicholas Rescher. *Many-valued Logic*. McGraw-Hill, New York, 1969.
22. Gernot Salzer. Optimal axiomatizations for multiple-valued operators and quantifiers based on semi-lattices. In *13th Int. Conf. on Automated Deduction (CADE'96)*, volume 1104 of *Lecture Notes in Artificial Intelligence*, pages 226–230. Springer, 1996.
23. Peter K. Schotch, Jorgen B. Jensen, Peter F. Larsen, and Edwin J. MacLellan. A note on three-valued modal logics. *Notre Dame Journal of Formal Logic*, 19:63–68, 1978.
24. Krister Segerberg. Some modal logics based on a three-valued logic. *Theoria*, 33:53–71, 1967.
25. S. K. Thomason. Possible worlds and many truth values. *Studia Logica*, 37:195–204, 1978.

Some Remarks on Completeness, Connection Graph Resolution and Link Deletion*

Reiner Hähnle[1], Neil V. Murray[2], and Erik Rosenthal[3]

[1] Dept. of Computer Science, Univ. of Karlsruhe, 76128 Karlsruhe, Germany,
reiner@ira.uka.de
[2] Dept. of Computer Science, State Univ. of New York, Albany, NY 12222, USA,
nvm@cs.albany.edu
[3] Dept. of Mathematics, Univ. of New Haven, 300 Orange Avenue, West Haven, CT
06516, USA, brodsky@charger.newhaven.edu

Abstract. A new completeness proof that generalizes the Anderson-Bledsoe excess literal argument is developed for connection-graph resolution. The technique also provides a simplified completeness proof for semantic resolution. Some observations about subsumption and about link deletion are made. Link deletion is the basis for connection graphs. Subsumption plays an important role in most resolution-based inference systems. In some settings—for example, connection graphs in negation normal form—both subsumption and link deletion can be quite tricky. Nevertheless, a completeness result that uses both is obtained in this setting.

1 Introduction

Robinson developed semantic tree arguments [12] to provide completeness proofs for resolution and related inference systems. These arguments were not entirely transparent in that they do not directly construct a proof in the deduction system. Rather, they work with an intermediate data structure—semantic trees. Some novices find the resulting proofs difficult to follow. The excess literal technique discovered by Anderson and Bledsoe [1], which is almost completely syntactic, is a considerable simplification. This technique, which is essentially an induction on the size of the formula, was the basis for the first completeness proofs of certain refinements of resolution.

In this paper we continue work begun in [7, 6] and describe how the excess literal technique can be adapted to provide concise completeness proofs for a variety of deduction methods including for some non-clausal systems. In addition to providing simplified proofs of known results the role of subsumption in non-clausal proof systems is discussed, and a completeness proof for non-clausal connection-graph resolution is obtained.

---

* This research was supported in part by the National Science Foundation under grants CCR-9404338 and CCR-9504349 and by the Deutsche Forschungsgemeinschaft within Schwerpunktprogramm Deduktion.

Kowalski [8] introduced connection-graph (cg) resolution in 1975. It keeps track of all links—complementary pairs of literals—occurring in a clause set and employs deletion of the links used in a resolution step to reduce the size of the search space.

Fewer links made completeness questionable, and the semantic tree and excess literal techniques seemed to be insufficient. After six years, Bibel [3] finally proved cg-resolution to be complete. The crucial advance he made was the notion of spanning and the observation that cg-resolution steps preserve spanning (see Section 3). His approach is similar in some respects to the one presented here, but it requires a non-trivial double induction and relies on the existence of refutations that must conform to an excessively rigid structure.

Link deletion is sometimes possible with inference rules that do not rely on clause form. Path dissolution [11] and the tableau method implicitly delete links. Link deletion is also possible with semi-resolution [9], but proving that the spanning property is preserved is highly non-trivial. Some insight into link deletion in negation normal form is provided in Section 5.3.

Stickel [13] defined and implemented non-clausal cg-resolution, but he did not give a completeness proof, suspecting that it "may be difficult." Indeed, if one tries to generalize Bibel's proof to the non-clausal case, one quickly faces seemingly insurmountable technicalities. In this paper, a variation of the Anderson-Bledsoe excess literal technique is employed to prove completeness of non-clausal cg-resolution (Theorem 4). In addition, we show that the same technique can be used to provide simplified proofs of other (known) results.

In Section 2, a variation of the Anderson-Bledsoe excess literal technique for proving completeness of resolution is described. The method is illustrated with a succinct, straightforward completeness proof for semantic resolution. In Section 3, the proof technique is used to provide a simplified proof of the completeness of clausal cg-resolution. Some new observations regarding cg-subsumption are also noted. Some prerequisites on NNF formulas are given in Section 4, and subsumption is generalized to NNF in Section 5.2. Completeness of non-clausal cg-resolution is then proven in Section 5.3.

2 A Simple Completeness Proof for Semantic Resolution

The work described here was inspired by the Anderson-Bledsoe [1] excess literal proof of the completeness of resolution. It is essentially an induction on the size of a sentence $\mathcal{S}$. An interesting variation of their proof can be obtained by applying the induction to the number of distinct atoms that appear in $\mathcal{S}$. A good example is the proof below that semantic resolution is complete. Older proofs of this result are widely recognized as quite opaque. Bachmair & Ganzinger's proof [2] is well structured, but does not yield a direct, syntactic construction.

The proof here is for the ground case; it lifts in the usual way. We make the standard convention that a set of clauses is interpreted as the conjunction of its members, and that a clause is a set of literals, interpreted as the disjunction of its members. Thus defined, duplicate clauses in a sentence and duplicate literals in a clause are automatically "merged" into one copy.

Semantic resolution is defined with respect to a given interpretation I. Let S be a set of clauses, and let $\mathcal{F}$ be the subset of S whose members are falsified by I. (Regardless of the choice of I, $\mathcal{F}$ is non-empty if S is unsatisfiable.) Let $\mathcal{T} = S - \mathcal{F}$. Obviously, I satisfies all clauses in $\mathcal{T}$.

To define semantic resolution, let $N = \{\overline{b_1}, \ldots, \overline{b_n}\} \cup C$ be a clause in $\mathcal{T}$ in which I satisfies the $\overline{b_i}$'s and falsifies the literals in C, and suppose the clauses $B_1, B_2, \ldots, B_n$ are in $\mathcal{F}$ with $b_k \in B_k$. Suppose further that for any $b' \in B_i - \{b_i\}$, $b' \neq b_j, 1 \leq j \leq n$. Then the clause $R = C \cup (\bigcup_{i=1}^{n} B_i - \{b_i\})$ is the *semantic resolvent* of the *nucleus clause* N and the *satellite clauses* $B_1, \ldots, B_n$. Obviously, I falsifies R. That R can be soundly inferred from N and the B_i's can easily be seen by noting that a sequence of binary resolutions between N and the B_i's produces R. Semantically, any interpretation that satisfies the parent clauses either satisfies one literal in C (and thus R) or else some $\overline{b_i}$. But then I falsifies b_i and must satisfy $B_i - \{b_i\}$ (and thus R). Observe that hyperresolution may be regarded as a special case of semantic resolution by considering the interpretation that assigns false to every atom.

We begin by recalling the *pure rule*: A literal in a set of clauses is said to be *pure* if its complement does not occur in any other clause. In that case, the clause containing the pure literal is also said to be pure. The proof of the next lemma is straightforward.

Lemma 1 (Pure Rule). Let S be an unsatisfiable clause set in which the clause C is pure. Then $S' = S - \{C\}$ is unsatisfiable.

Lemma 2. Let $S = \{C_0, C_1, C_2, \ldots, C_k\}$ be a minimally unsatisfiable set of clauses (i.e., no proper subset of S is unsatisfiable), and suppose $C_0 = \{p\} \cup \{q_1, \ldots, q_n\}$, where $n \geq 0$. Obtain S' from S by deleting every occurrence of p in S. Then S' is unsatisfiable, and every minimally unsatisfiable subset of S' contains $C_0' = \{q_1, \ldots, q_n\}$ but contains no clauses that contain $\overline{p}$.

Proof. Every clause in S' subsumes a clause in S, so S' is unsatisfiable. By the pure rule, any minimally unsatisfiable subset of S' cannot contain any clauses that contain $\overline{p}$. Also, since S is minimally unsatisfiable, there must be an interpretation I_0 that falsifies C_0 but satisfies every other clause in S. Thus, I_0 satisfies every clause in S' other than C_0'. But then C_0' must be in any minimally unsatisfiable subset of S'. □

Theorem 1. Semantic resolution is refutation complete for propositional logic.

Proof. Let $S = \{C_1, C_2, \ldots, C_m\}$ be an unsatisfiable set of clauses. We assume that S is minimally unsatisfiable; otherwise, restrict attention to a minimally unsatisfiable subset. We must show that, given an arbitrary interpretation I, there is a refutation of S using semantic resolution with respect to I.

Proceed by induction on the number n of distinct atoms in S. If there are none, then S contains the empty clause, and we are done.

So suppose that all unsatisfiable sets of clauses with at most n atoms can be refuted with semantic resolution, and assume that S has $n+1$ atoms including the atom p. Let $q = p$ if I falsifies p, $q = \bar{p}$ otherwise. Since S is minimal, the Pure Rule implies that p cannot be pure, so both q and $\bar{q}$ occur in S.

If S contains the unit clause $\{q\}$, fine; otherwise, remove all occurrences of q from S. This formula is unsatisfiable by Lemma 2. Consider a minimally unsatisfiable subset; by Lemma 2, every clause that had contained q in S is in this set and no clause containing $\bar{q}$ is present. By the induction hypothesis, there is a refutation $\mathcal{R}_q$ by semantic resolution. Now apply that refutation to S; call the resulting refutation $\mathcal{R}'_q$. The effect is to reintroduce q into some clauses, so that, with merging, the clause $\{q\}$ rather than the empty clause may be produced by $\mathcal{R}'_q$.

Nevertheless, each step is a semantic resolution step. The reason is that I falsifies q, and so, by reintroducing q, the membership in $\mathcal{F}$ or $\mathcal{T}$ of the resulting clauses is unchanged. Also, none of the clauses containing $\bar{q}$ are resolved upon, and the result is either the empty clause[1] or the clause $\{q\}$. This clause is the last semantic resolvent and is in $\mathcal{F}$.

Analogously, if we begin by deleting $\bar{q}$, a refutation $\mathcal{R}_{\bar{q}}$ by semantic resolution can be found. Let the proof that results from applying $\mathcal{R}_{\bar{q}}$ to S be denoted by $\mathcal{R}'_{\bar{q}}$. This proof yields either the empty clause or the unit $\{\bar{q}\}$. However, it may not be a semantic resolution proof with respect to I, because reintroducing $\bar{q}$ into a member of $\mathcal{F}$ produces a clause satisfied by I and thus in $\mathcal{T}$. But a semantic resolution proof can be constructed from $\mathcal{R}'_{\bar{q}}$ using the unit $\{q\}$ derived by $\mathcal{R}'_q$.

Consider an arbitrary step in $\mathcal{R}_{\bar{q}}$ with nucleus $N = \{\bar{b_1}, \ldots, \bar{b_n}, c_1, \ldots, c_m\}$ and satellites $B_k \in \mathcal{F}$ with $b_k \in B_k, 1 \leq k \leq n$. Suppose that $\bar{q}$ is reintroduced into some of these clauses. First, if $\bar{q}$ is in N, simply use $\{q\}$ as an additional satellite, and it will be resolved away. If $\bar{q}$ is also reintroduced into some B_k, the resulting clause $B_k \cup \{\bar{q}\}$ is a member of $\mathcal{T}$ and cannot be a satellite. But we may first use it as a nucleus clause in a semantic resolution with satellite $\{q\}$, and the semantic resolvent is simply B_k. This in turn can be used in the step that employed nucleus N, and so that step is unchanged. Note that this construction assures that the last step that produced $\bar{q}$ in $\mathcal{R}'_{\bar{q}}$ now produces the empty clause.

Combining $\mathcal{R}'_q$ and the modified $\mathcal{R}'_{\bar{q}}$ produces the required semantic resolution proof. □

The above induction is somewhat reminiscent of the Davis-Putnam-Loveland procedure [5]: Refutations are obtained from the induction hypothesis by removing all occurrences of a given atom. A simplified completeness proof for connected CNF tableaux can also be obtained with this technique [7, 6].

3 Connection Graphs

Let S be a set of clauses. A *link* in S is a pair of complementary literals from different clauses, and a *c-path* through S is a set containing exactly one literal

---

[1] In fact, this cannot happen because of minimality, but this is not really relevant.

occurrence from each clause in $\mathcal{S}$. A c-path may also be thought of as a maximal conjunction of literal occurrences in $\mathcal{S}$ and corresponds to a clause in a DNF equivalent of $\mathcal{S}$. Any satisfying interpretation for $\mathcal{S}$ will satisfy every literal on some c-path of $\mathcal{S}$. Obviously, if $\mathcal{S}$ is unsatisfiable, no c-path can be satisfied, and thus every c-path will contain a link.

A link set containing every link in $\mathcal{S}$ is said to be *full*, and $\mathcal{S}$ is said to be *spanned* by a set of links $\mathcal{L}$ if every c-path of $\mathcal{S}$ contains a member of $\mathcal{L}$. If $\mathcal{S}$ is unsatisfiable, then it is spanned by its links; note, however, that a link set $\mathcal{L}$ may span $\mathcal{S}$ and yet not be full. Intuitively, if $\mathcal{L}$ spans $\mathcal{S}$, then $\mathcal{L}$ contains enough information to demonstrate the unsatisfiability of $\mathcal{S}$. A *connection graph* for $\mathcal{S}$ consists of $\mathcal{S}$ along with a set $\mathcal{L}$ of links from $\mathcal{S}$; it is denoted $\mathcal{G}(\mathcal{S}, \mathcal{L})$. We write $\mathcal{G}$ when the clause and link set are obvious from context. We say that $\mathcal{G}$ (or $\mathcal{S}$) is *minimally spanned* if removal of any clause and its associated links produces a graph that is not spanned.

Example 1. Consider the connection graph $\mathcal{G}(\mathcal{S}, \mathcal{L})$ with unsatisfiable clause set $\mathcal{S}$ consisting of $C = \{p, \bar{q}\}$, $D = \{\bar{p}, \bar{q}\}$, $E = \{q\}$ and full link set $\mathcal{L} = \{\{p_C, \bar{p}_D\}\} \{\bar{q}_C, q_E\} \{\bar{q}_D, q_E\}\}$. $\mathcal{G}$ is minimally spanned.

3.1 The Pure Rule and Subsumption

Care is required with links in a connection graph since a complementary pair of literal occurrences may not be in the link set. Thus, if p is a literal in the clause C, we often use p_C to denote the occurrence of p in C. Care is also required for the notion of purity: Suppose $\mathcal{G}(\mathcal{S}, \mathcal{L})$ is spanned and contains the clause $C = A \cup \{p\}$. Suppose further that no link in $\mathcal{L}$ contains the literal p_C. Then the literal occurrence p_C and the clause C are said to be *pure* in $\mathcal{G}$. The next lemma is the pure rule for connection graphs; its proof is straightforward.

Lemma 3 (Pure Rule). *Let $\mathcal{G}(\mathcal{S}, \mathcal{L})$ be spanned and contain the clause $C = A \cup \{p\}$, where p_C is pure. Then $\mathcal{G}' = \mathcal{G}(\mathcal{S}', \mathcal{L}')$ is spanned, where $\mathcal{S}' = \mathcal{S} - \{C\}$ and $\mathcal{L}'$ is the result of removing from $\mathcal{L}$ all links that meet clause C.*

In the next lemma, and in subsequent developments, objects are removed from the clauses of a connection graph. In such situations, it is implicitly assumed that links associated with the removed objects are also removed.

Lemma 4. *Let $\mathcal{S} = \{C_0, C_1, C_2, \ldots, C_k\}$ be a minimally spanned set of clauses, and suppose $C_0 = \{p\} \cup \{q_1, \ldots, q_n\}$, where $n \geq 0$. Obtain $\mathcal{S}'$ from $\mathcal{S}$ by deleting p from C_0, Then $\mathcal{S}'$ is spanned, and every minimally spanned subset of $\mathcal{S}'$ contains $C_0' = \{q_1, \ldots, q_n\}$ but contains no clauses that contain an unlinked $\bar{p}$. More generally, p may be removed from all clauses in $\mathcal{S}$ to produce a spanned set of clauses with one fewer atom.*

Proof. Removing a literal from a clause in $\mathcal{S}$ simply removes the c-paths containing that literal, so $\mathcal{S}'$ is spanned. By the pure rule, any minimally spanned subset of $\mathcal{S}'$ cannot contain pure clauses.

Observe that not all c-paths through $\mathcal{S} - \{C_0\}$ are linked since $\mathcal{S}$ is minimally spanned. Thus, there is a linkless c-path P_0 through $\mathcal{S} - \{C_0\}$. Since $\mathcal{S}$ is spanned, there is a link connecting every literal in C_0 to a literal in P_0. Since P_0 is also a c-path (without links) through $\mathcal{S}' - \{C_0'\}$, $\mathcal{S}' - \{C_0'\}$ is not spanned, and C_0' is in any minimally spanned subset of $\mathcal{S}'$.

Finally, successive applications of the first part of the lemma can be used to produce a minimally spanned set of formulas in which p has been completely removed. □

The classical notion of subsumption carries over to the connection-graph setting in a spanning-preserving manner [3]. It must be modified because a link set may not be full. Suppose connection graph $\mathcal{G}(\mathcal{S}, \mathcal{L})$ contains clauses C and D, and suppose that $C \subseteq D$; i.e., C "classically" subsumes D. For each literal p_C in C, we denote by $\mathcal{L}_{p_C}$ the set of literal occurrences linked to p_C in $\mathcal{L}$. Suppose further that for every literal p_C, $\mathcal{L}_{p_C} \supseteq \mathcal{L}_{p_D}$. Then we say clause C *cg-subsumes* clause D.

Lemma 5 (Cg-subsumption). Suppose that $\mathcal{G}(\mathcal{S}, \mathcal{L})$ is spanned and contains clauses C and D, and suppose that C cg-subsumes D. Then $\mathcal{G}'(\mathcal{S}', \mathcal{L}')$ is spanned, where $\mathcal{S}' = \mathcal{S} - \{D\}$, and $\mathcal{L}'$ is the result of removing from $\mathcal{L}$ all links that meet clause D.

We will subsequently investigate and further develop cg-subsumption. However, the tools already described are sufficient to produce a quite elegant completeness proof for cg-resolution.

3.2 Cg-resolution

To define connection graph resolution, let $\mathcal{G}(\mathcal{S}, \mathcal{L})$ be a connection graph containing clauses $C = \{p\} \cup A$ and $D = \{\bar{p}\} \cup B$. Suppose further that $\mathcal{L}$ contains the link $L = \{p_C, \bar{p}_D\}$. Then we may cg-resolve clauses C and D to produce the connection graph $\mathcal{G}'(\mathcal{S}', \mathcal{L}')$, where $\mathcal{S}' = \mathcal{S} \cup \{E\}$, $E = A \cup B$, and $\mathcal{L}' = \mathcal{L} - \{L\} \cup INH$. The set INH consists of *inherited links*. Intuitively, each literal in E comes from one in C or from one in D and so inherits the links the corresponding literal had in the parent clause. When the same literal occurs in both parents, the occurrences are merged, and the resolvent inherits the links of both. Formally,

$$INH = \{\{p_E, \bar{p}_Q\} \mid \{p_C, \bar{p}_Q\} \in \mathcal{L} \text{ or } \{p_D, \bar{p}_Q\} \in \mathcal{L}\}$$

Example 2. In $\mathcal{G}(\mathcal{S}, \mathcal{L})$ from Example 1 we may cg-resolve clauses C and D via $L = \{p_C, \bar{p}_D\}$ with resolvent $F = \{\bar{q}\}$ and inherited link set $INH = \{\{\bar{q}_F, q_E\}\}$. The resulting link set $\mathcal{L}'$ spans $\mathcal{S}' = \mathcal{S} \cup \{F\}$, but $\mathcal{G}'(\mathcal{S}', \mathcal{L}')$ is not minimally spanned, since $\{E, F\} \subset \mathcal{S}'$ is spanned by INH.

The link set in the example is not full after only one cg-resolution step. Nevertheless, spanning is preserved. It is not difficult to prove but was quite difficult to see; it was the key observation that Bibel made that led to the first completeness proof of cg-resolution. Below a new proof is presented that is included here because it is shorter and simpler than Bibel's original proof, and because it will make it easier for the reader to follow the proof of the NNF case in Section 5.3. Note, however, that the new proof relies heavily on Bibel's lemma:

Lemma 6 (Bibel [3]). Suppose that a link in the spanned connection graph $\mathcal{G}(\mathcal{S}, \mathcal{L})$ is selected and cg-resolution produces the connection graph $\mathcal{G}'(\mathcal{S}', \mathcal{L}')$. Then $\mathcal{G}'$ is spanned, i.e., $\mathcal{S}'$ is spanned by $\mathcal{L}'$.

Theorem 2. Connection-graph resolution is refutation complete for propositional logic.

Proof. Let $\mathcal{G}(\mathcal{S}, \mathcal{L})$ be a connection graph spanned by $\mathcal{L}$; we must show that there is a refutation of $\mathcal{S}$ using cg-resolution. We proceed by induction on the number n of distinct atoms in $\mathcal{S}$. If there are none, then $\mathcal{S}$ contains the empty clause, and we are done. Otherwise, assume that all spanned connection graphs with at most n atoms can be refuted with cg-resolution, and suppose that $\mathcal{S}$ has $n + 1$.

If $\mathcal{S}$ is not minimally spanned, then restrict attention to a minimally spanned subset. Let p be any atom in $\mathcal{S}$; we begin by deriving the unit clause $\{p\}$ from $\mathcal{S}$. If $\mathcal{S}$ contains $\{p\}$, we have it; otherwise, by Lemma 4, we can remove all occurrences of p in $\mathcal{S}$, producing the spanned $\mathcal{G}'(\mathcal{S}', \mathcal{L}')$. By the induction hypothesis, there is a refutation of $\mathcal{S}'$. This refutation applied to $\mathcal{S}$ produces either the empty clause, in which case the proof is complete, or the unit clause p, as promised.

The unit $\{p\}$ cannot be pure because no occurrence of p in $\mathcal{S}$ was pure, and no resolution step involved any clause from $\mathcal{S}$ that contained $\bar{p}$. In particular, there are no new clauses containing $\bar{p}$. Let $C = \{\bar{p}\} \cup C'$ be a clause that is linked to p. The cg-resolvent of $\{p\}$ and C is C', which cg-subsumes C since no link to C' has been deleted. Thus, a minimally spanned subset of the resulting connection graph is a spanned clause set in which the number of occurrences of $\bar{p}$ is reduced.

We now iterate this process, successively reducing the number of occurrences of $\bar{p}$ while preserving the spanning property. Eventually a spanned graph with no occurrences of $\bar{p}$ is produced. By the pure rule, a minimally spanned subset contains no occurrences of p. Thus, we have a spanned connection graph with at most n distinct atoms. By the induction hypothesis, there is a refutation of the empty clause. □

Remark. In [3], Bibel uses a somewhat restricted inheritance rule to further reduce the number of links. The above proof applies with that inheritance rule virtually without modification.

3.3 Cg-subsumption

In the proof of Theorem 2, we made rather straightforward use of cg-subsumption. The requirements on the links of the clauses involved are necessary to ensure that spanning is preserved. In fact, whenever C subsumes D, C can be forced to cg-subsume D by performing enough cg-resolutions on D. For example, if the literal p occurs in both clauses, and if p_D is linked to $\overline{p_E}$ but p_C is not, then cg-resolve on the $\{p_D, \overline{p_E}\}$ link. This deletes the link; with enough cg-resolutions involving a subsumed clause, excess links that disallow cg-subsumption can be deleted, and the subsumed clause will be cg-subsumed.

This offers the advantage of deleting the subsumed clause, but the penalty is severe: A number of extra clauses are introduced into the search space. If these extra clauses turn out to be necessary for a proof, then the penalty is really no penalty at all. However, this seems unintuitive. On the other hand, if the extra cg-resolutions are avoided, the subsumed clause must be kept.

There may be a better approach. Suppose that in connection graph $\mathcal{G}$, clause C subsumes clause D but does not cg-subsume D. Suppose further that for each literal p_C in C such that $\mathcal{L}_{p_C} \not\supseteq \mathcal{L}_{p_D}$, we add to $\mathcal{L}_{p_C}$ exactly those links required to ensure that $\mathcal{L}_{p_C} \supseteq \mathcal{L}_{p_D}$. By cg-subsumption, D may now be deleted from the graph. We say that D has been removed by *augmented cg-subsumption*.

First observe that augmented cg-subsumption preserves spanning. Adding links to the subsuming clause cannot harm the spanning property, nor can the subsequent removal of the cg-subsumed clause. However, we would like to know that we have left completeness unaffected and not increased either the proof or the search space.

Investigation of the proof and search spaces is beyond the scope of this paper and left for future work. Completeness is, however, easily settled. First, cg-subsumption is a special case of augmented cg-subsumption; so the current proof of Theorem 2 goes through with the latter. But that proof could in fact be greatly simplified: All occurrences of literal p could be removed at once, and a derivation of (at least one) unit clause $\{p\}$ results. *All* other clauses containing p are then deleted by augmented cg-subsumption. As a result, *all* clauses with $\overline{p}$ are either already pure or become so after being resolved with the unit. Then the unit is pure. Voilá: The induction hypothesis provides a refutation.

4 Negation Normal Form

Formulas in *negation normal form* (NNF) are defined inductively: (i) Literals, true, false are NNF formulas; (ii) if $\mathcal{F}_1, \ldots, \mathcal{F}_m$ are NNF formulas, so are $\mathcal{F}_1 \wedge \cdots \wedge \mathcal{F}_m$ and $\mathcal{F}_1 \vee \cdots \vee \mathcal{F}_m$. We identify formulas that are equal up to associativity.

The *subformulas* of an NNF formula $\mathcal{G}$ are defined as follows: (i) Literals have only themselves as subformulas; (ii) if $\mathcal{G} = \mathcal{F}_1 \circ \cdots \circ \mathcal{F}_m$, where $\circ \in \{\wedge, \vee\}$, then for any $\{i_1, \ldots, i_r\} \subseteq \{1, \ldots, m\}$, $\mathcal{F}_{i_1} \circ \cdots \circ \mathcal{F}_{i_r}$ is a subformula of $\mathcal{G}$; (iii) the subformula relation is transitive.

If p and q are literals in an NNF formula $\mathcal{F}$, and if $p \in \mathcal{G}$ and $q \in \mathcal{H}$ where $\mathcal{G} \wedge \mathcal{H}$ is a subformula of $\mathcal{F}$, then p and q are said to be *c-connected*; if $\mathcal{G}$ and $\mathcal{H}$ are disjoined, then p and q are said to be *d-connected*. A *link* is a complementary pair of c-connected literals. Unless otherwise stated, we will assume that for any formula, the following simplification rules have been applied; in them, p is a literal and $\mathcal{S}$ is an arbitrary formula.

$$p \vee \cdots \vee p = p \wedge \cdots \wedge p = p \qquad p \vee \neg p = \text{true} \qquad p \wedge \neg p = \text{false}$$
$$\mathcal{S} \vee \text{false} = \mathcal{S} \wedge \text{true} = \mathcal{S} \qquad \mathcal{S} \wedge \text{false} = \text{false} \qquad \mathcal{S} \vee \text{true} = \text{true}$$

A *c-path through* $\mathcal{F}$ is a maximal set of mutually c-connected literals, and a *d-path through* $\mathcal{F}$ is a maximal set of mutually d-connected literals. The c-paths of a formula correspond to the clauses of one of its disjunctive normal form (DNF) equivalents. Similarly, the d-paths correspond to the clauses of a CNF equivalent. If $\mathcal{F} = \mathcal{G} \wedge \mathcal{H}$, and if P and Q are c-paths through $\mathcal{G}$ and $\mathcal{H}$, respectively, then $P \cup Q$ is a c-path through $\mathcal{F}$; it is denoted PQ. (If P and Q are d-paths through $\mathcal{G}$ and $\mathcal{H}$, respectively, then P and Q are each a d-path through $\mathcal{F}$.) The next lemma (from [10]) is easy to prove.

Lemma 7. A formula is satisfiable if and only if some c-path in it is satisfiable, and a c-path is satisfiable if and only if it does not contain a link.

We use *a set* of NNF formulas to denote the conjunction of the its members. In particular, we assume the formulas in the set to be either disjunctions or literals. Thus, complementary literals residing in different formulas are in fact c-connected and constitute a link. Links may also reside within a single formula. As in the clausal case, a spanned set of NNF formulas is *minimally spanned* if removing any member produces a set that is not spanned.

Our completeness arguments require the removal of syntactic objects from formulas in a spanning preserving way. This is more complicated for NNF formulas than it is for clause form, where all that is necessary is the removal of literal occurrences from the clauses in which they appear. One way to remove an object $\mathcal{P}$ from a spanned formula $\mathcal{S}$ to produce a spanned formula $\mathcal{S}_\mathcal{P}$ is to remove all c-paths containing $\mathcal{P}$. Since all c-paths of $\mathcal{S}$ contain a link, and since the c-paths of $\mathcal{S}_\mathcal{P}$ will be a subset of the c-paths of $\mathcal{S}$, $\mathcal{S}_\mathcal{P}$ will also be spanned. It turns out that a good choice for $\mathcal{S}_\mathcal{P}$ is the *c-path complement* of $\mathcal{P}$ in $\mathcal{S}$, denoted $CC(\mathcal{P}, \mathcal{S})$. We use a definition of $CC(\mathcal{P}, \mathcal{S})$ which is tailored to the special case required in this paper; for the general case, see [11].

Definition 1. *Let* $\mathcal{H}$ *be a subformula of the formula* $\mathcal{G}$. *Then the* c-path complement *of* $\mathcal{H}$ *with respect to* $\mathcal{G}$, *denoted* $CC(\mathcal{H}, \mathcal{G})$, *is the subformula that results from replacing* $\mathcal{H}$ *by* false *and making obvious simplifications.*

Alternatively, $CC(\mathcal{H}, \mathcal{G})$ can be simply characterized with the CE operator defined as follows. Let $\mathcal{P}$ be a subformula of an NNF formula $\mathcal{G}$. The *c-extension* $\mathcal{P}$, denoted $CE(\mathcal{P})$ is the largest subformula of $\mathcal{G}$ of the form $\mathcal{P} \wedge \mathcal{P}'$. The *d-extension* is denoted $DE(\mathcal{P})$ and is similarly defined. Observe that one of $DE(\mathcal{P})$ and $CE(\mathcal{P})$ must consist of $\mathcal{P}$ alone, and the other one cannot (unless $\mathcal{F} = \mathcal{P}$). The following Theorem is proved in [11].

Theorem 3. *If* $\mathcal{P}$ *is a subformula of formula* $\mathcal{S}$ *then* $CC(\mathcal{P}, \mathcal{S}) = \mathcal{S} - CE(\mathcal{P})$.

The pure rule is valid for NNF formulas [11] but requires some care. In essence, the NNF equivalent of the clause containing a given literal p is $DE(p)$.

Lemma 8 (Pure Rule for Negation Normal Form). If p is a pure literal in an unsatisfiable NNF formula S, then $S - DE(p)$ is unsatisfiable.

The previous lemma is adapted to the cg setting in a straightforward way:

Lemma 9 (Pure Rule for NNF Connection Graphs). If p is a pure literal in a spanned NNF formula S, then $S - DE(p)$ is spanned.

Proof. Removal of $DE(p)$ cannot harm spanning unless for some c-path Q, all the links on Q meet $DE(p)$. We may write Q as $Q'Q_{DE(p)}$. Note that $DE(p)$ is a disjunction, and $Q_{DE(p)}$ is a c-path through one of its disjuncts other than p. But then $Q'p$ is a linkless c-path through S, contrary to hypothesis. □

Lemma 10. Let $S = \{\mathcal{F}_0, \mathcal{F}_1, \mathcal{F}_2, \ldots, \mathcal{F}_k\}$ be a minimally spanned set of NNF formulas, and p a literal in $\mathcal{F}_0$. Obtain S' from S by removing $CE(p)$ for every occurrence of p in $\mathcal{F}_0$. Obtain S'' by applying the pure rule to S' (possibly repeatedly). $\mathcal{F}'_i$ *are the formulas in* S' *corresponding to* $\mathcal{F}_i$ *in* S.

1. *Both S' and S'' are spanned;*
2. $\mathcal{F}'_0$ *is a member of any minimally spanned subset of* S'.

Proof. 1. Removal of $CE(p)$ removes c-paths from the formula (Theorem 3). Since S is spanned, every c-path contains a link, and that is unchanged; S'' is spanned by Lemma 9.

2. Let P_0 be a linkless c-path through $\{\mathcal{F}_1, \ldots, \mathcal{F}_k\}$. Such a P_0 must exist since otherwise S would not be minimally spanned. However, since S is spanned, every extension of P_0 through $\mathcal{F}_0$ must contain a link. Since P_0 is also an unlinked c-path through $S' - \mathcal{F}'_0$, $S' - \mathcal{F}'_0$ is not spanned. Thus, any spanned subset of S' must contain $\mathcal{F}'_0$. □

5 Non-clausal Resolution

We begin by providing a precise definition of non-clausal (NC) resolution on literals for NNF formulas. Let $\mathcal{F}$ and $\mathcal{G}$ be arbitrary ground NNF formulas, where p is an atom occurring in both $\mathcal{F}$ and $\mathcal{G}$. We denote by $\mathcal{F}[p/\beta]$ the result of replacing all occurrences of p in $\mathcal{F}$ by β. If $\beta =$ true or $\beta =$ false, we assume that simplifications are performed. Then the *NC-resolvent* of $\mathcal{F}$ and $\mathcal{G}$ on the atom p is:

$$\mathcal{F}[p/\text{false}] \vee \mathcal{G}[p/\text{true}] .$$

5.1 Polarity

Although the definition above is symmetric with respect to $\mathcal{F}$ and $\mathcal{G}$, we have the following *polarity restriction*:

If p occurs only positively in $\mathcal{F}$ or only negatively in $\mathcal{G}$, then we need not consider the dual NC-resolvent

$$\mathcal{F}[p/\text{true}] \vee \mathcal{G}[p/\text{false}] .$$

The atom p occurs positively in $\mathcal{F}$ if it occurs as the literal p; p occurs negatively in $\mathcal{F}$ if it occurs in the literal $\bar{p}$. Note that p may occur both positively and negatively in $\mathcal{F}$.

5.2 NNF subsumption

In the discussion that follows, we will often refer to subsumption of d- and c-paths rather than of disjuncts and conjuncts. Paths are defined as sets of literal occurrences, but with regard to subsumption, we consider the literal set of a path P. In this way, no change in the basic definitions is needed. Clausal subsumption is generalized to the NNF case in a fairly straightforward manner.

Recalling that the d-paths of a formula correspond to the clauses of a CNF equivalent, we say that $\mathcal{F}$ *d-subsumes* $\mathcal{F}'$ if for every non-tautological d-path P' in $\mathcal{F}'$, there is a d-path P in $\mathcal{F}$, such that $P \subseteq P'$. The following lemma is easy.

Lemma 11. Let $\mathcal{S}$ be an unsatisfiable set of NNF formulas in which $\mathcal{F}$ d-subsumes $\mathcal{F}'$. Then $\mathcal{S} - \{\mathcal{F}'\}$ is unsatisfiable.

Clearly, if $\mathcal{F}$ d-subsumes $\mathcal{F}'$, then $\mathcal{F} \models \mathcal{F}'$. There is a dual (but not equivalent) syntactic characterization of subsumption for NNF formulas. We say that $\mathcal{F}$ *c-subsumes* $\mathcal{F}'$ if for every non-contradictory c-path P in $\mathcal{F}$, there is a c-path P' in $\mathcal{F}'$ such that $P \supseteq P'$. Of course, if $\mathcal{F}$ c-subsumes $\mathcal{F}'$, then $\mathcal{F} \models \mathcal{F}'$.

Lemma 12. Let $\mathcal{S}$ be an unsatisfiable set of NNF formulas in which $\mathcal{F}$ c-subsumes $\mathcal{F}'$. Then $\mathcal{S} - \{\mathcal{F}'\}$ is unsatisfiable.

To see that neither of these subsumption rules capture implication completely, consider $\mathcal{F}_1 = p$, $\mathcal{F}_2 = (p \vee q)$, $\mathcal{F}_3 = ((p \vee q) \wedge (r \vee \bar{r}))$. It is easy to see that $\mathcal{F}_1$ implies both $\mathcal{F}_2$ and $\mathcal{F}_3$. But $\mathcal{F}_1$ both d- and c-subsumes $\mathcal{F}_2$; $\mathcal{F}_1$ also d-subsumes $\mathcal{F}_3$ but does not c-subsume it.

The conditions of Lemmas 11 and 12 are quite strong; they are in fact necessary if we wish to delete an entire formula from a set of formulas. However, subsumption can more usefully be generalized further, so that more modest reductions can be captured. Lemma 13 below captures as a special case, Lemma 12. First, we define *c-reduction* for NNF formulas.

Given formulas $\mathcal{F}$ and $\mathcal{F}'$, suppose there is a subformula $\mathcal{H}'$ of $\mathcal{F}'$, such that $\mathcal{F}$ c-subsumes $\mathcal{H}'$. Then we say that $\mathcal{F}$ *c-reduces* $\mathcal{F}'$ via $\mathcal{H}'$.

Lemma 13. Let $\mathcal{S}$ be an unsatisfiable set of NNF formulas in which $\mathcal{F} \in \mathcal{S}$ c-reduces $\mathcal{F}'$ via $\mathcal{H}'$. Then $\mathcal{S}_{\mathcal{H}'} = (\mathcal{S} - \{\mathcal{F}'\}) \cup \{\mathcal{F}' - DE(\mathcal{H}')\}$ is unsatisfiable.

Proof. Suppose interpretation I satisfies $\mathcal{S}_{\mathcal{H}'}$. Then I satisfies some c-path P_I of $\mathcal{S}$, and P_I contains a c-path through $\mathcal{F}$. If P_I does not meet $DE(\mathcal{H}')$, then P_I is a c-path through $\mathcal{S}$, and I satisfies $\mathcal{S}$.

If P_I meets $DE(\mathcal{H}')$, then P_I may be written $PP_{DE(\mathcal{H}')}P_{\mathcal{F}}$, where $P_{DE(\mathcal{H}')}$ is P_I restricted to $DE(\mathcal{H}')$, and $P_{\mathcal{F}}$ is P_I restricted to $\mathcal{F}$. Since $\mathcal{F}$ c-subsumes $\mathcal{H}'$, $P_{\mathcal{F}} \supseteq P'$, for some c-path P' of $\mathcal{H}'$. Obviously, I satisfies the literals of P' and thus satisfies $PP'P_{\mathcal{F}}$. But this latter c-path is a c-path through $\mathcal{S}$, and so I satisfies $\mathcal{S}$. □

Of course, the dual notion of *d-reduction* is defined in the obvious way, and the dual of Lemma 13 holds. To adapt these notions to the connection-graph setting, it is most convenient to restrict attention to c-reduction.

Suppose $\mathcal{F}$ c-reduces $\mathcal{F}'$ via $\mathcal{H}'$. Suppose furthermore that for each non-contradictory c-path P' in $\mathcal{H}'$, each literal p is linked to a subset of the literals linked to occurrences of p on $P \supseteq P'$ for some non-contradictory c-path P in $\mathcal{F}$. Then we say that $\mathcal{F}$ *cg-c-reduces* $\mathcal{F}'$ via $\mathcal{H}'$.

Lemma 14. Let $\mathcal{S}$ be a spanned set of NNF formulas in which $\mathcal{F} \in \mathcal{S}$ cg-c-reduces $\mathcal{F}'$ via $\mathcal{H}'$. Then $\mathcal{S}_{\mathcal{H}'} = (\mathcal{S} - \{\mathcal{F}'\}) \cup \{\mathcal{F}' - DE(\mathcal{H}')\}$ is spanned.

Proof. Suppose for some c-path P through $\mathcal{S}$, every link on P meets $DE(\mathcal{H}')$. We may write $P = P_1 P_{DE(\mathcal{H}')} P_{\mathcal{F}}$, where $P_{DE(\mathcal{H}')}$ is the part of P in $DE(\mathcal{H}')$ and $P_{\mathcal{F}}$ is the part of P in $\mathcal{F}$. Since $\mathcal{F}$ c-subsumes $\mathcal{H}'$, $P_{\mathcal{F}} \supseteq P'$, for some c-path P' of $\mathcal{H}'$. The c-path $P_1 P' P_{\mathcal{F}}$ is a c-path of $\mathcal{S}$ and contains a link. But this link must be to P'. But then $P_{\mathcal{F}}$ is also linked to the same literal by the definition of c-reduction, contrary to the hypothesis that every link of P was to $DE(\mathcal{H}')$. □

As with augmented cg-subsumption for CNF formulas, we may define *augmented c-reduction* for NNF formulas. Suppose that in NNF connection graph $\mathcal{G}$, formula $\mathcal{F}$ c-reduces formula $\mathcal{F}'$ via $\mathcal{H}'$ but does not cg-c-reduce $\mathcal{F}'$. Suppose further that for each literal $p_{\mathcal{F}}$ on c-path $P_{\mathcal{F}}$, corresponding to literal $p_{\mathcal{F}'}$ on c-path $P_{\mathcal{F}'} \subseteq P_{\mathcal{F}}$, such that $\mathcal{L}_{p_{\mathcal{F}}} \not\supseteq \mathcal{L}_{p_{\mathcal{F}'}}$, we add to $\mathcal{L}_{p_{\mathcal{F}}}$ exactly those links required to ensure that $\mathcal{L}_{p_{\mathcal{F}}} \supseteq \mathcal{L}_{p_{\mathcal{F}'}}$. We may now apply cg-c-reduction. We say that $\mathcal{F}'$ has been *augmented cg-c-reduced*.

Of course, augmented cg-c-reduction preserves spanning. Adding links to the c-reducing formula cannot harm the spanning property, nor can the subsequent application of cg-c-reduction. As in the clausal case, we believe that completeness is unaffected, and neither the proof nor the search space have increased. As we have said, the NNF case is complex; an in depth investigation of completeness and of the proof and search spaces is beyond the scope of this paper and left for future work. In the next subsection, we do take some preliminary steps in this direction.

5.3 Non-clausal connection-graph resolution

To define cg-resolution for NNF formulas, let $\mathcal{G}(\mathcal{S}, \mathcal{L})$ be a connection graph containing NNF formulas $\mathcal{E}$ and $\mathcal{F}$, and suppose the atom p occurs positively in $\mathcal{E}$ and negatively in $\mathcal{F}$. Suppose further that $\mathcal{L}$ contains a non-empty set L of links of the form $\{p_{\mathcal{E}}, \bar{p}_{\mathcal{F}}\}$. Then we may cg-resolve formulas $\mathcal{E}$ and $\mathcal{F}$ to produce the connection graph $\mathcal{G}'(\mathcal{S}', \mathcal{L}')$, where $\mathcal{S}' = \mathcal{S} \cup \{\mathcal{H}\}$, and

$$\mathcal{H} = \mathcal{E}[p/\text{false}] \vee \mathcal{F}[p/\text{true}] .$$

The resulting set of links $\mathcal{L}'$ is defined as follows: If p occurs only positively in $\mathcal{E}$ and only negatively in $\mathcal{F}$, then $\mathcal{L}' = \mathcal{L} - L \cup INH$; otherwise, $\mathcal{L}' = \mathcal{L} \cup INH$. The set *INH* consists of *inherited links*. As in the CNF case, each literal in $\mathcal{H}$ comes from one in $\mathcal{E}$ or from one in $\mathcal{F}$ and so inherits the links the corresponding literal had in the parent formula. Formally,

$$INH = \{\{p_{\mathcal{H}}, \bar{p}_Q\} \mid \{p_{\mathcal{E}}, \bar{p}_Q\} \in \mathcal{L} \text{ or } \{p_{\mathcal{F}}, \bar{p}_Q\} \in \mathcal{L}\} .$$

Example 3. Consider the NNF connection graph consisting of four numbered formulas in Fig. 1.

The graph is fully linked except for $\{p_2, \bar{p}_4\}$, which is not present. (In the figure to the right, links are denoted by arcs connecting the linked literals). The remaining links span the graph.

If we cg-resolve on $\{p_3, \bar{p}_4\}$ (highlighted by the dashed arc), the resolvent is (false ∨ (true ∧ r)) ∨ (false ∨ q). Simplifying the latter yields (5) $r \vee q$.

Observe that the activated link must not be deleted, because p has mixed polarity in (3).

(1)
(2)
(3)
(4)

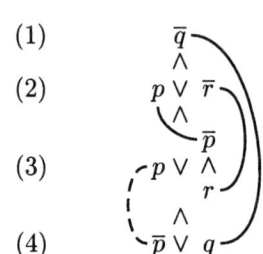

Fig. 1. NNF connection graph.

The careful reader will have noticed that whenever the atom resolved upon has mixed polarity in either parent formula, no links are deleted. This may at first seem surprising; however, past studies have shown that link deletion in the NNF setting is complicated. See, for example, [10]. More to the point, this restriction is crucial to the spanning-preservation lemma below. To see this, consider again Example 3 (Fig. 1). If the activated link were deleted, the c-path $\{\bar{q}_1, p_2, p_3, \bar{p}_4, r_5\}$ would be unlinked. In fact, successive applications of the Pure Rule would remove every formula from the graph, resulting in the (satisfiable!) empty conjunction.

Lemma 15. Suppose that a link $\{p_\mathcal{E}, \bar{p}_\mathcal{F}\}$ between formulas $\mathcal{E}$ and $\mathcal{F}$ in the NNF connection graph $\mathcal{G}(\mathcal{S}, \mathcal{L})$ is selected and cg-resolution produces the connection graph $\mathcal{G}'(\mathcal{S}', \mathcal{L}')$. Then $\mathcal{G}'$ is spanned, i.e., $\mathcal{S}'$ is spanned by $\mathcal{L}'$.

Proof. If p occurs negatively in $\mathcal{E}$ or positively in $\mathcal{F}$, then no links are deleted and the result is immediate. Otherwise, let the cg-resolvent be denoted by $\mathcal{H}$. Consider a c-path P through $\mathcal{G}$, for which the only links on P are in L. We must show that every extension of P through $\mathcal{H}$ is linked.

Observe that the cg-resolvent $\mathcal{H}$ is constructed as the disjunction of $\mathcal{E}'$ and $\mathcal{F}'$, where $\mathcal{E}'$ has all c-paths in $\mathcal{E}$ that do not contain p, and $\mathcal{F}'$ has all c-paths in $\mathcal{F}$ that do not contain $\bar{p}$. (All literals containing p are replaced by false, and this amounts to computing the c-path complement of those occurrences; see Theorem 3.) All such c-paths in $\mathcal{H}$ have isomorphic c-paths in either $\mathcal{E}$ or in $\mathcal{F}$ from which a link is inherited.

More precisely, let $P = P' P_\mathcal{E} P_\mathcal{F}$ be a c-path containing only links of the form $\{p_\mathcal{E}, \bar{p}_\mathcal{F}\}$, which are deleted in the cg-resolution. Without loss of generality, extend P through $\mathcal{E}'$ in $\mathcal{H}$ to $PP_\mathcal{H} = P' P_\mathcal{E} P_\mathcal{F} P_{\mathcal{E}'}$. Let $P^\mathcal{E}$ be the c-path in $\mathcal{E}$ isomorphic to $P_{\mathcal{E}'}$. Since $\mathcal{G}$ is spanned, the c-path $P' P^\mathcal{E} P_\mathcal{F}$ must have a link to $P^\mathcal{E}$. But this link is inherited on $P_{\mathcal{E}'}$. □

Theorem 4. Non-clausal connection-graph resolution is refutation complete for propositional logic.

Proof. Let $\mathcal{G}(\mathcal{S}, \mathcal{L})$ be an NNF formula spanned by $\mathcal{L}$; we must show that there is a refutation of $\mathcal{S}$ using cg-resolution. Proceed by induction on the number n

of distinct atoms in S. If there are none, then S contains false, and we are done. Otherwise, assume that all spanned connection graphs with at most n atoms can be refuted with cg-resolution, and suppose that S has $n+1$.

If S is not minimally spanned, then restrict attention to a minimally spanned subset. Let p be any atom in S; we begin by deriving the unit formula $\{p\}$ from S. If S contains the unit $\{p\}$, fine. Otherwise, successive applications of Lemma 10 produces a minimally spanned set of formulas S' in which all occurrences of $CE(p)$ and $DE(\bar{p})$ have been removed.

By the induction hypothesis, there is a refutation $\mathcal{R}$ by cg-resolution. Were we dealing with clauses, we could argue that if $\mathcal{R}$ is applied to S, either false or the literal p would be derived. However, in the NNF case, this is not obvious. The reason is that a given formula $\mathcal{F}_i$ of S may contain several occurrences of p and of $\bar{p}$; the c-extension of each occurrence of p and the d-extension of each occurrence of $\bar{p}$ are missing from the corresponding $\mathcal{F}'_i$'s in S'. The effect of $\mathcal{R}$ on each of these structures must be considered when applying $\mathcal{R}$ to S.

Consider first the c-extensions of p. The steps of $\mathcal{R}$ can change $\mathcal{F}_i$ outside $CE(p)$ and can change $CE(p)$ itself. Changes outside $CE(p)$ either leave it unchanged or delete it in its entirety due to a simplification rule. Changes inside $CE(p)$ can also delete it entirely. The other possibility is that what remains is a formula of the form $p \wedge \psi$. The point is, if $CE(p)$ is not completely deleted, then a formula of the form $p \wedge \psi$ remains.

The d-extensions of $\bar{p}$ are disjunctions. When $\bar{p}$ occurs in $\mathcal{F}_i$, $DE(\bar{p})$ is either $\mathcal{F}_i$ itself or a proper subformula of $\mathcal{F}_i$. In the former case, $\mathcal{F}_i$ would have been removed in forming S' and not have participated in $\mathcal{R}$. Otherwise, $DE(\bar{p})$ is one conjunct in some conjunction of $\mathcal{F}_i$. This conjunction is eliminated in $\mathcal{R}$ and will still be eliminated when $\mathcal{R}$ is applied to S, unless it also contains p; in that case the conjunction is simply one of the c-extensions of p that remain as discussed above.

In any case, the result of applying $\mathcal{R}$ to S is either false, in which case we are done, or a formula set S'' made up of formulas of the form $p \wedge \psi$. Observe that $S''[\text{false}/p]$ will reduce to false via the simplification rules since every $p \wedge \psi$ has this property.

Thus, $\mathcal{R}$ applied to S produces either false or the unit $\{p\}$. This unit cannot be pure, and we may proceed as in the CNF case: Resolve this unit with $\mathcal{F}'$'s that contain $\{\bar{p}\}$. The resolvent will cg-subsume the other parent formula. Thus we can produce a formula set containing fewer occurrences of $\bar{p}$. Repeating this process produces a formula set with one less atom, and the induction hypothesis gives us a refutation. □

We note that with augmented cg-c-subsumption, the above proof can be simplified in much the way that the proof for the CNF case was. It is worth mentioning that by using the distributive laws or d-path complement operators, a formula containing p both positively and negatively can be replaced by two, in each of which p occurs with only one polarity. In other words, we can force link deletion to be enabled, although it is not at all clear that doing so provides an advantage.

5.4 Strong vs. weak completeness

We close with a brief discussion of strong completeness (any sequence of resolution steps chosen according to some easily decidable condition and starting with an unsatisfiable formula does end with the empty formula) vs. weak completeness (*there is* a sequence of resolution steps ending with the empty formula). Strong completeness of clausal cg-resolution was recently proven [4]. Unfortunately, we cannot hope to obtain a proof of strong completeness easily from the method employed in the present paper. The reason is that it is inherently existential, as are many completeness arguments. The *existence* of a proof combined with some simple fairness criteria yields strong completeness trivially when link deletion is not present: The space of proofs that exist is invariant with respect to what step is chosen next. But with link deletion, some proofs that exist prior to a step no longer exist after that step. Others do, but we cannot be sure that they are not perpetually receeding over the horizon.

References

1. R. Anderson and W. Bledsoe. A linear format for resolution with merging and a new technique for establishing completeness. *JACM*, 17(3):525–534, July 1970.
2. L. Bachmair and H. Ganzinger. A theory of resolution. Technical Report MPI-I-97-2-005, Max-Planck-Institut für Informatik, Saarbrücken, 1997. To appear in: J. A. Robinson & A. Voronkov (eds.), *Handbook of Automated Reasoning*, Elsevier.
3. W. Bibel. On matrices with connections. *JACM*, 28:633–645, 1981.
4. W. Bibel and E. Eder. Decomposition of tautologies into regular formulas and strong completeness of connection graph resolution. *JACM*, 44(2):320–344, 1997.
5. M. Davis, G. Logemann, and D. Loveland. A machine program for theorem-proving. *CACM*, 5:394–397, 1962.
6. R. Hähnle, N. Murray, and E. Rosenthal. Completeness for linear regular negation normal form inference systems. In Z. W. Raś and A. Skowron, editors, *Found. of Intelligent Systems ISMIS'97, Charlotte/NC, USA*, volume 1325 of *LNCS*, pages 590–599. Springer-Verlag, 1997.
7. R. Hähnle, N. Murray, and E. Rosenthal. A remark on proving completeness. In D. Galmiche, editor, *Position Papers at Conf. on Analytic Tableaux and Related Methods, Nancy, France*, pages 41–47, 1997. Tech. Rep. 97-R-030, CRIN Nancy.
8. R. Kowalski. A proof procedure using connection graphs. *JACM*, 22(4):572–595, 1975.
9. N. Murray, A. Ramesh, and E. Rosenthal. The semi-resolution inference rule and prime implicate computations. In *Proc. Fourth Golden West International Conference on Intelligent Systems, San Fransisco, CA, USA*, pages 153–158, June 1995.
10. N. V. Murray and E. Rosenthal. Inference with path resolution and semantic graphs. *JACM*, 34(2):225–254, 1987.
11. N. V. Murray and E. Rosenthal. Dissolution: Making paths vanish. *JACM*, 40(3):504–535, 1993.
12. J. A. Robinson. The generalized resolution principle. In *Machine Intelligence*, volume 3, pages 77–93. Oliver and Boyd, Edinburgh, 1968.
13. M. E. Stickel. A nonclausal connection-graph resolution theorem-proving program. In D. Waltz, editor, *Proc. National Conference on Artificial Intelligence*, pages 229–233, Pittsburgh, PA, Aug. 1982. AAAI Press.

Simplification and Backjumping in Modal Tableau

Ullrich Hustadt and Renate A. Schmidt

Department of Computing, Manchester Metropolitan University,
Chester Street, Manchester M1 5GD, United Kingdom
{U.Hustadt, R.A.Schmidt}@doc.mmu.ac.uk

Abstract. This paper is concerned with various schemes for enhancing the performance of modal tableau procedures. It discusses techniques and strategies for dealing with the nondeterminism in tableau calculi, as well as simplification and backjumping. Benchmark results obtained with randomly generated modal formulae show the effect of combinations of different schemes.

1 Introduction

Usually the literature on theorem provers for modal logic confines itself to a description of the underlying calculus and methodology. Sometimes the description is accompanied with a consideration of the worst-case complexity of an algorithm based on the presented calculus or a small collection of benchmark results. Problems arising when implementing modal theorem provers and also considerations concerning optimisations towards increased efficiency have received much less attention, which, of course, is typical in a field under development. Sometimes the description of the theorem prover mentions some simplification rules [2] or discusses the use of structure sharing and use-checking [9]. Less attention has been paid to an empirical evaluation of the influence of such optimisations. But, recent work by Giunchiglia and Sebastiani [7], Horrocks [10], and Hustadt and Schmidt [11,12] has put increased emphasis on optimisation techniques for modal decision procedures.

In this paper we discuss various known techniques and strategies [7,9,10] and study their usefulness by experiments. We focus on two techniques which we think are instrumental for increased efficiency, namely *simplification* and *backjumping*. These techniques are well-known from other areas of computer science, like automated theorem proving in propositional logic, constraint solving and search. Our exposition concentrates on a modal **KE** tableau [3,4], but applies equally to standard tableau [6,8].

The paper is organised as follows. Section 2 recalls some basic notions and describes a standard tableau calculus for the basic modal logic **K**. Section 3 discusses the problems of dealing with the nondeterminism in tableau calculi. In Sections 4 and 5 we describe a simplification technique for modal tableau procedure and discuss the importance of backjumping and dependency-directed backtracking. Finally, Section 6 describes experiments which illustrate the effects in practice of the different optimisation techniques.

2 Basic notions

By definition, a *formula* of the basic modal logic **K** is a boolean combination of propositional and modal atoms. A *modal atom* is an expression of the form $\Box \psi$ where ψ is a formula of **K**. A *modal literal* is either a modal atom or its negation. We assume that $\phi \vee \psi$ is the abbreviation for $\neg(\neg \phi \wedge \neg \psi)$ and $\Diamond \psi$ for $\neg \Box \neg \psi$. $\top$ denotes a constant true proposition and $\bot$ a constant false proposition. $\overline{\phi}$ denotes the complementary formula of ϕ, for example $\overline{\neg p} = p$ and $\overline{\Box p} = \Diamond \neg p$. The following notation will be used: ϕ and ψ denote modal formulae, C and D denote multisets of modal formulae, $C; D$ denotes the multiset-union $C \cup D$, $C; \phi$ denotes $C \cup \{\phi\}$, and $\Box C$ denotes the multiset $\{\Box \phi \mid \phi \in C\}$.

$$(\wedge) \frac{C; \phi \wedge \psi}{C; \phi; \psi} \qquad (\vee) \frac{C; \phi \vee \psi}{C; \phi \mid C; \psi} \qquad (\neg) \frac{C; \neg \neg \phi}{C; \phi}$$

$$(\bot) \frac{C; \neg \phi; \phi}{\{\bot\}} \qquad (\theta) \frac{C; D}{C} \qquad (K) \frac{\Box C; \Diamond \phi}{C; \phi}$$

Fig. 1. Tableau rules for basic modal logic

Figure 1 describes the rules of a standard tableau system for basic modal logic as given by Goré [8] with a slight modification of the $(\bot)$ rule. The *numerator* of any rule is a set of formulae of which one or two are distinguished. For example, the distinguished formula of the numerator of the rule $(\wedge)$ is $\phi \wedge \psi$. Distinguished formulae are called *principal formulae*. The denominator of any rule is a list of sets of formulae. The rules $(\wedge)$, $(\vee)$, $(\neg)$, and (K) are the *elimination rules*, (θ) is the *thinning rule*, and $(\bot)$ is the *closure rule*. A *tableau* for a set C of formulae is a finite tree labelled with finite sets of modal formulae whose root is labelled with C. A tableau rule with numerator C is *applicable* to a node labelled with C. The steps for extending a tableau are the following.

1. Choose a leaf node N labelled with C, a rule R which is applicable to C, and a set of principal formulae D.
2. If D are the principal formulae of R, having k denominators $D_1, \ldots, D_k$, then create k successor nodes for N labelled with $D_1, \ldots, D_k$, respectively.

A branch in a tableau is *closed* if its end node contains $\bot$, otherwise it is *open*. A tableau is *closed* if all its branches are closed. The tableau calculus for basic modal logic by Fitting [6] uses the following refinement of the thinning rule (θ), called the *branch modification rule*:

$$(BM) \frac{D; \Box C; \Diamond \phi}{\Box C; \Diamond \phi} \quad \text{where } D \text{ contains no } \Box\text{-formula.}$$

3 Nondeterminism in tableau calculi

As usual when implementing a set of proof rules in a deterministic algorithm an important issue is how to deal with the nondeterminism present in the calculus. Several choices need to be made in extending a tableau, namely, which leaf node to continue with, which rule to apply, and to which principal formulae. The choices are don't care nondeterministic and don't know nondeterministic. A *don't care nondeterministic* choice is an arbitrary choice of one among multiple possible continuations for a computation which render the same result, for example, the nondeterminism of the ($\wedge$) and ($\neg$) rules, while a *don't know nondeterministic* choice is a choice among multiple possible continuations which do not necessarily render the same result, for example, the nondeterminism of the (K) rule. Techniques and strategies are required for dealing with nondeterministic choices, in a way that ensures soundness and completeness while delivering good performance.

For the completeness of the calculus it is sufficient to assume that all choices are don't know nondeterministic. However, a deterministic algorithm based on the assumption that all choices are don't know nondeterministic will be hopelessly inefficient, since it has to consider all possible continuations of all the don't care nondeterministic choices where the consideration of only one would suffice. Therefore, a clear distinction between don't care and don't know nondeterministic choices is necessary.

Even if we know which choices are don't care and which are don't know nondeterministic, the algorithm has to use a particular strategy to make these choices. It is well known for propositional decision procedures that different strategies lead to vastly different computational performance. For modal decision procedures such a body of knowledge does not seem to exist.

One of the problems is that applications of the (BM) and (K) rules have to be done don't know nondeterministically. Consider the tableau $\mathbf{T}_1$ given by the single node $\{\Box p, \Diamond p, \Diamond \neg p\}$. If we apply (BM) and (K) to the formula $\Diamond p$, we obtain $\mathbf{T}_2$ which cannot be closed. However, if we apply (BM) and (K) to the formula $\Diamond \neg p$ we obtain $\mathbf{T}_3$ which can be closed by an application of the closure rule. Within the framework of Fitting's tableau calculus it is impossible to avoid this don't know nondeterminism. Whenever there is more than one $\Diamond$-formula on a branch we systematically have to consider the application of the (K) rule to each of them (which can be done using backtracking).

$$\mathbf{T}_2: \quad \{\Box p, \Diamond p, \Diamond \neg p\} \qquad \mathbf{T}_3: \quad \{\Box p, \Diamond p, \Diamond \neg p\}$$
$$(BM),(K)\downarrow \qquad\qquad\qquad (BM),(K)\downarrow$$
$$\{p,p\} \qquad\qquad\qquad\qquad \{p, \neg p\}$$

It is not only relevant to which formula we apply the diamond elimination rule, but also at which state of our computation we do so, since the preceding application of the branch modification rule might delete information which is relevant for finding a closed tableau. Consider the tableau $\mathbf{T}_4$ given by the single node $\{\Box p \vee \Box q, \Diamond(\neg p \wedge \neg q)\}$. We can apply (BM) and (K) to the second formula in the tableau and obtain tableau $\mathbf{T}_5$.

$$\{\Box p \vee \Box q, \Diamond(\neg p \wedge \neg q)\}$$
$$(BM),(K)\downarrow$$
$$\{\neg p \wedge \neg q\}$$

It is not possible to obtain a closed tableau from $\mathbf{T}_5$. However, if we start by applying $(\vee)$ to the first formula in $\mathbf{T}_4$ and then (BM) and (K) to the second formula, we obtain the closed tableau $\mathbf{T}_6$.

$$\{\Box p \vee \Box q, \Diamond(\neg p \wedge \neg q)\}$$

$\{\Box p, \Diamond(\neg p \wedge \neg q)\}$ $\overset{(\vee)}{\longleftarrow}$ $\overset{(\vee)}{\longrightarrow}$ $\{\Box q, \Diamond(\neg p \wedge \neg q)\}$
$(BM),(K)\downarrow$ $(BM),(K)\downarrow$
$\{p, \neg p \wedge \neg q\}$ $\{q, \neg p \wedge \neg q\}$
$(\wedge)\downarrow$ $(\wedge)\downarrow$
$\{p, \neg p, \neg q\}$ $\{q, \neg p, \neg q\}$
$(\bot)\downarrow$ $(\bot)\downarrow$
$\{\bot\}$ $\{\bot\}$

There are several solutions how a deterministic algorithm can deal with this problem. The most obvious solution is to use backtracking. The states when branch modification is applicable are remembered, and we can apply the rule at the current state of our computation or we can delay the application. We explore one possibility and restore the current state of the computation unless a closed tableau was found.

The nondeterminism of the branch modification rule can be avoided altogether. We can delay the application of the rule until we are sure that we can find a closed tableau whenever it exists for the formula under consideration. This is the case, for example, if we delay the application of the branch modification rule until no further elimination rules for the boolean connectives are applicable on the current branch. The knowledge representation system $\mathcal{KRIS}$ uses this solution [1].

There is a trade-off between these two solutions, and we are not aware of any theoretical or empirical analysis of this trade-off. The performance of a tableau-based theorem prover depends on our ability to generate as few branches in a tableau as possible, and to close branches as soon as possible. So, it is desirable to apply the diamond elimination and branch modification rules as early as possible. However, if we fail to close the tableau and we have to go back to an earlier state of the computation, then a lot of computational effort has been wasted.

There are also solutions between the two extremes. For example, the modal theorem prover KSAT [7] proceeds as follows. Instead of delaying the application of the diamond elimination rule until no further applications of the elimination rules for boolean connectives are possible, it systematically applies all possible applications of the diamond elimination rule before a possible application of Davis-Putnam's propositional split rule. If none of the applications of the diamond elimination rule close the branch, it will continue with the intended application of the split rule.

Unlike the diamond elimination rule and branch modification rule, the application of the elimination rules for the boolean connectives can be performed don't care nondeterministically. However, this does not mean, as far as the efficiency of a modal theorem prover is concerned, all possible continuations are equally preferable. Already Smullyan [16, p. 28] noted that it is more efficient to give priority to applications of the conjunction elimination rule, that is, disjunction elimination should be delayed until no further application of conjunction elimination is possible. Even if we follow Smullyan's guideline it remains to decide in which order we apply the elimination rules to the formulae in a tableau. Three possible solutions can be found in existing systems:

(1) Select the first formula.
(2) Select the formula which contains the least number of occurrences of the disjunction operator (assuming all formulae are in negation normal form).
(3) Select the formula which has the smallest symbol weight.

For example, for $\{(p \vee r) \vee (q \vee r), \Diamond(p \wedge r) \vee \Diamond q, p \vee (q \vee r)\}$ strategy (1) will select the first formula, strategy (2) will select the second formula, and strategy (3) will select the last formula. While the literature on heuristics for selecting split variables in Davis-Putnam algorithms is extensive, very little is known about appropriate strategies for selecting formulae in tableau algorithms.

4 Simplification

This section describes a known simplification technique which helps reducing both the lengths of branches and the number of nonredundant branches.

D'Agostino [3] has shown that an algorithm for testing the satisfiability of propositional formulae based on the tableau rules of Figure 1 for the propositional connectives is not able to simulate the truth table method for propositional logic in polynomial time. He proposes the replacement of the disjunction elimination rule by the rules of Figure 2. The resulting calculus, restricted to the rules for propositional connectives, is called **KE**. The connectives $\wedge$ and $\vee$ are assumed to be commutative. The rule $(\vee S)$ is usually not explicit in presentations of **KE**, but implicit in the definition of *subsumed formulae* on a branch [15]. The calculus obtained by adding (θ) and (K) will be referred to as **MKE**. (Note that **MKE** is not related to the calculus $\Box$**KE** presented by Pitt and Cunningham [15] which is a free variable tableau method for modal logics.)

The truth table method *always* succeeds to prove that ϕ is a tautology using $\mathcal{O}(n \cdot 2^k)$ computation steps, where n is the length of ϕ and k is the number of

$$(PB)\frac{}{C;\phi \mid C;\overline{\phi}} \qquad (\vee E)\frac{C;\phi \vee \psi;\overline{\phi}}{C;\psi;\overline{\phi}} \qquad (\vee S)\frac{C;\phi \vee \psi;\phi}{C;\phi}$$

Fig. 2. Rules for disjunction elimination in **KE**

distinct variables in ϕ. D'Agostino and Mondadori [4, p. 303] show that there is a **KE**-refutation **T** of $\overline{\phi}$ with **T** having $\mathcal{O}(n \cdot 2^k)$ many nodes. However, their argument requires (i) applications of the (PB) rule which are not strongly analytic, and requires that (ii) after an application of the $(\wedge)$ rule to a conjunction $\varphi \wedge \psi$ one of φ or ψ has to be chosen *don't know* nondeterministically for further analysis, while the other subformula is prohibited from further analysis. For systems which are based on the **KE** calculus but do not adhere to restriction (ii) there is no guarantee that we find a **KE**-refutation of $\overline{\phi}$ using only $\mathcal{O}(n \cdot 2^k)$ computation steps. Consider the formula ϕ_1

$$\neg p \vee \neg q \vee (\neg \phi_1^1 \wedge \neg \phi_1^2) \vee \ldots \vee (\neg \phi_m^1 \wedge \neg \phi_m^2) \vee ((p \vee \neg \phi_{m+1}^1) \wedge (q \vee \neg \phi_{m+1}^2)),$$

where the ϕ_j^i, $1 \leq i \leq 2$, $1 \leq j \leq m+1$, are pairwise distinct boolean combinations of p and q, not identical to either p or q. A truth table method will consider the four possible truth assignments to p and q and will show that ϕ_1 is a tautology using $\mathcal{O}(|\phi_1|)$ computation steps. Note that the truth value of ϕ_1 under a truth assignment to p and q is actually independent of the truth value of the ϕ_j^i.

The tableau method based on **KE** will try to find a closed tableau for $\overline{\phi_1}$, that is

$$p \wedge q \wedge (\phi_1^1 \vee \phi_1^2) \wedge \ldots \wedge (\phi_m^1 \vee \phi_m^2) \wedge ((\neg p \wedge \phi_{m+1}^1) \vee (\neg q \wedge \phi_{m+1}^2)).$$

Following D'Agostino and Mondadori [4] we start by applying (PB) repeatedly with respect to the propositional variables p and q, to get the tableau of Figure 3. If we continue by applying conjunction elimination to $\overline{\phi_1}$ we are able to close all branches but the leftmost branch. On the leftmost branch we can not apply the $(\bot)$ rule, but have to proceed by further applications of the (PB) rule. Figure 4 shows that we can close the leftmost branch with only one application of the (PB) rule to the formula $\neg p \wedge \phi_{m+1}^1$ followed by applications of the $(\wedge)$, $(\vee E)$ and $(\bot)$ rules. The size of the tableau is $\mathcal{O}(m)$ which is within the upper bound given by D'Agostino and Mondadori. However, if we proceed by applications of the (PB) rule to the formulae $\phi_1^1, \ldots, \phi_m^1$, and finally to $\neg p \wedge \phi_{m+1}^1$, then we obtain a tableau of size $\mathcal{O}(m \cdot 2^m)$ which exceeds the upper bound. Figure 5 shows a part of the tableau we obtain in this case.

The formula selection strategies of the previous section do not ensure the tableau of Figure 4 is constructed instead of the one of Figure 5. Our proposed solution to this problem is the introduction of *simplification techniques* into tableau calculi.

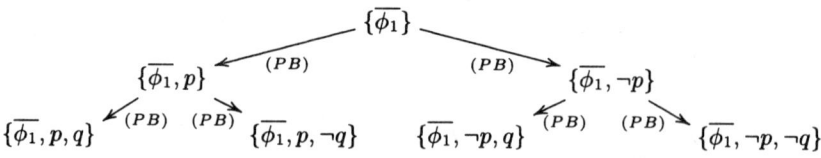

Fig. 3. First part of a tableau $\mathbf{T}_{10}$ for ϕ_1

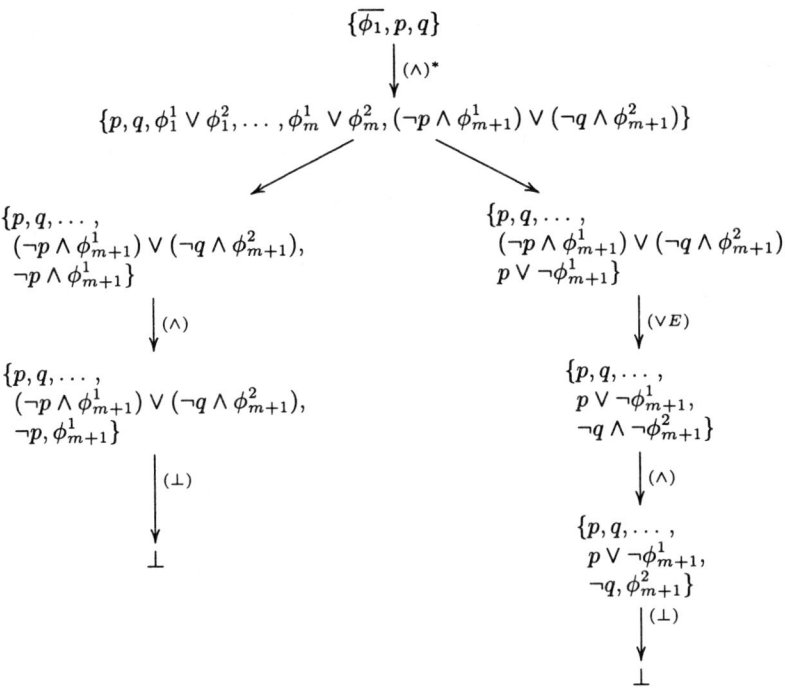

Fig. 4. Application of (PB) with respect to $p \vee \neg \phi^1_{m+1}$ in the leftmost branch of $\mathbf{T}_{10}$

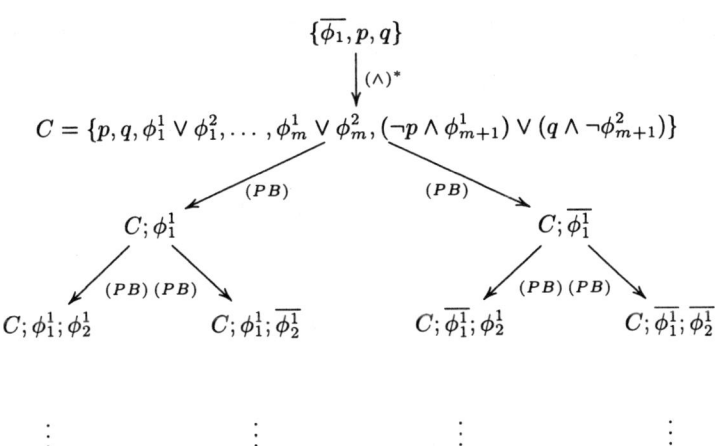

Fig. 5. Applications of (PB) with respect to $\phi^1_1, \ldots, p \vee \neg\phi^1_{m+1}$ in the leftmost branch of $\mathbf{T}_{10}$

$$\neg\phi \vee \phi \to \top \qquad \phi \vee \top \to \top \qquad \phi \vee \bot \to \phi \qquad \phi \vee \phi \to \phi$$
$$\neg\phi \wedge \phi \to \bot \qquad \phi \wedge \top \to \phi \qquad \phi \wedge \bot \to \bot \qquad \phi \wedge \phi \to \phi$$
$$\Box\top \to \top \qquad \Diamond\bot \to \bot \qquad \neg\top \to \bot \qquad \neg\bot \to \top$$

Table 1. Rewrite rules for modal formulae

Let $\phi[\psi/\omega]$ be the formula obtained from ϕ by replacing all occurrences of ψ which do not occur in the scope of a modality by the formula ω (either $\top$ or $\bot$). More precisely, $\phi[\psi/\omega]$ is defined by

$$\phi[\psi/\omega] = \begin{cases} \omega, & \text{if } \phi =_{AC} \psi \\ \overline{\omega} & \text{if } \phi =_{AC} \overline{\psi} \\ \neg\phi_1[\psi/\omega], & \text{else if } \phi = \neg\phi_1 \\ \phi_1[\psi/\omega] \wedge \phi_2[\psi/\omega], & \text{else if } \phi = \phi_1 \wedge \phi_2 \\ \phi_1[\psi/\omega] \vee \phi_2[\psi/\omega], & \text{else if } \phi = \phi_1 \vee \phi_2 \\ \phi, & \text{else.} \end{cases}$$

Here $=_{AC}$ denotes equality modulo associativity and commutativity of the connectives $\wedge$ and $\vee$, while $=$ denotes syntactic equality. Let $\phi\!\downarrow$ be the normal form of the formula ϕ obtained with the rewrite rules of Table 1. By $C[\psi/\omega]$ we denote the set $\{\phi[\psi/\omega] \mid \phi \in C\}$ and by $C\!\downarrow$ we denote the set $\{\phi\!\downarrow \mid \phi \in C\}$.

The definition $\phi[\psi/\omega]$ differs from the one given by Massacci [14] in the use of equality modulo associativity and commutativity. Massacci achieves this by replacing the binary connectives $\wedge$ and $\vee$ with set-oriented versions which are commutative, associative and idempotent. Furthermore, Massacci makes the implicit assumption that the replacement operation $\phi[\psi/\omega]$ is followed by a form of elimination of the $\top$ and $\bot$ symbols introduced, but does not specify how.

Let **SKE** and **MSKE** be the calculi of **KE** and **MKE** endowed with the following rule.

$$(S) \frac{C; \phi}{C[\phi/\top]\!\downarrow; \phi}$$

The rule can be applied don't care nondeterministically. One possible strategy is to apply (S) after an application of (PB) as follows.

$$\begin{array}{ccc} & C & \\ & \swarrow^{(PB)} \quad {}^{(PB)}\searrow & \\ C; \phi & & C; \overline{\phi} \\ \downarrow (S) & & \downarrow (S) \\ C[\phi/\top]\!\downarrow; \phi & & C[\overline{\phi}/\top]\!\downarrow; \overline{\phi} \end{array}$$

In this case, the $(\vee E)$ and $(\vee S)$ rules are obsolete.

Figure 6 shows an **SKE**-tableau for the formula ϕ_1. Since each of the final sets contains $\bot$, the tableau is closed. A procedure following the strategy exemplified in Figure 6, and restricting (PB) to propositional variables, corresponds exactly

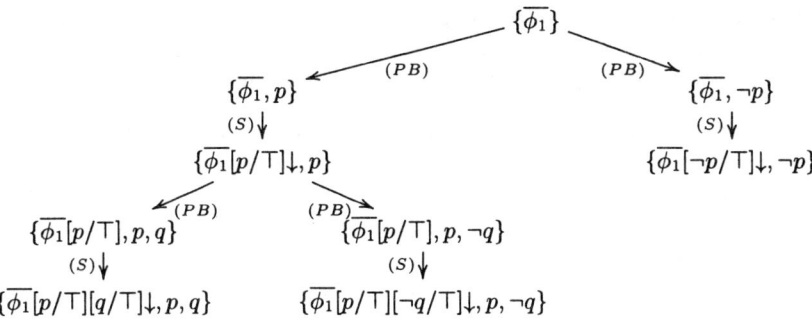

Fig. 6. A closed tableau for ϕ_1 using **SKE**

to a Davis-Putnam procedure for formulae which are not in clausal form. If we delay the application of the rule (S) until we have applied (PB) to all propositional variables occurring in the formula ϕ under consideration, we obtain a procedure which corresponds exactly to the truth table method.

We do not claim that simplification is a solution to the problem of choosing the next formula to be expanded, but adopting the case distinction mechanism of the truth table method together with simplification, a procedure can be devised which is no worse than the truth table method. Note that the rule (S) can also be added to standard tableau calculi. Simplification by (S) is a means of closing a branch as soon as possible. As a side effect the number of branches is also reduced, since superfluous applications of the (PB) rule can be avoided. In addition, branches are shorter and the counterexample obtainable from an open branch is simpler.

5 Backjumping

This section addresses how a tableau procedure can deal with the don't know nondeterminism inherent in the alternatives of the disjunction elimination rule by forms of backtracking.

Many procedures utilize chronological backtracking, that is, they go back to the most recent state before an application of disjunction elimination. The next example illustrates the drawbacks of this form of backtracking. Let ϕ_2 be a modal formula of the form

$$\neg \Box s \wedge \Box (p \vee r) \wedge (\Box \neg r \vee \Box q) \wedge (\neg \Box p \vee \Box r)$$
$$\wedge (\phi_1^1 \vee \phi_1^2 \vee \phi_1^3)$$
$$\ldots$$
$$\wedge (\phi_n^1 \vee \phi_n^2 \vee \phi_n^3)$$

where the ϕ_j^i, with $1 \leq i \leq 3$, $1 \leq j \leq n$, are modal literals different from the modal literals in the first four conjuncts of ϕ_2. Assume that ϕ_2 is satisfiable. Then:

1. $\Box\neg r$ is false in any model of ϕ_2, since $\Box\neg r$ and $\neg\Box s \wedge (\neg\Box p \vee \Box r)$ imply $\neg\Box p$ and $\Box(p \vee r) \wedge \Box\neg r \wedge \neg\Box p$ is not satisfiable.
2. A simplification step replacing $\Box\neg r$ by $\top$ in ϕ_2 does not affect the literal $\Box r$.

A procedure based on **MSKE** will start by applying conjunction elimination to ϕ_2. It is reasonable to continue with an application of the (S) rule to the units $\neg\Box s$ and $\Box(p \vee r)$. This will not close the tableau. Suppose we proceed by a sequence of applications of (PB) and (S). Let us assume that one of the first formulae to which we apply (PB) is $\Box\neg r$, followed by n modal formulae $\psi_1, \ldots, \psi_n$ chosen from $\phi_1^1, \ldots, \phi_n^3$, and finally $\neg\Box p$. Let us further assume that we construct the tableau in a depth-first way, considering the branch where $\Box\neg r$ is true first. As $\Box\neg r$ is false in any model, traversing the tree below this node, which has 2^n branches, is wasted. After closing the first and second branch we know that whenever $\neg\Box s$, $\Box(p \vee r)$, and $\Box\neg r$ are true on a branch we will be able to close it. However, this insight is not used by the procedure in order to skip the consideration of the branches generated by the n applications of the rule (PB) to the formulae $\psi_1, \ldots, \psi_n$.

This phenomenon is called *thrashing* [13]. Thrashing is the exploration of subtrees of the search tree which differ only in inessential features. Due to the rather complex constraints imposed by the modal formulae on a branch and due to strategies which delay the evaluation of these constraints, modal theorem provers are extremely vulnerable to thrashing.

Techniques that can be used to improve the backtracking behaviour of an algorithm are *backjumping* [5] and *dependency-directed backtracking* [17]. Backjumping backtracks to the last branching point of the search tree which is *relevant* to the failure on the current branch. For example, in the tableau

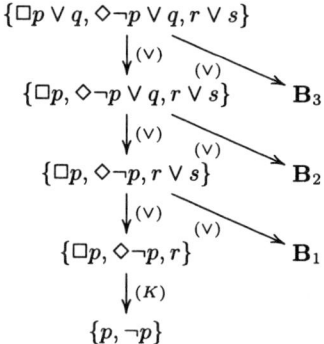

after closing the left-most branch backjumping enables us to backtrack past the last branching point, thus skipping $\mathbf{B}_1$, since the application of the $(\vee)$ rule to $r \vee s$ has not introduced a formula to the leftmost branch that has contributed to the derivation of p and $\neg p$.

Dependency-directed backtracking requires, in addition, the maintenance of assumption sets which contain all formulae which have contributed to the closure of a branch. Assumption sets are used to avoid the investigation of any branch which contains the formulae in one of the assumption sets. The assumption sets need to be transfered from one branch of the tableau to another. Furthermore, their number can grow exponentially. For this reason, the additional benefit of using dependency-directed backtracking instead of backjumping is often out of relation to the additional overhead.

Existing systems with backjumping are FaCT [10] and the Logics Workbench [9].

6 Empirical evaluation

This section describes an empirical analysis of implementations of the calculus **MKE**, and its extension **MSKE** with simplification as well as **MSKE** with backjumping. The procedures were implemented using SICStus Prolog Version 3.5. Common features of the procedures are:

1. Input formulae are transformed into negation normal form, and then into a normal form with respect to the rewrite rules of Table 1.
2. The connectives ∧ and ∨ are considered to be n-ary operators.
3. The rules (∧), (¬), and (⊥) are preferred over (PB) and $(\vee E)$. These in turn are preferred over (θ) and (K).
4. Our realisation of the (PB) rule is strongly analytic. It is applied only to the smallest disjunct of the selected formula, where the ordering on formulae is defined by a weighted symbol count with □ and ◇ assigned 7, ∧ and ∨ assigned 0, and ¬ and every propositional variable assigned 1. The intention is that propositional literal have smaller weight than modal literals of depth one which in turn have smaller weight than modal literals of depth two (on the assumption that clauses have maximal length 3).
5. The elimination rules are applied to formulae with smallest weight (the weights are those as specified in 4), that is, selection strategy (3) is used.

The procedure **MSKE** applies the (S) rule to any formula introduced by the (PB) rule.

The evaluation was done on a large set of formulae randomly generated as proposed by Giunchiglia and Sebastiani [7] and adopted by Hustadt and Schmidt [11]. The generated formulae are determined by five parameters: the number of propositional variables N, the number of modalities M, the number of modal subformulae per disjunction K, the number of modal subformulae per conjunction L, the modal degree D, and the probability P. Based on a given choice of parameters random modal KCNF formulae are defined inductively according to:

1. A *random (modal) atom* of degree 0 is a variable randomly chosen from the set of N propositional variables. A *random modal atom* of degree D, $D>0$, is with probability P a random modal atom of degree 0 or an expression of the form $\Box_i\phi$, otherwise, where $\Box_i$ is a modality randomly chosen from the set of M modalities and ϕ is a random modal KCNF clause of modal degree $D-1$.

2. A *random modal literal* (of degree D) is with probability 0.5 a random modal atom (of degree D) or its negation, otherwise.
3. A *random modal KCNF clause* (of degree D) is a disjunction of K random modal literals (of degree D).
4. Now, a *random modal KCNF formula* (of degree D) is a conjunction of L random modal KCNF clauses (of degree D).

It is important to note that in contrast to generating random 3SAT formulae, typically used for the evaluation of propositional decision procedures, generating a random modal 3CNF clause of degree 0 means randomly generating a *multiset* of three propositional variables and negating each member of the multiset with probability 0.5. This means the scheme we have just described allows for tautologous subformulae, like $p \vee \neg p \vee q$, and contradictory subformulae, like $\neg \Box (p \vee \neg p \vee p)$. Furthermore, most of the unsatisfiable random modal 3CNF formulae are trivially unsatisfiable. By definition, a random modal 3CNF formula ϕ is *trivially unsatisfiable* if the conjunction of the purely propositional clauses of ϕ is unsatisfiable. It turns out, these formulae are well suited to show the advantages and disadvantages of backjumping.

We use the parameter settings **PS0** ($N=5$, $M=1$, $K=3$, $D=2$, $P=0.5$), **PS1** ($N=10$, $M=1$, $K=3$, $D=2$, $P=0.5$), and **PS2** ($N=4$, $M=1$, $K=3$, $D=1$, $P=0.0$). **PS2** was generated in accordance with the guidelines of [11], which means, for all occurrences of $\Box \phi$ in a random modal 3CNF formula of degree 1, ϕ has to be a nontautologous clause containing exactly three differing literals.

The tests were conducted by proceeding as follows. We take one of the parameter settings which fixes all parameters except L, the number of clauses. The parameter L ranges from N to $40N$ for **PS0** and **PS1** and from N to $30N$ for **PS2**. For each value of the ratio L/N a set of 100 random modal KCNF formulae of degree D is generated. For small L the generated formulae are more likely to be satisfiable and for larger L the generated formulae are more likely to be unsatisfiable. For **PS2**, already for $L/N=30$ all formulae are unsatisfiable, so there is no need to increase the ratio beyond 30. For each generated formula ϕ we measure the time needed by one of the tableau procedures to determine the satisfiability of ϕ. There is a upper limit for the CPU time consumed. As soon as this limit is reached, the computation for ϕ is stopped. Our tests were run on a Sun Ultra 1/170E with 196MB main memory using a time-limit of 1000 CPU seconds.

Figures 7–10 depict the outcome of our experiments in the form of percentile graphs. Formally, the $Q\%$-*percentile* of a set of values is the value V such that $Q\%$ of the values is smaller or equal to V and $(100-Q)\%$ of the values is greater than V. The median of a set coincides with its 50%-percentile.

Figure 7 illustrates the importance of design decision 2. Without accounting for associativity of $\vee$ and $\wedge$, the computational behaviour of **MKE** is drastically worse. This can be attributed to the phenomenon that, whenever the (PB) rule introduces a formula ϕ on a branch, modal clauses like $\psi_1 \vee (\phi \vee \psi_2)$ on the branch become true and need not be considered. However, $(\vee S)$ is not applicable when not regarding $\vee$ as being associative. Instead, **MKE** will apply the (PB)

rule to ψ_1. This leads to increased thrashing. Interestingly, the associativity assumption is not important for **MSKE**.

When assuming $\vee$ is an n-ary operator inside **MKE**, then the rules ($\vee E$) and ($\vee S$) have essentially the same effect as simplification in **MSKE** on KCNF formulae. This is illustrated by the graphs of Figures 8 and 9, for **MKE** and **MSKE** on **PS1**. Only for the parameter setting **PS2** we see the advantage of using simplification.

The effect of enhancing **MSKE** with backjumping is apparent in Figure 10. Even though a slow down is evident for the 50%–70%-percentiles on **PS1**, the 80%–90%-percentiles show a significant improvement. The slow down is due to the additional computational overhead of managing the information required for performing backjumping. Nevertheless, there are almost no samples in our benchmark suite, for which **MSKE** with backjumping fails to terminate within the given time limit.

It should be stressed that the advantages of the optimisation techniques become most apparent on hard test formulae which are difficult to come by. The usefulness of the optimisation techniques discussed in this paper on other problem sets, like the benchmark collection of the Logics Workbench, requires further investigation.

Our discussion was limited to optimisations of modal **KE** procedures. Similar results can obtained for standard tableau-based procedures with the same enhancements. It is open, however, how enhanced **KE** and standard tableau procedures compare.

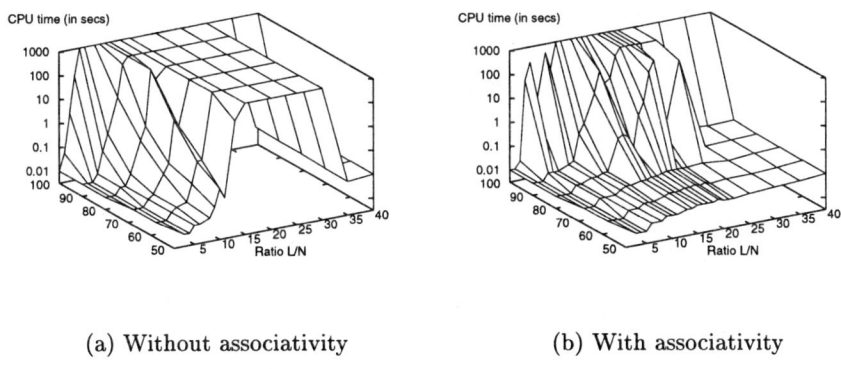

(a) Without associativity   (b) With associativity

Fig. 7. Graphs for **MKE** on **PS0**

Acknowledgements

We thank the anonymous referees for their comments. This research was conducted while the authors were employed at the Max-Planck-Institut für Infor-

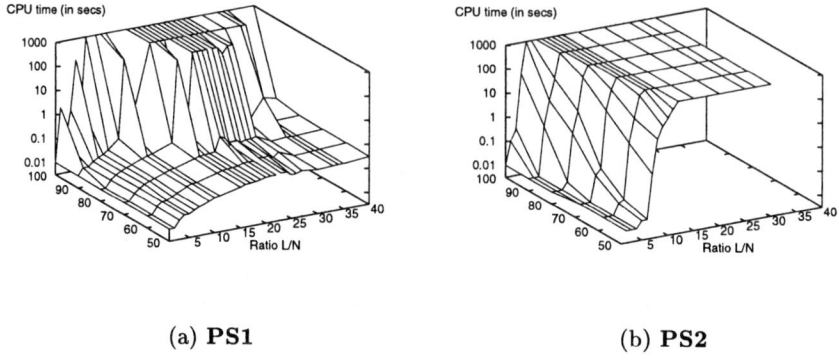

(a) **PS1**   (b) **PS2**

Fig. 8. Graphs for **MKE**

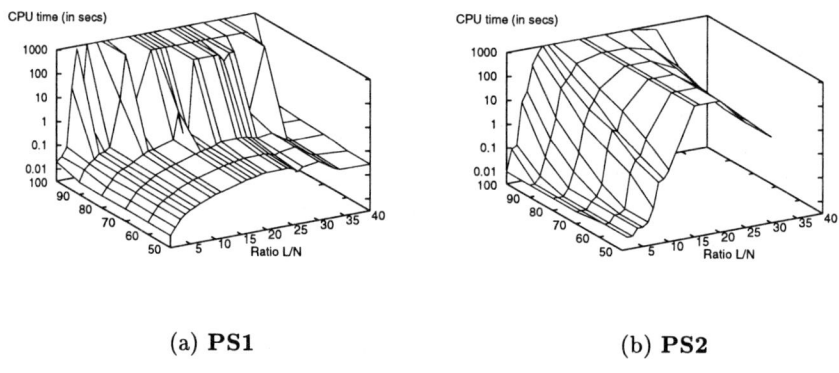

(a) **PS1**   (b) **PS2**

Fig. 9. Graphs for **MSKE**

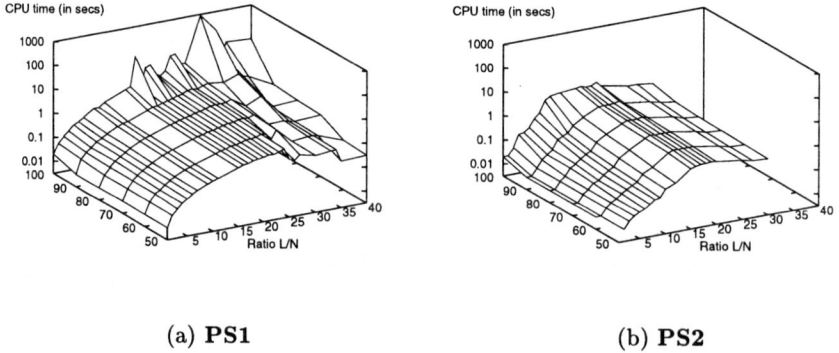

(a) **PS1**   (b) **PS2**

Fig. 10. Graphs for **MSKE** with backjumping

matik in Saarbrücken, Germany. The work was funded by the MPG and the DFG through grant Ga 261/8-1 and the TRALOS-Project.

References

1. F. Baader and B. Hollunder. A terminological knowledge representation system with complete inference algorithms. In H. Boley and M. M. Richter, editors, *Proc. of the Intern. Workshop on Processing Declarative Knowledge (PDK '91)*, LNAI 567, pages 67–86. Springer, 1991.
2. L. Catach. Tableaux: A general theorem prover for modal logics. *Journal of Automated Reasoning*, 7(4):489–510, 1991.
3. M. D'Agostino. Are tableaux an improvement on truth-tables? *Journal of Logic, Language, and Information*, 1:235–252, 1992.
4. M. D'Agostino and M. Mondadori. The taming of the cut. Classical refutations with analytic cut. *Journal of Logic and Computation*, 4(3):285–319, 1994.
5. R. Dechter. Enhancement schemes for constraint processing: Backjumping, learning, and cutset decomposition. *Artifical Intelligence*, 41(3):273–312, 1989/90.
6. M. Fitting. *Proof Methods for Modal and Intuitionistic Logics*, volume 169 of Synthese Library, Studies in Epistemology, Logic, Methodology, and Philosophy of Science. D. Reidel Publishing Company, 1983.
7. F. Giunchiglia and R. Sebastiani. Building decision procedures for modal logics from propositional decision procedures: Case study of modal K. In M. A. McRobbie and J. K. Slaney, editors, *Proc. of the 13th Intern. Conf. on Automated Deduction (CADE-13)*, LNAI 1104, pages 583–597. Springer, 1996.
8. R. Goré. Tableau methods for modal and temporal logics. Technical Report TR-ARP-15-95, Automated Reasoning Project, Australian National University, Canberra, Australia, November 1995.
9. A. Heuerding, G. Jäger, S. Schwendimann, and M. Seyfried. The Logics Workbench LWB: A snapshot. *Euromath Bulletin*, 2(1):177–186, 1996.
10. I. Horrocks. Optimisation techniques for expressive description logics. Technical Report UMCS-97-2-1, Department of Computer Science, University of Manchester, Manchester, UK, February 1997.
11. U. Hustadt and R. A. Schmidt. On evaluating decision procedures for modal logic. Research report MPI-I-97-2-003, Max-Planck-Institut für Informatik, Saarbrücken, Germany, February 1997.
12. U. Hustadt and R. A. Schmidt. On evaluating decision procedures for modal logic. In M. E. Pollack, editor, *Proc. of the Fifteenth Intern. Joint Conf. on Artificial Intelligence (IJCAI'97)*, pages 202–207. Morgan Kaufmann, 1997.
13. A. K. Mackworth. Constraint satisfaction. In S. C. Shapiro, editor, *Encyclopedia of Artificial Intelligence*, pages 205–211. John Wiley & Sons, 1987.
14. F. Massacci. Simplification with renaming: A general proof technique for tableau and sequent-based provers. Technical Report 424, Computer Laboratory, University of Cambridge, Cambridge, UK, May 1997.
15. J. Pitt and J. Cunningham. Theorem proving and model building with the calculus ke. *Journal of the IGPL*, 4(1):129–150, February 1996.
16. R. M. Smullyan. *First-order logic*. Springer, New York, 1968.
17. R. M. Stallman and G. J. Sussman. Forward reasoning and dependency-directed backtracking in a system for computer-aided circuit analysis. *Artifical Intelligence*, 9:135–196, 1977.

Free Variable Tableaux for a Logic with Term Declarations

P.J. Martín, A. Gavilanes, J. Leach

Dep. de Sistemas Informáticos y Programación. Universidad Complutense de Madrid
e-mail: {pjmartin, agav, leach}@eucmos.sim.ucm.es [*]

Abstract. We study free variable tableau methods for logics with term declarations. We show how to define a substitutivity rule preserving the soundness of the tableaux and we prove that some other attempts lead to unsound systems. Based on this rule, we define a sound and complete free variable tableau system and we show how to restrict its application to close branches by defining a sorted unification calculus.

1 Introduction

Recently in various areas of artificial intelligence, the importance of hybrid systems of representation, in which sub-systems that employ different languages and inference mechanisms co-exist, has been manifested. Amongst them, the logics that integrate sort information in their specific structure have a special relevance in the field of automatic deduction. Furthermore, it is well known that the use of sorts in the universe of discourse can produce a drastic reduction in the search space, which would involve in itself more efficient deductions.

In some cases, besides the improvement of efficiency, the necessity of the use of sorts in the explicit reasoning has been manifested when dealing with taxonomic information. In this way the order sorted logics arise [Coh 87] [Wal 87] [Sch 89] [Fri 91], in which deductions take note of the sort hierarchy. In a lot of them, the framework of the deduction systems is substitutional, which means that the information about sorts is only used when the process of unification obtains substitutions.

However, the use of order sorted logics, where the information about sorts co-exists with the information about individuals within the same formal framework [Wei 91] [GLMN 96] turns out to be also interesting. One can say therefore that sorts are dynamic in opposition to the static behaviour that they maintain in the previous logics. Following this line, it is possible to generalize the dynamism by declaring the function and predicate symbols on the same level as the sort hierarchy and the information about individuals. In this way the so called *logics with term declarations* arise as hybrid systems of representation that include, in a unique formalism, a classic many sorted logic together with all the information that it entails (relations between sorts, and sort declarations for function and predicate symbols).

In this paper we study free variable tableau methods for logics with term declarations. A critical point of this work is determining which substitutions are well-sorted, in the sense that the (static) sort of a variable and the (dynamic) sort of the substituting term are the same. A right concept of well-sortedness preserves the soundness of substitutivity in tableaux while a wrong one produces unsound tableau systems. We prove the latter occurs in the only paper we know about a similar dynamic logic [Wei 95].

---

[*]Research partially supported by the ESPRIT BR Working Group 6028 CCLII.

We present a first tableau method and prove its soundness with respect to our notion of well-sorted substitutivity, and we show how to improve the method by restricting the substitutivity rule to close branches. It implies the definition of a calculus for solving unification problems, with respect to term declaration theories. We show this calculus is sound and complete, and we use it to prove the completeness of an improved free variable tableau method.

This paper is organized as follows. Section 2 presents the Logic with Term Declarations, explaining its main syntactical and semantical features. Section 3 studies our well-sortedness concept. In Section 4, a ground tableau method is shown, and it is extended to a free variable version in Section 5. Finally, Section 6 presents a sorted unification calculus, which is used in Section 7 for defining a tableau system where the substitutivity rule is only applied to close branches. Due to lack of space, some proofs have been omitted. They can be found in [GLMN 97b].

2 The Logic With Term Declarations LTD

The logic LTD uses sorts in a dynamic way, not having function and constant symbols static declarations in the signature. This means that we can not infer the sort of any term from its structure, neither syntactic, nor semantically. Instead of it, LTD takes advantage of sorts using a new formula constructor ($t \in s$) to declare that the term t belongs to the sort s.

For example if we have available the sorts nat and int, to respectively denote natural and integer numbers, the function $+$, that adds two integers, can be declared by the formula $\forall x^{int} \forall y^{int}(x^{int} + y^{int} \in int)$. In fact, this declaration could be done in a logic with static sorts, putting down the function symbol $+ : int \times int \to int$ in the signature, but then the behavior of $+$ could not be specified anymore. In our approach, we are able to declare the function symbol $+$ only when such information is required. Even more, we can refine the behavior of $+$; for example, we can overload the function $+$, expressing that when applied to naturals it results to be another natural, with the formula $\forall x^{nat} \forall y^{nat}(x^{nat} + y^{nat} \in nat)$.

Another advantage of using term declarations, comes from its combination with sorted variables for expressing relations between sorts. For example we can express that natural are integer numbers by $\forall x^{nat}(x^{nat} \in int)$, or that the intersection of natural and integer numbers is not empty by $\exists x^{nat}(x^{nat} \in int)$.

On the other hand, since predicate symbols denote boolean functions, dynamic declarations make no sense for them. However, as we do not know anything about the sort of a term, we can apply predicates to every term.

A signature Σ for LTD consists of a finite set S of sorts s, together with sets of constants $\mathcal{C}$, function symbols $\mathcal{F}$ and symbols of predicate $\mathcal{P}$, the last ones with associated arity for each of its elements. Next, we define terms and formulas for a signature Σ and a sorted family of countable sets of variables $X = (X^s)_{s \in S}$.

Definition 1 *The sets of Σ-terms $T(\Sigma)$ and Σ-formulas $F(\Sigma)$ are defined by:*

$t ::= x^s (\in X^s) \mid c\ (\in \mathcal{C}) \mid f(t_1, \ldots, t_n)\ (f^n \in \mathcal{F};\ t_1, \ldots, t_n \in T(\Sigma))$
$\varphi ::= t \in s \mid P(t_1, \ldots, t_n)\ (P^n \in \mathcal{P}; t_1, \ldots, t_n \in T(\Sigma)) \mid \neg \varphi \mid \varphi \wedge \varphi' \mid \exists x^s \varphi.$

A $\in$-*theory*, or simply a *theory*, is a set of *term declarations* $t \in s$. Usual first-order formulas are defined by their classical abbreviations ($\vee, \to, \leftrightarrow, \forall$).

Regarding semantics, we need domains to represent every sort appearing in the signature; so the structures are built by families of domains. Considering that we do not have static declarations, our domains can be empty.

Definition 2 *A Σ-structure $\mathcal{D}$ is a tuple consisting of:*
1. *A total domain D containing a family of domains $\{D^s | s \in S\}$.*
2. *Sets $\{c^{\mathcal{D}} \in D | c \in \mathcal{C}\}, \{f^{\mathcal{D}} : D^n \to D | f^n \in \mathcal{F}\}$ and $\{P^{\mathcal{D}} : D^n \to \{\underline{t}, \underline{f}\} | P^n \in \mathcal{P}\}$ of interpretations of constants, function symbols and predicate symbols.*

A valuation for $\mathcal{D}$ is a sorted function $\rho = (\rho^s)_{s \in S}$ such that ρ^s is a finite map from X^s to D^s, for every $s \in S$. We will denote ρ^s by $[\rho^s(x_1^s)/x_1^s, \ldots, \rho^s(x_n^s)/x_n^s]$, where $dom(\rho^s) = \{x_1^s, \ldots, x_n^s\}$ is the domain of ρ^s, and $dom(\rho) = \bigcup_{s \in S} dom(\rho^s)$ is the domain of ρ. Note that a valuation for $\mathcal{D}$ verifies that $\rho^s = [\;]$ for any $s \in S$ such that $D^s = \emptyset$. Finally we will denote by $\rho[d/x^s]$ the valuation that coincides with ρ except for x^s, whose value is d.

A Σ-interpretation is a pair $\langle \mathcal{D}, \rho \rangle$ composed of a Σ-structure and a valuation for it. The semantic value of a term t in $\langle \mathcal{D}, \rho \rangle$ is defined if its variables appear in $dom(\rho)$. In this case we denote it by $[\![t]\!]_\rho^{\mathcal{D}}$ which is an element of D defined in the usual way. For example, if $f^n \in \mathcal{F}$ then $[\![f(t_1, \ldots, t_n)]\!]_\rho^{\mathcal{D}} = f^{\mathcal{D}}([\![t_1]\!]_\rho^{\mathcal{D}}, \ldots, [\![t_n]\!]_\rho^{\mathcal{D}})$ whenever the variables of $t_1, \ldots, t_n$ are in $dom(\rho)$.

Formulas in LTD are interpreted in a bivalued way when its free variables belong to the domain of the valuation.

Definition 3 *The boolean value of a Σ-formula φ in a Σ-interpretation $\langle \mathcal{D}, \rho \rangle$, such that the free variables of φ belong to $dom(\rho)$, is an element of $\{\underline{t}, \underline{f}\}$, denoted by $[\![\varphi]\!]_\rho^{\mathcal{D}}$ and defined by:*

- $[\![t \in s]\!]_\rho^{\mathcal{D}} = \begin{cases} \underline{t} & \text{if } [\![t]\!]_\rho^{\mathcal{D}} \in D^s \\ \underline{f} & \text{otherwise.} \end{cases}$

- $[\![P(t_1, \ldots, t_n)]\!]_\rho^{\mathcal{D}} = P^{\mathcal{D}}([\![t_1]\!]_\rho^{\mathcal{D}}, \ldots, [\![t_n]\!]_\rho^{\mathcal{D}})$.

- *The semantics of $\neg$ and $\wedge$ is the usual.*

- $[\![\exists x^s \varphi]\!]_\rho^{\mathcal{D}} = \begin{cases} \underline{t} & \text{if there is } d \in D^s \text{ such that } [\![\varphi]\!]_{\rho[d/x^s]}^{\mathcal{D}} = \underline{t} \\ \underline{f} & \text{otherwise.} \end{cases}$

In the sequel, when we write $[\![t]\!]_\rho^{\mathcal{D}}$ (resp. $[\![\varphi]\!]_\rho^{\mathcal{D}}$), we assume that the free variables of t (resp. φ) are in $dom(\rho)$. Note that this assumption trivially holds when dealing with ground terms (resp. sentences). Actually in this case, no valuation is needed, so structures are enough to interpret these terms (resp. formulas) and then we simplify the notation by writing $[\![t]\!]^{\mathcal{D}}$ (resp. $[\![\varphi]\!]^{\mathcal{D}}$). The concepts of *model* and *logical consequence* are defined as usual and represented by using the symbol $\models$.

In order to show the expressive power of LTD we present two examples.

Example 4 *We can prove the formula $\exists x^{s''}(x^{s''} \in s')$, expressing that the intersection of the sorts s' and s'' is not empty, as a logical consequence of the set of formulas $\{a \in s, f(a) \in s'', \forall x^s (f(x^s) \in s')\}$. Note that in a static many-sorted logic without equality this formula could not be expressed because we can not represent the identification of an element of two different sorts.*

Example 5 *Structures with the s-domain empty are syntactically characterized with the formula $\forall x^s (x^s \notin s)$.*

3 Well-Sorted Substitutions

First-order logic satisfies *Substitution Lemma*. This lemma states that the interpretation of a substituted formula $\varphi[t_1/x_1, \ldots, t_n/x_n]$ is equal to the interpretation of the formula φ, but properly changing the valuation of the variables $x_1, \ldots, x_n$ by the interpretation of the terms $t_1, \ldots, t_n$. In tableau methods, this result is needed to assure soundness of the γ-rule, and also when dealing with free variable tableau versions.

Since we expect this kind of result in our logic, we have to find out which substitutions satisfy it. The difficult point in *LTD* is that variables are sorted and when a substitution $[t_1/x_1^{s_1}, \ldots, t_n/x_n^{s_n}]$ is applied, the interpretation of some introduced term t_i may not belong to the sort s_i of the replaced variable. So we must use substitutions only in contexts guaranteeing $[\![t_i]\!]_\rho^\mathcal{D} \in D^{s_i}$, for every $1 \leq i \leq n$. A theory will be the syntactic context that provides enough sort information about the terms to assure the previous property.

Definition 6 (Well-Sorted Substitution) *A substitution $[t_1/x_1^{s_1}, \ldots, t_n/x_n^{s_n}]$ is well-sorted w.r.t. the theory $\mathcal{L}$ if $t_i \notin X^{s_i} \Longrightarrow (t_i \in s_i) \in \mathcal{L}$, for every $1 \leq i \leq n$.*

Lemma 7 *Let $\mathcal{L}$ be a theory, $\langle \mathcal{D}, \rho \rangle$ a Σ-interpretation satisfying $\mathcal{L}$ and $\tau \equiv [t_1/x_1^{s_1}, \ldots, t_n/x_n^{s_n}]$ a well-sorted substitution w.r.t. $\mathcal{L}$ such that $(t_i \in s_i) \notin \mathcal{L} \Longrightarrow t_i \in dom(\rho)$, for every $1 \leq i \leq n$. Then $[\![t_i]\!]_\rho^\mathcal{D} \in D^{s_i}$, for every $1 \leq i \leq n$.*

Lemma 8 (Substitution for terms and formulas) *Let $\mathcal{L}$ be a theory, $\langle \mathcal{D}, \rho \rangle$ a Σ-interpretation satisfying $\mathcal{L}$ and $\tau \equiv [t_1/x_1^{s_1}, \ldots, t_n/x_n^{s_n}]$ a well-sorted substitution w.r.t. $\mathcal{L}$ such that $(t_i \in s_i) \notin \mathcal{L} \Longrightarrow t_i \in dom(\rho)$, for every $1 \leq i \leq n$. For any term t (resp. formula φ) such that $free(t) - \{x_1^{s_1}, \ldots, x_n^{s_n}\}$ (resp. $free(\varphi) - \{x_1^{s_1}, \ldots, x_n^{s_n}\}$)[1] is included in $dom(\rho)$, it holds:*

1. $[\![t\tau]\!]_\rho^\mathcal{D} = [\![t]\!]_{\rho[[\![t_1]\!]_\rho^\mathcal{D}/x_1^{s_1}, \ldots, [\![t_n]\!]_\rho^\mathcal{D}/x_n^{s_n}]}^\mathcal{D}$
2. $[\![\varphi\tau]\!]_\rho^\mathcal{D} = [\![\varphi]\!]_{\rho[[\![t_1]\!]_\rho^\mathcal{D}/x_1^{s_1}, \ldots, [\![t_n]\!]_\rho^\mathcal{D}/x_n^{s_n}]}^\mathcal{D}$

4 The Ground Tableau Method

In this section we outline a ground tableau method for *LTD* as a basis for the free variable tableau version. We will suppose that Σ has been extended to a signature $\overline{\Sigma}$, with a countable set of constants. A tableau for a set of sentences is a tree growing and branching by the application of expansion tableau rules, according to the patterns of the formulas labelling its nodes. For conjunction and disjunction formulas, branches are enlarged or split, respectively, as in classical first-order tableaux, using the rules α and β [Fit 96]. For quantified formulas we have the following expansion rules:

$$\gamma) \quad \frac{\neg \exists x^s \varphi \quad t \in s}{\neg \varphi[t/x^s]} \qquad \delta) \quad \frac{\exists x^s \varphi}{\varphi[c/x^s] \quad c \in s}$$

In γ, t is a ground term. In δ, c is a new constant not occurring in the branch.

Note that these two new rules are similar to the classical first-order ones, but here the (dynamic) sort information is managed, using $(t \in s)$ in the case of γ or introducing $(c \in s)$ in the case of δ, while in classical tableaux the (static) sort information is given by the signature and no explicit reference is required by the corresponding rules.

---

[1] *free* supplies the set of free variables of a term, a formula or a set of formulas.

Definition 9 *A branch B of a tableau is closed if it contains an atomic contradiction, that is φ and $\neg\varphi$ (φ atomic) appear in B. A tableau is closed if all its branches are closed.*

Theorem 10 (Soundness and Completeness) *[GLMN 97a] For every set of Σ-sentences Φ, Φ has a closed tableau if and only if Φ is not satisfiable.*

5 Free Variable Tableaux

Now we will assume that the extended signature $\overline{\Sigma}$ also contains a countable set of function symbols. When dealing with free variables, the sorted variable occurring in a γ-rule is not replaced by a ground term, but by a new free variable of the same sort. So, in *LTD* we get the following two rules:

$$\gamma') \quad \frac{\neg \exists x^s \varphi}{\neg \varphi[y^s/x^s]} \qquad \delta') \quad \frac{\exists x^s \varphi}{\varphi[f(x_1^{s_1},\ldots,x_n^{s_n})/x^s]} \\ f(x_1^{s_1},\ldots,x_n^{s_n}) \in s$$

In γ', y^s is a (new) free variable. In δ', f is a new function symbol applied to the free variables $(x_1^{s_1},\ldots,x_n^{s_n})$ occurring in the branch.

The free variables of a tableau may be substituted, but as argued in Section 3, they can be replaced only by terms interpreted in the corresponding domain. This condition is assured for a branch when the substitution is well-sorted w.r.t. the theory included in this branch. However substitutions are applied to the whole tableau. Requiring well-sortedness w.r.t. the theory of every tableau branch leads us to a too strong condition, since a variable in the domain of a substitution may not occur free in every branch. It will be enough to consider well-sortedness of substitutions w.r.t. a branch, when they are restricted to the free variables of the branch. This property is formalized in the following definition.

Definition 11 *The substitution τ is well-sorted w.r.t. a tableau $\mathcal{T}$ with branches $B_1,\ldots,B_n$, if $\tau|_{free(B_i)}$[2] is well-sorted w.r.t. the theory included in B_i, for every $1 \leq i \leq n$.*

Definition 12 (Substitutivity Rule) *If $\mathcal{T}$ is a free variable tableau and τ is an idempotent substitution well-sorted w.r.t. $\mathcal{T}$ then $\mathcal{T}\tau$ is a free variable tableau.*

This rule will be called *sub* and we denote by $\mathcal{S}1$ the tableau system composed of α, β, γ', δ' and *sub* together with a closure definition, where the concepts of closed tableau branch and closed tableau are defined as for ground tableaux.

The sense in which $\mathcal{S}1$ preserves soundness has to be made more precise because of the existence of empty domains. For example, the formula $\forall x^s(\neg P(a) \wedge P(a))$ is satisfiable in structures with empty s-domain, nevertheless the method $\mathcal{S}1$ is able to build a closed tableau for it. In order to prevent these cases, we prove soundness as follows.

---

[2]This is the restriction of τ to the free variables of B_i.

Definition 13 *A free variable tableau $\mathcal{T}$ is satisfiable if there exists a structure $\mathcal{D}$ such that for every valuation ρ for $\mathcal{D}$, verifying free($\mathcal{T}$)$\subseteq$ dom(ρ), there exists a branch B with $\langle \mathcal{D}, \rho \rangle \models B$.*

Lemma 14 *Let $\mathcal{T}$ be a free variable tableau satisfiable in a structure that for every sort s has a non-empty s-domain. If the tableau $\mathcal{T}'$ is built by applying one of the S1-rules to $\mathcal{T}$, then $\mathcal{T}'$ is satisfiable in a structure with non-empty s-domain for every sort s.*

Proof. We only prove the case *sub*. Let τ be an idempotent substitution well-sorted w.r.t. $\mathcal{T}$. If $\mathcal{D}$ is the structure that models $\mathcal{T}$, we prove that it also makes $\mathcal{T}' = \mathcal{T}\tau$ satisfiable. Let ρ be a valuation such that $free(\mathcal{T}\tau) \subseteq dom(\rho)$; we extend ρ in order to make it defined in $\mathcal{T}$, which is possible because $\mathcal{D}$ has non-empty domains. This new valuation ρ' operates as ρ in $\mathcal{T}'$.

By hypothesis, there exists a branch B_1 in $\mathcal{T}$ such that $\langle \mathcal{D}, \rho' \rangle \models B_1$. If we define $\tau_1 \equiv \tau|_{free(B_1)}$, then from the well-sortedness of τ_1 w.r.t. B_1 and Lemma 8 we can deduce that, for all φ such that $free(\varphi) \subseteq dom(\rho')$, $[\![\varphi\tau_1]\!]_{\rho'}^{\mathcal{D}} = [\![\varphi]\!]_{\rho'\tau_1}^{\mathcal{D}}$ [3].

By hypothesis again, there exists B_2 in $\mathcal{T}$ such that $\langle \mathcal{D}, \rho'\tau_1 \rangle \models B_2$. If $B_1 = B_2$ then we have finished because $[\![B_2\tau]\!]_{\rho'}^{\mathcal{D}} = [\![B_2\tau_1]\!]_{\rho'}^{\mathcal{D}} = [\![B_2]\!]_{\rho'\tau_1}^{\mathcal{D}} = \underline{t}$[4]. Otherwise if $\tau_2 \equiv \tau|_{free(B_2)-free(B_1)}$, from the well-sortedness of τ_2 w.r.t. B_2 and Lemma 8 we can deduce that for all φ such that $free(\varphi) \subseteq dom(\rho'\tau_1)$, $[\![\varphi\tau_2]\!]_{\rho'\tau_1}^{\mathcal{D}} = [\![\varphi]\!]_{\rho'\tau_1\tau_2}^{\mathcal{D}}$. Note that $B_2\tau = B_2\tau_1\tau_2$, because τ is idempotent; so the process can be continued.

We prove that $\mathcal{T}\tau$ is satisfiable by repeating this procedure until we reach a branch already used. This will be the case because $\mathcal{T}$ has a finite number of branches. ∎

Theorem 15 (Soundness of S1) *For every set of Σ-sentences Φ, if Φ has a closed tableau then Φ is not satisfiable in structures where every sort s has a non-empty domain.*

Proof. Suppose that Φ is satisfiable in a structure $\mathcal{D}$ with non-empty domains and that Φ has a closed tableau $\mathcal{T}$. Then by Lemma 14, $\mathcal{T}$ is satisfiable in a certain structure $\mathcal{D}'$ with non-empty domains. So, we can build a valuation ρ defined for every free variable of $\mathcal{T}$ such that there exists a branch B with $\langle \mathcal{D}', \rho \rangle \models B$. By semantics, we get contradiction because B is closed. ∎

Observe that our claim for idempotence in the rule *sub* is critical for assuring soundness, as we notice from the proof of Lemma 14. If *sub* rule only demands well-sortedness the system turns to be unsound, as the following example shows.

Example 16 *Let Σ be a signature with sorts s and s', constant symbols a and b, and predicate symbols Q and P. Let $\mathcal{D}$ be a Σ-structure such that $a^{\mathcal{D}} \in D^s \cap D^{s'}$, $b^{\mathcal{D}} \in D^s - D^{s'}$, $P^{\mathcal{D}}$ is true in $D^{s'}$ but false in $D^s - D^{s'}$, $Q^{\mathcal{D}}$ is false in $D^{s'}$ but true in $D^s - D^{s'}$. Then the set $\{a \in s, b \in s, \neg Q(a), \neg P(b), \forall x^s(Q(x^s) \vee (x^s \in s' \wedge \forall y^{s'} P(y^{s'})))\}$ can be checked to be satisfiable in $\mathcal{D}$, in spite of having a closed tableau. In effect, applying α, β and γ' properly, the following sketch $\mathcal{T}$ of tableau can be built:*

---

[3] $\rho\tau$ stands for the valuation ρ but interpreting every replaced variable x of τ as $[\![\tau(x)]\!]_{\rho}^{\mathcal{D}}$.
[4] The interpretation of a branch is the conjunction of the interpretations of its formulas.

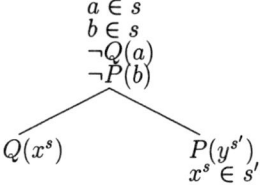

Note that $[a/x^s, x^s/y^{s'}]$ is well-sorted w.r.t. $\mathcal{T}$, and if we apply it to $\mathcal{T}$ we get $\mathcal{T}'$:

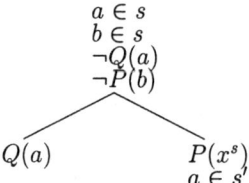

$\mathcal{T}'$ can be closed if we apply the substitution $[b/x^s]$ which is well-sorted w.r.t. $\mathcal{T}'$.

Let us now make some reflections about some other possible presentations of a well-sorted substitution. In the table below, we present four different forms for defining a well-sorted (WS) substitution $\tau \equiv [t_1/x_1^{s_1}, \ldots, t_n/x_n^{s_n}]$ w.r.t. a theory $\mathcal{L}$.

Name	Definition
RWS	$\forall i (t_i \notin X^{s_i} \longrightarrow (t_i \in s_i) \in \mathcal{L}\tau)$
UWS	$\forall i (t_i \in T_{\mathcal{L}}(s_i))$
WeiWS	τ is UWS w.r.t. $\mathcal{L}$ $\exists \tau'(\tau' \leq^5 \tau \wedge dom(\tau') \subseteq free(\mathcal{L}) \cup dom(\tau) \wedge \tau'$ is RWS w.r.t. $\mathcal{L})$
RUWS	τ is RWS and UWS w.r.t. $\mathcal{L}$

$T_{\mathcal{L}}(s)$ stands for the set of terms of sort s occurring in the theory $\mathcal{L}$, when $\mathcal{L}$ is closed under sorted substitution, that is, using the formulas of $\mathcal{L}$ as universally (U) quantified. The names of the previous well-sortedness concepts derive from their definition, so RWS stands for *rigid* [6], UWS stands for *universal*, WeiWS stands for the definition used in [Wei 95], and RUWS stands for both *universal and rigid*.

The important fact about these definitions is that none of them satisfies Lemma 7 and so they lead to unsound tableau systems. Before proving it, let us remark that, according to our previous analysis, each of these four definitions can adopt two forms. We can either require the substitution to be well-sorted w.r.t. every branch (total substitutivity rules), or demand the well-sortedness of the replaced term only w.r.t. the branch where the variable occurs free (loose substitutivity rules). In all cases, we obtain unsound substitutivity rules that make them not applicable to tableaux. We only prove the two cases of WeiWS. The other ones can be found in [GLMN 97b].

Example 17 *Let Σ be a signature with sorts s and s', constant symbol a, and function symbol f. Let $\mathcal{D}$ be the Σ-structure presented in the figure below, where arrows represent the definition of $f^{\mathcal{D}}$.*

---

[5] This means that τ' is a particular case of τ.
[6] Note that our well-sorted substitutions can also be seen as rigid.

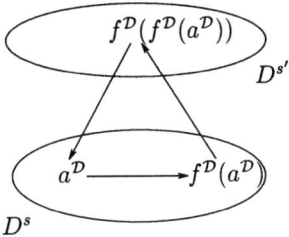

Then the set $\{a \in s, \neg f(f(a)) \in s, \forall x^s(f(f(x^s)) \in s \vee (f(x^s) \in s \wedge \forall y^s(y^s \in s)))\}$ can be checked to be satisfiable in $\mathcal{D}$, but it has a closed tableau. Applying α, β and γ' properly, the following sketch $\mathcal{T}$ of tableau can be built:

$$
\begin{array}{c}
a \in s \\
\neg f(f(a)) \in s \\
\diagup \qquad \diagdown \\
f(f(x^s)) \in s \quad \begin{array}{c} f(x^s) \in s \\ y^s \in s \end{array}
\end{array}
$$

Then the substitution $\tau \equiv [a/x^s, f(f(a))/y^s]$ is total WeiWS w.r.t. $\mathcal{T}$, and so loose WeiWS, and it can be used to close the tableau.

Theorem 18 (Completeness of $\mathcal{S}1$) *For every set of Σ-sentences Φ, if Φ is not satisfiable then Φ has a closed free variable tableau.*

Proof. As Φ is not satisfiable, then it has a closed ground tableau $\mathcal{T}$. Now we show how we can systematically build in $\mathcal{S}1$ a free variable tableau $\mathcal{T}'$ from the rules used in $\mathcal{T}$, such that $\mathcal{T} = \mathcal{T}'$.

- Every time we apply α or β in $\mathcal{T}$ we apply them in $\mathcal{T}'$.

- Every time we apply γ in $\mathcal{T}$ we apply γ' and sub in $\mathcal{T}'$, building $\mathcal{T}'$ like $\mathcal{T}$, as follows. If we use $(t \in s)$ and $\neg \exists x^s \varphi$ in $\mathcal{T}$, then we introduce $\neg \varphi[y^s/x^s]$ in $\mathcal{T}'$ and apply the substitution $[t/y^s]$. This is possible because $\mathcal{T}'$ is built like $\mathcal{T}$, and then $(t \in s)$ occurs in the branch and t is ground (so the substitution is trivially idempotent and obviously well-sorted).

- Every time we apply δ in $\mathcal{T}$ we apply δ' in $\mathcal{T}'$, using the same constant symbol. Note that this is possible because after each of these steps $\mathcal{T}'$ remains ground, so the function symbol introduced by δ' is a constant. ∎

6 Sorted Unification

As in classical first-order tableaux [Fit 96], it is not convenient to apply the rule sub in an unrestricted way, because in that case we would not improve the ground version. Therefore, we study its application only for closing branches. In this setting we need an unification calculus in order to find well-sorted unifiers for potentially complementary literals occurring in a branch.

Our calculus has to be strong enough to find well-sorted unifiers in a free variable tableau whenever its related ground version is closed. This idea is outlined in the following example.

Example 19 *Let $\mathcal{T}$ be the closed ground sketch of tableau presented on the left below. On the right, let $\mathcal{T}'$ be the free variable tableau built as $\mathcal{T}$.*

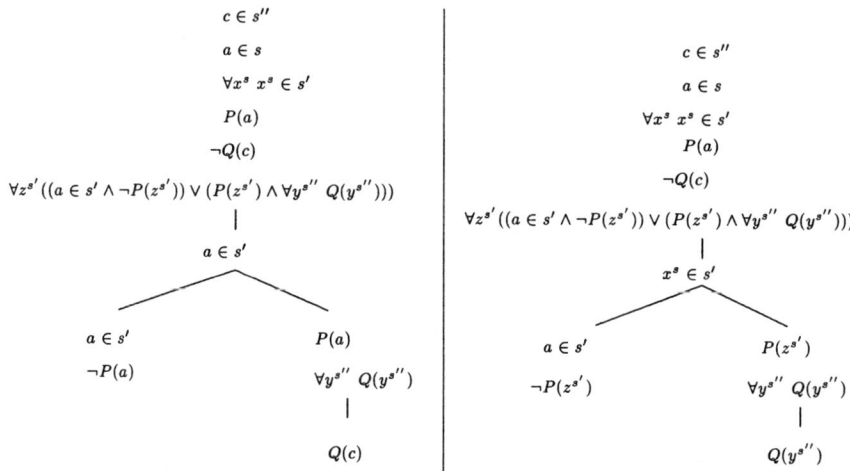

The sequence of unitary substitutions $\sigma \equiv [c/y^{s''}] [x^s/z^{s'}] [a/x^s]$ relates both tableaux. We have used it instead of $\tau \equiv [c/y^{s''}, a/z^{s'}, a/x^s]$ to express more clearly the order in the γ-rule applications to $\mathcal{T}$.

Sequences remark the idea of an existing order in the free variables substitutions of $\mathcal{T}'$, which in its turn corresponds to the order used in the γ-applications to $\mathcal{T}$. We introduce the concept of well-sorted sequence because the replacement of a certain variable can affect to the well-sortedness of another one.

Definition 20 *Let $\sigma \equiv \sigma_1 \ldots \sigma_n$, $\mathcal{L}$ and $\mathcal{T}$ be a sequence of unitary substitutions, a theory and a tableau, respectively. We say that σ is well-sorted w.r.t. $\mathcal{L}$ (resp. $\mathcal{T}$) if σ_i is well-sorted w.r.t. $\mathcal{L}\sigma_1 \ldots \sigma_{i-1}$ (resp. $\mathcal{T}\sigma_1 \ldots \sigma_{i-1}$), for every $1 \leq i \leq n$.*

Note that in the example σ is a sequence of unitary substitutions well-sorted w.r.t. $\mathcal{T}'$ and then is applicable, by Theorem 15. However, we would not be allowed to apply directly τ above to $\mathcal{T}'$, because this substitution is not well-sorted w.r.t. $\mathcal{T}'$, since $a \in s'$ does not belong to the second branch.

Using the same formulas that close $\mathcal{T}$, for closing $\mathcal{T}'$, our sorted unification calculus should be able to find out a unifier of the following two unification problems, one for each branch:

1. $z^{s'} \simeq a$ (branch B_1).   2. $y^{s''} \simeq c$ (branch B_2).

A unifier must be found because σ is a well-sorted sequence of unitary substitutions and so both problems have solution.

In the next subsection the sorted calculus $\mathcal{C}$ is presented. It tries to solve each problem separately using the theory presented in the respective branch. However the unifiers it obtains could not be well-sorted w.r.t. the whole tableau, so we have to check well-sortedness before applying the unifier. Such a test does not belong to the calculus itself but to the closure rule of the tableau method (cfr. *unif* rule in Definition 29). In the example, although the calculus $\mathcal{C}$ can solve the first problem (branch B_1) through $[a/z^{s'}]$, this substitution can not be applied to $\mathcal{T}'$ because it is not well-sorted w.r.t. the branch B_2.

6.1 The Sorted Calculus $\mathcal{C}$

The sorted calculus $\mathcal{C}$ begins with a set of equations Γ to be unified and a theory $\mathcal{L}$ that has to be observed. It works by building a well-sorted sequence of unitary substitutions w.r.t. $\mathcal{L}$ by non-deterministic application of its rules. Every time a new element of the sequence is obtained, we record it immediately, but it is never applied directly to build a unique substitution because we need to keep the order in the applications. The unification problem is solved when the set of equations is empty.

There are twelve rules in $\mathcal{C}$, the six standard rules for syntactic unification (tautology, decomposition, orientation, application, decomposition failure and cycle failure, cfr. [Wei 95]) and the following new six sorted rules:

The Sorted Rules of $\mathcal{C}$

1. Extraction
$$\frac{x^s \simeq y^s,\ \Gamma \qquad \sigma_1 \ldots \sigma_n}{\Gamma \qquad \sigma_1 \ldots \sigma_n[y^s/x^s]}$$

2. Functional Weakening
$$\frac{x^s \simeq f(t_1,\ldots,t_n),\ \Gamma \qquad \sigma_1 \ldots \sigma_n}{y^{s'} \simeq f(t_1,\ldots,t_n),\ \Gamma \qquad \sigma_1 \ldots \sigma_n[y^{s'}/x^s]}$$

In $\mathcal{L}\sigma_1\ldots\sigma_n$ there are elements of the form: $y^{s'} \in s$, $y^{s''} \in s', \ldots, f(u_1,\ldots,u_n) \in s^r$

3. Functional Failure
$$\frac{x^s \simeq f(t_1,\ldots,t_n),\ \Gamma \qquad \sigma_1 \ldots \sigma_n}{\text{FAIL}}$$

In $\mathcal{L}\sigma_1\ldots\sigma_n$ there are not elements $y^{s'} \in s$, $y^{s''} \in s', \ldots, f(u_1,\ldots,u_n) \in s^r$

4. Variable Weakening
$$\frac{x^s \simeq y^{s'},\ \Gamma \qquad \sigma_1 \ldots \sigma_n}{t \simeq y^{s'},\ \Gamma \qquad \sigma_1 \ldots \sigma_n[t/x^s]}$$

$s \neq s'$, $(t \in s) \in \mathcal{L}\sigma_1\ldots\sigma_n$, $t \notin X^s$, $x^s \notin free(t)$

5. Variable Orientation
$$\frac{x^s \simeq y^{s'},\ \Gamma \qquad \sigma_1 \ldots \sigma_n}{y^{s'} \simeq x^s,\ \Gamma \qquad \sigma_1 \ldots \sigma_n}$$

$s \neq s'$, $(t \in s') \in \mathcal{L}\sigma_1\ldots\sigma_n$, $t \notin X^{s'}$, $y^{s'} \notin free(t)$

6. Variable Weakening Failure
$$\frac{x^s \simeq y^{s'},\ \Gamma \qquad \sigma_1 \ldots \sigma_n}{\text{FAIL}}$$

$s \neq s'$ and in $\mathcal{L}\sigma_1\ldots\sigma_n$ there are not elements neither of the form:

$$\begin{cases} u^{s_{x_1}} \in s, u^{s_{x_2}} \in s_{x_1}, \ldots, u^{s''} \in s_{x_r} \\ v^{s_{y_1}} \in s', v^{s_{y_2}} \in s_{y_1}, \ldots, v^{s''} \in s_{y_k} \end{cases}$$

nor

$$\begin{cases} u^{s_{x_1}} \in s, u^{s_{x_2}} \in s_{x_1}, \ldots, f(\ldots) \in s_{x_r} \\ v^{s_{y_1}} \in s', v^{s_{y_2}} \in s_{y_1}, \ldots, f(\ldots) \in s_{y_k} \end{cases}$$

Now let us make some comments about the calculus $\mathcal{C}$. Firstly we suppose that the application of the standard rules has always preference w.r.t. the sorted rules. In particular, this means that if *functional weakening* is applied to the equation $x \simeq f(t_1,\ldots,t_n)$ then x does not occur in any other equation.

Concerning the sorted rules we comment the following. Only three rules in $\mathcal{C}$ (*extraction*, and *functional* and *variable weakening*) extract unitary substitutions and build sequences.

The *functional weakening* rule is used when we try to unify a variable x^s and a functional term t. If there exists a chain of declarations in the current theory $\mathcal{L}\sigma_1...\sigma_n$ expressing that another term with the same root symbol as t belongs to the sort s, then we append the first link of the chain to the sequence of unitary substitutions. If the chain has a single element $(f(u_1,...,u_n) \in s) \in \mathcal{L}\sigma_1...\sigma_n$ then, we apply the rule directly, without introducing extra variables. In both cases, x^s must not occur in any term of the chain. Furthermore if the chain is not unitary, we suppose $s \neq s'$, excluding not intelligent applications of the rule.

The *variable weakening* rule is required to unify two variables x^s and $y^{s'}$ of different sorts. The rule binds x^s to a term, not containing x^s, having a term declaration of sort s in the current theory.

The *variable weakening failure* rule is used when we try to unify two variables of different sorts and the current theory does not contain chains of term declarations neither expressing there is a common subsort of s and s' nor finishing in terms of subsorts of s and s', with the same functional root symbol.

Definition 21 *Let* Γ, Γ', $\sigma_1\sigma_2...\sigma_n$, $\sigma_1\sigma_2...\sigma_{n'}$ *and* $\mathcal{L}$ *be two sets of equations, two sequences of unitary substitutions with* $n' \in \{n, n+1\}$ *and a theory, respectively. We say that the pair* $\langle \Gamma', \sigma_1...\sigma_{n'}\rangle$ *is $\mathcal{C}$-accessible from* $\langle \Gamma, \sigma_1...\sigma_n\rangle$*, written* $\langle \Gamma, \sigma_1...\sigma_n\rangle \vdash_\mathcal{C} \langle \Gamma', \sigma_1...\sigma_{n'}\rangle$*, if we can obtain the pair* $\langle \Gamma', \sigma_1...\sigma_{n'}\rangle$ *from* $\langle \Gamma, \sigma_1...\sigma_n\rangle$ *using a $\mathcal{C}$-rule. We say that $\mathcal{C}$ unifies Γ w.r.t. $\mathcal{L}$ by the sequence of unitary substitutions* $\sigma_1\sigma_2...\sigma_n$ *if there exists a chain of the form* $\langle \Gamma, \emptyset\rangle \vdash_\mathcal{C} ... \vdash_\mathcal{C} \langle \emptyset, \sigma_1...\sigma_n\rangle$.

The calculus $\mathcal{C}$ is sound in the sense that, given a solvable set of equations Γ and a theory $\mathcal{L}$, it only builds sequences well-sorted w.r.t. $\mathcal{L}$ unifying Γ. Furthermore, every sequence defines an idempotent substitution so, by the soundness of the *sub* rule, we can apply it safely. The idempotence derives from the fact that the sequences that the $\mathcal{C}$-calculus obtains are triangular.

Definition 22 *[Kog 95] A sequence of unitary substitutions* $[t_1/x_1{}^{s_1}]...[t_n/x_n{}^{s_n}]$ *is triangular if*
  1. $free(t_i) \cap \{x_1{}^{s_1},...,x_i{}^{s_i}\} = \emptyset$, *for every* $1 \leq i \leq n$
  2. $x_i \neq x_j$, $1 \leq i < j \leq n$.

Theorem 23 (Soundness) *Let* $\Gamma, \mathcal{L}$ *and* $\sigma \equiv \sigma_1...\sigma_n$ *be a set of equations, a theory and a sequence of unitary substitutions, respectively. If $\mathcal{C}$ unifies Γ w.r.t. $\mathcal{L}$ by $\sigma_1...\sigma_n$ then:*
  (i) σ is a well-sorted sequence w.r.t. $\mathcal{L}$,
  (ii) σ is triangular,
  (iii) σ unifies Γ.

Note that in the previous theorem there is no mention of failure. In the calculus $\mathcal{C}$ failure does not mean total failure but a partial one, in the sense that when we apply a rule and the produced set of equations is not unifiable, we must do backtracking.

6.2 Completeness

Completeness means that if there exists a well-sorted unifier as a sequence of unitary substitutions, then $\mathcal{C}$ finds a *most general well-sorted unifier*. Considering that we are interested in a tableau system, we can restrict ourselves to a particular set of unifiers; we only need to subsume the well-sorted sequences of unitary substitutions that could be inferred from a closed ground tableau. These sequences are ground[7] and they can be characterized by the concept of *hyperwell-sortedness*.

Definition 24 *A sequence of unitary substitutions $[t_1/x_1{}^{s_1}] \ldots [t_n/x_n{}^{s_n}]$ is hyperwell-sorted w.r.t. $\mathcal{L}$ if $(t_i \in s_i) \in \mathcal{L}$, for every $1 \leq i \leq n$.*

In a hyperwell-sorted sequence, the order of the substitutions is not relevant in the sense that the declaration of the replaced term explicitly appears in the theory. In fact, it can be easily proved that if a triangular sequence is hyperwell-sorted w.r.t. a theory $\mathcal{L}$ then it is also well-sorted w.r.t. $\mathcal{L}$.

We prove the completeness of $\mathcal{C}$ in two steps, first with respect to theories and then, with respect to tableaux. In both cases, in the completeness theorem, we only consider hyperwell-sorted sequences and we implicitly introduce in the theorem the concept of most general unifier.

Theorem 25 (Completeness for a theory) *Let $\sigma_1 \ldots \sigma_n$ be a ground triangular hyperwell-sorted sequence w.r.t. a theory $\mathcal{L}$. Let Γ be a set of equations that is unified by $\sigma_1 \ldots \sigma_n$. Then the calculus $\mathcal{C}$ unifies Γ w.r.t. $\mathcal{L}$ by $\tau_1 \ldots \tau_p$ and there exists another sequence $\theta_1 \ldots \theta_k$ such that:*
  1. *$\theta_1 \ldots \theta_k$ is triangular,*
  2. *$\theta_1 \ldots \theta_k$ is hyperwell-sorted w.r.t. $\mathcal{L}\tau_1 \ldots \tau_p$,*
  3. *$\tau_1 \ldots \tau_p \theta_1 \ldots \theta_k = \sigma_1 \ldots \sigma_n$.*

Sketch of proof. The idea is to build a transformation system $\mathfrak{S}$ of triples:

$$\langle \Gamma, \tau_1 \ldots \tau_p, \eta_1 \ldots \eta_r \rangle \vdash_{\mathfrak{S}} \langle \Gamma', \tau_1 \ldots \tau_{p'}, \zeta_1 \ldots \zeta_s \rangle$$

such that:

(i) $\langle \Gamma, \tau_1 \ldots \tau_p \rangle \vdash_{\mathcal{C}} \ldots \vdash_{\mathcal{C}} \langle \Gamma', \tau_1 \ldots \tau_{p'} \rangle$.

(ii) If $\eta_1 \ldots \eta_r$ is a triangular hyperwell-sorted sequence w.r.t. $\mathcal{L}\tau_1 \ldots \tau_p$ which unifies Γ and verifies $\tau_1 \ldots \tau_p \eta_1 \ldots \eta_r = \sigma_1 \ldots \sigma_n$, then $\zeta_1 \ldots \zeta_s$ is a triangular hyperwell-sorted sequence w.r.t. $\mathcal{L}\tau_1 \ldots \tau_{p'}$ which unifies Γ' and verifies $\tau_1 \ldots \tau_{p'} \zeta_1 \ldots \zeta_s = \sigma_1 \ldots \sigma_n$.

Starting with $\langle \Gamma, \emptyset, \sigma_1 \ldots \sigma_n \rangle$ and decreasing step-by-step the complexity of the triples w.r.t. a well-founded order, the system $\mathfrak{S}$ reaches a triple with minimum complexity satisfying $\Gamma = \emptyset$. In these conditions, we can build a sequence of the form:

$$\langle \Gamma, \emptyset, \sigma_1 \ldots \sigma_n \rangle \vdash_{\mathfrak{S}} \ldots \vdash_{\mathfrak{S}} \langle \emptyset, \tau_1 \ldots \tau_p, \theta_1 \ldots \theta_k \rangle$$

and so:
  a) $\mathcal{C}$ unifies Γ w.r.t. $\mathcal{L}$ by $\tau_1 \ldots \tau_p$
  b) $\theta_1 \ldots \theta_k$ is a triangular hyperwell-sorted sequence w.r.t. $\mathcal{L}\tau_1 \ldots \tau_p$ verifying $\sigma_1 \ldots \sigma_n = \tau_1 \ldots \tau_p \theta_1 \ldots \theta_k$. ∎

---

[7] When required, sequences of unitary substitutions will be analyzed and compared through the composition they define.

The completeness of $\mathcal{C}$ w.r.t. a theory can be lifted to a tableau. First we define how to extend the concept of hyperwell-sortedness to tableaux.

Definition 26 *A sequence of unitary substitutions $\sigma \equiv [t_1/x_1{}^{s_1}]\ldots[t_n/x_n{}^{s_n}]$ is hyperwell-sorted w.r.t. a tableau $\mathcal{T}$ if for every branch B of $\mathcal{T}$, the subsequence of σ that only replaces free variables of B, written $\sigma|_{free(B)}$, is hyperwell-sorted w.r.t. the theory included in B.*

As with simple theories, we can prove the same result relating well and hyperwell-sortedness. Therefore a triangular hyperwell-sorted sequence w.r.t. a tableau $\mathcal{T}$ is also well-sorted w.r.t. $\mathcal{T}$.

If $\mathcal{T}$ is a closed ground tableau and $\mathcal{T}'$ is its related free variable version, then we can easily deduce from $\mathcal{T}$ different triangular hyperwell-sorted sequences w.r.t. $\mathcal{T}'$. In Example 19, the two sequences $[c/y^{s''}][x^s/z^{s'}][a/x^s]$ and $[x^s/z^{s'}][a/x^s][c/y^{s''}]$ are triangular hyperwell-sorted w.r.t. $\mathcal{T}'$. However, only the first one reflects the idea of a bottom-up ordering in the introduction of variables by application of γ'-rule (the last variable introduced by γ' is the first variable appearing in the sequence). In order to precise this ordering and specify the sequences that will be lifted, we introduce the following concept.

Definition 27 *A sequence of unitary substitutions $[t_1/x_1{}^{s_1}]\ldots[t_n/x_n{}^{s_n}]$ is well-ordered w.r.t. a tableau $\mathcal{T}$ if for every $1 \leq i < j \leq n$ such that there exists a branch B of $\mathcal{T}$ with $x_i, x_j \in free(B)$, it holds $x_i \in free(B') \Longrightarrow x_j \in free(B')$, for every branch B' of $\mathcal{T}$.*

So in a well-ordered sequence, if two variables x and y appear in a branch, and x appears before than y in the sequence then, y has been introduced by γ'-application in every branch where x occurs. This means that the substitution $[x^s/z^{s'}][a/x^s][c/y^{s''}]$ of Example 19 is not well-ordered w.r.t. $\mathcal{T}'$ because $z^{s'}$ and $y^{s''}$ occur in the second branch of $\mathcal{T}'$, but only $z^{s'}$ occurs in the first branch.

From now on, we will only deal with well-ordered sequences. Following the ideas explained in Example 19, it can be easily deduced triangular hyperwell-sorted and well-ordered sequences w.r.t. a closed ground tableau.

In order to get completeness, a result like Theorem 25 is not strong enough because it only assures well-sortedness w.r.t. the initial theory; now we have to be sure that the obtained $\mathcal{C}$-sequence unifying Γ is well-sorted w.r.t. the whole tableau $\mathcal{T}$, which means that its application is sound, by Theorem 15. So completeness is stated as follows.

Theorem 28 (Completeness for a tableau) *Let $\sigma_1 \ldots \sigma_n$ be a ground triangular hyperwell-sorted and well-ordered sequence w.r.t. a tableau $\mathcal{T}$. Let Γ be a set of equations, obtained from a given branch B of $\mathcal{T}$, that is unified by $\sigma_1 \ldots \sigma_n$. Then the calculus $\mathcal{C}$ unifies Γ w.r.t. the theory included in B by the sequence $\tau_1 \ldots \tau_p$, which is well-sorted w.r.t. to $\mathcal{T}$, and there exists another sequence $\theta_1 \ldots \theta_k$ such that:*
  1. *$\theta_1 \ldots \theta_k$ is triangular,*
  2. *$\theta_1 \ldots \theta_k$ is hyperwell-sorted and well-ordered w.r.t. $\mathcal{T}\tau_1 \ldots \tau_p$,*
  3. *$\tau_1 \ldots \tau_p \theta_1 \ldots \theta_k = \sigma_1 \ldots \sigma_n$.*

7 Free Variable Tableaux with Sorted Unification

In this section we take advantage of our sorted unification calculus $\mathcal{C}$, presenting the tableau system $\mathcal{S}2$. This system is composed of the rules α, β, γ', δ' and the following unification rule *unif*. The concepts of closed branch and closed tableau in $\mathcal{S}2$ are defined as in $\mathcal{S}1$.

Definition 29 (Unification Rule) *Let B be a branch of a free variable tableau $\mathcal{T}$ containing two potentially complementary literals φ and $\neg\varphi'$. If $\mathcal{C}$ unifies the set $\{\varphi \simeq \varphi'\}$ w.r.t. B by the sequence $\sigma_1 \ldots \sigma_n$ and such a sequence is well-sorted w.r.t. $\mathcal{T}$, then $\mathcal{T}\sigma_1 \ldots \sigma_n$ is a free variable tableau.*

Note the importance of using sequences of unitary substitutions instead of a unique substitution. Structuring $\mathcal{C}$-unifiers in unitary substitutions determines whether the $\mathcal{C}$-unifier is well-sorted or not w.r.t. the whole tableau.

Theorem 30 (Soundness of $\mathcal{S}2$) *For every set of Σ-sentences Φ, if Φ has a closed tableau then Φ is not satisfiable in structures with non-empty domains, for every sort.*

Completeness of $\mathcal{S}2$ is obtained by a lifting lemma expressing that if we can close a ground tableau, then we can close its related free variable version.

Lemma 31 (Lifting Lemma) *Let $\mathcal{T}$ be a free variable tableau and $\sigma_1 \ldots \sigma_n$ a ground triangular hyperwell-sorted and well-ordered sequence w.r.t. $\mathcal{T}$. Let $\mathcal{T}_b$ be a ground tableau such that $\mathcal{T}\sigma_1 \ldots \sigma_n = \mathcal{T}_b$. Then if $\mathcal{T}_b$ is closed, $\mathcal{T}$ can be closed.*

Proof. It is enough to prove that we can close a branch B of $\mathcal{T}$ using *unif* via a sequence $\tau_1 \ldots \tau_p$, and that there exists a ground triangular hyperwell-sorted and well-ordered sequence $\theta_1 \ldots \theta_k$ w.r.t. $\mathcal{T}\tau_1 \ldots \tau_p$ such that $\mathcal{T}\tau_1 \ldots \tau_p \theta_1 \ldots \theta_k = \mathcal{T}_b$.

As $\mathcal{T}_b$ is closed then in B we have two literals of the form $P(t)$ and $\neg P(t')$ (w.l.o.g. we suppose that P is unary) such that $t\sigma_1 \ldots \sigma_n = t'\sigma_1 \ldots \sigma_n$. By Theorem 28, there exists a sequence $\tau_1 \ldots \tau_p$ such that $\mathcal{C}$ unifies the set $\{t \simeq t'\}$ w.r.t. B by $\tau_1 \ldots \tau_p$. This theorem also assures that such a sequence is well-sorted w.r.t. the whole tableau – then we can apply *unif* – and that there exists a ground triangular hyperwell-sorted and well-ordered sequence $\theta_1 \ldots \theta_k$ w.r.t. $\mathcal{T}\tau_1 \ldots \tau_p$ such that $\tau_1 \ldots \tau_p \theta_1 \ldots \theta_k = \sigma_1 \ldots \sigma_n$. Then $\mathcal{T}\tau_1 \ldots \tau_p \theta_1 \ldots \theta_k = \mathcal{T}_b$. ∎

Theorem 32 (Completeness of $\mathcal{S}2$) *For every set of Σ-sentences Φ, if Φ is not satisfiable then Φ has a closed free variable tableau.*

8 Conclusions

We have presented the logic with term declarations *LTD*. This is an order-sorted logic which extends the classical first-order logic by introducing a new formula constructor $t \in s$, allowing the dynamic declaration of the term t as an element of sort s.

In [GLMN 97a] we have presented a ground tableau method for an extension of *LTD* with a new kind of formula $t \sqsubseteq t'$, expressing that the term t is less or equal than the term t' in a preordered domain.

This time we have studied free-variable tableau versions for LTD. The first question to be solved is how to define sound substitutions of variables in tableaux. As far as we know, the only proposal of a similar tableau method [Wei 95] uses an unsound substitutivity rule. We have proved that this one and some other possible attempts to define such a rule fall into error, while our concept of well-sorted substitution σ w.r.t. a theory of term declarations $\mathcal{L}$ avoids unsoundness, by requiring idempotence to σ and the explicit declaration $t \in s$ in $\mathcal{L}$, for every substituted variable $[t/x^s]$ of σ. This notion entails a certain component of rigidity in the sense that demanding $(t \in s) \in \mathcal{L}$ assures that the interpretation of t belongs to s, when the valuation of its variables is fixed. So the variables of t do not behave as universally quantified, but as constants.

A free-variable tableau version $\mathcal{S}1$ based on this substitutivity rule is proved sound and complete. However there is no improvement w.r.t. the ground tableau version. So we have studied how to restrict its application to the closure of branches and we have defined a calculus $\mathcal{C}$ for unifying equations w.r.t. a set of term declarations. This calculus is sound and complete not only w.r.t. term declaration theories, but also w.r.t. tableaux. In both cases, substitutions are structured into unitary components $[t_i/x_i]$; a sequence of such unitary substitutions reflects the idea of an order in the introduction of free variables by applications of γ'-rule. Finally a free-variable tableau version $\mathcal{S}2$ based on this unification rule is proved also complete.

Acknowledgments: We are greatly indebted to Susana Nieva for helpful discussions and comments.

References

[Coh 87]   A. G. Cohn. *A more expressive formulation of many sorted logic*. Journal of Automated Reasoning 3, 113–200, 1987.

[Fit 96]   M. Fitting. *First-Order Logic and Automated Theorem Proving*. Second edition. Springer, 1996.

[Fri 91]   A. M. Frisch. *The substitutional framework for sorted deduction: fundamental results on hybrid reasoning*. Artificial Intelligence 49, 161–198, 1991.

[GLMN 96]   A. Gavilanes, J. Leach, P. J. Martín, S. Nieva. *Reasoning with preorders and dynamic sorts using free variable tableaux*. Proc. AISMC-3. LNCS 1138, 365–379, 1996.

[GLMN 97a]   A. Gavilanes, J. Leach, P. J. Martín, S. Nieva. *Ground semantic tableaux for a logic with preorders and dynamic declarations*. TR DIA 50/97. 1997.

[GLMN 97b]   A. Gavilanes, J. Leach, P. J. Martín. *Free variable tableaux for a logic with term declarations*. TR DIA 69/97. 1997.

[Kog 95]   E. Kogel. *Rigid E-unification simplified*. Proc. 4th International Workshop TABLEAUX'95. LNCS 918, 17–30, 1995.

[Sch 89]   M. Schmidt-Schauss. *Computational Aspects of an Order Sorted Logic with Term Declarations*. LNAI 395. Springer, 1989.

[Wal 87]   C. Walther. *A Many-sorted Calculus based on Resolution and Paramodulation*. Research Notes in Artificial Intelligence. Pitman, 1987.

[Wei 91]   C. Weidenbach. *A sorted logic using dynamic sorts*. MPI-I-91-218, 1991.

[Wei 95]   C. Weidenbach. *First-order tableaux with sorts*. J. of the Interest Group in Pure and Applied Logics 3(6), 887–907, 1995.

Simplification*
A General Constraint Propagation Technique for Propositional and Modal Tableaux

Fabio Massacci

Dipartimento di Informatica e Sistemistica,
Università di Roma "La Sapienza"
via Salaria 113, I-00198 Roma, Italy
massacci@dis.uniroma1.it

Abstract. Tableau and sequent calculi are the basis for most popular interactive theorem provers for formal verification. Yet, when it comes to automatic proof search, tableaux are often slower than Davis-Putnam, SAT procedures or other techniques. This is partly due to the absence of a bivalence principle (viz. the cut-rule) but there is another source of inefficiency: the lack of constraint propagation mechanisms.
This paper proposes an innovation in this direction: the rule of *simplification*, which plays for tableaux the role of subsumption for resolution and of unit for the Davis-Putnam procedure.
The simplicity and generality of simplification make possible its extension in a uniform way from propositional logic to a wide range of modal logics. This technique gives an unifying view of a number of tableaux-like calculi such as DPLL, KE, HARP, hyper-tableaux, BCP, KSAT.
We show its practical impact with experimental results for random 3-SAT and the industrial IFIP benchmarks for hardware verification.

1 Introduction

It is a widespread belief that methods based on the sequent calculus (such as tableaux) are hopeless for "real life" satisfiability and validity search. Even for decidable problems with a natural appeal for tableaux, such as modal or propositional logics, experimental results have shown that other algorithms outperform them by orders of magnitude [6, 12, 16, 27, 33].

The key question is whether such a gap is inherent to tableau methods or something is simply missing. The answer is not only of theoretical interest but has an extreme relevance for the development of formal verification tools.

Indeed, variants of the sequent calculus are the main techniques used by interactive theorem provers, such as Isabelle [22], PVS [26] or HOL [13]. Those provers have successfully tackled hardware and software verification and often require to prove some properties in decidable sub-theories such as propositional

---

* An comprehensive review of the various results on simplification, including the first order case can be found in [18].

logic (e.g. a N-bits binary adder) or fragments of arithmetics [26]. If tableaux are "hopeless by nature", then non sequent-based systems (outside the prover) should be consulted as oracles. For instance efficient algorithms for propositional logics as one-off "derived rules" for HOL have been proposed in [15].

To fill the computational gap, a number of works have improved the effectiveness of tableau method by switching from ground calculi to free variables versions [23, 11] with smart skolemisation techniques [1], adding ad-hoc rules for modus ponens and tollens [24, 7, 16, 21], imposing regularity conditions [7, 17], using controlled form of cut [17, 7], or factoring and merging [20, 32], exploiting universal variables [4, 3], incorporating features of hyper-resolution [2]. In some cases [6] the quest for an efficient implementation has lead from an original sequent calculus to a (seemingly different) Davis-Putnam procedure.

We propose a general technique, which subsumes a number of these approaches, and whose intuition is due to a comparison between tableaux and their "historical competitors".

Since the very beginning, the Davis-Putnam-Longeman-Loveland (DPLL for short) procedure [9, 8], and resolution [25] included rules for the simplification of the formulae to be proved unsatisfiable (e.g. unit or subsumption) *without* changing the basic inference mechanism.

The implementation of this simplification procedures is almost a research field in itself [31, 33] and difficult problems could hardly (if at all) be solved by either resolution or DPLL without using them. On the contrary, even in the modern texts on tableaux [11] everything is proved by "first principles".

We advocate that the lack of *rules for constraint propagation* is one of the main cause for the computational gap[1]. Hence we need a simple theoretical innovation: an operation that *plays for the sequent calculus the same role of unit for DPLL and subsumption for resolution*. We call it *simplification* and discuss its application to propositional, and modal logics.

The simplicity and flexibility of simplification makes it possible its application to a wide range of logics and logical formalism, as soon as there is sequent calculus. The only difficult feature is the enhancement of simplification as we "upgrade" the logic. Its effectiveness can enhance the computational power of interactive and automatic tableau based provers.

The introduction of simplification for the sequent calculus provides an *unifying perspective* of many (tableau based) deduction techniques and "explains away" the characteristic of the DPLL procedure (or KSAT-procedures of [12] which are based on DPLL) or KE. For instance the first order version in [8] can be "re-interpreted" as a tableau á la Smullyan "plus" propositional simplification. Such an interpretation can also explain the comparative inefficiency of first-order DPLL versus propositional DPLL (see further [18]).

We show the practical impact of the operation of simplification on the performance of tableau methods: a simple Prolog implementation shows essentially the same easy-hard-easy computation pattern for random 3-SAT shown by DPLL

---

[1] This has also been confirmed for modal logics by recent experimental studies [16] which clarified the gap between KSAT and tableau based procedures.

[6] even with a a simpler notion of simplification than the full fledged used by DPLL. An extensive experimental analysis has also been carried on the industrial IFIP (non clausal) benchmarks for hardware verification [5].

In the rest of the paper we present the intuitions behind the operation of simplification (2) and introduce some preliminary notation (§3). Next, we present the calculus for propositional logics (§4), show how other approaches are subsumed (§5) and present the extension to modal logics (§6). Finally, we discuss the experimental results (§7), and conclude (§8).

2 Principles and Intuitions

We assume a basic knowledge of the sequent calculus (see [28] for an introduction). In the sequel formulae are denoted by A, B, C, sets of formulae by Γ, Δ and sequents by $\Gamma \Longrightarrow \Delta$.

Simplification can be explained by comparing it with subsumption: bottom-up methods (resolution) explore the search space by generating new information and to simplify the search they *retain information* to discard (newly generated) irrelevant facts. Top-down methods (tableaux) work by breaking existing information and therefore the best way to delete irrelevant information should be *anticipating* the outcome of the search.

For example, consider the following α-rule:

$$\frac{\Gamma, A, B \Longrightarrow \Delta}{\Gamma, A \wedge B \Longrightarrow \Delta} \ (\alpha)$$

After this rule a tableau will continue to break down connectives until a branch closure rule can be applied. It looks for something like $\Gamma, C \Longrightarrow \Delta, C$.

Yet, after the α-step, there is some information that the calculus does not use to anticipate the outcome of its search: when all formulae in $\Gamma \cup \{A, B\}$ are true then A must be true, no matter what A is (equally for B). We can use this information about A to *simplify Γ and Δ before reducing other connectives.*

The simplest way is syntactic search: look for all exact[2] occurrences of A in Γ and replace it with the constant $\top$; next perform some boolean operations to eliminate $\top$ and $\bot$ from the formula; only afterwards continue with other rules.

The computational pay-off may seem doubtful: some time must be spent for scanning the formula replacing the occurrences with $\top$; eliminate boolean constant and so on. So what could be the gain?

Intuition 1. The sequent proof of $\Gamma \Longrightarrow \Delta$ is potentially of size $O(2^{|\Gamma \cup \Delta|})$ [30]. If can reduce $\Gamma \Longrightarrow \Delta$ to $\Gamma^* \Longrightarrow \Delta^*$ so that its size decreases, if only by 1, we would reduce the potential search space at least by half.

We trade off some polynomial processing for an exponential gain: scanning the formula and looking for "exact" copies can be done in polynomial time.

---
[2] With modal connectives or variables it is not so simple but this is just the intuition.

Hence, if there is at least another occurrence of A (or B) in Γ, it always pays off to replace this occurrence with $\top$: the resulting sequent will be smaller and thus the proof tree below will also be (exponentially) smaller.

We would like simplification to be

- *locally applicable* to single formulae;
- a *sound and complete* admissible rule for the underlying calculus;
- substantially a *rewriting rule*;
- a rule requiring *polynomial time*;
- such that *the resulting formulae are (potentially) smaller*.

Soundness is obvious. We need completeness only if we impose that simplification is applied before other rules. If rules can be applied in any order, incomplete simplification would be an "unsafe" rule over which we may backtrack [22].

Local applicability is also a key property and leads to rules like "simplify B using A". The advantage of this definition and the rules we propose is that when we "upgrade" the logic we do not have (in general) to change the rules but only to specify the behavior of the operation for the *new* connectives we have introduced. The flexibility and modularity makes it easy also to upgrade current implementations with these techniques.

3 Notation and Terminology

For an introduction to tableaux for propositional logic see [11, 28] and [10, 14] for propositional modal logics.

We construct propositional formulae A, B, C from a set of propositional atoms $p, q \in \mathcal{P}$ and the boolean constants $\top$ and $\bot$ with the logical connectives $\wedge, \vee, \neg, \supset$ etc. Propositional modal logics are obtained by adding the unary operators $\Box$ (necessity) and $\Diamond$ (possibility).

For sake of modularity we use signed formulae and the α, β, ν, π notation of Smullyan [28] and Fitting [10]. Thus a *signed formula*, denoted by φ and ψ, is a pair $\mathtt{t}.A$ or $\mathtt{f}.A$. In particular $\mathtt{t}.A$ ($\mathtt{f}.A$) is a formula with positive (resp. negative) sign. If A is a propositional letter we say that it is a signed atom. The intuitive interpretation is that A is assumed to be respectively true and false. The *conjugate* $\overline{\varphi}$ of a signed formula φ is obtained by switching $\mathtt{t}.$ with $\mathtt{f}.$ and vice versa. The uniform classification of formulae is given in Table 1.

Sets of formulae are represented by Γ and Δ and a *sequent*, written using Gentzen notation, is the pair $\Gamma \Longrightarrow \Delta$. With signed formula the sequent $\Gamma \Longrightarrow \Delta$ is represented as a set $\mathcal{S}$ of signed formulae as follows: $\mathcal{S} = \{\mathtt{t}.A \mid A \in \Gamma\} \cup \{\mathtt{f}.A \mid A \in \Delta\}$. We abbreviate $\mathcal{S} \cup \{\varphi\}$ as $\mathcal{S}, \varphi$. In this set-oriented framework the usual structural rules such as weakening and contraction become redundant.

4 Propositional Logic

The rules and axioms of the calculus are given in Fig. 1. The only new rule is denoted by $(simp)$, and we call it *local simplification*. The use of signed formulae makes the definition of rule $(simp)$ extremely compact.

α	α_1	α_2
t.$A \wedge B$	t.A	t.B
f.$A \vee B$	f.A	f.B
f.$A \supset B$	t.A	f.B

α	α_1, α_2
t.$\neg A$	f.A
f.$\neg A$	t.A

β	β_1	β_2
f.$A \wedge B$	f.A	f.B
t.$A \vee B$	t.A	t.B
t.$A \supset B$	f.A	t.B

ν	ν_0
t.$\Box A$	t.A
f.$\Diamond A$	f.A

π	π_0
f.$\Box A$	f.A
t.$\Diamond A$	t.A

Table 1. Uniform notation

$$\frac{\mathcal{S}, \alpha_1, \alpha_2}{\mathcal{S}, \alpha} \; (\alpha) \qquad \frac{\mathcal{S}, \beta_1 \quad \mathcal{S}, \beta_2}{\mathcal{S}, \beta} \; (\beta) \qquad \frac{\mathcal{S}, \psi[\varphi], \varphi}{\mathcal{S}, \psi, \varphi} \; (simp) \qquad \frac{\mathcal{S}}{\mathcal{S}, \varphi} \; (thin)$$

$$\frac{}{\mathcal{S}, \mathtt{f}.\top} \; (\top) \qquad \frac{}{\mathcal{S}, \mathtt{t}.\bot} \; (\bot) \qquad \frac{}{\mathcal{S}, \varphi, \overline{\varphi}} \; (Ax) \qquad \frac{\mathcal{S}, \varphi \quad \mathcal{S}, \overline{\varphi}}{\mathcal{S}} \; (cut)$$

Fig. 1. Propositional Sequent Rules with Simplification

Rule (*thin*) is not necessary and rule (*cut*) can be eliminated without hindering soundness and completeness [28, 11], although it may impact the computational complexity [7]. We introduce them to show that some particular forms of simplification are derived by using them in combination with (*simp*).

The definition of sequent tree and proof are standard [11, 28]: a *sequent tree* is a dyadic tree where each node is labelled by a sequent, the root is labelled with the sequent we are trying to prove and, for each node, its children are labelled with the corresponding consequents of a rule of the calculus. A *proof* is a sequent tree where each leaf is labelled with an axiom.

The next tool of our machinery, is the definition of the proper operation of simplification $\psi[\varphi]$. Signs will be left unchanged by simplification and we have just to specify the progression over formulae. In the following definition $\sharp$ is any binary (propositional) connective.

$$A[\varphi] \mapsto \begin{cases} \top & \text{if } \varphi = \mathtt{t}.A \\ \bot & \text{elseif } \varphi = \mathtt{f}.A \\ \neg(B[\varphi]) & \text{elseif } A = \neg B \\ B[\varphi] \sharp C[\varphi] & \text{elseif } A = B \sharp C \\ A & \text{otherwise} \end{cases} \qquad (1)$$

By $A = B$ we mean, at this preliminary stage, syntactic identity. So $A \wedge B$ is different from $B \wedge A$. Further extensions, where the commutative properties of the connectives are considered, are described in the following subsections.

There is a number of observations worth making:

1. we do not only reduce branching, we get smaller and simpler formulae
2. the operation of simplification is parametric w.r.t. the connectives used;

$$\cfrac{\cfrac{A,(A \supset B) \vee (A \supset C), D \Longrightarrow A}{A,(A \supset B) \vee (A \supset C) \Longrightarrow D \supset A} \; \alpha_\supset \quad \cfrac{\cfrac{A, B \Longrightarrow B, C}{A, A \supset B \Longrightarrow B, C} \; MP \quad \cfrac{A, C \Longrightarrow B, C}{A, A \supset C \Longrightarrow B, C} \; MP}{\cfrac{A,(A \supset B) \vee (A \supset C) \Longrightarrow B, C}{A,(A \supset B) \vee (A \supset C) \Longrightarrow B \vee C} \; \alpha_\vee} \; \beta_\vee}{\cfrac{\cfrac{A,(A \supset B) \vee (A \supset C) \Longrightarrow (D \supset A) \wedge (B \vee C)}{A \Longrightarrow (A \supset B) \vee (A \supset C) \supset (D \supset A) \wedge (B \vee C)} \; \alpha_\supset}{\Longrightarrow A \supset ((A \supset B) \vee (A \supset C) \supset (D \supset A) \wedge (B \vee C))} \; \alpha_\supset} \; \beta_\wedge$$

$$\cfrac{\cfrac{\cfrac{A, B \vee C \Longrightarrow B \vee C}{A \Longrightarrow B \vee C \supset B \vee C} \; \alpha_\supset}{A \Longrightarrow ((A \supset B) \vee (A \supset C) \supset (D \supset A) \wedge (B \vee C))} \; simp_{t.A}}{\Longrightarrow A \supset ((A \supset B) \vee (A \supset C) \supset (D \supset A) \wedge (B \vee C))} \; \alpha_\supset$$

Fig. 2. HARP modus ponens versus simplification

3. the simplifying formula φ needs not be a literal nor have any normal form;
4. without restriction, simplification can proceed recursively down any level of nesting of propositional connectives.

The last step is one of the key differences with modus ponens and tollens applied in HARP [21], the KE β^c-rules [7] and the $\vee - simp_{0,1}$-rule used in [16]. Indeed we will show that they are particular forms of simplification where the recursion is stopped at the main connective of ψ.

Theorem 1. *Simplification (simp) is a sound and complete (invertible) rule for propositional logic.*

Proposition 1. *For propositional logic the size of $\psi[\varphi]$ is never larger than the size of ψ. It is strictly smaller whenever φ occurs as a subformula in ψ.*

An example is shown in Fig 2 where we compare simplification with the rule of Modus-Ponens in HARP. To simplify the notation we assume the usual associativity precedence of $\wedge$ over $\vee$ and of those two connectives over $\supset$ and use the $\Gamma \Longrightarrow \Delta$ notation in the obvious way.

Of course one may argue that the pattern A and $(A \supset B) \vee (A \supset C)$ could be easily recognized but it is also easy to change it substantially; for instance as $(A \supset B) \vee ((A \wedge \neg B) \vee D \supset C)$. This formula can still be proved with local simplification without any branching whereas HARP's proof is harder than the one shown in Fig. 2. Moreover the point of simplification is that we do not want to remember a lot of particular patterns.

There is another advantage: with open branches *simplification leads to fewer and smaller counter-models*, i.e. it sets the truth value of fewer propositions. For example one may try $\mathtt{t}.p \wedge (r \wedge \neg(q \supset p) \supset q)$. "Normal" tableaux have four open branches: the first (left-to-rigth) is $\{\mathtt{t}.p, \mathtt{f}.r\}$, then $\{\mathtt{t}.p, \mathtt{f}.q\}$, followed by $\{\mathtt{t}.p\}$ and $\{\mathtt{t}.p, \mathtt{t}.q\}$. The application of simplification yields only the model $\{\mathtt{t}.p\}$.

5 Subsuming Other Approaches

A wide variety of rules used by tableau provers turn out to be restricted and shallow forms of simplification. Here we consider the case for the unit of DPLL [9], the β^c-rules of KE [7],the Modus ponens and tollens of HARP [21], hyper-tableau expansions [2] and boolean constraint propagation [19].

First consider DPLL unit: "let l be a unit clause, delete every clause which contains l and delete any occurrence of $\bar{l}$ from the remaining clauses" [8].

Suppose we have $\Gamma, l \Longrightarrow \emptyset$ where Γ is set of clauses and we apply simplification to all formulae of Γ. By applying the boolean reductions $\bot \lor l_1 \lor \ldots l_n \mapsto l_1 \lor \ldots \lor l_n$ and $\top \lor l_1 \lor \ldots l_n \mapsto \top$, and finally $\top \land C_1 \land \ldots \land C_n \mapsto C_1 \land \ldots \land C_n$ we obtain the same result of the unit rule.

Fact 1. *The* DPLL-*unit rule is derived by the sequent calculus with the* (simp)-*rule restricted to signed atoms, when S is a set of clauses with positive sign.*

Boolean constraint propagation is an inference system where the unit rule is the only rule of inference. Hence we subsume directly also the BCP system [19].

Then, we simply need to restrict the application of the cut rule to atomic proposition to make the following observation:

Fact 2. *The* DPLL *procedure is derived by the sequent calculus with rules* (simp), (cut) *and* (thin) *restricted to signed atoms, when S is a set of clauses with positive sign.*

The reason for the successes of the split-rule of DPLL or the folding operation with unit lemmata [17] is indeed their combination with (hidden) unit simplification: we add a unit literal (the negation of the left-hand side), we apply simplification (unit) and then use thinning to eliminate the literal. So *the size of the right subtree is never larger*.

Equally, the β^c-rules of KE from [7] are a generalization to signed formulae of the modus ponens and modus tollens of HARP [21] and the $\lor - simp_{0,1}$ in [16]. Reversed bottom-up from their original tableau-notation they are as follows:

$$\frac{S, \beta_2, \overline{\beta_1}}{S, \beta, \overline{\beta_1}} \beta_1^c \qquad \frac{S, \beta_1, \overline{\beta_2}}{S, \beta, \overline{\beta_2}} \beta_2^c$$

It can be easily shown, by cases, that they are extremely restricted form of simplification. For instance consider the following instance of β_1^c:

$$\frac{\Gamma, B \longrightarrow \Delta, \Lambda}{\Gamma, A \lor B \Longrightarrow \Delta, A} \text{ t.}A^c \qquad \frac{\Gamma, (A \lor B)[\texttt{t}.A] \Longrightarrow \Delta, A}{\Gamma, A \lor B \Longrightarrow \Delta, A} \text{ } simp_{\texttt{f}.A}$$

If we applied the full-fledge version of simplification we would have had:

$$(A \lor B)[\texttt{f}.A] \mapsto A[\texttt{f}.A] \lor B[\texttt{f}.A] \mapsto \bot \lor B[\texttt{f}.A] \mapsto B[\texttt{f}.A]$$

Thus, while KE only yields B, the full fledged use of simplification would have given us $B[\texttt{f}.A]$. If A occurred somewhere in B simplification would have reduced B and, in general, shortened the proof further.

Fact 3. *The β_i^c rules for KE are instances of the simplification rule, restricted to β-formulae and such that recursive simplification stops at the main connective.*

As a simpler corollary we get:

Fact 4. *Modus ponens, tollens etc. of HARP and the $\vee - simp_{0,1}$ rules in [16] are instances of the simplification rule (simp), when recursion stops at the main connective of S formulae and it is only applied to formulae of particular forms.*

The extension step of hyper-tableaux [2] can also be subsumed:

Fact 5. *The expansion rule of propositional hyper tableaux [2] is derived by the sequent calculus with (simp) restricted to atoms, a restriction on the application of the β-rules (β-formulae must not contains negated atoms), when S is a set of clauses with positive sign.*

The next restriction that is imposed on clausal tableau proofs or model elimination techniques is *regularity* [2, 7, 17, 32, 30]: a literal should never appear twice on a branch. In the setting of propositional logic and clausal theorem proving (such as [2, 17]) this is equivalent to saying that a clause is never selected for the extension of a branch in the tableau if the extension step (an n-ary β-rule) will yield a literal to appear twice. It is a system to constrain the search space and in particular the non-determinism in the selection of clauses for reduction.

Since we use sequents as sets, and are not restricted to clausal normal form we rephrase it in more general terms: a proof is regular iff all subformulae introduced by α, β rules (i.e. $\beta_1, \beta_2, \alpha_1, \alpha_2$) do not occur already in the sequent S before the application of the rule.

Fact 6. *If simplification is given precedence over other rules then all propositional tableaux proofs are regular.*

We can devise more powerful forms of simplification if we replace $\wedge$ and $\vee$, which are commutative, associative and idempotent, with set-oriented versions $\bigwedge \{\}$ and $\bigvee \{\}$. For instance $A \wedge A \wedge B \wedge (C \vee D \vee E) \mapsto \bigwedge \{A, B, \bigvee \{C, D, E\}\}$. This is already a simplification since we eliminate duplicates. This topic is described in more details in [18]. In the experiments mentioned in §7 this is obtained by using prolog lexicographically ordered lists (without repetitions).

It has been sometimes argued that the use of the name TABLEAU in [6] was somehow misleading since it was an efficient implementation of DPLL. In the light of the results of this paper the choice of the name was indeed correct.

6 Modal Logics from K to S5

A wide range of tableau and sequent calculi have been proposed for modal logics (see e.g. [10]). For simplicity, we use the Gentzen-like formulation without adding extra-logical symbols. These techniques can be adapted to prefixed systems [10].

The rules are shown in Fig. 3. Different logics can be obtained by changing the operation $\mathcal{S}^*$. The same figure list the cases for the major modal logics

Modal Rules	Logic	Composition of S^*	Logic	Composition of S^*
$\dfrac{S^*, \pi_0}{S, \pi}\ \pi$	K	$\{\nu_0 \mid \nu \in \mathcal{S}\}$	T	$\{\nu_0 \mid \nu \in \mathcal{S}\}$
	K4	$\{\nu_0, \nu \mid \nu \in \mathcal{S}\}$	S4	$\{\nu \mid \nu \in \mathcal{S}\}$
	K45	$\{\nu_0, \nu, \pi \mid \nu, \pi \in \mathcal{S}\}$	S5	$\{\nu, \pi \mid \nu, \pi \in \mathcal{S}\}$
$\dfrac{S, \nu_0}{S, \nu}\ \nu$				

Rule ν is only used for logics T, S4, S5

Fig. 3. Modal Sequent Rules

of knowledge and belief[3] [10,14]. Deontic variants are obtained by waiving the requirement to have a π-formula (and the corresponding π_0) in the π-rule.

We do not need to change our general framework nor to impose clausal normal form, as done in KSAT [12] to apply propositional DPLL. We inherit all propositional rules including simplification. We simply need to specify how $A[\varphi]$ will behave in presence of $\square$ and $\diamond$.

The simplest way is to consider $\square B$ or $\diamond B$ as atoms ("modal atoms") and apply boolean simplification. This is already sufficient for a speed-up [12,16].

Fact 7. *Basic KSAT is an instance of the modal sequent calculus for logic K with propositional simplification, S contains only modal-CNF formulae with positive sign and with only the $\square$ operator.*

This restriction is such that KSAT may not even simplify directly $\square p \wedge \diamond \neg p$ without reduction to normal form.

To enhance simplification we look inside modal atoms. We skip boolean rules as they remain unchanged. The new (interesting) cases in the simplification of $A[\varphi]$ are those where φ is a modal formula i.e. either a ν formula or a π-formula. The uniform notation gives extremely elegant rules.

Modal-K simplification is the simplest:

$$A[\nu] \mapsto \begin{cases} \square(B[\nu_0]) & \text{if } A = \square B \\ \diamond(B[\nu_0]) & \text{if } A = \diamond B \\ \ldots & \text{as for propositional logic} \end{cases}$$

The idea is that if we want to simplify $\square A$ with $[\mathbf{t}.\square C]$ (a ν-formula) then we can exploit the properties of the modal operators and transform this operation in the simplification of A with $[\mathbf{t}.C]$ (the corresponding ν_0-formula).

In a nutshell, while propositional connectives leave the simplifying formula unchanged in the recursive steps (see §4 and reduction (1)), the semantics of the modal operators is such that we must do the recursion on the subformulae of both the simplified and the simplifying formula.

---

[3] With this sequent calculus á la Fitch, analytic cut is not eliminable for strong completeness of the logic S5. One has to resort to prefixed systems to avoid it [10].

For instance suppose we have $\Diamond(\Box p \supset \Box(p \supset r) \vee \Diamond \neg r)$ and want to simplify it with the ν-formula $\mathsf{t}.\Box\Box p$. Then we get the following computation:

$$(\Diamond(\Box p \supset \Box(p \supset r) \vee \Diamond \neg r))[\mathsf{t}.\Box\Box p]$$
$$\mapsto_K \Diamond((\Box p \supset \Box(p \supset r) \vee \Diamond \neg r)[\mathsf{t}.\Box p])$$
$$\mapsto_{bool} \Diamond((\Box p)[\mathsf{t}.\Box p] \supset (\Box(p \supset r))[\mathsf{t}.\Box p] \vee (\Diamond \neg r)[\mathsf{t}.\Box p])$$
$$\mapsto_K \Diamond(\top \supset \Box((p \supset r)[\mathsf{t}.p] \vee \Diamond(\neg r[\mathsf{t}.p])$$
$$\mapsto_{bool} \Diamond(\Box r \vee \Diamond \neg r)$$

Modal-K4 simplification exploits directly the properties of transitivity (represented by the axiom $\Box A \supset \Box\Box A$):

$$A[\nu] \mapsto \begin{cases} \Box(B[\nu][\nu_0]) & \text{if } A = \Box B \\ \Diamond(B[\nu][\nu_0]) & \text{if } A = \Diamond B \\ \ldots & \text{as for propositional logic} \end{cases}$$

This means that for any logic containing the transitivity axiom we can first simplify B with ν and then simplify the result with ν_0. The presence of the transitivity adds a further level of simplification with respect to logic K.

Notice that the recursive call to simplification always terminates since the nesting of modal operators decreases in the simplified formula. For instance one may try the previous example with $\mathsf{t}.\Box p$ as the simplifying formula.

Modal-K5 simplification is the following

$$A[\nu] \mapsto \Box(B[\nu_0]) \text{ if } A = \Box B$$
$$A[\nu] \mapsto \Diamond(B[\nu_0]) \text{ if } A = \Diamond B$$
$$A[\pi] \mapsto \Box(B[\pi]) \text{ if } A = \Box B$$
$$A[\pi] \mapsto \Diamond(B[\pi]) \text{ if } A = \Diamond B$$

The rule can be understood by considering the corresponding axiom, $\Diamond A \supset \Box \Diamond A$: it tells that possibility-like formulae can propagate inside modalities.

The calculus for K45 uses the simplification procedures of both K5 and K4:

$$A[\varphi] \mapsto \begin{cases} \Box(B[\nu][\nu_0]) & \text{if } A = \Box B \text{ and } \varphi = \nu \\ \Diamond(B[\nu][\nu_0]) & \text{if } A = \Diamond B \text{ and } \varphi = \nu \\ \Box(B[\pi]) & \text{if } A = \Box B \text{ and } \varphi = \pi \\ \Diamond(B[\pi]) & \text{if } A = \Diamond B \text{ and } \varphi = \pi \\ \ldots & \text{as for propositional logic} \end{cases}$$

Theorem 2. *The rules for L-Modal simplification, where L is one of K, T, K4, S4, K45, S5, is a sound and complete (invertible) rule.*

Proposition 2. *For the propositional modal logics K, T, K4, S4, K45, S5, the size of $\psi[\varphi]$ is never larger than the size of ψ.*

7 Experimental Analysis

We performed experiments for both propositional and modal logics For propositional logics we have used the random 3-SAT and the IFIP benchmarks[4].

---

[4] The benchmarks are available at http://www.dis.uniroma1.it/~massacci/ifip both in Isabelle and prolog format.

We have used a tableau prover (Beatrix) implemented in sicstus prolog. Beatrix has been designed in the spirit of lean tableau theorem proving [4] with few modifications. At first, we used set oriented connectives implemented with prolog ordered lists without repetitions, e.g. the formula $a_1 \wedge a_2 \wedge (b_1 \vee b_2)$ is represented by the prolog list [conj,a1,a2,[disj,b1,b2]]. The benchmarks are easier to represent in this way. Second we used prolog lexicographic ordering among formulae, reduced conjunctions before disjunctions and small formulae before large ones. Finally, before applying an α or β rule to a formula φ we use φ to simplify all other formulae in the current sequent. Formula are reduced off-line in negation normal form (without any optimizations) as in leanTAP. The machine used was a workstation Sun SuperSPARK (time is in seconds).

To make the comparison more interesting we used two β-rules: the standard rule of Fig. 1 and the asymmetric rule for a limited form of cut [7, 17]:

$$\frac{\mathcal{S}, \beta_1 \quad \mathcal{S}, \beta_2}{\mathcal{S}, \beta} \; Dir \qquad \frac{\mathcal{S}, \beta_1 \quad \mathcal{S}, \overline{\beta_1}, \beta_2}{\mathcal{S}, \beta} \; Lem$$

A simple implementation is already sufficient for interesting conclusions wrt the relative gains of simplification and lemmaizing. In the following tables the normal β rule is denoted by "Dir" and the lemmaizing version by "Lem".

Remark 1. None of the problems listed here could be solved by Beatrix without simplification in any reasonable amount of time (one night), *no matter the branching rule*. So analytic cut does not necessarily helps for hard problems.

Thus, we do not show the benchmark results for Beatrix without simplification (in practice an enhanced version of leanTAP) since it would be an empty column except for the easiest formulae (first line of the table) of the 3-SAT benchmark.

Table 2 shows the result for the standard random distribution of 3-SAT[5]: $3-sat(V,C)$ means that samples had C clauses, with 3 literals selected uniformly among V variables and each literal negated with probability 0.5.

A number of experimental studies [6, 12] have shown that DPLL (and indeed most complete satisfiability checking procedures) exhibits an easy-hard-easy computation pattern as the ratio between the number of clauses and variables increases, with the hardest formulae around $C/V = 4.2 - 4.3$. This phenomena has been sometimes attributed to the "semantical" branching described by the splitting rule of DPLL. On the contrary we have found out the following:

Fact 8. *For random 3-CNF local simplification produces already the easy-hard-easy pattern. Only in the transition region lemmaizing improves the performance.*

Apart from the CNF benchmark we have tested our implementation on the IFIP benchmarks for hardware verification [5]. These benchmarks are known to be hard for DPLL [29] and are formulated in non clausal form, with an extensive use of equivalences ($\equiv$) and exclusive or ($\oplus$). Each problem requires to prove

---

[5] After $C = 10 \times V$ a linear relation between C and running time is due to inefficient list management by prolog.

Table 2. Benchmark on Random 3-SAT

Problem	Dir	Lem	Problem	Dir	Lem
3-sat(32,96)	.3	.2	3-sat(64,192)	1.4	1.0
3-sat(32,128)	3.9	1.2	3-sat(64,256)	334.6	38.4
3-sat(32,136)	6.1	1.8	3-sat(64,272)	554.3	56.4
3-sat(32,144)	6.9	2.1	3-sat(64,288)	1,050.9	72.0
3-sat(32,160)	8.2	2.4	3-sat(64,320)	568.6	60.0
3-sat(32,192)	7.7	2.6	3-sat(64,384)	240.3	39.4
3-sat(32,224)	6.0	2.2	3-sat(64,448)	139.3	30.6
3-sat(32,256)	5.3	2.3	3-sat(64,512)	74.3	23.5
3-sat(32,288)	4.9	2.3	3-sat(64,576)	69.6	21.8
3-sat(32,320)	4.8	2.4	3-sat(64,640)	59.6	20.4
3-sat(32,640)	5.9	3.7	3-sat(64,1280)	38.2	19.8
3-sat(32,1280)	10.5	6.5	3-sat(64,2560)	52.3	24.9

$$Spec = \begin{cases} car_1 \equiv (a_1 \wedge b_1), \\ som_1^s \equiv (a_1 \oplus b_1), \\ som_2^s \equiv (a_2 \oplus b_2 \oplus car_1), \\ cout^s \equiv (((a_2 \vee b_2) \wedge car_1) \vee (a_2 \wedge b_2)), \end{cases}$$

$$Impl = \begin{cases} cout_1 \equiv (b_1 \wedge a_1)), \\ som_1^i \equiv \neg((\neg a_1 \wedge \neg b_1) \vee (a_1 \wedge b_1)), \\ som_2^i \equiv \neg((((\neg a_2 \wedge \neg b_2) \vee (a_2 \wedge b_2)) \wedge \neg cout_1) \vee \\ \vee (cout_1 \wedge \neg((\neg a_2 \wedge \neg b_2) \vee (a_2 \wedge b_2)))), \\ cout^i \equiv ((a_2 \wedge cout_1) \vee (b_2 \wedge cout_1) \vee (a_2 \wedge b_2)) \end{cases}$$

$$Spec \cup Impl \implies (som_1^s \equiv som_1^i) \wedge (som_2^s \equiv som_2^i) \wedge (cout^s \equiv cout^i)$$

Fig. 4. A two-bit ripple adder rip02.be

that the outputs of two circuits are equivalent i.e. the specification is equivalent to the implementation.

An example is shown in Fig. 4 as a sequent to be proved. It is a simple 2-bits binary ripple adder: it takes four bits of input (a_i and b_i) and returns three bits of output: the sums som_i and the carry $cout$. The two circuits are described by two sets of equivalences, each equivalence defining the output of a sub-circuit. The "upper" part in Fig. 4 describes the specification and the "lower" part describes the implementation. Using this description we must derive a (sequent) proof of the equivalence of the respective outputs.

In practice all have been solved except large multipliers, which are hard also for BDDs, timed out after 10 minutes (for a comparison between DPLL and OBDDs see [29]). Satisfiable formulae (marked by *) are solved within few milliseconds. What is more interesting here is the comparison with a standard

Table 3. Simplification versus Isabelle

Problem	fast_tac	ifip_tac
ex2	13.1	1.3
transp	43.8	0.2
risc	-	9.8
counter	-	68.8
hostint1	-	96.5
mul	-	130.9

Problem	Dir	Lem
ex2	0.0	0.0
transp	0.0	0.0
risc	0.4	0.6
counter	0.1	0.1
hostint1	0.3	0.2
mul	0.4	0.2

Prob.	fast_tac	ifip_tac
rip02	2848.2	1.6
rip04	-	994.5
rip06	-	-
rip08	-	-

Prob.	Dir	Lem
rip02	0.0	0.0
rip04	0.5	0.6
rip06	3.0	3.0
rip08	18.2	18.4

Table 4. Run times on IFIP Benchmarks

Problem	Dir	Lem	Problem	Dir	Lem	Problem	Dir	Lem	Problem	Dir	Lem
counter	0.1	0.2	ztwaalf1	0.8	0.8	add1	24.4	12.2	alupla20	784.0	618.1
d3(*)	0.1	0.1	ztwaalf2	0.8	0.4	alu	21.3	7.1	dc2	149.7	12.5
dk27	2.2	2.3	mjcg_no (*)	0.0	0.0	dk17	19.8	3.0	sqn	297.7	11.2
ex2	0.0	0.0	mjcg_yes	1.8	1.1	f51m	21.4	5.7	z9sym	166.8	9.8
hostint1	0.2	0.2	mp2d	3.7	1.1	table	13.9	2.8	mul03	-	20.1
misg	0.7	1.0	rip02	0.0	0.0	vg2	12.8	7.0	pitch	-	5.7
mul	0.2	0.2	rip04	0.4	0.5	x1dn	13.4	7.2	rd73	-	30.4
risc	0.4	0.5	rip06	2.9	2.9	z4	2.9	2.3	rom2	-	2.5
transp	0.0	0.0	rip08	18.2	18.4	z5xpl	11.5	4.1	root	-	33.7

interactive prover such as Isabelle [22]. This is shown in Table 3, where $fast\_tac$ is the standard automatic tactic used in Isabelle, and $ifip\_tac$ applies $fast\_tac$ after a pre-processing step which eliminates the equivalences by substituting each defined proposition with the corresponding formula (sub-circuit). Notice that also this preprocessing step, which eliminates abbreviations and leaves only "natural" (in)equivalences (at the price of an increase in size), is not enough.

The scaling factor for problems such as $rip0n.be$ is more interesting than the absolute values. Absolute values simply tell us about the relative efficiency of lean proving in prolog vs big provers in ML, somehow strengthening the argument of [4]. The gap in the scaling factor (a factor of 1000 for Isabelle in the passage from $rip0n$ to $rip0n + 2$ is matched by a factor of 10 for Beatrix) tells that the use of simplification can substantial increase the tractability threshold for tableau calculi. Further benchmarks are shown in Table 4.

The experiments on modal logics lead to the same results noted in [16] on the gap between "standard" tableaux and KSAT: after the introduction of (limited) simplification, the gap claimed by Giunchiglia & Sebastiani [12] disappears.

Intuition 2. Simplification is a must (and a win) for efficient tableau provers, no matter if you use cut or not.

8 Conclusion

In this paper we have shown how the introduction of a simple technique called simplification may boost the deduction capabilities of tableau methods. It plays the same role of unit for DPLL and subsumption for resolution and provides a uniform framework which subsumes a number of techniques for the improvement of tableau-based methods.

Its flexibility and its simplicity make it possible its incorporation into sequent based interactive theorem provers used for hardware and software verification. Moreover there is no need of black-boxes since the computational effectiveness of DPLL and KSAT can now be compiled into standard sequent-based tactics.

While propositional and modal logics can be uniformly treated without much changes, the effective use of simplification for first order logic requires more advanced techniques. This topic is further discussed in [18].

At the end, there is a criticism we may have to face: after adding all these simplifications to sequent calculi are not we abandoning the calculus itself?

The answer is a question: is resolution with subsumption no longer resolution?

Acknowledgments

This work have been developed while the author was visiting the Computer Laboratory at the University of Cambridge.

I would like to acknowledge many fruitful discussions with L. Paulson, L. Carlucci Aiello, F. Donini, F. Giunchiglia, R. Goré, J. Harrison, F. Pirri and R. Sebastiani, the members of the AI group in Rome and the Automatic Reasoning group in Cambridge. Comments from the referees were helpful.

This work has been financially supported by ASI, CNR, and MURST grants.

References

1. M. Baaz and C. Fermüller. Non-elementary speedups between different versions of tableaux. In *Proc. of the 4th Workshop on Theorem Proving with Analytic Tableaux and Related Methods*, LNAI 918, pages 217–230. Springer-Verlag, 1995.
2. P. Baumgartner, U. Furbach, and I. Niemelä. Hyper tableaux. In *Proc. JELIA-96*, LNAI 1126, pages 1–17. Springer-Verlag, 1996.
3. B. Beckert. Semantic tableaux with equality. *Journal of Logic and Computation*, 7(1):38–58, 1997.
4. B. Beckert and J. Possega. leanTAP: Lean tableau-based deduction. *JAR*, 15(3):339–358, 1995.
5. L. J. Claesen, editor. *Formal VLSI Correctness Verification: VLSI Design Methods*, volume II. Elsevier Science, 1990.
6. J. Crawford and L. Auton. Experimental results on the crossover point in satisfiability problems. In *Proc. of the 11th National Conference on Artificial Intelligence (AAAI-93)*, pages 21–27. AAAI Press/The MIT Press, 1993.
7. M. D'Agostino and M. Mondadori. The taming of the cut. *JLC*, 4(3):285–319, 1994.

8. M. Davis, G. Longemann, and D. Loveland. A machine program for theorem-proving. *Comm. of the ACM*, 5(7):394–397, 1962.
9. M. Davis and H. Putnam. A computing procedure for quantificational theory. *JACM*, 7(3):201–215, 1960.
10. M. Fitting. *Proof Methods for Modal and Intuitionistic Logics*. Reidel, 1983.
11. M. Fitting. *First Order Logic and Automated Theorem Proving*. Springer-Verlag, 1990.
12. F. Giunchiglia and R. Sebastiani. Building decision procedures for modal logics from propositional decision procedures - the case study of modal K. In *Proc. CADE-96*, LNAI 1104, pages 583–597. Springer-Verlag, 1996.
13. M. Gordon and T. Melham. *Introduction to HOL*. Cambridge University Press, 1993.
14. J. Halpern and Y. Moses. A guide to completeness and complexity for modal logics of knowledge and belief. *AIJ*, 54:319–379, 1992.
15. J. Harrison. Binary decision diagrams as a HOL derived rule. *The Computer Journal*, 38:162–170, 1995.
16. U. Hustadt and R. Schmidt. On evaluating decision procedure for modal logic. In *Proc. of IJCAI-97*, 1997.
17. R. Letz, K. Mayr, and C. Goller. Controlled integration of the cut rule into connection tableau calculi. *JAR*, 13(3):297–337, 1994.
18. F. Massacci. Simplification with renaming: A general proof technique for tableau and sequent-based provers. Tech. Rep. 424, Computer Laboratory, Univ. of Cambridge (UK), 1997.
19. D. McAllester, Truth maintenance. In *Proc. of AAAI-90*, pages 1109-1116, 1990.
20. N. V. Murray and E. Rosenthal. On the computational intractability of analytic tableau methods. *Bulletin of the IGPL*, 2(2):205–228, 1994.
21. F. Oppacher and E. Suen. HARP: a tableau-based theorem prover. *JAR*, 4:69–100, 1988.
22. L. C. Paulson. *Isabelle: A Generic Theorem Prover*, volume 828 of *Lecture Notes in Computer Science*. Springer-Verlag, 1994.
23. D. Prawitz. An improved proof procedure. *Theoria*, 26:102–139, 1960.
24. W. Quine. *Methods of Logic*. Harvard Univ. Press, iv ed., 1983.
25. J. A. Robinson. A machine oriented logic based on the resolution principle. *JACM*, 12(1):23–41, 1965.
26. J. Rushby. Automated deduction and formal methods. In *Proc. of CAV-96*, LNCS 1102, pages 169–183. Springer-Verlag, 1996.
27. B. Selman, H. Levesque, and D. Mitchell. A new method for solving hard satisfiability problems. In *Proc. of AAAI-92*, 1992.
28. R. M. Smullyan. *First Order Logic*. Springer-Verlag, 1968. Republished by Dover, 1995.
29. T. Uribe and M. E. Stickel. Ordered binary decision diagrams and the Davis-Putnam procedure. In *Proc. of the 1st Internat. Conf. on Constraints in Computational Logics*, LNCS 845, Springer-Verlag, 1994.
30. A. Urquhart. The complexity of propositional proofs. *Bull. of Symbolic Logic*, 1(4):425–467, Dec 1995.
31. A. Voronkov. The anatomy of vampire (implementing bottom-up procedures with code trees). *JAR*, 15(2):237–265, 1995.
32. K. Wallace and G. Wrightson. Regressive merging in model elimination tableau-based theorem provers. *Bull. of the IGPL*, 3(6):921–937, 1995.
33. H. Zhang and M. E. Stickel. An efficient algorithm for unit-propagation. In *Proc. of AI-MATH-96*, 1996.

A Tableaux Calculus for Ambiguous Quantification*

Christof Monz and Maarten de Rijke

ILLC, University of Amsterdam, Plantage Muidergracht 24, 1018 TV Amsterdam,
The Netherlands. E-mail: {christof, mdr}@wins.uva.nl

Abstract. Coping with ambiguity has recently received a lot of attention in natural language processing. Most work focuses on the semantic representation of ambiguous expressions. In this paper we complement this work in two ways. First, we provide an entailment relation for a language with ambiguous expressions. Second, we give a sound and complete tableaux calculus for reasoning with statements involving ambiguous quantification. The calculus interleaves partial disambiguation steps with steps in a traditional deductive process, so as to minimize and postpone branching in the proof process, and thereby increases its efficiency.

1 Introduction

Natural language expressions can be highly ambiguous, and this ambiguity may have various faces. Well-known phenomena include lexical and syntactic ambiguities. In this paper we focus on representing and reasoning with a different source of ambiguity, namely quantificational ambiguity, as exemplified in (1).

(1) a. Every man loves a woman.
    b. Every boy doesn't see a movie.

The different readings of (1.a) correspond to the two logical representations in

(2) a. $\forall x \, (man(x) \to \exists y \, (woman(y) \land love(x, y)))$.
    b. $\exists y \, (woman(y) \land \forall x \, (man(x) \to love(x, y)))$.

We refer the reader to [KM93,DP96] for extensive discussions of these and other examples of quantificational ambiguity. All we want to observe here is this. Examples like (1.a) have a preferred reading namely the wide-scope reading represented by (2.a)). Additional linguistic or non-linguistic information, or the context, may overrule this preference. For instance, if (1.a) is followed by (3), then the second reading (2.b) is preferred. But if (1.a) occurs in isolation, then the first reading (2.a) is preferred.

(3) But she is already married.

---

* The research in this paper was supported by the Spinoza project 'Logic in Action' at the University of Amsterdam.

Clearly, if we want to process a discourse from left to right and take the context of an expression into account, our semantic representation for (1.a) must initially allow for both possibilities. And, similarly, any reasoning system for ambiguous expressions needs to be able to integrate information that helps the disambiguation process within the deductive process.

Although the problem of ambiguity and underspecification has recently enjoyed a considerable increase in attention from computational linguists, computer scientists and logicians (see, for instance, [DP96]), the focus has mostly been on semantic aspects, and deductive reasoning with ambiguous sentences is still in its infancy.

The aim of this paper is to present a tableaux calculus for reasoning with expressions involving ambiguous quantification. An important feature of our calculus is that it integrates two processes: disambiguation and deductive reasoning. The calculus operates on semantic representations of natural language expressions. These representations contain both ambiguous and unambiguous subparts, and an important feature of our representations is that they represent all possible disambiguations of an ambiguous statement in such a way that unambiguous subparts are shared as much as possible. As we will explain below, compact representations of this kind will allow us to keep ambiguities 'localized' — a feature which has important advantages from the point of view of efficiency.

In setting up a deductive system for ambiguous quantification we have had two principal desiderata. First, although this is not the topic of the present paper, we aim to implement the calculus as part of a computational semantics work bench; this essentially limits our options to resolution and tableaux based calculi. Second, to incorporate information arising from the disambiguation process within a proof system, the proofs themselves need to be incremental in the sense that at any stage we have a 'partial' proof that can easily be extended to cope with novel information. We believe that a tableaux style calculus has clear advantages over resolution based systems in this respect.

The paper is organized as follows. A considerable amount of work goes into setting up semantic representations and a mechanism for for recording ambiguities and disambiguations in such a way that it interfaces rather smoothly with traditional deductive proof steps. This work takes up Sections 2 and 3. Then, in Section 4 we present two tableaux calculi, one which deals with fully disambiguated representations of ambiguous natural language expressions, and a more interesting one in which traditional tableaux style deduction is interleaved with partial disambiguation. Section 5 contains a detailed example, and Section 6 provides conclusions and suggestions for further work.

2 Representing Ambiguity

Lexical ambiguities can be represented pretty straightforwardly by putting the different readings into a disjunction. (Cf. [Dee96,KR96] for further elaboration.) It is also possible to express quantificational ambiguities by a disjunction, but quite often this involves much more structure than in the case of lexical ambiguities, because quantificational ambiguities are not tied to a particular atomic expression. For instance, the only way to represent the ambiguity of (1.a) in a disjunctive manner is (4).

(4) $\forall x\, (man(x) \rightarrow \exists y\, (woman(y) \land love(x,y)))$
$\lor\ \exists y\, (woman(y) \land \forall x\, (man(x) \rightarrow love(x,y)))$

Obviously, there seems to be some redundancy, because some subparts appear twice. If we put indices at the corresponding subparts, as in (5) below, we see that these subparts are not proper expressions of first-order logic, except subpart k.

(5) $\underline{\forall x\, (man(x) \rightarrow_i \underline{\exists y\, (woman(y) \land_j \underline{love(x,y)}_k))}}$
$\lor\ \underline{\exists y\, (woman(y) \land_j \underline{\forall x\, (man(x) \rightarrow_i \underline{love(x,y)}_k))}}$

The difference between the readings lies not in the material used, both readings are built from the parts i, j and k, but in the order these are put together.

A reasonable way to represent improper expressions like i and k is to abstract over those parts that are missing in order to yield a proper expression of first-order logic. [Bos95] calls these missing parts *holes*. Roughly speaking, they are variables over occurrences of first-order formulas. To distinguish the occurrence of an expression from its logical content, it is necessary to supplement first-order formulas with labels. Holes may be subject to constraints; for instance, the semantic representations of verbs have to be in the scope of its arguments, because otherwise it may happen that the resulting disambiguations contain free variables. So we do not want to permit disambiguations like $\forall x\, (man(x) \rightarrow love(x,y) \land \exists y\, (woman(y)))$. These constraints are expressed by a partial order on the labels.

Definition 1 (Underspecified Representation). *For $i \in \mathbb{N}$, let h_i a new atomic symbol, called a hole. A formula φ is an h-formula, or a formula possibly containing holes, if it is built up from holes and atomic formulas from first-order logic using the familiar boolean connectives and quantifiers.*

Next, we specify the format of an underspecified representation UR of a natural language expression. An underspecified representation is a quadruple $\langle LHF, L, H, C\rangle$ consisting of

1. *A set of labeled h-formulas LHF.*
2. *The set of labels L occurring in LHF.*
3. *The set of holes H occurring in LHF.*
4. *A set of order-constraints C of the form $k \leq k'$, meaning that k has to be a subexpression of k', where $k, k' \in L \cup H$ and C is closed under reflexivity, antisymmetry and transitivity.*

An obvious question at this point is, how does one associate a UR with a given natural language expression? We will not address this issue here, but we will assume that there exists some mechanism for arriving at UR's, see for example [Kön94]. For notational convenience we write $UR(S)$ for the underspecified representation, associated with a sentence S. By way of example, we reconsider (4) and obtain the following underspecified representation:

(6) $\langle\{l_0 : h_0, l_1 : \forall x\, (man(x) \to h_1), l_2 : \exists y\, (woman(y) \land h_2), l_3 : love(x,y)\},$
$\{l_0, l_1, l_2, l_3\},$
$\{h_0, h_1, h_2, h_3\}\rangle,$
$closure(\{l_1 \leq h_0, l_2 \leq h_0, l_3 \leq h_1, l_3 \leq h_2\})$

There are two possible sets of instantiations, ι_1 and ι_2, of the holes h_0, h_1, h_2, h_3 in (6) which obey the constraints in (6): $\iota_1 = \{h_0 := l_1, h_1 := l_2, h_2 := l_3\}$ and $\iota_2 = \{h_0 := l_2, h_2 := l_1, h_1 := l_3\}$.

It is also possible to view UR's as upper semi-lattices, as it is done in [Rey93]:

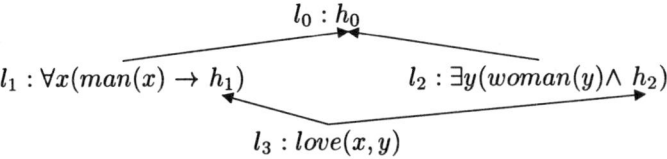

For each instantiation of the holes there is a corresponding substitution $\sigma(\iota)$ which is like ι but $h := \varphi \in \sigma(\iota)$ iff there is a l, such that $l : \varphi \in LHF$ and $h := l \in \iota$.

The next step is to define an extension of the language of first-order logic, $\mathcal{L}$, in which both standard (unambiguous) expressions occur side by side with the above underspecified representations. The resulting language of the language of underspecified logic, or $\mathcal{L}^u$ for short, is the language in which we will perform deduction.

Definition 2 (Underspecified Logic). *A formula φ is a formula of our underspecified logic $\mathcal{L}^u$, or a u-formula, that is, a formula possibly containing underspecified representations, if it is built up from underspecified representations and the usual atomic formulas from standard first-order logic using the familiar boolean connectives and quantifiers.*

Example 1. As an example of a more complex u-formula consider the semantic representation of *if every boy didn't sleep and John is a boy, then John didn't sleep.*

$$\left(\begin{array}{c} l_0 : h_0 \\ l_1 : \neg h_1 \quad\quad l_2 : \forall x\,(boy(x) \to h_2) \\ l_3 : sleep(x) \end{array}\right) \land boy(j)\,) \to \neg sleep(j)$$

Definition 3 (Total Disambiguations). *To define the total disambiguation $\delta(\varphi)$ of a u-formula φ, we need the following notion of a join.*

Given an underspecified representation $\langle LHF, L, H, C\rangle$ and $k, k', k'' \in L \cup H$ and $k'' \leq k, k' \in C$ then k'' is the join of k and k', $k \sqcup k' = k''$, only if there is no $k''' \in L \cup H$ and $k''' \leq k, k' \in C$ and $k''' > k'' \in C$.

Then, by $\delta(\varphi)$ we denote the set of total disambiguations of the u-formula φ, where for all $d \in \delta(\varphi)$, $d \in \mathcal{L}$. For complex u-formulas δ is defined recursively:

1. $\delta(\langle LHF, L, H, C \rangle) = $ the set of $LHF\sigma(\iota)$ such that
   (i) ι is an instantiation and $\sigma(\iota)$ is the corresponding substitution
   (ii) $H\iota = L$
   (iii) for all $l, l' \in L$, if $l \sqcup l'$ is defined, then $l \leq l' \in closure(C\iota)$ or $l' \leq l \in closure(C\iota)$
2. $\delta(\neg\varphi) = \{ \neg d \mid d \in \delta(\varphi) \}$
3. $\delta(\varphi \circ \psi) = \{ d \circ d' \mid d \in \delta(\varphi), d' \in \delta(\psi) \}$, where $\circ \in \{\wedge, \vee, \rightarrow\}$
4. $\delta(\mathcal{Q}x\varphi) = \{ \mathcal{Q}xd \mid d \in \delta(\varphi) \}$, where and $\mathcal{Q} \in \{\forall, \exists\}$.

If $l \leq l' \notin C$ and $l' \leq l \notin C$, then it does not have to be case that there is a scope ambiguity between quantifiers belonging to l and l'. For instance, if l and l' belong to different conjuncts, they are not ordered to each other. The restriction that $l \sqcup l'$ has to be defined excludes this.

Example 2. To illustrate the purpose of this restriction see the underspecified representation for *every man who doesn't have a car rides a bike*

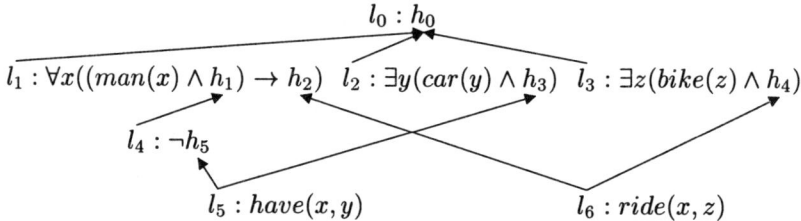

Although l_3 and l_4 are not related to each other, it cannot happen that l_3 is in the scope of l_4, because the negation must be a subformula of the antecedent of l_1, whereas l_3 might have scope over l_1 as a whole or might be in the scope of the succedent of l_1. More generally, this is due to the fact that l_3 and l_4 do not have to share a subformula, i.e., $l_3 \sqcup l_4$ is not defined.

3 Semantics of Underspecified Formulas

In the previous section we introduced a formalism that allows for a compact semantic representation of ambiguous expressions. Now we want to see what the validity conditions of these underspecified representations are, and how they interact with the classical logical connectives.

If an ambiguous sentence S with $\delta(UR(S)) = \{d_1, d_2\}$ is uttered, and we want to check, whether S is valid, we simply have to see whether all of its disambiguations are valid. That is, it must be the case that $\models d_1$ and $\models d_2$. If, on the other hand, an ambiguous sentence S with $\delta(UR(S)) = \{d_1, d_2\}$ is claimed to be false, things are different. Here it is not sufficient that either $\not\models d_1$ or $\not\models d_2$; one has to be sure that *all* disambiguations are false, i.e., $\not\models d_1$ and $\not\models d_2$. To model this distribution of falsity, van Eijck and Jaspars [EJ96] use the notions of a countermodel and a falsification relation $\dashv$. Roughly, if only unambiguous expressions appear as premises or consequences $\dashv$ corresponds to $\not\models$, but if at least one underspecified expression appears as premise or consequence, we have to define the (counter-) consequence relation appropriately.

Definition 4. *We define the underspecified consequence relation $\models_u$ and underspecified falsification relation $\dashv_u$ for $\mathcal{L}^u$ and an arbitrary model M.*

1. $M \models_u \varphi$ iff $M \models \varphi$, if φ is an unambiguous expression.
   $M \dashv_u \varphi$ iff $M \not\models \varphi$, if φ is an unambiguous expression.
2. $M \models_u UR$ iff $M \models d$, for all $d \in \delta(UR)$.
   $M \dashv_u UR$ iff $M \not\models d$, for all $d \in \delta(UR)$.
3. $M \models_u \neg\varphi$ iff $M \dashv_u \varphi$
   $M \dashv_u \neg\varphi$ iff $M \models_u \varphi$
4. $M \models_u \varphi \wedge \psi$ iff $M \models_u \varphi$ and $M \models_u \psi$
   $M \dashv_u \varphi \wedge \psi$ iff $M \dashv_u \varphi$ or $M \dashv_u \psi$
5. $M \models_u \varphi \vee \psi$ iff $M \models_u \varphi$ or $M \models_u \psi$
   $M \dashv_u \varphi \vee \psi$ iff $M \dashv_u \varphi$ and $M \dashv_u \psi$
6. $M \models_u \varphi \to \psi$ iff $M \dashv_u \varphi$ or $M \models_u \psi$
   $M \dashv_u \varphi \to \psi$ iff $M \models_u \varphi$ and $M \dashv_u \psi$
7. $M \models_u \forall x \varphi$ iff $M \models_u \varphi[a]$, for all $a \in D(M)$.
   $M \dashv_u \forall x \varphi$ iff $M \dashv_u \varphi[a]$, for some $a \in D(M)$.
8. $M \models_u \exists x \varphi$ iff $M \models_u \varphi[a]$, for some $a \in D(M)$.
   $M \dashv_u \exists x \varphi$ iff $M \dashv_u \varphi[a]$, for all $a \in D(M)$.

Example 3. We now give an example demonstrating the convenience of having the falsification relation.

In our setting of ambiguous expressions, some familiar classical tautologies are no longer valid. For instance, if A is ambiguous and B unambiguous we do not want $(A \wedge B) \to A$ because the two occurrences of A may be disambiguated in different ways. For instance, if $\delta(A) = \{d_1, d_2\}$, then $\models_u (A \wedge B) \to A$ iff $\models_u (d_1 \wedge B) \to d_1$, $\models_u (d_1 \wedge B) \to d_2$, $\models_u (d_2 \wedge B) \to d_1$ and $\models_u (d_2 \wedge B) \to d_2$. If we were to model falsity by $\not\models$, applying the definitions would yield:

$\models_u (A \wedge B) \to A$ iff $\not\models_u A \wedge B$ or $\models_u A$
$\qquad$ iff $\not\models_u A$ or $\not\models_u B$ or $\models_u A$
$\qquad$ iff $\not\models d_1$ or $\not\models d_2$ or $\not\models B$ or ($\models d_1$ and $\models d_2$).

The latter is classically valid, and it would therefore make the classical tautology valid. On the other hand, if we model falsity by $\dashv_u$ we manage to avoid this, as $\dashv_u$ distributes over disambiguations of A, whereas $\not\models$ does not:

$\models_u (A \wedge B) \to A$ iff $\dashv_u A \wedge B$ or $\models_u A$
$\qquad$ iff $\dashv_u A$ or $\dashv B$ or $\models_u A$
$\qquad$ iff ($\not\models d_1$ and $\not\models d_2$) or $\dashv B$ or ($\models d_1$ and $\models d_2$).

Definition 5. *Let $\varphi_1, \ldots, \varphi_n, \psi$ be $\mathcal{L}^u$-formulas, possibly containing underspecified representations. We define relation of underspecified consequence $\models_u$ as follows:*

$$\varphi_1, \ldots, \varphi_n \models_u \psi \text{ iff}$$
$$\text{for all } d_1 \in \delta(\varphi_1), \ldots, d_n \in \delta(\varphi_n)$$
$$\text{and for all } d' \in \delta(\psi) \text{ it holds that}$$
$$d_1, \ldots, d_n \models d'.$$

The underlying intuition is that if someone utters a statement of the form *if S then S'*, where S and S' are ambiguous sentences with $\delta(UR(S)) = \{d_1, d_2\}$, $\delta(UR(S')) = \{d'_1, d'_2\}$, then we do not know exactly what the speaker had in mind by uttering this. So to be sure that this was a valid utterance, one has to check whether it is valid for every possible combination of disambiguations, i.e., whether each of $d_1 \models d'_1$, $d_1 \models d'_2$, $d_2 \models d'_1$, and $d_2 \models d'_2$ is a valid classical consequence.

Unfortunately, this definition of entailment is not a conservative extension of classical logic. Even the reflexivity principle $A \models A$ fails. For instance, if we take $\delta(UR(S)) = \{d_1, d_2\}$, then $UR(S) \models_u UR(S)$ iff $d_1 \models d_1, d_1 \models d_2, d_2 \models d_1$, and $d_2 \models d_2$, i.e. iff $\models d_1 \leftrightarrow d_2$. As we will show below, this has some clear consequences for our calculus, especially the closure conditions. We refer the reader to [Dee96,Jas97] for alternative definitions of the ambiguous entailment relation.

4 An Underspecified Tableaux Calculus

The differentiation between consequence and falsification can be nicely modeled in a labeled tableaux calculus, where the nodes in the tableaux tree are of the form $T : \varphi$ or $F : \varphi$, meaning that we want to construct a model or countermodel for φ, respectively. Tableaux calculi are especially well suited, because the notion of a countermodel is implicit in the notion of an open tableaux tree, where one constructs a countermodel for a formula.

But what does it mean, if we not only allow first-order formulas to appear in a tableaux proof but as also u-formulas? According to the semantic definitions in Section 3, a proof for a u-formula is simply a proof for each of its disambiguations (in a classical tableaux calculus $\mathcal{TC}$). In the following two subsections we first introduce a calculus $\mathcal{TC}_u$ which integrates the mechanism of disambiguation in its deduction rules, and thereby allows one to postpone the disambiguation until it is really needed. $\mathcal{TC}_u$ nicely shows how ambiguity and branching of tableaux trees correspond to each other. But $\mathcal{TC}_u$ still makes no use of the compact representation of underspecified representations, introduced in Section 2. Therefore, we give a modified version of $\mathcal{TC}_u$, called $\mathcal{TC}_{up}$, which also allows us to reason within an underspecified representation.

Our tableaux calculi are based on the labeled free-variable tableaux calculus, see for instance [Fit96] for a general introduction to tableaux calculi.

4.1 Reasoning with Total Disambiguations

The definitions of the logical connectives in section 3 allow us to treat logical connectives occurring in u-formulas in the same way as in a tableaux calculus for classical logic $\mathcal{TC}$, as long as they do not occur inside of a UR. Here it is necessary to disambiguate the UR first, and then apply the rules in the normal way.

Example 4. If we try to deduce $(A \land B) \to A$, with $\delta(A) = \{d_1, d_2\}$ and B unambiguous, we have to prove each of $\vdash_{TC} (d_1 \land B) \to d_1$, $\vdash_{TC} (d_1 \land B) \to d_2$, $\vdash_{TC} (d_2 \land B) \to d_1$ and $\vdash_{TC} (d_2 \land B) \to d_2$. This leads to the following classical labeled tableaux proof trees.

$$
\begin{array}{cccc}
(a) & (b) & (c) & (d) \\
F : (d_1 \land B) \to d_1 & F : (d_1 \land B) \to d_2 & F : (d_2 \land B) \to d_1 & F : (d_2 \land B) \to d_2 \\
| & | & | & | \\
T : d_1 \land B & T : d_1 \land B & T : d_2 \land B & T : d_2 \land B \\
| & | & | & | \\
F : d_1 & F : d_2 & F : d_1 & F : d_2 \\
| & | & | & | \\
T : d_1 & T : d_1 & T : d_2 & T : d_2 \\
| & | & | & | \\
T : B & T : B & T : B & T : B
\end{array}
$$

At least structurally, the above proof trees are the same. It does not matter whether they contain underspecified representations. This suggests a natural strategy: to postpone disambiguation and merge those parts of the trees that are similar.

$$
\begin{array}{c}
(1)\ F : (A \land B) \to A \\
| \\
(2)\ T : A \land B \\
| \\
(3)\ F : A \\
| \\
(4)\ T : A \\
| \\
(5)\ T : B \\
\diagup \quad \diagdown \\
(6)\ F : d_1 \qquad (7)\ F : d_2 \\
\diagup \diagdown \quad\quad \diagup \diagdown \\
(8)\ T : d_1 \ \ (9)\ T : d_2 \ \ (10)\ T : d_1 \ \ (11)\ T : d_2
\end{array}
$$

This is a much more compact representation. Again, since A is ambiguous, (3) and (4) do not allow one to close the branch, because reflexivity is not a valid principle in our ambiguous setting.

The deduction rules for our underspecified tableaux calculus for totally disambiguated expressions $\mathcal{TC}_u$ are given in Table 1. Besides the last two rules $(T_u : UR)$ and $(F_u : UR)$, all rules are stated in a standard way and need no further explanation. The purpose of the last two rules is to disambiguate UR's and to start a new branch for each of its disambiguations. This implements the idea of postponing disambiguation, because disambiguation applies now only to UR's and not to any u-formula.

Theorem 1. *Let $\varphi \in \mathcal{L}^u$. Then $\vdash_{\mathcal{TC}_u} \varphi$ iff $\vdash_{\mathcal{TC}} d$, for all $d \in \delta(\varphi)$.*

Corollary 1. *Let $\varphi \in \mathcal{L}^u$. Then $\vdash_{\mathcal{TC}_u} \varphi$ iff $\models_u \varphi$.*

Table 1. Deduction rules of the underspecified tableaux calculus $\mathcal{TC}_u$

$$\frac{T_u : \varphi \wedge \psi}{\begin{array}{c} T_u : \varphi \\ T_u : \psi \end{array}} \; (T_u : \wedge) \qquad\qquad \frac{F_u : \varphi \wedge \psi}{F_u : \varphi \;\mid\; F_u : \psi} \; (F_u : \wedge)$$

$$\frac{T_u : \varphi \vee \psi}{T_u : \varphi \;\mid\; T_u : \psi} \; (T_u : \vee) \qquad\qquad \frac{F_u : \varphi \vee \psi}{\begin{array}{c} F_u : \varphi \\ F_u : \psi \end{array}} \; (F_u : \vee)$$

$$\frac{T_u : \varphi \to \psi}{F_u : \varphi \;\mid\; T_u : \psi} \; (T_u : \to) \qquad\qquad \frac{F_u : \varphi \to \psi}{\begin{array}{c} T_u : \varphi \\ F_u : \psi \end{array}} \; (F_u : \to)$$

$$\frac{T_u : \neg \varphi}{F_u : \varphi} \; (T_u : \neg) \qquad\qquad \frac{F_u : \neg \varphi}{T_u : \varphi} \; (F_u : \neg)$$

$$\frac{T_u : \forall x \varphi}{T_u : \varphi[x/X]} \; (T_u : \forall) \qquad\qquad \frac{F_u : \forall x \varphi}{F_u : \varphi[x/f(X_1,\ldots,X_n)]} \; (F_u : \forall)^\dagger$$

$$\frac{T_u : \exists x \varphi}{T_u : \varphi[x/f(X_1,\ldots,X_n)]} \; (T_u : \exists)^\dagger \qquad\qquad \frac{F_u : \exists x \varphi}{F_u : \varphi[x/X]} \; (F_u : \exists)$$

$$\frac{T_u : UR}{T_u : d_1 \;\mid\; \ldots \;\mid\; T_u : d_n} \; (T_u : UR)^\ddagger \qquad\qquad \frac{F_u : UR}{F_u : d_1 \;\mid\; \ldots \;\mid\; F_u : d_n} \; (F_u : UR)^\ddagger$$

†Where $X_1, \ldots, X_n$ are the free variables in φ.
‡Where $d_1, \ldots, d_n \in \delta(UR)$.

4.2 Reasoning with Partial Disambiguations

From a computational point of view $(T_u : UR)$ and $(F_u : UR)$ are not optimal, since they cause a lot of branchings of the tableaux tree. Also, total disambiguation is not the appropriate means for underspecified reasoning, because the advantage of the compact representation, namely avoiding redundancy, gets lost. So $\mathcal{TC}_u$ is appropriate for dealing with formulas containing UR's but not for reasoning inside the UR's themselves.

Sometimes it is not necessary to compute all disambiguations, because there exists a strongest (weakest) partial disambiguation. If such a strongest (weakest) disambiguation does exist, it suffices to verify (falsify) this one, because it entails (is entailed by) all other disambiguations. But what are the circumstances under which a strongest (weakest) disambiguation exists?

Before we can determine a strongest (weakest) reading, we have to resolve the relative position of negative contexts and quantifiers. To this end we define positive and negative contexts (see also [TS96]).

Definition 6. *A u-formula φ is a positive context for a subformula ξ of φ, notation: $con^+(\varphi,\xi)$, iff*

$$\varphi ::= \xi \mid \psi \wedge \chi[\xi] \mid \chi[\xi] \wedge \psi \mid \psi \vee \chi[\xi] \mid \chi[\xi] \vee \psi \mid \psi \to \chi[\xi] \mid \forall x\chi[\xi] \mid \exists x\chi[\xi]$$

where ξ occurs in χ and $con^+(\chi,\xi)$ holds, or $\varphi ::= \neg\chi[\xi] \mid \chi[\xi] \to \psi$, where ξ occurs in χ and $con^-(\chi,\xi)$ holds.

A u-formula φ is a negative context for a subformula ξ of φ, $con^-(\varphi,\xi)$, iff

$$\varphi ::= \psi \wedge \chi[\xi] \mid \chi[\xi] \wedge \psi \mid \psi \vee \chi[\xi] \mid \chi[\xi] \vee \psi \mid \psi \to \chi[\xi] \mid \forall x\chi[\xi] \mid \exists x\chi[\xi],$$

where ξ occurs in χ and $con^-(\chi,\xi)$ holds, or $\varphi ::= \neg\chi[\xi] \mid \chi[\xi] \to \psi$, where ξ occurs in χ and $con^+(\chi,\xi)$ holds.

To apply the tableaux rules to a formula ψ it is necessary to know whether ψ occurs positively in a superformula φ — then we have to apply a T-rule —, or negatively — then we have to apply an F-rule. In an underspecified representation it may happen that a formula occurs positively in one disambiguation and negatively in another. We call formulas of this kind *indefinite*, and in this case we cannot apply a tableaux rule.

Definition 7. *Given an underspecified representation $\langle LHF, C, L, H \rangle$, a labeled h-formula $l : \varphi[h] \in LHF$ is* definite *if for every $l' : \psi[h'] \in LHF$, such that $con^-(\psi,h')$ holds and $h \sqcup h'$ defined, then it holds that $l \leq h' \in C$ or $l' \leq h \in C$. It is called* indefinite *otherwise.*

Why do we consider definite formulas? Intuitively, we need to know which quantifier we are actually dealing with when we are trying to find a strongest (weakest) reading. Formulas can be made more definite by using the rules for partial negation resolution given in Table 2. Roughly, we obtain more definite h-formulas within a given underspecified representation by adding further constraints which let indefinite h-formulas become definite by using one of the rules of *partial negation resolution* as specified in Table 2, which are generalizations of the method of partial disambiguation in [KR96]. These rules reduce the number of indefinite h-formulas occurring in an underspecified representation by creating partial disambiguations in which the indefinite h-formula has scope over (or is in the scope of one of) the h-formulas inducing the indefiniteness; in Table 2 this is $l_m : \varphi_m[h_n]$, where $con^-(\varphi_m, h_n)$ holds and $h_k \sqcup h_n$ is defined. Solid lines between two labels or holes, k, k', indicate immediate scope relation, dashed lines are the transitive closure of solid lines. For instance, let $\varphi_j = \forall x(\varphi)$ and $\varphi_m = \neg h_n$, we do not know, whether $\forall x$ binds x universally or existentially, because it can appear above or under the negation. Applying $(T_u : \pi)$ yields the two possible cases, namely $\forall x(\varphi)$ occurring above (left branch) or under (right branch) the negation.

To put it differently, suppose that $l_m : \varphi_m[h_n]$ is the only h-formula, which causes indefiniteness of $l_j : \varphi_j$ in an application of $(T_u : \pi)$, then the rule for left partial disambiguation labels $l_j : \varphi_j$ with T_u, because now it has scope over

Table 2. Tableaux rules for partial negation resolution

$$
\frac{\begin{array}{c} T_u : h_i \\ l_j : \varphi_j[h_k] \quad \ldots \quad l_l : \varphi_l \\ l_m : \varphi_m[h_n] \end{array}}{\begin{array}{c|c} T_u : h_i & T_u : h_i \\ l_j : \varphi_j[h_k] \ \ldots \ l_l : \varphi_l & \ldots \ l_l : \varphi_l \\ l_m : \varphi_m[h_n] & l_m : \varphi_m[h_n] \\ & l_j : \varphi_j[h_k] \end{array}} \ (T_u : \pi)
$$

$$
\frac{\begin{array}{c} F_u : h_i \\ l_j : \varphi_j[h_k] \quad \ldots \quad l_l : \varphi_l \\ l_m : \varphi_m[h_n] \end{array}}{\begin{array}{c|c} F_u : h_i & F_u : h_i \\ l_j : \varphi_j[h_k] \ \ldots \ l_l : \varphi_l & \ldots \ l_l : \varphi_l \\ l_m : \varphi_m[h_n] & l_m : \varphi_m[h_n] \\ & l_j : \varphi_j[h_k] \end{array}} \ (F_u : \pi)
$$

the negative context, and the rule for right partial disambiguation labels $l_j : \varphi_j$ with F_u, because it is in the scope of the negative context.

Our complete set of deduction rules for underspecified representations is given by combining Tables 2 and 3. This set defines our tableaux calculus, $\mathcal{TC}_{up}$.

Observe that there are three sets of rules in Table 3. The first set deals with ordinary logical connectives only. The second group are so-called interface rules; roughly speaking, they control the flow of information between traditional tableaux reasoning and disambiguation. Reasoning within an underspecified representation starts at its top-hole and compares all its daughters, i.e., those formulas that appear immediately in its scope. A similar interface is needed for h-formulas. The logical connectives in complex h-formulas are also treated with the T/F-rules, but for treating holes we need to know what material goes into them. For holes having only one daughter, it is possible to apply the normal tableaux rules to this daughter, see $(T_u : \uparrow)$ and $(F_u : \uparrow)$.

As to the rules in the third group, these are designed to partially construct the weakest or strongest readings of u-formulas, respectively. Both $(T_u : \forall)$ and $(F_u : \exists)$ presuppose that $l_j : \exists x \varphi[h_l]$ or $l_j : \forall x \varphi[h_l]$ occurs definite, otherwise we

Table 3. Set of deduction and interface rules of $\mathcal{TC}_{up}$

$$\frac{T:\varphi \wedge \psi}{\begin{array}{c}T:\varphi\\T:\psi\end{array}} \;(T:\wedge) \qquad \frac{F:\varphi\wedge\psi}{F:\varphi \;\mid\; F:\psi}\;(F:\wedge)$$

$$\frac{T:\varphi\vee\psi}{T:\varphi\;\mid\;T:\psi}\;(T:\vee) \qquad \frac{F:\varphi\vee\psi}{\begin{array}{c}F:\varphi\\F:\psi\end{array}}\;(F:\vee)$$

$$\frac{T:\varphi\to\psi}{F:\varphi\;\mid\;T:\psi}\;(T:\to) \qquad \frac{F:\varphi\to\psi}{\begin{array}{c}T:\varphi\\F:\psi\end{array}}\;(F:\to)$$

$$\frac{T:\neg\varphi}{F:\varphi}\;(T:\neg) \qquad \frac{F:\neg\varphi}{T:\varphi}\;(F:\neg)$$

$$\frac{T:\forall x\varphi}{T:\varphi[x/X]}\;(T:\forall) \qquad \frac{F:\forall x\varphi}{F:\varphi[x/f(X_1,\ldots,X_n)]}\;(F:\forall)^{\dagger}$$

$$\frac{T:\exists x\varphi}{T:\varphi[x/f(X_1,\ldots,X_n)]}\;(T:\exists)^{\dagger} \qquad \frac{F:\exists x\varphi}{F:\varphi[x/X]}\;(F:\exists)$$

$$\frac{T:UR}{T_u:h_0}\;(T:UR) \qquad \frac{F:UR}{F_u:h_0}\;(F:UR)$$

with $l_i:\varphi_i \;\ldots\; l_n:\varphi_n$ premises.

$$\frac{T:h_i}{T_u:h_i}\;(T:h) \qquad \frac{F:h_i}{F_u:h_i}\;(F:h)$$

with $l_i:\varphi_i \;\ldots\; l_n:\varphi_n$ premises.

$$\frac{T_u:h_i \;\leftarrow\; l_j:\varphi}{T:\varphi}\;(T_u:\uparrow) \qquad \frac{F_u:h_i \;\leftarrow\; l_j:\varphi}{F:\varphi}\;(F_u:\uparrow)$$

$$\frac{T_u:h_i \;\leftarrow\; l_j:\forall x\varphi[h_l]\; l_k:\mathcal{Q}y\psi\;\ldots\;l_n:\varphi_n}{T_u:h_i \;\leftarrow\; l_j:\forall x\varphi[h_l]\;\ldots\;l_n:\varphi_n,\; l_k:\mathcal{Q}y\psi}\;(T_u:\forall)^{\ddagger}$$

$$\frac{F_u:h_i \;\leftarrow\; l_j:\exists x\varphi[h_l]\; l_k:\mathcal{Q}y\psi\;\ldots\;l_n:\varphi_n}{F_u:h_i \;\leftarrow\; l_j:\exists x\varphi[h_l]\;\ldots\;l_n:\varphi_n,\; l_k:\mathcal{Q}y\psi}\;(F_u:\exists)^{\ddagger}$$

†Where $X_1,\ldots,X_n$ are the free variables in φ.
‡Where $\mathcal{Q}\in\{\forall,\exists\}$, l_j is definite, and $\forall x\varphi[h]$ and $\exists x\varphi[h]$ are *special* (see below).

would not be able to tell what the quantificational force of $l_j : \exists x \varphi$ or $l_j : \forall x \varphi$ is. So, before applying the rules it may be necessary to apply partial negation resolution as presented in Table 2 first so as to make $l_j : \forall x \varphi[h_l]$ definite. There is an important restriction on the applicability of the rules $(T_u : \forall)$ and $(F_u : \exists)$: to guarantee soundness of the rules, the formulas $\forall x\, \varphi[h]$ and $\exists x\, \phi[h]$ in l_j should be *special*. Here $\forall x\, \varphi[h]$ is special if it is of the form $\forall x\, (\chi_1 \to h)$ or $\forall x\, (\chi_1 \wedge h \to \chi_2)$, while $\exists x\, \varphi[h]$ is special if it is of the form $\exists x\, (\chi_1 \wedge h)$.

To conclude this section, we briefly turn to soundness and completeness. First, now that our tableaux may have different kinds of labelings (there are T/F-nodes and T_u/F_u-nodes), we need to specify what it means for a tableaux to close. We say that a branch b *closes* if there are two nodes $T : \varphi$ and $F : \psi$ belonging to b, such that φ and ψ are atomic formulas of $\mathcal{L}$ and φ and ψ are unifiable. In particular, it is not possible to close a tableau with two nodes $T : \varphi$ and $F : \psi$ containing holes or underspecified representations.

Next, what do soundness and completeness mean in our ambiguous setting? Sound and complete with respect to which semantics or system? We have opted to state soundness and completeness with respect to tableaux provability of all total disambiguations.

Theorem 2 (Soundness and Completeness). *Let $\varphi \in \mathcal{L}^u$. Then $\vdash_{\mathcal{TC}_{up}} \varphi$ if, and only if, for all $d \in \delta(\varphi)$ $\vdash_{\mathcal{TC}} d$*

Proof (Sketch). The soundness part ('only if') boils down to a proof that the T_u/F_u rules do not introduce any information that would not have been available by totally disambiguating first. The restrictions on the rules $(T_u : \forall)$ and $(F_u : \exists)$ that were discussed above allow us to establish this.

Proving completeness ('if') is in some way easier: any open branch in a (completely developed) tableau for $\mathcal{TC}_{up}$ corresponds to a (completely developed) open branch in a tableau proof for $\mathcal{TC}_u$. See [MR98] for the details.

5 An Example

Consider the sentence *every boy doesn't see a movie* appearing as a premise in a tableau. Because displaying derivations in our calculus is very space-consuming, we can only give the beginning of one of its branches, which is given in Figure 1. Each box corresponds to a node in a tableau tree. Because in (1) $l_1 : \forall x\, (boy(x) \to h_1)$ occurs indefinite, it is necessary to apply partial negation resolution first. The total disambiguation of the left branching would be

$$\{\forall x\, (boy(x) \to \exists y\, (movie(y) \wedge \neg see(x,y))),$$
$$\forall x\, (boy(x) \to \neg \exists y\, (movie(y) \wedge see(x,y))),$$
$$\exists y\, (movie(y) \wedge \forall x\, (boy(x) \to \neg see(x,y)))\},$$

That is, formulas in which the universal quantifier has scope over the negation, disregarding the existential quantifier. Now $(T_u : \forall)$ is applicable and the universal quantifier is given wide scope in (4), corresponding to the readings

Fig. 1. Part of a proof in $\mathcal{TC}_{up}$

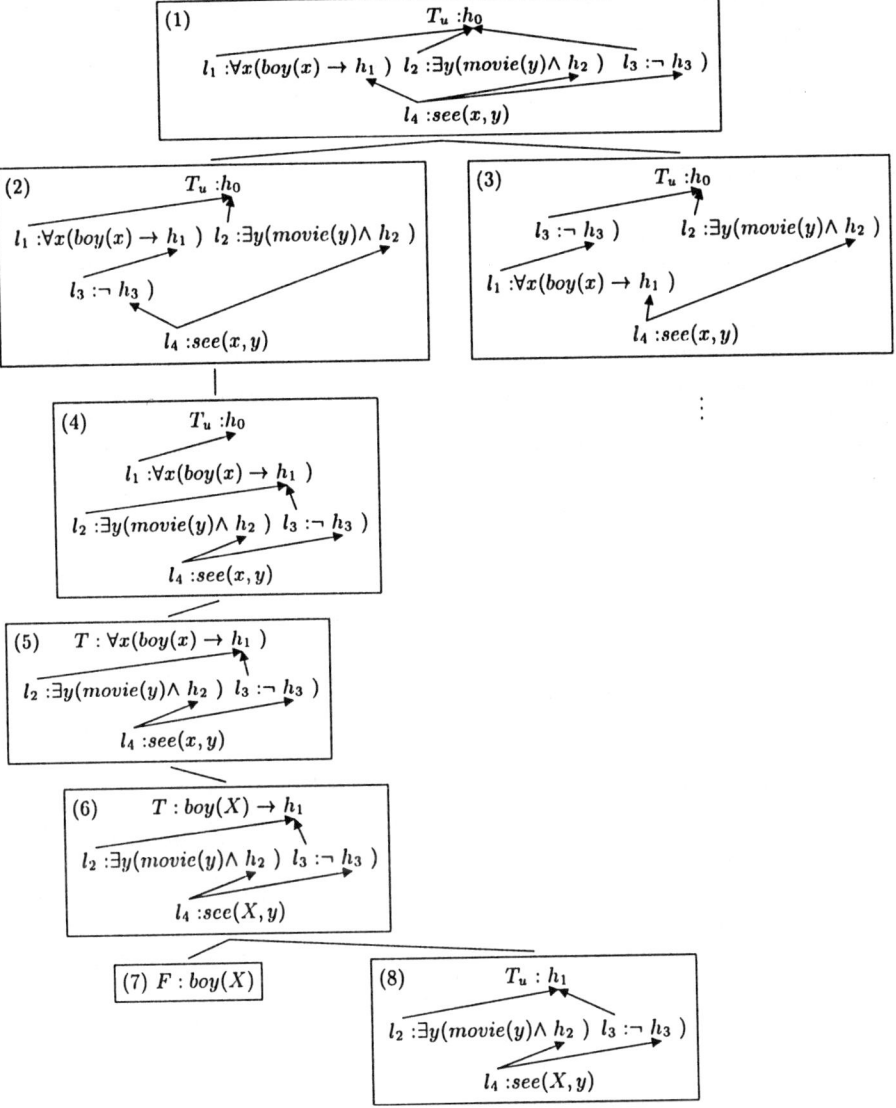

$\forall x\,(boy(x) \rightarrow \exists y(movie(y) \wedge \neg see(x,y)))$ and $\forall x\,(boy(x) \rightarrow \neg \exists y\,(movie(y) \wedge see(x,y)))$. Because h_0 has only one daughter, the normal tableaux rules for logical connectives can be applied to it. So we instantiate x with a free variable X and apply $(T:\rightarrow)$, which causes a branching of the proof tree, where (7) is a non-ambiguous literal with which we can try to close a tableaux branch. In (8) h_1 is the top-node to which the underspecified tableaux rules can be applied.

6 Conclusion

In this paper we have presented a tableaux calculus for reasoning with ambiguous quantification. We have set up a representation formalism that allows for a smooth interleaving of traditional deduction steps with disambiguation steps.

Our ongoing work focuses on two aspects. First, we are adding rules for coping with additional forms of ambiguity to the calculus, such as ambiguity of binary connectives. Second, we are in the process of implementing the calculus $\mathcal{TC}_{up}$; as part of this work new and interesting theoretical issues arise, such as 'proof optimization': for reasons of efficiency it pays to postpone disambiguations as long as possible, but to be able to apply some of the rules expressions need to be definite and for this reason early disambiguation may be required. What is the best way of reconciling these two demands?

References

[Bos95] J. Bos. Predicate logic unplugged. In P. Dekker and M. Stokhof, editors, *Proc. 10th Amsterdam Colloquium*. ILLC, University of Amsterdam, 1995.

[Dee96] K. van Deemter. Towards a logic of ambiguous expressions. In Peters and Deemter [DP96].

[DP96] K. van Deemter and S. Peters, editors. *Semantic Ambiguity and Underspecification*. CSLI Publications, 1996.

[EJ96] J. van Eijck and J. Jaspars. Ambiguity and reasoning. Technical Report CS-R9616, Centrum voor Wiskunde en Informatica, Amsterdam, 1996.

[Fit96] M. Fitting. *First-Order Logic and Automated Theorem Proving*. Springer-Verlag New York, 2nd edition, 1996.

[Jas97] J. Jaspars. Minimal logics for reasoning with ambiguous expressions. CLAUS-Report 94, University of Saarbrücken, 1997.

[KM93] H. S. Kurtzman and M. C. MacDonald. Resolution of quantifier scope ambiguities. *Cognition*, 48:243–279, 1993.

[Kön94] E. König. A study in grammar design. Arbeitspapier des Sonderforschungsbereich 340 no. 54, Institut für Maschinelle Sprachverarbeitung, 1994.

[KR96] E. König and U. Reyle. A general reasoning scheme for underspecified representations. In H.-J. Ohlbach and U. Reyle, editors, *Logic and Its Applications. Festschrift for Dov Gabbay*. Kluwer Academic Publishers, 1996.

[MR98] C. Monz and M. de Rijke. Reasoning with ambiguous expressions. Unpublished manuscript, 1998.

[Rey93] U. Reyle. Dealing with ambiguities by underspecification: Construction, representation, and deduction. *Journal of Semantics*, 10(2):123–179, 1993.

[TS96] A. S. Troelstra and H. Schwichtenberg. *Basic Proof Theory*. Cambridge University Press, 1996.

From Kripke Models to Algebraic Counter-valuations

Sara Negri and Jan von Plato

Department of Philosophy
PL 24, Unioninkatu 40 B
00014 University of Helsinki, Finland
{negri,vonplato}@helsinki.fi

Abstract. Starting with a derivation in the refutation calculus **CRIP** of Pinto and Dyckhoff, we give a constructive algebraic method for determining the values of formulas of intuitionistic propositional logic in a counter-model. The values of compound formulas are computed pointwise from the values on atoms, in contrast to the non-local determination of forcing relations in a Kripke model based on classical reasoning.

1 Introduction

Systems of terminating sequent calculi for intuitionistic propositional logic were first given in Dyckhoff (1992) and in Hudelmaier (1992). These calculi have the property that bottom-up proof search of provable sequents always terminates, a feature obtained through a refinement of the left implication rule of the usual cut-free sequent calculi for intuitionistic propositional logic (see Troelstra and Schwichtenberg 1996 for standard versions of these calculi).

In Pinto and Dyckhoff (1995), a related *refutation calculus* **CRIP** was given, for showing underivability of a sequent $\Gamma \Rightarrow \Delta$. They proved that for intuitionistic propositional logic, either the sequent $\Gamma \Rightarrow \Delta$ is derivable in Dyckhoff's calculus **LJT***, or the *antisequent* $\Gamma \not\Rightarrow \Delta$ is derivable in **CRIP**. For the latter case, a method was given for constructing a Kripke counter-model. A related method was developed by Stoughton (1996) for producing small Kripke counter-models.

We shall here propose an algebraic method for computing the values of compound formulas in a counter-model. The method is constructive, and can replace the determination of forcing of compound formulas in a Kripke model. In the latter, classical reasoning on the meta-level is used; Our method, instead, uses a direct pointwise computation from values on atomic formulas.

In Kripke trees as well as in Heyting algebras, there is no internal notion for expressing that, say, an element is strictly above another one in the partial order, but this can only be seen by looking "from the outside", if at all. We propose a structure, that of a *positive* Heyting algebra, that internalizes the intuitive situation. This is done by requiring a relation $a \not\leq b$, read as "a exceeds b", with properties such that the usual partial order comes out as a negation,

$a \leqslant b \equiv \sim a \not\leqslant b$. This is quite analogous to the definition of an equality relation as a negation of apartness. Next, we define a formula A to be *invalid* if there exists a valuation v to a positive Heyting algebra such that $v(\top) \not\leqslant v(A)$. If not, A is defined as valid, and we have for all valuations v that $v(\top) \leqslant v(A)$. In von Plato (1997), it is shown that this initially perhaps surprising definition of intuitionistic validity as a negative notion coincides with the usual definition. Further, with positive Heyting algebras we can express and prove *soundness* of rules of refutation, by showing that if there is a counter-valuation for the premises, there is a counter-valuation for the conclusion.

The paper is organized as follows: We introduce the algebraic semantics of refutation, and then present the calculus **CRIP**. In Section 5, we show how to construct an algebraic counter-model parallel to the construction of a Kripke counter-model. The key step is the operation of combining (positive) Heyting algebras that corresponds to the gluing of Kripke models. In Section 6, we show how the valuations in positive Heyting algebras are computed, and in Section 7 we give some examples; These show concretely how the values of compound formulas are computed from values on atoms, instead of the non-local and classical determination of forcing in a Kripke model.

2 Positive partial order, lattices and Heyting algebras

We assume given a set with a primitive relation $a \not\leqslant b$, to be read a *exceeds* b, and satisfying the axioms of *irreflexivity* and *splitting*:

PPO1. $\sim a \not\leqslant a$,   PPO2. $a \not\leqslant b \supset a \not\leqslant c \vee c \not\leqslant b$.

A set with such a relation is called a *positive partial order*. Observe that the relation is not a partial order, for transitivity does not in general hold, but a relation whose negation is a partial order, defined by:

Definition 1. $a \leqslant b \equiv \sim a \not\leqslant b$.

This *weak partial order* relation is reflexive by PPO1 and transitive by contraposition of PPO2. As there will be no need for a primitive notion of equality, we define equality by $a = b \equiv a \leqslant b \;\&\; b \leqslant a$. Thus, our weak partial order is what is sometimes called a quasi-ordering.

We can further define an apartness relation by $a \neq b \equiv a \not\leqslant b \vee b \not\leqslant a$. It has the usual properties, and its negation coincides with equality defined above. Strict partial order can be defined by $a < b \equiv b \not\leqslant a \;\&\; \sim a \not\leqslant b$ and it is irreflexive and transitive.

A *positive lattice* is obtained by adding meet and join operations and the following axioms to a positive partial order.

MTI $\sim a \wedge b \not\leqslant a$, $\sim a \wedge b \not\leqslant b$,   JNI $\sim a \not\leqslant a \vee b$, $\sim b \not\leqslant a \vee b$,

MTU $c \not\leqslant a \wedge b \supset c \not\leqslant a \vee c \not\leqslant b$,   JNU $a \vee b \not\leqslant c \supset a \not\leqslant c \vee b \not\leqslant c$.

Positive Heyting algebras result from adding to a positive lattice a third construction $a \rightarrow b$, to be called *Heyting arrow*, with the axioms

PHI $\quad \sim (a{\to}b)\wedge a \not\leq b,$ $\qquad\qquad$ PHU $\quad c \not\leq a{\to}b \supset c\wedge a \not\leq b.$

The first axiom validates modus ponens, the second, a constructive uniqueness principle, identifies implication as the supremum of anything that together with a gives b.

Here we use positive Heyting algebras with a bottom element 0. This is obtained by the principle

PHB $\quad \sim 0 \not\leq a,$

and a top element 1 is now defined by $1 = 0{\to}0$.

Each of the positive structures is constructively stronger than the corresponding usual structure, because of the presence of splitting and the uniqueness axioms. But if we define partial order through the negation of excess, the usual axioms for partial order, lattices and Heyting algebras are obtained by taking the negative axioms for excess and the contrapositions of the positive ones. For instance, the axioms for partial order defined negatively are PPO1 and

$$\sim a \not\leq c \ \& \sim c \not\leq b \supset \sim a \not\leq b,$$

and the ones to be added for lattices are MTI, JNI, and

$$\sim c \not\leq a \ \& \sim c \not\leq b \supset \sim c \not\leq a\wedge b,$$
$$\sim a \not\leq c \ \& \sim b \not\leq c \supset \sim a\vee b \not\leq c,$$

and for Heyting algebras PHB, PHI, and

$$\sim c\wedge a \not\leq b \supset \sim c \not\leq a{\to}b.$$

If a formula in which all atoms are negated is proved in the theory of positive Heyting algebras, then it can be proved in the theory with the above axioms. This conservativity result is proved in Negri (1997) by means of a cut-free sequent calculus for the theory of positive Heyting algebras. (The ideas and methods of this proof require too much space to be summarized here).

We say that a map ϕ from a positive Heyting algebra H_1 to a positive Heyting algebra H_2 is a *homomorphism of positive Heyting algebras* if it reflects the excess relation and preserves meet, join, Heyting arrow and bottom, that is, for all $a, b \in H_1$ we have

1. $\phi(a) \not\leq \phi(b)$ implies $a \not\leq b$,

2. $\phi(a\wedge b) = \phi(a)\wedge\phi(b)$,

3. $\phi(a\vee b) = \phi(a)\vee\phi(b)$,

4. $\phi(a{\to}b) = \phi(a){\to}\phi(b)$,

5. $\phi(0_1) = 0_2$,

where 0_1 and 0_2 are the bottom elements of H_1 and H_2, respectively.

If a map ϕ reflects the excess relation, then by contraposition it is monotone with respect to the partial order defined through negation of excess. As a consequence, the conditions 2–5 can be weakened into the following:

2'. $\sim \phi(a) \wedge \phi(b) \not\leq \phi(a \wedge b)$,

3'. $\sim \phi(a \vee b) \not\leq \phi(a) \vee \phi(b)$,

4'. $\sim \phi(a) \rightarrow \phi(b) \not\leq \phi(a \rightarrow b)$,

5'. $\sim \phi(0_1) \not\leq 0_2$.

An *isomorphism of positive Heyting algebras* is a bijective homomorphism of positive Heyting algebras.

The following lemma will be used in the proof of proposition 11:

Lemma 2. *If H_1 and H_2 are positive Heyting algebras and $\phi : H_1 \to H_2$ and $\psi : H_1 \to H_2$ are maps that reflect the excess relation and are inverses of each other, then ϕ is an isomorphism of positive Heyting algebras.*

Proof. We prove 2', the other conditions being dealt with similarly.

By bijectivity, we have $a \wedge b = \psi \phi(a \wedge b)$, and thus also $\psi \phi(a) \wedge \psi \phi(b) = \psi \phi(a \wedge b)$. By monotonicity of ψ we have $\sim \psi(\phi(a) \wedge \phi(b)) \not\leq \psi \phi(a) \wedge \psi \phi(b)$, and therefore $\sim \psi(\phi(a) \wedge \phi(b)) \not\leq \psi \phi(a \wedge b)$. By monotonicity of ϕ and the fact that ϕ is the inverse of ψ, we obtain $\sim \phi(a) \wedge \phi(b) \not\leq \phi(a \wedge b)$.

3 Algebraic semantics of refutation

We shall show that positive Heyting algebras lead to a natural formal semantics of refutation, corresponding precisely to the usual algebraic semantics for derivability. A *valuation* is, as usually, a homomorphism $v :$ Form $\to H$ from the set of formulas Form (here of intuitionistic propositional logic) to a positive Heyting algebra H, satisfying the equations

$v(A \& B) = v(A) \wedge v(B)$,

$v(A \vee B) = v(A) \vee v(B)$,

$v(A \supset B) = v(A) \rightarrow v(B)$,

$v(\bot) = 0$.

Let Γ range over finite sets of formulas. We shall write $v(\Gamma)$ for the meet of the values of formulas in Γ, with $v(\Gamma) = 1$ in case Γ is empty.

Definition 3. *A formula A is* invalid *under Γ, written $\Gamma \not\models A$, if there is a valuation v to a positive Heyting algebra such that $v(\Gamma) \not\leq v(A)$. In this case we say that v is a* counter-valuation *to Γ, A.*

In particular, a formula A is invalid, denoted by $\not\models A$, if there is a valuation v to a positive Heyting algebra such that $1 \not\leq v(A)$. We can also define *consistency* of Γ internally, by requiring that there is a valuation v for which $v(\Gamma) \not\leq 0$. This is most naturally written as $\Gamma \not\models$ (or, equivalently, $\Gamma \not\models \bot$).

Definition 4. $\Gamma \models A$ *if and only if not $\Gamma \not\models A$.*

We shall say that A is valid under Γ, or a logical consequence of Γ. In particular, A is valid if not $\not\vDash A$, and Γ is inconsistent if not $\Gamma \not\vDash$.

We emphasize that this order of concepts is essential for reasoning constructively. If a classical meta-logic is used, validity can equally be taken as the basic notion.

It follows from our definition that $\Gamma \vDash A$ if and only if for all valuations v to positive Heyting algebras, $v(\Gamma) \leqslant v(A)$. In particular, we have that a formula A is valid if and only if $v(A) = 1$ for all valuations. This is just like the standard definition of validity for intuitionistic logic except that it refers to positive Heyting algebras, and as shown in von Plato (1997), the new notion of validity coincides with the old one. To give a brief example of a proof of validity, let us show $\vDash A\&B \supset A$. So assume there is a valuation v such that $1 \not\leqslant v(A\&B \supset A)$. Then $1 \not\leqslant v(A) \wedge v(B) \to v(A)$, so by PHU, $v(A) \wedge v(B) \not\leqslant v(A)$ which gives a contradiction by MTI. So for all valuations v we have $\sim 1 \not\leqslant v(A\&B \supset A)$, that is, $1 \leqslant v(A\&B \supset A)$. Observe that the proof is constructive: no *reductio ad absurdum* is used, but the negative definition of validity.

In von Plato (1997), details of the application of positive Heyting algebras to intuitionistic propositional logic can be found. For example, it is shown that the Lindenbaum algebras of intuitionistic propositional logic have the structure of positive Heyting algebras, from which completeness relative to these algebras follows.

4 Refutation calculi

For us, a refutation calculus is a system of syntactic rules for showing refutability. Refutability is a positive notion, in contrast to the weak negative notion of underivability.

We shall here make use of the calculus **CRIP** of Pinto and Dyckhoff (1995), with the role of falsum in the rules made explicit (Roy Dyckhoff, personal communication November 1997). In the rules below, an *antisequent* is an expression of form $\Gamma \not\Rightarrow \Delta$ where Γ, Δ are finite multisets of formulas. The rules of **CRIP**, from Pinto and Dyckhoff (1995, p. 227), are to be read as follows: We start from an antisequent $\Gamma \not\Rightarrow \Delta$ at the bottom, and infer sufficient conditions upwards. If we reach *axioms* in all leaves of the upward-growing tree, the refutation was successful. We can then read the tree top-down as a derivation of the initial antisequent as a theorem of **CRIP**, and, therefore, as a nontheorem of intuitionistic propositional logic. If not, the sequent $\Gamma \Rightarrow \Delta$ is derivable in the multisuccedent calculus **LJT*** of Dyckhoff (1992).

In the rules of **CRIP** below, we use $P, Q, R, \ldots$ for atomic formulas and $A, B, C, \ldots$ for arbitrary formulas. Two of the rules have conditions, and in them, an *atomic implication* is one with an atom as antecedent.

CRIP:

$$\overline{P_1 \supset B_1, \ldots, P_k \supset B_k, \Gamma \not\Rightarrow \Delta} \quad axiom$$

$$\frac{A, B, \Gamma \not\Rightarrow \Delta}{A\&B, \Gamma \not\Rightarrow \Delta} \ (1) \qquad \frac{\Gamma \not\Rightarrow \Delta, A}{\Gamma \not\Rightarrow \Delta, A\&B} \ (2) \qquad \frac{\Gamma \not\Rightarrow \Delta, B}{\Gamma \not\Rightarrow \Delta, A\&B} \ (3)$$

$$\frac{A, \Gamma \not\Rightarrow \Delta}{A \vee B, \Gamma \not\Rightarrow \Delta} \ (4) \qquad \frac{B, \Gamma \not\Rightarrow \Delta}{A \vee B, \Gamma \not\Rightarrow \Delta} \ (5) \qquad \frac{\Gamma \not\Rightarrow \Delta, A, B}{\Gamma \not\Rightarrow \Delta, A \vee B} \ (6)$$

$$\frac{P, B, \Gamma \not\Rightarrow \Delta}{P, P \supset B, \Gamma \not\Rightarrow \Delta} \ (7) \qquad \frac{C \supset B, D \supset B, \Gamma \not\Rightarrow \Delta}{(C \vee D) \supset B, \Gamma \not\Rightarrow \Delta} \ (8)$$

$$\frac{C \supset (D \supset B), \Gamma \not\Rightarrow \Delta}{(C\&D) \supset B, \Gamma \not\Rightarrow \Delta} \ (9) \qquad \frac{B, \Gamma \not\Rightarrow \Delta}{(C \supset D) \supset B, \Gamma \not\Rightarrow \Delta} \ (10)$$

$$\frac{C_1, D_1 \supset B_1, \Gamma_1 \not\Rightarrow D_1 \ldots C_n, D_n \supset B_n, \Gamma_n \not\Rightarrow D_n \quad \Gamma', E_1 \not\Rightarrow F_1 \ldots \Gamma', E_m \not\Rightarrow F_m}{\Gamma' \not\Rightarrow E_1 \supset F_1, \ldots, E_m \supset F_m, \Delta} \ (11)$$

where we use the abbreviations:

$\Gamma' = (C_1 \supset D_1) \supset B_1, \ldots, (C_n \supset D_n) \supset B_n, \Gamma$,
$\Gamma_i = \Gamma' - (C_i \supset D_i) \supset B_i$.

$$\frac{\Gamma \not\Rightarrow \Delta}{\bot \supset B, \Gamma \not\Rightarrow \Delta} \ (12)$$

The conditions in *axiom* are that Γ contains only atomic formulas, Δ contains only atomic formulas or $\bot$, Γ and Δ are disjoint, and each P_i is atomic and not in Γ.

The restrictions in rule (11) are: Each formula in Γ is either atomic or an atomic implication, no antecedent of an atomic implication is equal to an atom in Γ, Δ contains only atoms or $\bot$, Γ and Δ are disjoint.

The *axiom*-rule is a special case of rule (11), with $m = n = 0$. The conditions and rule (12) are amendments to Pinto and Dyckhoff (1995). It is possible to avoid adding rule (12) if in rule (11) Γ is permitted to contain implications with $\bot$ as antecedent.

5 Construction of counter-valuations

We show how to construct positive Heyting algebras serving as codomains of counter-valuations for the nontheorems of intuitionistic propositional logic. We use the calculus **CRIP** and the construction of Kripke counter-models from derivations of antisequents in **CRIP**, to obtain the construction of positive Heyting algebras and counter-valuations.

We start by recalling the construction of a Heyting algebra out of a Kripke model (for more details, see Fitting 1969). Let K be a Kripke model, with a reflexive and transitive relation $\leq$ and a forcing relation $\Vdash$ between elements w of K and formulas, with the usual properties.[1] The algebraic model $H(K)$ corresponding to K is the collection of the upward closed subsets[2] of K, with ordering given by subset inclusion. The meet and join operations are intersection and union, respectively. The top element 1 is the whole set K, the bottom is the empty set. The K-valuation $v(P)$ of an atomic formula P is the set of nodes of the Kripke model forcing P,

Definition 5. $v(P) \equiv \{w \in K \mid w \Vdash P\}$.

We have (Fitting 1969, p. 24):

Proposition 6. $H(K)$ is a Heyting algebra, with $v(A) = 1$ iff $K \Vdash A$.

For propositional logic, finite Kripke models suffice for the construction of counter-models. These are discrete structures, with a decidable partial order.

Whereas finite sets have a decidable membership, subfinite sets, i.e., subsets of a finite set, do not necessarily have a decidable membership. We therefore define the Heyting algebra associated to a finite Kripke tree to consist of *finite subsets* of the Kripke tree. Then the associated Heyting algebra has a decidable order, and is indeed a positive Heyting algebra with the excess relation defined by

$$U \not\leq V =_{df} (\exists u \in U)(u \notin V).$$

We therefore have

Proposition 7. *If K is a finite Kripke tree, $H(K)$ is a positive Heyting algebra.*

The following representation of elements of $H(K)$ will be useful:

Lemma 8. *If K is a finite Kripke tree, then every element of $H(K)$ can be uniquely represented as*

$$\bigcup_{a \in F} \uparrow a$$

where $\uparrow a = \{b \in K \mid a \leq b\}$, F is a finite subset of K, and any two distinct elements of F are incomparable.

---

[1] By well known results (see Troelstra and van Dalen 1988, ch. 2.6) we can consider Kripke models as represented by trees, and call a Kripke tree the lattice structure of a Kripke model.

[2] Recall that a subset U of S is *upward closed* if, whenever $x \in S$ and $a \leq x$ for some $a \in U$, then $x \in U$.

Proof. Immediate.

In the construction of Kripke models, an essential step is the *gluing* of a finite number of Kripke models $K_1, \ldots, K_n$. The resulting Kripke model has an initial world w_0 with immediate successors given by the initial worlds of the n given Kripke models. The forcing relation can be modified by the forcing of certain atoms in the new root w_0.

We shall denote by $g(K_1, \ldots, K_n)$ the Kripke tree obtained by gluing of $K_1, \ldots, K_n$. Our next task is to find the operation on (positive) Heyting algebras corresponding to gluing, that is, the operation ∘ solving up to positive Heyting algebra isomorphism the equation

$$H(g(K_1, \ldots, K_n)) = \circ(H(K_1), \ldots, H(K_n)).$$

For the sake of simplicity, we consider the case of $n = 2$ only, but what follows generalizes to any finite number in an obvious way.

Before giving the general construction, we discuss two examples:

Example 9. Let K_1 be the singleton-set Kripke tree. Then $H(K_1)$ is the (positive) Heyting algebra consisting of two elements

Observe that when one draws diagrams of this kind, one neatly places the points apart, even though in the theories based on partial order there is no internal notion to express the visual effect.

Example 10.

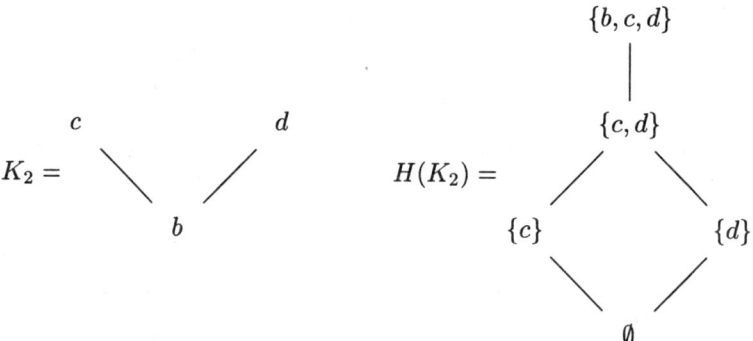

Observe that K_2 is itself the gluing of two Kripke trees of the first kind, so $H(K_2) \cong H(K_1) \circ H(K_1)$ where ∘ is the operation to be determined.

If we glue together K_1 and K_2 we obtain

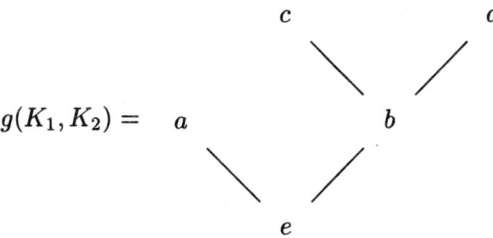

and the corresponding positive Heyting algebra is (with explicit labels omitted)

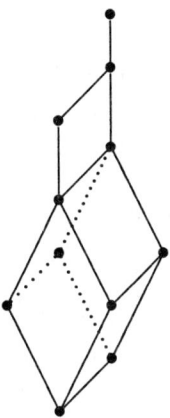

The **general construction** behind these examples is as follows: Given two positive Heyting algebras H_1 and H_2, with respective top elements 1_1 and 1_2, let $H_1 \times H_2$ be their Cartesian product with excess relation defined by

$$(a_1, a_2) \not\leq (b_1, b_2) \equiv a_1 \not\leq b_1 \vee a_2 \not\leq b_2$$

and component-wise meet, join and Heyting arrow, and let $\widehat{H_1 \times H_2}$ be the lattice obtained by adding an "extra-top" element 1 and extending the excess relation by posing $\sim (a,b) \not\leq 1$ for $a \in H_1$, $b \in H_2$ and $1 \not\leq (1_1, 1_2)$. It is clear that in the examples we have

$$H(K_2) = H(g(K_1, K_1)) \cong \widehat{H(K_1) \times H(K_1)},$$

$$H(g(K_1, K_2)) \cong \widehat{H(K_1) \times H(K_2)}.$$

Indeed we have, in full generality, that the extra-topped Cartesian product is the operation on Heyting algebras corresponding to the gluing of Kripke models:

Proposition 11. *Let K_1 and K_2 be finite Kripke trees. Then*

$$H(g(K_1, K_2)) \cong \widehat{H(K_1) \times H(K_2)}.$$

Proof. By lemma 8, an element in $H(g(K_1, K_2))$ is either $1 = \uparrow w_0$, where w_0 is the root of $g(K_1, K_2)$, or

$$\bigcup_{a \in F_1} \uparrow a \; \cup \; \bigcup_{b \in F_2} \uparrow b$$

where F_1 and F_2 are finite subsets of K_1 and K_2. The maps

$$\phi : H(g(K_1, K_1)) \to \widehat{H(K_1) \times H(K_1)}$$
$$\uparrow w_0 \mapsto 1$$
$$\bigcup_{a \in F_1} \uparrow a \; \cup \; \bigcup_{b \in F_2} \uparrow b \mapsto (\bigcup_{a \in F_1} \uparrow a, \bigcup_{b \in F_2} \uparrow b)$$

$$\psi : \widehat{H(K_1) \times H(K_1)} \to H(g(K_1, K_1))$$
$$1 \mapsto \uparrow w_0$$
$$(\bigcup_{a \in F_1} \uparrow a, \bigcup_{b \in F_2} \uparrow b) \mapsto \bigcup_{a \in F_1} \uparrow a \; \cup \; \bigcup_{b \in F_2} \uparrow b$$

reflect the excess relation and are inverses of each other, therefore by lemma 2 they give an isomorphism between $H(g(K_1, K_2))$ and $\widehat{H(K_1) \times H(K_2)}$.

We adopt from Pinto and Dyckhoff (1995) the following:

Definition 12. *A Kripke tree is a strong counter-model to a sequent $\Gamma \Rightarrow \Delta$ if in its initial world all the formulas in Γ are forced and none of the formulas in Δ are forced.*

Our corresponding algebraic notion is:

Definition 13. *A positive Heyting algebra H with a valuation v is an algebraic counter-model to a sequent $\Gamma \Rightarrow \Delta$ if for all A in Γ, we have $v(A) = 1$ and $1 \not\leq \bigvee_{B \in \Delta} v(B)$.*

Lemma 14. *If K is a strong counter-model to $\Gamma \Rightarrow \Delta$, then $H(K)$, with the K-valuation as defined in 5, is an algebraic counter-model to $\Gamma \Rightarrow \Delta$.*

As an aside, we recall from Pinto and Dyckhoff (1995) that a Kripke tree is a *counter-model* to a sequent $\Gamma \Rightarrow \Delta$ if it has a node in which all formulas in Γ are forced and none of the formulas in Δ are forced. If K is a counter-model to $\Gamma \Rightarrow \Delta$ then the positive Heyting algebra $H(K)$ with the K-valuation has the property that

$$\bigwedge_{A \in \Gamma} v(A) \not\leq \bigvee_{B \in \Delta} v(B)$$

and we call such a positive Heyting algebra with a valuation satisfying the above property a *weak algebraic counter-model*. The relation between algebraic counter-models and weak algebraic counter-models parallels the relation between strong counter-models and counter-models, that is, every algebraic counter-model is a weak algebraic counter-model but not conversely.

Theorem 15. *If $\Gamma \not\Rightarrow \Delta$ is derivable in* **CRIP**, *then there is an algebraic counter-model to the sequent $\Gamma \Rightarrow \Delta$.*

Proof. Consider the derivation in **CRIP** of the antisequent $\Gamma \not\Rightarrow \Delta$. For each step of the construction of the Kripke counter-model as given in proposition 1 of Pinto and Dyckhoff (1995), there is a corresponding step of construction of a positive Heyting algebra and an algebraic counter-valuation, given as follows:

–To the Kripke tree consisting of a single world there corresponds the positive Heyting algebra consisting of two elements. All atoms that are forced are evaluated into the top element, the others into the bottom.

–To the gluing of $n > 1$ Kripke trees K_i there corresponds the extra-topped Cartesian product of Heyting algebras. The atoms forced in the new root are evaluated into the extra top, the other atoms P into $(v_1(P), \ldots, v_n(P))$ where $v_i(P)$ is the K-valuation of $H(K_i)$. In the special case of an application of rule (11) with just one premise, and in all other rules, no gluing of Kripke models is performed, and correspondingly, no extra-topped Cartesian product is taken: an algebraic counter-model for the premise is also a counter-model for the conclusion.

By proposition 11, the Heyting algebra H resulting from this construction is isomorphic to the Heyting algebra $H(K)$ associated to the resulting Kripke tree. Moreover, by lemma 14, H with the K-valuation of $H(K)$ is an algebraic counter-model to $\Gamma \Rightarrow \Delta$.

6 Computation of counter-valuations

The proof of theorem 15 prescribes how to construct an algebraic counter-model starting from a successful **CRIP** refutation. The positive Heyting algebra that serves as codomain of the valuation is defined inductively: The starting points are the two-element Heyting algebras, serving as counter-models for the axioms, and given $n > 1$ positive Heyting algebras that serve as counter-models for the n premises of rule 11, the counter-model for the conclusion is obtained by taking their extra-topped Cartesian product; The construction also gives the valuation for atomic formulas. The evaluation of compound formulas can then be done in a component-wise fashion, but before that a remark on the Cartesian product of positive Heyting algebras is in order:

If $H_1, \ldots, H_n$ are positive Heyting algebras, then the set given by their Cartesian product with excess relation given by

$$(a_1, \ldots, a_n) \not\leq (b_1, \ldots, b_n) \equiv a_1 \not\leq b_1 \vee \ldots \vee a_n \not\leq b_n$$

and meet, join and Heyting arrow defined component-wise, is a positive Heyting algebra.

Let H be the extra-topped Cartesian product of the positive Heyting algebras $H_1, \ldots, H_n$ and let v be a valuation on atoms. Then for all formulas A, $v(A)$ is either 1 or $(a_1, \ldots, a_n)$, where $a_i \in H_i$. For the sake of simplicity we can also denote by a vector $(t_1, \ldots, t_n)$ the extra-top and extend the excess relation and the meet and join operations of H_i by stating that $t_i \not\leq a_i$ for all $a_i \in H_i$ and by posing $t_i \wedge a_i = a_i$ and $t_i \vee a_i = t_i$. Then valuations can be computed componentwise, with some care for implication. So assume that $v(B) = (b_1, \ldots, b_n)$ and $v(C) = (c_1, \ldots, c_n)$ have been computed. We then have:

$$v(B \& C) = v(B) \wedge v(C) = (b_1 \wedge c_1, \ldots, b_n \wedge c_n),$$

$$v(B \vee C) = v(B) \vee v(C) = (b_1 \vee c_1, \ldots, b_n \vee c_n).$$

For $v(B \supset C)$ we distinguish three cases:

If $\sim v(B) \not\leq v(C)$, then $v(A) = (t_1, \ldots, t_n)$,

if $v(B) \not\leq v(C)$ and $v(B) = (t_1, \ldots, t_n)$ then $v(A) = (c_1, \ldots c_n)$,

if $v(B) \not\leq v(C)$ and $(t_1, \ldots, t_n) \not\leq v(B)$ then $v(A) = (b_1 \to c_1, \ldots b_n \to c_n)$.

The evaluation is algorithmic and no use of reasoning on the meta-level is needed, whereas in Kripke models the computation of the values of compound formulas uses classical reasoning on the meta-level.

7 Some examples of algebraic counter-models

Example 16. $(P \supset Q) \vee (Q \supset P)$, with P and Q distinct atoms:

The antisequent $\not\Rightarrow (P \supset Q) \vee (Q \supset P)$ has the following **CRIP** derivation:

$$\cfrac{\cfrac{\overline{P \not\Rightarrow Q}^{\ axiom} \quad \overline{Q \not\Rightarrow P}^{\ axiom}}{\not\Rightarrow P \supset Q, Q \supset P}^{(11)}}{\not\Rightarrow (P \supset Q) \vee (Q \supset P)}^{(6)}$$

The Kripke counter-model is obtained by gluing the single-world Kripke models K_1 and K_2, with $K_1 \Vdash P$ and $K_2 \Vdash Q$. The algebraic counter-model is obtained by taking the extra-topped Cartesian product of the corresponding positive Heyting algebras of two elements

$$\begin{array}{cc} 1 = v_1(Q) & 1 = v_2(P) \\ | & | \\ 0 = v_1(P) & 0 = v_2(Q) \end{array}$$

that is,

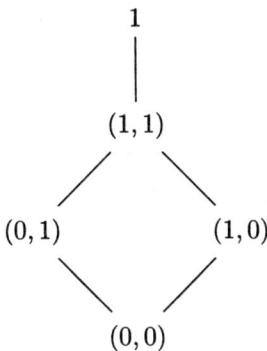

The valuation of the atoms is

$$v(Q) = (v_1(Q), v_2(Q)) = (1,0), \qquad v(P) = (v_1(P), v_2(P)) = (0,1)$$

and now we can compute

$$v(P \supset Q) = (1,0), \quad v(Q \supset P) = (0,1), \quad v((P \supset Q) \vee (Q \supset P)) = (1,1),$$

so that

$$1 \not\leq v((P \supset Q) \vee (Q \supset P)).$$

Example 17. $(P \supset Q \vee R) \supset (P \supset Q) \vee (P \supset R)$, with P, Q, R distinct atoms:
The **CRIP** derivation is

$$\dfrac{\dfrac{\dfrac{\overline{R, P \not\Rightarrow Q}\ \text{axiom}}{Q \vee R, P \not\Rightarrow Q}\ (5)}{P \supset Q \vee R, P \not\Rightarrow Q}\ (7) \quad \dfrac{\dfrac{\overline{Q, P \not\Rightarrow R}\ \text{axiom}}{Q \vee R, P \not\Rightarrow R}\ (4)}{P \supset Q \vee R, P \not\Rightarrow R}\ (7)}{\dfrac{\dfrac{P \supset Q \vee R \not\Rightarrow P \supset Q, P \supset R}{P \supset Q \vee R \not\Rightarrow (P \supset Q) \vee (P \supset R)}\ (6)}{\not\Rightarrow (P \supset Q \vee R) \supset (P \supset Q) \vee (P \supset R)}\ (11)}\ (11)$$

We get the Kripke counter-model by gluing the two single-world Kripke models, K_1 forcing R and P, and K_2 forcing Q and P. Observe that the lower instance of rule (11) does not require any gluing. Thus, the corresponding algebraic model is as in example 16, with

$$v(P) = (1,1) \qquad v(Q) = (0,1), \qquad v(R) = (1,0).$$

Therefore

$$v(Q \vee R) = (1,1), \qquad v(P \supset Q) = (0,1), \qquad v(P \supset R) = (1,0),$$

thus

$$v(P \supset Q \vee R) = 1 \not\leq v((P \supset Q) \vee (P \supset R)) = (1,1).$$

Example 18. $(\sim P \supset Q \vee R) \supset (\sim P \supset Q) \vee (\sim P \supset R)$, P, Q, R distinct atoms:
The **CRIP** derivation is

$$\dfrac{\dfrac{\overline{P \not\Rightarrow \bot}\ axiom}{\bot \supset Q \vee R, P \not\Rightarrow \bot}\ (12) \quad \dfrac{\dfrac{\overline{R, \sim P \not\Rightarrow Q}\ axiom}{Q \vee R, \sim P \not\Rightarrow Q}\ (5)}{\sim P \supset Q \vee R, \sim P \not\Rightarrow Q}\ (10) \quad \dfrac{\overline{Q, \sim P \not\Rightarrow R}\ axiom}{Q \vee R, \sim P \not\Rightarrow R}\ (4)}{\dfrac{\sim P \supset Q \vee R \not\Rightarrow \sim P \supset Q, \sim P \supset R}{\dfrac{\sim P \supset Q \vee R \not\Rightarrow (\sim P \supset Q) \vee (\sim P \supset R)}{\not\Rightarrow (\sim P \supset Q \vee R) \supset (\sim P \supset Q) \vee (\sim P \supset R)}\ (11)}\ (6)}\ (11)$$

We construct the Kripke counter-model to the end-antisequent by gluing the three Kripke trees forcing, respectively, P, R, and Q. The corresponding positive Heyting algebra is the extra-topped cube

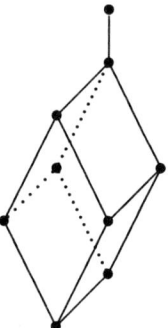

where $v(P) = (1,0,0)$, $v(R) = (0,1,0)$, $v(Q) = (0,0,1)$. We can now illustrate the ease by which the values of compound formulas are determined in the algebraic semantics, by simple computation from values of atomic formulas:

$v(\sim P) = (1,0,0) \rightarrow (0,0,0) = (0,1,1)$
$v(Q \vee R) = (0,1,1)$
$v(\sim P \supset Q \vee R) = (0,1,1) \rightarrow (0,1,1) = 1$
$v(\sim P \supset Q) = (0,1,1) \rightarrow (0,0,1) = (1,0,1)$
$v(\sim P \supset R) = (0,1,1) \rightarrow (0,1,0) = (1,1,0)$
$v((\sim P \supset Q) \vee (\sim P \supset R)) = (1,0,1) \vee (1,1,0) = (1,1,1)$
$v((\sim P \supset Q \vee R) \supset (\sim P \supset Q) \vee (\sim P \supset R)) = 1 \rightarrow (1,1,1) = (1,1,1)$.

8 Concluding remarks

We have given an algebraic semantics of refutation and replaced the determination of forcing of formulas in a Kripke model by a straightforward component-wise computation. Kripke models have been used only for showing the correctness of the construction, that parallels the construction of a Kripke counter-model out of a **CRIP** derivation. In a further work we plan to study the direct construction of counter-valuations avoiding Kripke models altogether.

Positive Heyting algebras and the definition of validity as a negative notion have been here introduced for systematic reasons, even if they could have been avoided in the case of intuitionistic propositional logic because of decidability. We hope to extend the algebraic semantics and counter-valuation construction to intuitionistic predicate logic and expect that the use of positive Heyting algebras will result in a computationally stronger semantics as compared to Kripke models.

Implementation of our algorithm of counter-model construction should present no particular difficulties.

We are indebted to Roy Dyckhoff for his useful comments and advice. Thanks are due to Paul Taylor for his package for drawing diagrams in LaTeX.

References

Dyckhoff, R. (1992) Contraction-free sequent calculi for intuitionistic logic, *The Journal of Symbolic Logic*, vol. 57, pp. 795–807.

Dyckhoff, R. and L. Pinto (1996) Implementation of a loop-free method for construction of counter-models for intuitionistic propositional logic, CS/96/8, Computer Science Division, St Andrews University.

Fitting, M. (1969) *Intuitionistic Logic, Model Theory and Forcing*, North-Holland, Amsterdam

Hudelmaier, J. (1989) *Bounds for cut elimination in intuitionistic propositional logic*, PhD thesis, University of Tübingen.

Hudelmaier, J. (1992) Bounds for cut elimination in intuitionistic propositional logic, *Archive for Mathematical Logic*, vol. 31, pp. 331–354.

Negri, S. (1997) Sequent calculus proof theory of intuitionistic apartness and order relations, *Archive for Mathematical Logic*, to appear.

Pinto, L. and R. Dyckhoff (1995) Loop-free construction of counter-models for intuitionistic propositional logic, in Behara et al. eds., *Symposia Gaussiana*, Conf. A, pp. 225–232, de Gruyter, Berlin.

von Plato, J. (1997) Positive Heyting algebras, manuscript

Stoughton, A. (1996) Porgi: a Proof-Or-Refutation Generator for Intuitionistic propositional logic. *CADE-13 Workshop on Proof Search in Type-Theoretic Languages*, Rutgers University, pp. 109–116.

Troelstra, A. S. and D. van Dalen (1988) *Constructivism in Mathematics*, vol. 1, North-Holland, Amsterdam.

Troelstra, A. S. and H. Schwichtenberg (1996) *Basic Proof Theory*, Cambridge University Press.

Deleting Redundancy in Proof Reconstruction

Stephan Schmitt[1] Christoph Kreitz[2]

[1] Fachgebiet Intellektik, Fachbereich Informatik, Darmstadt University of Technology
Alexanderstr. 10, 64283 Darmstadt, Germany
steph@informatik.tu-darmstadt.de

[2] Department of Computer Science, Cornell University
Ithaca, NY 14853, USA
kreitz@cs.cornell.edu

Abstract. We present a framework for eliminating redundancies during the reconstruction of sequent proofs from matrix proofs. We show that search-free proof reconstruction requires knowledge from the proof search process. We relate different levels of *proof knowledge* to *reconstruction knowledge* and analyze which redundancies can be deleted by using such knowledge. Our framework is uniformly applicable to classical logic and all non-classical logics which have a matrix characterization of validity and enables us to build adequate conversion procedures for each logic.

1 Introduction

Automated theorem proving in non-classical logics has become important in many branches of Artificial Intelligence and Computer Science. As a result, the resolution principle [14] and the connection method [1, 2], which both have led to efficient theorem provers for classical logic [22, 9, 3], have been extended to characterizations of logical validity in modal logics, intuitionistic logic, and fragments of linear logic [20, 10, 21, 19, 7]. These characterizations are the foundation of efficient and uniform proof search procedures for all these logics [12, 13, 7] which are used as inference engines in automatic program development systems [8, 4] and other problem-oriented applications [5].

In many applications of theorem proving it is not sufficient to show *that* a theorem is valid. The need for further processing (e.g. generating programs from proofs) or a deeper understanding of the proof requires that proof details can be presented in a comprehensible form. On the other hand, the efficiency of automated proof methods strongly depends on a compact and machine-oriented characterization of logical validity. This makes it necessary to *reconstruct* a sequent proof, a natural deduction proof, or even a proof in a semi-natural mathematical language from an automatically generated machine proof.

As a complement to existing matrix-based proof search methods we have developed a uniform procedure for transforming classical and non-classical matrix proofs back into sequent style systems [16, 17, 7]. This procedure creates a sequent proof for a given formula by traversing its formula tree in an order which respects a *reduction ordering* induced by the matrix proof. It selects an appropriate sequent rule for each visited node by consulting tables which represent the peculiarities of the different logics. At nodes which cause the sequent proof to branch the reduction ordering has to be divided appropriately and certain redundancies need to be eliminated in order to ensure completeness.

Redundancy elimination is the most crucial aspect of proof reconstruction if matrix proofs shall be converted into sequent proofs *without additional search*. We will show that the complexity of eliminating redundancy strongly depends on the amount of *proof knowledge* made available by the proof search method. If the procedure has to rely only on the matrix characterization then additional search becomes necessary, but redundancies can be eliminated in polynomial time in the size of the matrix proof if the history of the proof search is known.

In this paper we shall present a detailed analysis of possible redundancies in a reduction ordering and of the *proof reconstruction knowledge* which is necessary to delete them. We shall study different levels of proof knowledge and their effects on the proof reconstruction process. We will introduce *prefixed connections* as a logic-independent concept which allows us to extract conditions for extending the elimination of redundancies to a maximal level. Our result can be used as a general framework for building efficient and complete proof reconstruction procedures for non-classical logics if the proof search method is known.

In Section 2 we give a brief summary of matrix characterizations and proof reconstruction in non-classical logics. Section 3 classifies redundancies in matrix proofs and the requirements for eliminating them. In Section 4 we discuss proof knowledge available from the *extension procedure* [2, 12] and the resulting redundancy elimination methods. In Section 5 we investigate the complexity, adequate completeness, and correctness of the refined proof reconstruction method.

2 Preliminaries

Matrix characterizations of logical validity were introduced for classical logic [1, 2] and later extended to intuitionistic and modal logics [21] and fragments of linear logic [7]. On this basis an efficient proof search procedure has been elaborated [12, 7] which captures all these logics in a uniform way. A uniform procedure for converting matrix proofs into sequent proofs has been developed in [17, 7].

2.1 Matrix Calculi for Non-classical Logics

In matrix proofs a formula F is represented by its formula tree $\ll$ whose nodes are called *positions*. Each position x of $\ll$ refers to a unique subformula F_x of F. The root w of $\ll$ represents F itself while its leaves (or *atomic positions*) refer to the atoms of F. Because of the corresponding subformula relation $\ll$ is called the *tree ordering* of F. Each position x is associated with a *polarity* $pol(x) \in \{0,1\}$, a *principal type* $Ptype(x)$, and its operator $op(x)$. The polarity determines whether F_x will appear in the succedent of a sequent proof ($pol(x)=0$) or in its antecedent. $F_x^{pol(x)}$ denotes the *signed formula* at position x. The principal type $Ptype(x)$ is the formula type of F_x according to the tableaux classification in [21, 6]. Principal types are a compact and logic-independent way to express proof-relevant properties of a formula [17]. In the following we will only consider the types α, β, and *atom*. Two atomic positions x and y are *α-related* ($x \sim_\alpha y$) or *β-related* ($x \sim_\beta y$) if their greatest common predecessor in $\ll$ is has principal type α (or β). A *non-normal form matrix* of F is a two-dimensional representation of the atomic positions of $\ll$ such that β-related positions are placed on top of each other while α-related are written side by side. A *path* p through F is a *maximal* subset of atomic positions which are pairwisely not β-related.

A *connection* is a subset $\{c_1, c_2\}$ of a path p such that the atoms F_{c_1} and F_{c_2} have the same predicate symbol but different polarities. It is *complementary* if F_{c_1} and F_{c_2} can be unified by some *combined substitution* $\sigma = \langle \sigma_Q, \sigma_L \rangle$. σ_Q is the usual quantifier substitution while σ_L, used to analyze non-permutabilities of sequent rules in a non-classical logic $\mathcal{L}$, unifies the prefixes of the connected atoms. The *prefix* of an atom F_x is a string consisting of special positions in $\ll$ between the root w and x. A set of connections $\mathcal{C}$ *spans* a formula F (or is a *spanning mating*) if each path contains a complementary connection $c \in \mathcal{C}$. The substitution σ induces a relation $\sqsubset$ on the positions of $\ll$ such that $(x, a) \in \sqsubset$ iff $\sigma(x) = a$ and a is not a variable. σ is *admissible* if the *reduction ordering* $\lhd = (\ll \cup \sqsubset)^+$ is irreflexive and some additional global conditions hold. Finally, multiple uses of subformulae in a matrix proof are represented by a *combined multiplicity* $\mu = \langle \mu_Q, \mu_L \rangle$ of the positions x in $\ll$. Using these concepts logical validity can be uniformly characterized as follows (see [21, 7] for details).

Theorem 1. *A formula F is valid wrt. a logic $\mathcal{L}$ iff there exists a multiplicity μ, an admissible substitution σ, and a set of connections $\mathcal{C}$ which spans F.*

Proof search procedures based on the matrix characterization of logical validity are generalizations of the *extension method* [2] to non-normal form matrices and non-classical logics [12, 7]. They consist of a general path-checking algorithm and a uniform and efficient algorithm for prefix unification [11].

In the following we shall use the reduction ordering $\propto^*$, a slight technical modification of $\lhd$ (see [17]) as starting point for proof reconstruction and consider only those aspects of σ which are encoded in $\propto^*$. The following example illustrates the matrix characterization of logical validity in intuitionistic logic.

Example 1. Consider $F_1 \equiv \neg A \lor \neg B \Rightarrow \neg B \lor \neg A$ and its intuitionistic matrix-proof, represented by the reduction ordering $\propto^*$ on the left hand side of Figure 1. The name α_i, β_i, or a_i for a position x encodes its principal type while its main operator $op(x)$ and the polarity $pol(x)$ are written beside it. There are two paths through F_1, $p_1 = \{a_1, a_3, a_4\}$ and $p_2 = \{a_2, a_3, a_4\}$. The two connections $\{a_1, a_4\}$ and $\{a_2, a_3\}$ (depicted at atomic positions) span F_1 wrt. some intuitionistic admissible substitution σ which induces the relation $\sqsubset = \{(\alpha_6, \alpha_2), (\alpha_5, \alpha_3)\}$ (indicated by curved arrows).

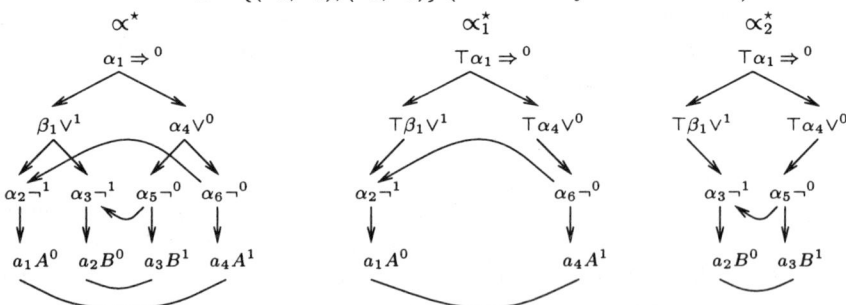

Fig. 1. Reduction ordering $\propto^*$ for $\neg A \lor \neg B \Rightarrow \neg B \lor \neg A$

2.2 Proof Reconstruction in Non-classical Logics

An algorithm for converting a matrix proof of a formula F into a sequent proof essentially has to traverse the reduction ordering $\propto^*$ while constructing a sequent rule at each visited position x. We focus on conversion into multiple conclusion sequent calculi (cf. [6]) where a sequent $\Gamma \vdash \Delta$ is described by *associated sets* of signed formulae: $S_\Delta = \{F_x^0 \mid F_x \in \Delta\}$, $S_\Gamma = \{F_x^1 \mid F_x \in \Gamma\}$, and $S = S_\Delta \cup S_\Gamma$. The main operator $op(x)$ and polarity $pol(x)$ uniquely describe the sequent rule necessary to reduce the sequent formula $F_x^{pol(x)} \in S$. The subformulae resulting from applying this rule to $F_x^{pol(x)}$ are determined by the set $succ(x)$ of immediate successors of x in $\ll$. The induced relation $\sqsubset$ encodes the non-permutabilities of sequent rules in a logic $\mathcal{L}$ and "blocks" certain positions x: rule construction for x will be delayed until all its predecessors wrt. $\sqsubset$ have been visited first.

At a β-position x a sequent proof branches into two independent subproofs. Accordingly, the reduction ordering $\propto^*$ must be split into suborderings $\propto_1^*, \propto_2^*$ and conversion continues separately on each of them. The operation $split\,(\propto^*, x)$, developed in [17] and illustrated in Figure 1, first divides $\propto^*$ and then eliminates components of each $\propto_i^*$ which are no longer relevant for the corresponding sequent subproof. Proof reconstruction terminates when all branches of the sequent proof have been closed by converting a connection from the matrix proof. Because of the uniformity of the conversion procedure the technical details are subtle. In the following we give a rather informal account of traversal and splitting and refer to [17, 18, 7] for a complete and algorithmic presentation.

Traversal of $\propto^*$. Each position in $\ll$ has to be visited and marked as *solved* if it is not blocked. A position x is *open* (i.e. elegible to be visited next) if its immediate predecessor $pred(x)$ is already solved but x is not. After x has been solved the set P_o of all open positions is updated to $P_o' = (P_o \setminus \{x\}) \cup succ(x)$. If x is a β-position, then $split\,(\propto^*, x)$ (see below) divides $\propto^*$ into $\propto_1^*, \propto_2^*$, recomputes the corresponding sets P_o^1, P_o^2 and each $\propto_i^*$ is traversed recursively. If two solved positions form a complementary connection then the conversion of $\propto_i^*$ terminates. P_o is initialized as $P_o = \{w\}$ where w is the root of $\propto^*$.

Example 2. Consider the formula F_1 from Example 1 and its matrix proof represented by the reduction ordering in Figure 1. We begin by initializing $P_o = \{\alpha_1\}$. We solve α_1 and construct the sequent rule $\Rightarrow r$ obtained from the main operator $\Rightarrow$ and polarity 0. Updating P_o yields $P_o = \{\beta_1, \alpha_4\}$. Solving α_4 next leads to applying $\vee r$ to $F_{\alpha_4}^0$. At β_1 we create $\vee l$ and the sequent proof branches: $split\,(\propto^*, \beta_1)$ divides $\propto^*$ into the suborderings $\propto_1^*$ and $\propto_2^*$ depicted in Figure 1 (where a $\top$ marks the already solved positions). Recomputing the sets of open positions results in $P_o^1 = \{\alpha_2, \alpha_6\}$ and $P_o^2 = \{\alpha_3, \alpha_5\}$.

We continue by traversing $\propto_1^*$. The position α_2 is blocked by α_6 since the corresponding sequent rule $\neg r$ has to be applied before $\neg l$ which belongs to α_2. We must solve α_6 to unblock α_2. The atomic position a_4 is next but no sequent rule will be generated since its connection partner a_1 has not been solved yet. We solve α_2 and complete the subproof by applying the axiom rule based on the connection $\{a_1, a_4\}$. Converting the subordering $\propto_2^*$ works as before. The resulting sequent proof is shown to the right.

$$\dfrac{\dfrac{\dfrac{\dfrac{\dfrac{A \vdash A}{\neg A, A \vdash}\,ax.\,(a_1, a_4)}{\neg A \vdash \neg B, \neg A}\,\neg r(\alpha_6)}{\neg A \vee \neg B \vdash \neg B, \neg A}\quad \dfrac{\dfrac{\dfrac{B \vdash B}{\neg B, B \vdash}\,ax.\,(a_2, a_3)}{\neg B \vdash \neg B, \neg A}\,\neg r(\alpha_5)}{\neg A \vee \neg B \vdash \neg B \vee \neg A}\,\vee l(\beta_1)}{\dfrac{\neg A \vee \neg B \vdash \neg B \vee \neg A}{\vdash \neg A \vee \neg B \Rightarrow \neg B \vee \neg A}\,\vee r(\alpha_4)}\,\Rightarrow r(\alpha_1)$$

Splitting at β-positions. The main modification of the reduction ordering during proof reconstruction occurs when traversal has reached a β-position x and $\propto^*$ has to be divided into $\propto_1^*$ and $\propto_2^*$. If $\{x_1, x_2\}$ are the successors of x then $F_{x_1}^{pol(x_1)}$ will move to the left subproof of the corresponding sequent proof and $F_{x_2}^{pol(x_2)}$ to the right one. Since the set of open positions P_o encodes the actual sequent, only one of the two successors of x will be added to each P_o^i, i.e. $P_o^i = (P_o \setminus \{x\}) \cup \{x_i\}$. Formally splitting is based on the following definitions.

Definition 1. *Let x be a position of $\propto^*$ and $\ll^x$ the subtree ordering with root x and position set $pos(x)$, including the pair $(pred(x), x) \in \ll^x$. The restriction of $\propto^*$ involving positions from $pos(x)$ is defined as $t^x := \ll^x \cup \sqsubset^x$, where $\sqsubset^x := \{(x_1, x_2) \in \sqsubset \mid x_1 \in pos(x) \lor x_2 \in pos(x)\}$. If C is the connection set of $\propto^*$ then $C^x := \{\{c_1, c_2\} \in C \mid c_1 \in pos(x)\}$.*

Definition 2. *Let x be of type β and $succ(x) = \{x_1, x_2\}$. The β-split of $\propto^*$ at x is defined by β-$split(\propto^*, x) := [\propto_1^*, \propto_2^*]$, where $\propto_1^* = \propto^* \setminus t^{x_2}$ and $\propto_2^* = \propto^* \setminus t^{x_1}$. For the subrelations and connections we have $\sqsubset_i := \sqsubset \setminus \sqsubset^{x_j}$, $\ll_i := \ll \setminus \ll^{x_j}$, and $C_i := C \setminus C^{x_j}$ where $i \neq j \in \{1, 2\}$.*

After a β-*split* certain redundancies need to be deleted from each $\propto_i^*$. This improves the efficieny of the reconstruction process and is necessary for ensuring its completeness when dealing with non-classical logics. We will discuss this now.

3 Classifying Redundancy in Matrix Proofs

Usually, the order in which a reduction ordering $\propto^*$ can be traversed while respecting the 'blocks' induced by $\sqsubset$ is not unique. In Example 2 we could have visited α_1, α_4, and then α_5 instead of β_1. This, however, would not lead to a successful sequent proof since applying the $\neg r$ rule corresponding to α_5 causes the deletion of the formula $\neg A$ which is relevant for completing the proof. Thus the reduction ordering $\propto^*$ is *not complete* wrt. rule non-permutabilities of the (non-standard) sequent calculus. In [17] we have introduced the concept of *wait*-labels which are dynamically assigned to special positions of $\propto^*$ during conversion and make $\propto^*$ complete wrt. all non-classical logics under consideration.

In intuitionistic logic an open position $x \in P_o$ is blocked by such a *wait*-label (denoted by $wait[x] = \top$) iff applying the corresponding sequent rule to F_x^0 would delete proof-relevant formulae. In Example 2 *wait*-labels must be assigned to α_5 and α_6 after solving α_1, α_4 since reducing $F_{\alpha_5}^0$ would delete $F_{\alpha_6}^0$ and vice versa. Hence β_1 must be solved next by applying β-$split(\propto^*, \beta_1)$. But the resulting suborderings $\propto_i^*$ contain redundancies which would create a deadlock. In $\propto_1^* = \propto^* \setminus t^{\alpha_3}$, for instance, the open positions are $P_o^1 = \{\alpha_2, \alpha_5, \alpha_6\}$ where α_5, α_6 are blocked by *wait*-labels and α_2 is blocked because of $(\alpha_6, \alpha_2) \in \sqsubset$. $F_{\alpha_5} = \neg B$ is not relevant for the subproof represented by $\propto_1^*$ and can safely be deleted by applying $\neg r$ to $F_{\alpha_6} = \neg A$. $wait[\alpha_6] = \top$ should not longer hold as well.

Since *wait*-labels shall prevent only the deletion of *proof-relevant* sequent formulae, we have to remove outdated *wait*-labels in order to guarantee completeness. We will solve this problem by *redundancy deletion* after β-*splits*, the identification of proof-relevant positions and elimination of redundant ones from the α_i^*. This procedure strongly depends on the amount of *proof knowledge* made available by the proof search process, which leads to *reconstruction knowledge* about relevant and redundant positions.

3.1 Literal Purity

The minimal proof knowledge available after proof search is the set of connections $\mathcal{C}$ and the substitution σ which induces the relation $\sqsubset$. It leads to a generalized *purity reduction* (cf. [2]): an atomic position x of α^* is called *pure* if it is not connected. Complementarity of paths will not depend on x or any literal in the same "clause" and the whole tree containing these literals is redundant.

Definition 3. *A position k with $|succ(k)| \geq 2$ is a β-node if $Ptype(k) = \beta$ and otherwise a Θ-node. The greatest predecessor k of a position x in $\ll$ with $|succ(k)| \geq 2$ is called the associated node of x. We write $k \ll^\beta x$ if k is a β-node and $k \ll^\Theta x$ otherwise.*

If x is pure after a β-split and $k \ll^\beta x$ then the whole subtree with root k can be eliminated from α^* and the predecessor position of k inherits the purity property. If $k \ll^\Theta x$ then only the branch containing x can be deleted whereas k and all other branches have to remain in α^* (usually k is no longer a Θ-node afterwards). Combining these two reductions yields the function (β, Θ)-*purity* which will be applied to each subrelation α_i^* after β-*split*.

Definition 4. *Let $P_r = \{ b \mid succ(b) = \emptyset \land \forall c \in \mathcal{C}. b \notin c \}$ be the set of pure leaf positions in α^*. Let $succ_j^+(x) := \{succ_j(x)\} \cup succ^+(succ_j(x))$ where $succ^+(x)$ is the set of all successors of x in $\ll$ and $succ_j(x)$ is a selection function with $succ_j(x) = x_j$ if $succ(x) = \{x_1,..,x_n\}$. The (β, Θ)-purity reduction is defined as:*

    function (β, Θ)-*purity* $(\alpha^*, \mathcal{C})$: *reduction-ordering*
      while $P_r \neq \emptyset$ **do**
        select $b \in P_r$; $P_r := P_r \setminus \{b\}$; let k be the associated node of b
        if $k \ll^\beta b$ **then** $\alpha^* := \alpha^* \setminus t^k$;   % β-*purity*
               $\boxed{\mathcal{C} := \mathcal{C} \setminus \mathcal{C}^k;}$
               $P_r := \{ b \mid succ(b) = \emptyset \land \forall c \in \mathcal{C}. b \notin c \}$
        else compute s where $b \in succ_s^+(k)$; $\alpha^* := \alpha^* \setminus t^{succ_s(k)}$ % Θ-*purity*

To ensure completeness (β, Θ)-*purity* must be integrated into β-*split* as follows.

Definition 5. *The* split*-operation at a β-node x is defined by* $split(\alpha^*, x) = [\alpha_{1'}^*, \alpha_{2'}^*]$, *where* $\alpha_{i'}^* = (\beta, \Theta)$-*purity* $(\alpha_i^*, \mathcal{C}_i)$ *and* $[\alpha_1^*, \alpha_2^*] = \beta$-*split* (α^*, x).

Consider again Example 2. After applying β-*split*(α^*, β_1) the position a_3 (a_4) becomes pure in α_1^* (α_2^*). Since $a_4 \ll_1^\Theta a_3$ ($a_4 \ll_2^\Theta a_4$) applying Θ-*purity* deletes t^{α_5} in α_1^* (t^{α_6} in α_2^*). Thus all *wait*-labels can be removed and traversal can proceed with each of the resulting suborderings $\alpha_{i'}^*$ (see Figure 1).

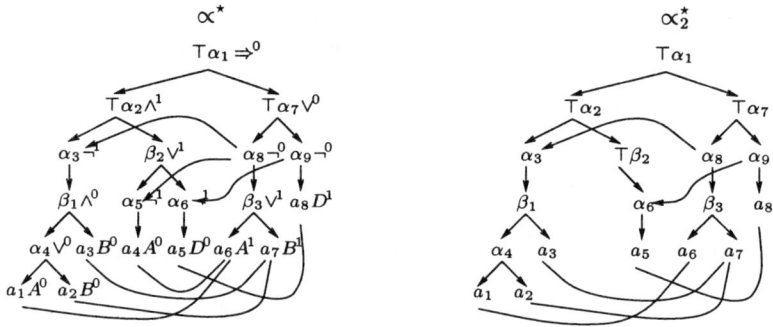

Fig. 2. Reduction orderings for $\neg((A \vee B) \wedge B) \wedge (\neg A \vee \neg D) \Rightarrow \neg(A \vee B) \vee \neg D$

3.2 The Decomposition Problem in α^*

The (β, Θ)-*purity* reduction is suitable for dealing with *first level redundancy* where proof relevance is determined by being connected. Pure positions, however, are not the only redundancies that may occur during proof reconstruction. After β-*split* and (β, Θ)-*purity* one of the resulting suborderings may consist of several "isolated" subrelations which do not have connections between each other. In this case, only *one* of these subrelations is sufficient for making all paths complementary. Nevertheless the purity principle may not apply if all leaves in the othere subrelations are connected and hence assumed to be proof-relevant.

Example 3. Consider $F_2 \equiv \neg((A \vee B) \wedge B) \wedge (\neg A \vee \neg D) \Rightarrow \neg(A \vee B) \vee \neg D$ and the reduction ordering α^* resulting from its intuitionistic matrix proof in Figure 2 (left hand side). For proof reconstruction we solve the positions $\alpha_1, \alpha_2, \alpha_7$ and split at β_2 which corresponds in the following proof fragment:

$$\dfrac{\dfrac{\neg((A \vee B) \wedge B), \neg A \vdash \neg(A \vee B), \neg D \quad \neg((A \vee B) \wedge B), \neg D \vdash \neg(A \vee B), \neg D}{\neg((A \vee B) \wedge B), \neg A \vee \neg D \vdash \neg(A \vee B), \neg D}}{\vdash \neg((A \vee B) \wedge B) \wedge (\neg A \vee \neg D) \Rightarrow \neg(A \vee B) \vee \neg D} \vee l \atop \Rightarrow r, \wedge l, \vee r$$

In the subrelation α_2^* resulting from splitting (Figure 2, right hand side), which corresponds to the right sequent after $\vee l$, the set of open positions is $P_o^2 = \{\alpha_3, \alpha_6, \alpha_8, \alpha_9\}$. The positions α_3 and α_6 are blocked by $\sqsubset$ whereas α_8 and α_9 are blocked by *wait*-labels since reducing $F_{\alpha_8}^0 = \neg(A \vee B)^0$ would delete $F_{\alpha_9}^0 = \neg D^0$ and vice versa. Furthermore, all atomic positions are connected and (β, Θ)-*purity* is not applicable. Thus proof reconstruction would run into a deadlock. But α_2^* has been *decomposed* into two "isolated" subrelations $\{t^{\alpha_3}, t^{\alpha_8}\}$ and $\{t^{\alpha_6}, t^{\alpha_9}\}$ and the *wait*-labels could be removed if we could determine which subrelation suffices for constructing the proof.

Deadlocks of above kind occur in intuitionistic logic and in all modal logics considered in [17] where additional *wait*-labels are required for proof reconstruction. In linear logic, *wait*-labels do not cause deadlocks and proof reconstruction has not to deal with this kind of redundancy [7]. Finding the appropriate isolated subrelation is the *decomposition problem in* α^* which we will formalize now.

Definition 6. *Let P_o be the set of open positions, P_a the set of atomic positions which are solved but connected, $P_u = P_o \cup P_a = \{x_1, .., x_n\}$ the set of usable positions, and $T_u = \{t^{x_1}, .., t^{x_n}\}$. The connection relation $R_C \subseteq T_u \times T_u$ is defined by $R_C = \{ (t^{x_i}, t^{x_j}) \mid 1 \leq i, j \leq n \wedge \exists \{c_i, c_j\} \in \mathcal{C}. c_i \in pos(x_i) \wedge c_j \in pos(x_j) \}$. By R_C^* we denote the transitive closure of R_C.*

Let P_r be the set of pure positions It is easy to see that R_C^* defines an equivalence relation on T_u if $P_r = \emptyset$. In this case we will write $\sim_C$ instead of R_C^*. The equivalence class $[t^x] \in T_u/\sim_C$ is defined by $[t^x] := \{t^y \mid t^y \sim_C t^x\}$.

Definition 7. *Let $P_r = \emptyset$ and $T_u/\sim_C = \{[t^{x_1}], \ldots, [t^{x_n}]\}$. The decomposition problem in $\propto^*$ is the problem of selecting the proof-relevant $[t^{x_i}] \in T_u/\sim_C$.*

Definition 8. *A reduction ordering $\propto^*$ is called a deadlock iff $P_r = \emptyset$ and for all $x \in P_o$ either $(y,x) \in \sqsubset$ for some unsolved position y, or $wait[x] = \top$.*

In Example 3 we have a decomposition problem in $\propto_2^*$ after $split(\propto^*, \beta_2)$ since $T_u/\sim_C = \{[t^{\alpha_8}], [t^{\alpha_9}]\} = \{\{t^{\alpha_3}, t^{\alpha_8}\}, \{t^{\alpha_6}, t^{\alpha_9}\}\}$. In addition, $\propto_2^*$ is a deadlock. The same situation is caused by α-*reduction* when solving α_7 after β_2.

In general, proof reconstruction will require a solution for a decomposition problem if $\propto^*$ is also a deadlock. This may only occur in non-classical logics and is characterized by the following lemma (see [18] for a proof).

Lemma 1. *Let $P_r = \emptyset$. If $\propto^*$ is a deadlock then there exists $\{w_1, w_2\} \subseteq P_o$ such that $wait[w_1] = wait[w_2] = \top$ and $[t^{w_1}] \neq [t^{w_2}]$.*

Thus deadlocks can only occur if there is also a decomposition problem $T/\sim_C = \{[t^{w_1}], [t^{w_2}]\}$. Completeness of proof reconstruction can only be guaranteed if the decomposition problem can be either avoided or solved since otherwise proof-relevant formulae might be deleted. A complete solution of the decomposition problem consists of establishing a selection function $f_{\sim_C}$ which chooses the only relevant class $[t^{x_i}]$ from $\propto^*$. A solution is called *adequate* if $f_{\sim_C}$ can be realized without any additional search. In Example 3 there is a deadlock in $\propto_2^*$ and $[t^{\alpha_8}] \neq [t^{\alpha_9}]$ holds for the two *wait*-labeled positions α_8, α_9. $f_{\sim_C}$ should select $[t^{\alpha_9}]$ since $[t^{\alpha_8}]$ does not lead to a proof.

Decomposition problems cannot be avoided during proof reconstruction. Nor can selection functions which are both complete and adequate be characterized for their solution. Completeness can be achieved only at the expense of adequateness, by searching all selections $[t^{x_i}]$ until proof reconstruction succeeds. We call this kind of redundancy *second level redundancy* since they cannot be solved adequately without the use of additional proof knowledge.

4 Integrating Proof Knowledge into Proof Reconstruction

Constructing adequate solutions for a decomposition problem requires additional knowledge about the proof search method and the proof knowledge it has gathered while developing the matrix-proof. In the following we characterize the knowledge that must be provided by the proof search method and assume this method to be based on the usual *extension procedure* [1, 2, 12]. We will encode the proof history of this procedure as *reconstruction knowledge* in the form of *prefixed connections* and derive a refinement of β-*split* and (β, Θ)-*purity* such that $T_u/\sim_C$ will always consist of a single class. This makes it possible to avoid decomposition problems and deadlocks during the reconstruction process.

4.1 Prefixed Connections

The extension procedure checks the complementarity of sets of paths by following connections. It keeps track of the order in which connections have been followed and uses *active paths* to denote the sets of paths with the same initial subpath which have not yet been proven complementary. The history of constructing a matrix-proof can be expressed by *directed connections*, i.e. pairs $(a, b) \in \mathcal{C}$ of atomic positions (instead of sets $\{a, b\}$), together with their *active path context*.

Definition 9. *An active path from a matrix proof $\propto^*$ is a a sequence $P_i = (a_i^1, b_i^1) \circ \ldots \circ (a_i^{m_i}, b_i^{m_i})$ of pairs of atomic positions such that each a_i^j is β-related to b_i^{j-1} in $\propto^*$, The set $\{P_1, \ldots, P_n\}$ of all active paths from a matrix proof $\propto^*$ will be denoted by $\mathcal{P}$. $at(P_i) = a_i^1 \cdots a_i^{m_i}$ is called the* atom string *of path P_i.*

Each connection $c \in \mathcal{C}$ can be related to the set of active subpaths in which it occurs. These subpaths will be represented by a set of *prefixes* $Pre(c)$ such that the cardinality of $Pre(c)$ encodes the multiple use of c in the matrix proof.

Definition 10. *Let $c = (a, b) \in \mathcal{C}$ be a connection, $\{P_1, \ldots, P_m\} \subseteq \mathcal{P}$ be the set of active paths in which c occurs. The set of* prefixes *assigned to c is defined by $Pre(c) = \bigcup_{i=1}^{m} \{q_i a \mid q_i a \preceq at(P_i)\}$.[1] $Pre(c)$ is also called the* paths context *of c. For a set of connections $\mathcal{C}' \subseteq \mathcal{C}$ we define $Pref(\mathcal{C}') = \bigcup_{c \in \mathcal{C}'} Pre(c)$.*

The basic property of the extension procedure is that *all* paths through an established connection are known to be complementary. Since active paths are explored depth-first we know that two connections used in the matrix proof cannot have common prefixes, i.e. $Pre(c_1) \cap Pre(c_2) = \emptyset$ if $c_1 \neq c_2$. From now on we shall illustrate matrix proofs and prefixed connections using the two-dimensional matrix representation of a formula (see section 2.1), although the reduction ordering $\propto^*$ remains to be the basic data structure for proof reconstruction.

Example 4. Consider the following (classical) non-normal form matrix which has been proven valid using the extension procedure with start clause $\{A^1\}$.

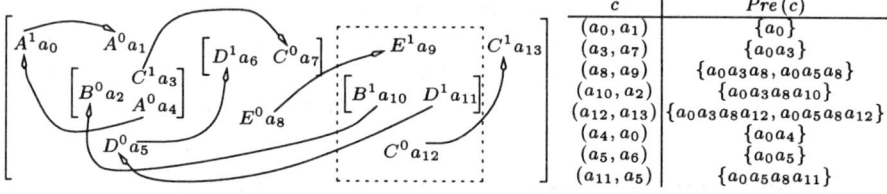

To obtain a unique indexing we have written atomic positions x from $\propto^*$ besides the atoms F_x. There are five active paths $\mathcal{P} = \{P_1, P_2, P_3, P_4, P_5\}$ with

$P_1 = (a_0, a_1) \circ (a_3, a_7) \circ (a_8, a_9) \circ (a_{10}, a_2)$ $P_4 = (a_0, a_1) \circ (a_5, a_6) \circ (a_8, a_9) \circ (a_{11}, a_5)$
$P_2 = (a_0, a_1) \circ (a_3, a_7) \circ (a_8, a_9) \circ (a_{12}, a_{13})$ $P_5 = (a_0, a_1) \circ (a_5, a_6) \circ (a_8, a_9) \circ (a_{12}, a_{13})$
$P_3 = (a_0, a_1) \circ (a_4, a_0)$

The atom strings $at(P_i) = p_i$ are $p_1 = a_0 a_3 a_8 a_{10}$, $p_2 = a_0 a_3 a_8 a_{12}$, $p_3 = a_0 a_4$, $p_4 = a_0 a_5 a_8 a_{11}$, and $p_5 = a_0 a_5 a_8 a_{12}$. Each connection $c \in \mathcal{C}$ receives a set of prefixes $Pre(c)$ (see next to the matrix). The connections (a_{12}, a_{13}) and (a_8, a_9) have two prefixes due to their multiple occurrence in the path contexts P_2, P_5 and P_1, P_2, P_4, P_5 respectively.

---

[1] $q \preceq p$ denotes that q is an initial substring of p. ϵ denotes the empty string.

4.2 Splitting with Prefixed Connections

Prefixed connections encode multiple occurrences of active subpaths in a matrix proof. The active paths encode possible split structures of sequent proofs wrt. the axioms determined by the connections. Splitting with prefixed connections basically derives a classification of the active paths (i.e. prefixes) which have been interrupted during β-split. From this, the operation β-split can be refined to extend redundancy deletion along these "interrupted subpaths".

Definition 11. *Let y be a position in α^*, $\ll^y$ its subtree, At^y the corresponding set of atomic positions, and C^y the set of directed connections in $\ll^y$. We divide C^y into the set of* entry connections $C_{en}^y = \{(c,d) \in C^y \mid d \in At^y\}$ *and the set of* extension connections $C_{ex}^y = \{(c,d) \in C^y \mid c \in At^y\}$. *Accordingly, $Pref(C^y)$ is divided into the set of* entry- *and* extension prefixes: $Pref(C_{en}^y)$ *and* $Pref(C_{ex}^y)$.

$C_{en}^y \cap C_{ex}^y$ is nonempty iff $(c,d) \in C^y$ is a connection and $\{c,d\} \subseteq At^y$. Splitting at a β-position causes, at a first level, the elimination of a submatrix At^y with connections C^y and, at a second level, *interruptions* of active subpaths. By separating entry- and extension connections after splitting we can identify prefixes of connections which depend on deleted connections and determine the interrupted active subpaths. Prefixes depending on extension connections cannot contribute to a subproof any longer and are *forward redundant*. Prefixes depending on entry connections are *backward redundant* towards a minimal subprefix.

In order to eliminate this kind of redundancy, we must respect these dependencies while deleting connections during β-split and (β, Θ)-purity. In a first step we will eliminate prefixes from $\mathcal{C}$ which are redundant wrt. the deleted connection set C^y. For this we need a copy of the *original* matrix proof, i.e. the reduction ordering and connection set. Moreover, we define a concept of α-related subproofs in a matrix in order to capture non-normal form reduction steps.

Definition 12. *Let α_o^*, C_o denote the original matrix proof. For a prefix qx_q let At_o^z be the* minimal submatrix *in α_o^* such that $q \in Pref(C_o^z)$ denotes the entry connection and $qx_q \in Pref(C_o^z)$ the extension connection in At_o^z. The set of α-related subproofs of qx_q is determined by the prefix set $T_o^{qx_q} = \{t \in Pref(C_o^z) \mid qx_q \prec t\}$. The set of atomic positions which are involved by $T_o^{qx_q}$ is defined as:*

$$P_o^{qx_q} = \{a \in At_o^z \mid \exists c=(c_1,c_2) \in C_o^z. \exists t \in T_o^{qx_q}. a=c_1 \vee a=c_2 \wedge t \in Pre(c)\}$$

During the conversion process, the set of α-related subproofs of qx_q is denoted by $T^{qx_q} \subseteq T_o^{qx_q}$. The corresponding atomic positions are given by $P^{qx_q} \subseteq P_o^{qx_q}$. Backward redundancy starts the deletion process at an entry prefix p and terminates at a minimal subprefix q of p which is determined either by a non-normal form reduction step in α^* or by the original matrix proof α_o^*.

Definition 13. *Let $D \subseteq Pref(C_o)$, $p \in D$, and $M_{qx_q} \subseteq T_o^{qx_q}$. q is called the* minimal subprefix *of p wrt. D, M_{qx_q} iff $q \in Pref(C_o) \cup \{\epsilon\}$, $q \prec p$, and either*
1. *q is the* maximal *initial substring of p for which there are $t, qx_q \in D$ such that $qx_q \preceq p \preceq t$ and $t \in M_{qx_q}$, or*
2. *q is the* maximal *initial substring of p for which $q \notin D$.*

For abbreviation, we write $q \prec_{min} p$ wrt. D, M_{qx_q}.

For a position set P we abbreviate $\forall y \in P. x \sim_\alpha y$ by $x \sim_\alpha P$ (assume $x \sim_\alpha \emptyset, x \sim_\alpha x$).

Definition 14. *Let p be an entry prefix and q its minimal subprefix. The operations for deleting backward and forward redundancy are defined by:*
$del_{en}(p,q) = \{s \mid qx'_q \preceq s \wedge p \not\prec s \wedge x'_q \sim_\alpha P^{qx_q}\}$ and $del_{ex}(p) = \{s \mid p \preceq s\}$.

$del_{en}(p,q)$ covers all active subpaths s which have an entry connection with a prefix p as a *direct* or as an *indirect* subgoal. Suppose that y is the β-position for splitting. A connection with prefix p uses an atom from At^y as entry point and will become *open ended* or "interrupted" after deletion of At^y. Hence, no split ordering in a sequent proof can close the subpaths s between the α-related subproofs of qx_q and all s with $p \not\prec s$, using the remaining connections from $\propto^*$.

$del_{ex}(p)$ describes all active subpaths s which depend on an extension connection with prefix p. Each such connection uses an atom from At^y as extension point and will receive an *open beginning* after the deletion of At^y. Again, the subpaths s cannot be closed with the given connections.

In the following we extend these deletion operations to all entry / extension prefixes $Pref(\mathcal{C}^y_{en}) / Pref(\mathcal{C}^y_{ex})$ of a submatrix At^y which has to be deleted next. This extension is defined stepwisely since the result of one deletion step may influence the remaining candidate sets for the next step. The resulting prefix set depends on the order of prefix selection during this iteration and does not lead to *unique* subproofs after splitting.

Definition 15. *Let $\mathcal{C}$ be the connection set of $\propto^*$ and y be a position in $\propto^*$. Let $Pref(\mathcal{C}^y_{en})$ the set of entry prefixes and $Pref(\mathcal{C}^y_{ex})$ be the set of extension prefixes corresponding to $\mathcal{C}^y$. We set $D^0 = Pref(\mathcal{C})$, $P^0_{en} = Pref(\mathcal{C}^y_{en})$, $P^0_{ex} = Pref(\mathcal{C}^y_{ex})$, $M^0_r = T^r$ for all $r \in Pref(\mathcal{C})$, and $Q^0 = \emptyset$. Then we define the following iteration:*
$D^i = D^{i-1} \setminus S^i$, $\quad P^i_{en} = D^i \cap P^{i-1}_{en}$, $\quad M^i_r = D^i \cap M^{i-1}_r$ for all $r \in D^i$,
$\quad\quad\quad\quad\quad\quad P^i_{ex} = D^i \cap P^{i-1}_{ex}$, $\quad Q^i = (Q^{i-1} \cup U^i) \setminus S^i$
until $i = n$ such that $P^n_{en} = P^n_{ex} = \emptyset$. During iteration we have:
$$S^i = \begin{cases} del_{en}(p,q) & \text{for a } p \in P^{i-1}_{en} \text{ with } q \prec_{min} p \text{ wrt. } D^{i-1}_{en}, M^{i-1}_{qx_q} \\ del_{ex}(p) & \text{for a } p \in P^{i-1}_{ex} \end{cases}$$
$U^i = \{qx_q \mid qx_q \notin D^i \wedge y \not\prec_\beta x_q \wedge M^{i-1}_{qx_q} \neq \emptyset \wedge M^i_{qx_q} = \emptyset\}$

The redundant prefixes are given by the sets S^i, depending if deletion of backward redundancy $p \in P^{i-1}_{en}$ or forward redundancy $p \in P^{i-1}_{ex}$ has been performed. The sets U^i denote *unblocked redundancies* caused during deletion of S^i. They consists of already deleted prefixes qx_q whose α-related subproofs M_{qx_q} have blocked deletion of backward redundancy in former steps. If step i results in a complete deletion of $M^i_{qx_q}$, the blocked backward deletion wrt. qx_q has to be continued. If $y \sim_\beta x_q$ (where y is the splitting node) continuation of backward deletion has to be performed by β-*split* in order to retain soundness. The sets U^i will be collected in a set Q^j and regularly updated modulo the actual deletion set S^j. This operation forces the sets S^j to deal with prefixes which have not necessarily been deleted from D^{j-1} in the *actual* step. The concept ensures a robust treatment of ordering dependencies in which prefixes will be selected from the candidate sets P^i_{en} and P^i_{ex}. In order to characterize a complete prefix deletion we integrate the elimination of unblocked redundancies as follows:

Definition 16. *We start with the sets* D^n, Q^n, *and* M_r^n *for all* $r \in D^n$. *Then*
$$D^{n+j} = D^{n+(j-1)} \setminus S^j \qquad M_r^{n+j} = D^{n+j} \cap M_r^{n+(j-1)} \text{ for all } r \in D^{n+j},$$
$$Q^{n+j} = (Q^{n+(j-1)} \cup U^j) \setminus S^j$$
until $j = m$ *such that* $Q^{n+m} = \emptyset$. *Again, we use during iteration:*
$$S^j = del_{en}(p, q) \text{ for a } p \in Q^{n+(j-1)} \text{ with } q \prec_{min} p \text{ wrt. } D^{n+(j-1)} \cup \{p\}, M_{qx_q}^{n+(j-1)}$$
$$U^j = \{qx_q \mid qx_q \notin D^{n+j} \wedge y \not\prec_\beta x_q \wedge M_{qx_q}^{n+(j-1)} \neq \emptyset \wedge M_{qx_q}^{n+j} = \emptyset\}$$
pref_del$(\mathcal{C}, \mathcal{C}^y)$ *denotes the complete operation* prefix deletion *in* $\mathcal{C}$ *wrt. a connection set* $\mathcal{C}^y$ *and yields a set* $\mathcal{C}'$ *with* $\text{Pref}(\mathcal{C}') = D^{n+m}$.

Obviously, $\text{Pref}(\mathcal{C}^y) = \emptyset$ after applying *pref_del*$(\mathcal{C}, \mathcal{C}^y)$ since $s = p$ satisfies the conditions of Definition 14, for all $p \in \text{Pref}(\mathcal{C}^y)$. A connection $c \in \mathcal{C}$ is called *redundant* if it does not occur within at least one active subpath, i.e. if $\text{Pre}(c) = \emptyset$.

Definition 17. *Let* y *be a position in* α^* *and* $\mathcal{C}^y \subseteq \mathcal{C}$ *be the connection set corresponding to* At^y. *Then* connection deletion *in* $\mathcal{C}$ *wrt.* $\mathcal{C}^y$ *is defined by* $con\_del(\mathcal{C}, \mathcal{C}^y) := \mathcal{C}' \setminus \{c \mid \text{Pre}(c) = \emptyset\}$, *where* $\mathcal{C}' = pref\_del(\mathcal{C}, \mathcal{C}^y)$.

Redefining the split operation. An extended elimination of redundancies during proof reconstruction is realized by redefining the *split* operation at β-positions. Connection deletion during β-*split* and (β, Θ)-*purity* is determined by $\mathcal{C} \setminus \mathcal{C}^k$ wrt. the deleted submatrix At^k. The refined connection deletion will remove all interrupted subpaths from $\mathcal{C}$ such that $\mathcal{C}^k \subseteq \{c \in \mathcal{C}' \mid \text{Prec}(c) = \emptyset\}$ will hold. This incremental process guarantees redundancy deletion on the "low" connection level and on the "higher" level of paths contexts. As a consequence, redundancy will be deleted already when it *occurs* and not when it becomes visible in form of a decomposition problem in α^*.

The refined connection deletion requires us to modify some definitions since $\mathcal{C}$ is now a set of directed connections. We redefine $\mathcal{C}^x$ in Definition 1 as $\{(c_1, c_2) \in \mathcal{C} \mid c_1 \in pos(x) \vee c_2 \in pos(x)\}$. In Definition 2 we replace $\mathcal{C}_i := \mathcal{C} \setminus \mathcal{C}^{x_j}$ by $\mathcal{C}_i := con\_del(\mathcal{C}, \mathcal{C}^{x_j})$. Definition 4 is modified by extending $b \in c$ to directed connections $c = (c_1, c_2)$ when defining the set P_r and removing the boxed part in (β, Θ)-*purity* since the refinements of β-*split* avoid further prefix and connection deletions. Finally, the *split* operation from Definition 5 uses the new β-*split* and (β, Θ)-*purity*. The following example illustrates the refined operations.

Example 5. Consider the proof history from the extension proof of Example 4. During proof reconstruction we split at the clause $\{E^1, \{B^1, D^1\}, C^0\}$ (dashed boxed) by solving a β-position x in α^* with $succ(x) = \{x_1, a_{12}\}$. We have $At^{x_1} = \{a_9, a_{10}, a_{11}\}$, $\mathcal{C}_{en}^{x_1} = \{(a_8, a_9)\}$ with $\text{Pref}(\mathcal{C}_{en}^{x_1}) = \{a_0 a_3 a_8, a_0 a_5 a_8\}$, and $\mathcal{C}_{ex}^{x_1} = \{(a_{10}, a_2), (a_{11}, a_5)\}$ with $\text{Pref}(\mathcal{C}_{ex}^{x_1}) = \{a_0 a_3 a_8 a_{10}, a_0 a_5 a_8 a_{11}\}$. Similarly, $At^{a_{12}} = \{a_{12}\}$, $\mathcal{C}_{ex}^{a_{12}} = \{(a_{12}, a_{13})\}$, and $\text{Pref}(\mathcal{C}_{ex}^{a_{12}}) = \{a_0 a_3 a_8 a_{12}, a_0 a_5 a_8 a_{12}\}$. Applying β-*split*(α^*, x) results in

$\alpha_1^* = \alpha^* \setminus t^{x_1}$ (i.e. deleting literals E^1, B^1, D^1) and $\mathcal{C}_1 = con\_del(\mathcal{C}, \mathcal{C}^{x_1})$,
$\alpha_2^* = \alpha^* \setminus t^{a_{12}}$ (i.e. deleting literal C^0) and $\mathcal{C}_2 = con\_del(\mathcal{C}, \mathcal{C}^{a_{12}})$.

We illustrate the connection deletion wrt. α_1^*. The operation $pref\_del(\mathcal{C}, \mathcal{C}^{x_1})$ starts with an initialization according to the first row of the table below. We begin with the entry prefix $a_0 a_3 a_8$ and its minimal subprefix $q = a_0$, $x_q = a_3$. The first deletion set is given by S^1 where the α-related subproof $M_{a_0 a_3}^0$ prevents deletion of $a_0 a_5$ (see

Definition 14). The second step selects $a_0 a_5 a_8$ with minimal subprefix a_0 and $x_q = a_5$. The whole iteration yields $D^4 = \{a_0, a_0 a_3 a_8 a_{12}, a_0 a_5 a_8 a_{12}\}$, $Q^4 = \{a_0 a_3, a_0 a_5\}$ and $M_r^4 = \emptyset$ for all $r \in D^4$. Prefix deletion in $\propto_1^*$ has to be completed by eliminating the unblocked redundancies Q^4 according to Definition 16. (see the right hand side of the table). We select $a_0 a_3$ with its minimal subprefix $q = \varepsilon$ and $x_q = a_0$. Recall, that S^5 also contains all prefixes from Q^4 which were not contained in D^4. Thus, after one step, we obtain $Q^5 = \emptyset$ and terminate with $pref\_del(C, C^{x_1}) = C'$ where $Pref(C') = D^5 = \{a_0 a_3 a_8 a_{12}\}$.

i	select	D^i	P^i_{en}	P^i_{ex}	S^i
0	–	$Pref(C)$	$\{a_0 a_3 a_8, a_0 a_5 a_8\}$	$\{a_0 a_3 a_8 a_{10}, a_0 a_5 a_8 a_{11}\}$	$\emptyset$
1	$a_0 a_3 a_8$	$D^0 \setminus S^1$	$\{a_0 a_5 a_8\}$	$\{a_0 a_3 a_8 a_{10}, a_0 a_5 a_8 a_{11}\}$	$\{a_0 a_3 a_8, a_0 a_3, a_0 a_4\}$
2	$a_0 a_5 a_8$	$D^1 \setminus S^2$	$\emptyset$	$\{a_0 a_3 a_8 a_{10}, a_0 a_5 a_8 a_{11}\}$	$\{a_0 a_5 a_8, a_0 a_5\}$
3	$a_0 a_3 a_8 a_{10}$	$D^2 \setminus S^3$	$\emptyset$	$\{a_0 a_5 a_8 a_{11}\}$	$\{a_0 a_3 a_8 a_{10}\}$
4	$a_0 a_5 a_8 a_{11}$	$D^3 \setminus S^4$	$\emptyset$	$\emptyset$	$\{a_0 a_5 a_8 a_{11}\}$

i	$M^i_{a_0 a_3}$	$M^i_{a_0 a_5}$	U^i	Q^i	i	select	D^i	Q^i	U^i	S^i
0	$\{a_0 a_3 a_8 a_{10}\}$	$\{a_0 a_5 a_8 a_{11}\}$	$\emptyset$	$\emptyset$	4	–	D^4	Q^4	U^4	$\emptyset$
1	$\{a_0 a_3 a_8 a_{10}\}$	$\{a_0 a_5 a_8 a_{11}\}$	$\emptyset$	$\emptyset$	5	$a_0 a_3$	$D^4 \setminus S^5$	$\emptyset$	$\emptyset$	$\{a_0, a_0 a_5 a_8 a_{12}\} \cup Q_4$
2	$\{a_0 a_3 a_8 a_{10}\}$	$\{a_0 a_5 a_8 a_{11}\}$	$\emptyset$	$\emptyset$						
3	$\emptyset$	$\{a_0 a_5 a_8 a_{11}\}$	$\{a_0 a_3\}$	U^3						
4	$\emptyset$	$\emptyset$	$\{a_0 a_5\}$	$Q^3 \cup U^4$						

Finally, β-split performs connection deletion wrt. $\propto_1^*$: $con\_del(C, C^{x_1}) = \{(a_{12}, a_{13})\}$ with $Pre((a_{12}, a_{13})) = \{a_0 a_3 a_8 a_{12}\}$. The refined split-operation terminates on $\propto_1^*$ with applications of the refined (β, Θ)-purity. The set P_r is initialized with all atomic positions of $\propto_1^*$, except a_{12} and a_{13}. Additional connection deletions are not required and all subrelations depending on iterative updating of P_r will be deleted from $\propto_1^*$ (see Definition 4). The resulting relevant submatrix corresponding to $\propto_{1'}^*$ is shown below:

$$\begin{bmatrix} C^0 a_{12} & C^1 a_{13} \end{bmatrix}$$

The example can easily be extended such that the "conventional" application of β-split would cause a decomposition problem in $\propto_{1'}^*$.

5 Complexity, Correctness, and Completeness

We will now show that the refinements of the β-split operation which we introduced in sections 3 and 4 do in fact lead to an adequate and complete proof reconstruction procedure. Since a traversal of $\propto^*$ can be completed in polynomial time if no deadlocks are present, it suffices to prove that the refined split operation has a polynomial complexity in the size of the matrix proof (adequateness) and that splitting with prefixed connections deletes all redundancies from a matrix proof (completeness). The proofs of the following lemmata and theorems can be found in the first author's technical report [15].

Complexity. The *size of a matrix proof* is usually defined as the number of inference steps for testing a mating C to be spanning. This measure is equivalent to the number of active paths $|\mathcal{P}|$ in a matrix proof which may increase exponentially in the size of $\propto^*$, i.e. the number of positions.

Lemma 2. *Let $|\mathcal{P}|$ be the size of a matrix proof, n the maximal length of paths $P_i \in \mathcal{P}$, and k the maximal length of (non-normal form) clauses in $\propto^*$. Then $|\mathcal{P}| \in \mathcal{O}(k^n)$ and n is polynomial wrt. the size of $\propto^*$.*

The refinements of the operation β-*split* are based on the deletion of prefixes wrt. a prefix set $Pref(\mathcal{C}^y)$. Testing the basic elimination conditions (Definition 14) as well as computing the relevant sets during the iteration process (Definition 15 and 16) can be done in polynomial time in the size of $|Pref(\mathcal{C})|$. But this this complexity is polynomial wrt. the size of the matrix proof $|\mathcal{P}|$.

Lemma 3. *Let $\mathcal{C}$ be the spanning mating from a matrix proof, $Pref(\mathcal{C})$ its prefix set, and k, n as in Lemma 2. Then $|Pref(\mathcal{C})| \in \mathcal{O}(k^n)$.*

The number of purity applications within (β, Θ)-*purity* is determined by the possible updates of the set P_r, which is polynomial wrt. the size of $\propto^\star$. Lemma 3 also states that integration of proof knowledge $Pref(\mathcal{C})$ into the proof reconstruction process requires polynomial (space) complexity wrt. $|\mathcal{P}|$. From this we conclude the following adequateness theorem:

Theorem 2. *The refined split operation $split(\propto^\star, x)$ at a β-node x using prefixed connections is polynomial wrt. the size of the matrix proof $|\mathcal{P}|$.*

In special cases the size $|\mathcal{P}|$ of a matrix proof and the size $|\mathcal{C}|$ of its spanning mating may differ exponentially. If k, n are defined as above and all active paths $P_i \in \mathcal{P}$ share the same connections from $\mathcal{C}$, we find examples such that $|\mathcal{C}| < n \cdot k$ and $|\mathcal{P}| \in \mathcal{O}(k^n)$. Hence, the additional search complexity on a decomposition problem when using spanning matings $\mathcal{C}$ may be transformed into an exponential representation requirement for $|\mathcal{P}|$ when integrating prefixed connections. However, the complexity of a matrix proof is reflected more adequately when taking its size $|\mathcal{P}|$ into account. Furthermore, conversion with prefixed connections avoids redundant steps in the resulting sequent proof which cannot be guaranteed when using matings together with additional search.

Correctness and Adequate Completeness. The correctness of the conversion procedure using prefixed connections is obvious. The split operation with refined connection deletion cuts subrelations from $\propto^\star$ without violating the relation $\sqsubset$ in the *remaining* part of $\propto^\star$. Incorrect applications of sequent rules wrt. rule non-permutabilities are impossible. Thus, reconstructing a sequent proof from $\propto^\star$ implies that the input formula is valid wrt. the selected logic.

In order to prove completeness we show that no redundant connections will survive in the $\propto_i^\star$ after applications of the refined split operation.

Lemma 4. *Let $\propto^\star$ represent an extension proof with connection set $\mathcal{C}$. Let x be a β-node and $\mathcal{C}_i' \subseteq \mathcal{C}$ be the connection sets of $\propto_i^\star$, after executing the refined split operation at x. Then a connection $c \in \mathcal{C}$ is proof-relevant wrt. $\propto_i^\star$ iff $c \in \mathcal{C}_i'$.*

The occurrence of a decomposition problem $T_u/\sim_\mathcal{C} = \{[t^{a_1}], .., [t^{a_n}]\}$ is always based on redundant connections since only a single $[t^{x_i}]$ is proof-relevant. From Lemma 4 it also follows that $T_u/\sim_\mathcal{C} = \{[t^a]\}$ for $a \in P_u$. According to Lemma 1 this implies that $\propto^\star$ will be deadlock free during the whole proof reconstruction process which leads to the concluding theorem.

Theorem 3. *Let x be a β-node. The extended split operation $split(\propto^\star, x)$ is correct and adequately complete for redundancy deletion in $\propto^\star$.*

6 Conclusion and Future Work

We have presented a method for eliminating redundancies during a conversion of matrix proofs into sequent proofs. Our approach refines the proof reconstruction procedure presented in [17, 7] and covers classical and intuitionistic logic, the modal logics $K, K4, D, D4, T, S4, S5$, and fragments of linear logic. For obtaining *adequate* (search-free) *completeness* of proof reconstruction, we have classified two levels of redundancy. We have shown that adequate solutions require additional knowledge from proof search in the matrix calculus. Assuming the usual extension proof search strategy we have introduced *prefixed connections* as a means for representing a proof history. We have integrated this concept into the proof reconstruction procedure and shown that the refined procedure will always generate a sequent proof in polynomial time wrt. the size of the matrix proof.

In the future we will make use of the uniformity of our approach and combine it with existing proof procedures for non-classical logics in order to guide derivations in interactive proof development systems. We will also be able to extend our approach to additional logics, such as larger fragments of linear logic, as soon as matrix characterizations and proof procedures have been developed for them. Apart from this we will generalize the concept of prefixed connections to other proof strategies. For example proof histories from tableau based proof procedures [13] may be expressed in terms of extension proofs in order to combine tableau provers with our proof reconstruction procedure as well.

References

1. W. BIBEL. On matrices with connections. *Journal of the ACM*, 28, p. 633–645, 1981.
2. W. BIBEL. *Automated theorem proving*. Vieweg, 1987.
3. W. BIBEL, S. BRÜNING, U. EGLY, T. RATH. Komet. *CADE-12*, LNAI 814, pp. 783–787. 1994.
4. W. BIBEL, D. KORN, C. KREITZ, F. KURUCZ, J. OTTEN, S. SCHMITT, G. STOLPMANN. A Multi-Level Approach to Program Synthesis. 7th *LOPSTR Workshop*, LNCS , 1998.
5. W. BIBEL, D. KORN, C. KREITZ, S. SCHMITT. Problem-Oriented Applications of Automated Theorem Proving. *DISCO-96*, LNCS 1128, pp. 1–21, 1996.
6. M. C. FITTING. *Proof Methods for Modal and Intuitionistic Logic.* D. Reidel, 1983.
7. C. KREITZ, H. MANTEL, J. OTTEN, S. SCHMITT. Connection-based proof construction in Linear Logic. *CADE-14*, 1997.
8. C. KREITZ, J. OTTEN, AND S. SCHMITT. Guiding Program Development Systems by a Connection Based Proof Strategy. 5th *LOPSTR Workshop*, LNCS 1048, pp. 137–151. 1996.
9. R. LETZ, J. SCHUMANN, S. BAYERL, W. BIBEL. SETHEO: A high-performance theorem prover. *Journal of Automated Reasoning*, 8:183–212, 1992.
10. H. J. OHLBACH. A resolution calculus for modal logics. PhD Thesis, Univ. Kaiserslautern, 1988.
11. J. OTTEN, C. KREITZ. T-string-unification: unifying prefixes in non-classical proof methods. 5th *TABLEAUX Workshop*, LNAI 1071, pp. 244–260, 1996.
12. J. OTTEN, C. KREITZ. A uniform proof procedure for classical and non-classical logics. *KI-96: Advances in Artificial Intelligence*, LNAI 1137, pp. 307–319, 1996.
13. J. OTTEN. ileanTAP: An intuitionistic theorem prover. 6th *TABLEAUX Workshop*, 1997.
14. J. A. ROBINSON. A machine-oriented logic based on the resolution principle. *Journal of the ACM*, 12(1):23–41, 1965.
15. S. SCHMITT. Avoiding Redundnacy for Proof Reconstruction in Classical and Non-Classical Logics. Technical Report, TU-Darmstadt, 1997.
16. S. SCHMITT, C. KREITZ. On transforming intuitionistic matrix proofs into standard-sequent proofs. 4th *TABLEAUX Workshop*, LNAI 918, pp. 106–121, 1995.
17. S. SCHMITT, C. KREITZ. Converting non-classical matrix proofs into sequent-style systems. *CADE-13*, LNAI 1104, pp. 418–432, 1996.
18. S. SCHMITT, C. KREITZ. A uniform procedure for converting non-classical matrix proofs into sequent-style systems. Technical Report AIDA-96-01, TU-Darmstadt, 1996.
19. T. TAMMET A Resolution Theorem Prover for Intuitionistic Logic *CADE-13*, LNAI 1104, pp. 2–16, 1996.
20. L. WALLEN. Matrix proof methods for modal logics. *IJCAI-87*, p. 917–923. 1987.
21. L. WALLEN. *Automated deduction in non-classical logics*. MIT Press, 1990.
22. L. WOS ET. AL. Automated reasoning contributes to mathematics and logic. *CADE-10*, LNCS 449, p. 485–499. 1990.

A New One-Pass Tableau Calculus for PLTL

Stefan Schwendimann

Institut für Informatik und
angewandte Mathematik
University of Berne
Neubrückstr.10
CH-3012 Bern
E-mail: schwendi@iam.unibe.ch
Phone: +41 31 6313317

Abstract. The paper presents a one-pass tableau calculus $PLTL_T$ for the propositional linear time logic PLTL. The calculus is correct and complete and unlike in previous decision methods, there is no second phase that checks for the fulfillment of the so-called eventuality formulae. This second phase is performed *locally* and is incorporated into the rules of the calculus. Derivations in $PLTL_T$ are cyclic trees rather than cyclic graphs. When used as a basis for a decision procedure, it has the advantage that only one branch needs to be kept in memory at any one time. It may thus be a suitable starting point for the development of a parallel decision method for PLTL.

1 Introduction

Temporal logic has proved to be a useful formalism for reasoning about execution sequences of programs. It can be employed to formulate and verify properties of concurrent programs, protocols and hardware (see for instance [1], [13], [14]). A prominent variant is the propositional linear time logic PLTL where the decision problem is known to be PSPACE-complete [15]. In most of the previous publications the decision algorithm itself has been presented as a 2-phase procedure:

1. A tableau procedure that creates a graph.
2. A procedure that checks whether the graph fulfills all eventuality formulae.

The second phase usually leads to an analysis of the strongly connected components (SCC) of the graph (see e.g. [16]). Typical descriptions of this 2-phase method can be found in [17] and [9] where, in both cases, the second phase is not treated formally.

The tableau method presented in [12] is claimed to be incremental, where 'incremental' means that only reachable nodes are created (this is also true for [17] and [9]). However, it is essentially still a 2-phase procedure. The focus there is on providing a refined method for linear temporal logic with past time operators.

The above methods can treat the verification problem directly as a logical implication '*spec* → *prop*', where *spec* is the PLTL formula representing a specification and *prop* the formula representing a property to be verified. The essence of the problem is to show the validity of this implication in PLTL.

An alternative approach uses state-based methods (also referred to as 'model checking'). One possibility is to translate both the specification (e.g. of a protocol) and the negation of the property into labeled generalized Büchi automata, where the property automaton is also generated by a tableau-like procedure. A second phase then checks whether the language accepted by the synchronous product of the two automata is empty. Once again, in general, this involves an SCC analysis. In [7] it is claimed that the check for emptiness can be done 'on-the-fly' during the generation of the product: the tableau-like procedure builds the property graph in a depth-first manner choosing only successors that 'match' the current state of the protocol. Validity can also be checked using this method. However, it is not clear from the description whether the procedure remains 'on-the-fly' when there is no protocol to 'match'. In [2] it is shown how a generalized Büchi automaton can be transformed into a classical Büchi automaton for which the emptiness check reduces to a simple cycle detection scheme. So in the area of state-based methods similar attempts have been made to intermix the two phases and to avoid a standard SCC analysis.

Here we present a one-pass tableau calculus which checks locally, on-the-fly, for the fulfillment of eventuality formulae on a branch-by-branch basis. No second phase is required. It can also be used for an incremental depth-first search where only reachable states are created. Derivations in this calculus result in (cyclic) tree-like structures rather than general graphs. Thus, the analysis of strongly connected components reduces to the detection of 'isolated subtrees', a task which is very simple and which can therefore be incorporated easily into the calculus. The new aspects basically consist of:

1. A branch-based loop check that ensures termination.
2. A part that synthesizes the essential information gleaned from expanding the subtrees of a node.

The 2-phase methods require the creation of a fully expanded tableau, which is often exponential in the size of the initial formula. Since our method involves only one pass and is complete, we can stop as soon as a (counter-) model is detected, thus, (sometimes) avoiding a fully expanded tableau. A further advantage is that only one branch of the derivation tree needs to be considered at any stage. Therefore, the calculus $PLTL_T$ is a natural analogue of the tableau and Gentzen-style sequent calculi for various modal logics, for instance K, KT and S4 (see e.g. [6], [8], [3]), where derivations are also trees, where it is always sufficient to consider one branch at any one time and where a check for loops is sometimes required to guarantee termination (see e.g. [11]).

While the two phases of the previous methods are an obstacle for parallelization, the branch-by-branch treatment offers natural possibilities for concurrent search. Of course, at the end, the resultant parts would need to be combined, but until then the processors could work independently on different subtrees without extra-communication.

There is of course a caveat. Since a naive derivation in PLTL$_T$ essentially unfolds a graph into a tree, the run-time may be significantly higher, especially for examples where the graphs have (relatively) few nodes and many edges. Clearly, the calculus must be applied in combination with suitable pruning and caching techniques. Algorithmic aspects, however, are beyond the scope of this paper. We will focus on the new definitions and the key lemmata and theorems. Simpler observations are stated as propositions without proofs.

2 Syntax

In the following we deal with an extension $\mathcal{L}$ of the language for classical propositional logic. It comprises: 1. Countably many propositional variables $p_0, p_1, \ldots$. 2. The propositional constants true and false. 3. The connectives $\neg, \wedge, \vee, \mathsf{X}$ (neXt time), F (sometime), G (generally), $\mathcal{U}$ (until), and $\mathcal{B}$ (before). As auxiliary symbols we have parentheses and commas. The formulae of $\mathcal{L}$ are inductively defined: 1. The propositional variables and constants are formulae. 2. If A and C are formulae, then $(\neg A)$, $(\mathsf{X}\,A)$, $(\mathsf{F}\,A)$, $(\mathsf{G}\,A)$, $(A \wedge C)$, $(A \vee C)$, $(A\,\mathcal{U}\,C)$, and $(A\,\mathcal{B}\,C)$ are formulae.

The set of propositional variables is denoted by Var and the set of all formulae by Fml. As metavariables for propositional variables we use P, Q, and as metavariables for formulae A, C, D, possibly with subscripts. Propositional variables are also called *positive literals*; if P is a propositional variable then $\neg P$ is a *negative literal*. As metavariable for positive literals we use P and as metavariable for literals M, possibly with subscripts. In order to increase readability, we omit outer parentheses and define the unary connectives to take precedence over all binary connectives. For example, we write $\mathsf{F}\,(p_7\,\mathcal{U}\,p_1) \wedge (p_0\,\mathcal{B}\,\neg \mathsf{X}\,p_1)$ for the formula $((\mathsf{F}\,(p_7\,\mathcal{U}\,p_1)) \wedge (p_0\,\mathcal{B}\,(\neg(\mathsf{X}\,p_1))))$.

3 Semantics

Definition 1. *A PLTL-model is a pair $\langle S, L \rangle$, where S is an infinite sequence of states $(s_i)_{i \in \mathbb{N}} = s_0\,s_1 \ldots$ and $L : S \to \mathrm{Pow}(\mathrm{Var})$ is a function which assigns to each state a set of propositional variables. L is called a 'labeling'.*

Definition 2. *Let $\mathcal{M} = \langle S, L \rangle$ be a PLTL-model, $s_i \in S$, and $A \in \mathcal{L}$. The relation '$\mathcal{M}$ satisfies A at state s_i', formally $\mathcal{M}, s_i \models A$, is inductively defined:*

1. $\mathcal{M}, s_i \models \mathsf{true}$ and $\mathcal{M}, s_i \not\models \mathsf{false}$.
2. $\mathcal{M}, s_i \models P$ iff $P \in L(s_i)$.
3. $\mathcal{M}, s_i \models \neg A$ iff $\mathcal{M}, s_i \not\models A$.
4. $\mathcal{M}, s_i \models A \wedge C$ iff $\mathcal{M}, s_i \models A$ and $\mathcal{M}, s_i \models C$.
5. $\mathcal{M}, s_i \models A \vee C$ iff $\mathcal{M}, s_i \models A$ or $\mathcal{M}, s_i \models C$.
6. $\mathcal{M}, s_i \models \mathsf{X}\,A$ iff $\mathcal{M}, s_{i+1} \models A$.

7. $\mathcal{M}, s_i \models \mathsf{G}\, A$ iff $\mathcal{M}, s_{i+j} \models A$ for all $j \geq 0$.
8. $\mathcal{M}, s_i \models \mathsf{F}\, A$ iff there exists a $j \geq 0$ such that $\mathcal{M}, s_{i+j} \models A$.
9. $\mathcal{M}, s_i \models A\,\mathcal{U}\,C$ iff there exists a $j \geq 0$ such that $\mathcal{M}, s_{i+j} \models C$ and $\mathcal{M}, s_{i+k} \models A$ for all $0 \leq k < j$.
10. $\mathcal{M}, s_i \models A\,\mathcal{B}\,C$ iff for all $j \geq 0$ with $\mathcal{M}, s_{i+j} \models C$ there exists a $0 \leq k < j$ with $\mathcal{M}, s_{i+k} \models A$.

If $\mathcal{M}, s_i \models A$ for all $s_i \in S$, we write $\mathcal{M} \models A$. A formula A is **PLTL-satisfiable** iff there exists a PLTL-model $\mathcal{M} = \langle S, L \rangle$ and a state $s_i \in S$ such that $\mathcal{M}, s_i \models A$. A formula A is **PLTL-valid** iff $\mathcal{M} \models A$ for all PLTL-models $\mathcal{M} = \langle S, L \rangle$. Then we write $\mathsf{PLTL} \models A$.

Formulae which contain the symbol $\neg$ only immediately before positive literals are called formulae in *negation normal form*. The PLTL-valid equivalences $(\neg \mathsf{X}\, A \leftrightarrow \mathsf{X}\, \neg A)$, $(\neg \mathsf{G}\, A \leftrightarrow \mathsf{F}\, \neg A)$, $(\neg (A\,\mathcal{U}\,C) \leftrightarrow (\neg A\,\mathcal{B}\,C))$, and $(\neg (A\,\mathcal{B}\,C) \leftrightarrow (\neg A\,\mathcal{U}\,C))$ allow us to push the negation inwards and to obtain for any formula an equivalent formula in negation normal form. In the following we restrict ourselves to formulae in negation normal form.

Definition 3. *The complement $\overline{A}$ of a formula A in negation normal form is inductively defined as follows. 1. $\overline{\mathsf{true}} := \mathsf{false}$ and $\overline{\mathsf{false}} := \mathsf{true}$. 2. $\overline{P} := \neg P$ and $\overline{\neg P} := P$. 3. $\overline{A \wedge C} := (\overline{A} \vee \overline{C})$ and $\overline{A \vee C} := (\overline{A} \wedge \overline{C})$. 4. $\overline{\mathsf{G}\, A} := \mathsf{F}\, \overline{A}$ and $\overline{\mathsf{F}\, A} := \mathsf{G}\, \overline{A}$. 5. $\overline{A\,\mathcal{B}\,C} := \overline{A}\,\mathcal{U}\,C$ and $\overline{A\,\mathcal{U}\,C} := \overline{A}\,\mathcal{B}\,C$.*

Definition 4. *We classify the formulae in negation normal form: 1. Propositional constants, literals and formulae of the form $\mathsf{X}\, A$ are called* elementary. *2. All other formulae are called* non-elementary *and can be represented either as α-formulae (conjunctions) or as β-formulae (disjunctions) according to the following tables:*

α	α_1	α_2
$A \wedge C$	A	C
$\mathsf{G}\, A$	A	$\mathsf{X}\,\mathsf{G}\, A$
$A\,\mathcal{B}\,C$	$\overline{C}$	$A \vee \mathsf{X}\,(A\,\mathcal{B}\,C)$

β	β_1	β_2
$A \vee C$	A	C
$\mathsf{F}\, D$	D	$\mathsf{X}\,\mathsf{F}\, D$
$C\,\mathcal{U}\,D$	D	$C \wedge \mathsf{X}\,(C\,\mathcal{U}\,D)$

β-formulae of the form $\mathsf{F}\, D$ and $C\,\mathcal{U}\,D$ are also called eventuality *formulae or* eventualities *for short; in order for these formulae to hold at a certain state in a model, there must be a future state where D 'eventually' holds.*

In the following we use α, α_1, α_2 to denote an α-formula and its conjuncts and β, β_1, β_2 to denote a β-formula and its disjuncts. Moreover, we assume for the rest of the paper that there are no formulae of the form $\mathsf{F}\, D$; they can be written as $\mathsf{true}\,\mathcal{U}\,D$.

Definition 5. *We define the closure $cl(A)$ for any formula A in negation normal form: 1. A is in $cl(A)$. 2. If $\neg P$ is in $cl(A)$, then P is in $cl(A)$. 3. If $\mathsf{X}\, B$ is in $cl(A)$, then B is in $cl(A)$. 4. If α is in $cl(A)$, then α_1 and α_2 are in $cl(A)$. 5. If β is in $cl(A)$, then β_1 and β_2 is in $cl(A)$.*

The closure of a formula is essentially the set of all subformulae augmented with the α_2 and β_2 parts of the temporal connectives. It is also called the Fischer-Ladner closure [5]. Before we turn to Hintikka structures for PLTL, we define some properties for more general 'labeling' functions which assign to states sets of formulae rather than sets of variables.

Definition 6. *Let S be a (possibly finite) sequence of states $s_0 \, s_1 \, \ldots$, L a function $L : S \to \text{Pow}(\text{Fml})$, and $s_i \in S$.*

1. *Propositional consistency properties:*
   (PC0) false *is not in $L(s_i)$.*
   (PC1) *If a literal M is in $L(s_i)$, then its complement $\overline{M}$ is not in $L(s_i)$.*
   (PC2) *If α is in $L(s_i)$, then α_1 and α_2 are in $L(s_i)$.*
   (PC3) *If β is in $L(s_i)$, then β_1 or β_2 is in $L(s_i)$.*
2. *Local consistency property:*
   (LC) *If $\mathsf{X} A$ is in $L(s_i)$ and s_i is not the last state if S is finite, then A is in $L(s_{i+1})$.*

We say that L fulfills one of the above properties if the respective condition is satisfied for all states s_i of the sequence S.

In the next definition we describe the set of eventualities that are not 'satisfied' in a sequence of states.

Definition 7. *Let S be a (possibly finite) sequence of states $s_0 \, s_1 \, \ldots$ and $L : S \to \text{Pow}(\text{Fml})$ a labeling. Then the set $\text{open}(S, L)$ of eventualities is defined as:*

$$\text{open}(S, L) := \{C \,\mathcal{U}\, D \mid \exists i \, (C \,\mathcal{U}\, D \in L(s_i)) \text{ and } \forall j \geq i \, (D \notin L(s_j))\}.$$

The following definition of a (pre-)Hintikka structure can be found in the literature (e.g. [4]).

Definition 8. *A pre-Hintikka structure $\mathcal{H}$ is a pair $\langle S, L \rangle$, where S is a sequence of states $(s_i)_{i \in \mathbb{N}} = s_0 \, s_1 \, \ldots$ and $L : S \to \text{Pow}(\text{Fml})$ is a labeling function that fulfills the properties (PC0-3) and (LC).*

By restricting the labeling function L to variables, we can associate with each pre-Hintikka structure $\mathcal{H} = \langle S, L \rangle$ a model $\mathcal{M}_\mathcal{H} := \langle S, L{\upharpoonright}\text{Var}\rangle$.

Definition 9. *We say that a pre-Hintikka structure $\mathcal{H} = \langle S, L \rangle$ is a Hintikka structure if $\text{open}(S, L) = \emptyset$, that is, if we have for any state s_i and any eventuality $C \,\mathcal{U}\, D$: If $C \,\mathcal{U}\, D \in L(s_i)$, then there exists a $j \geq i$ with $D \in L(s_j)$.*

$\mathcal{H}$ is said to be a (pre-)Hintikka structure for a formula A if $A \in L(s_0)$. We say that $\mathcal{H}$ is a complete (pre-)Hintikka structure for A if for all i: $L(s_i) = \{C \mid C \in cl(A) \text{ and } \mathcal{M}_\mathcal{H}, s_i \models C\}$.

Note that any Hintikka structure for A can be made into a complete Hintikka structure for A by adding to $L(s_i)$ all formulae of the closure that are satisfied at s_i. The following standard theorem relates the existence of Hintikka structures to the existence of models.

Theorem 1. *A formula A in negation normal form is* PLTL-*satisfiable iff there exists a Hintikka structure for A.*

Proof. See for instance [9].

In the following we deal with a set W of words over an alphabet S. We write ws for the concatenation of a word w and a single element $s \in S$. Similarly, we write ww' for the concatenation of the two words w and w'. w and w' may also be the empty word. Now we introduce a new type of structures which are essentially trees with loops on their branches.

Definition 10. *A* loop tree *is a tuple* $\mathcal{T} = \langle W, S, L, R \rangle$ *where:*

1. S *is a finite set.*
2. W *is a finite set of finite words over S where:*
   (a) *If* $w = s_0 s_1 \ldots s_k \in W$, *then* $s_i \neq s_j$ *for all* $0 \leq i < j \leq k$.
3. R *is a binary relation on W with the following properties:*
   (a) $(w, ws) \in R$ *for all* $w, ws \in W$.
   (b) *If* $w \in W$ *and* $ws \notin W$ *for all* $s \in S$, *then there exists a word* $w' \in W$ *such that* w' *is a prefix of* w *and* $(w, w') \in R$.
   (c) *If* $(w, w') \in R$, *then either* w' *is of the form* ws *or* w' *is a prefix of* w.
4. $L : W \to \text{Pow}(\text{Fml})$ *is a labeling function with the property:* $L(ws) = L(w's)$ *for all* $ws, w's \in W$.

The set S can be viewed as a set of nodes and the words W as directions how to reach these nodes. The conditions say that a word should contain a node only once, and that words which cannot be extended are related to a prefix. This means that we basically have a tree-like structure with loops back on the branches where at the end of each branch we have at least one loop back. The arrows in Fig. 1 correspond to the relation R. The labeling is controlled by the last node of a word. A word is essentially the last node plus the information how it is reached. Therefore words will also be called states.

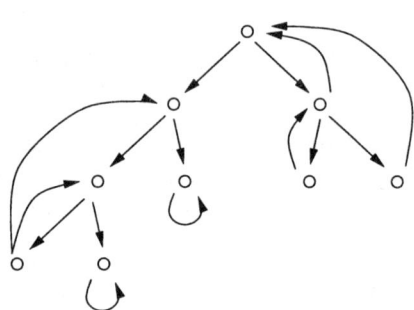

Fig. 1. Example of a loop tree.

Definition 11. *Let* $\mathcal{T} = \langle W, S, L, R \rangle$ *be a loop tree.*

1. *If* $ws \in W$ *and* $w \notin W$, *then* ws *is called a* root *of* $\mathcal{T}$.
2. *A* path *through* $\mathcal{T}$ *is a finite or infinite R-sequence of states* $w_0, \ldots, w_i, w_{i+1}, \ldots$, *where* $(w_i, w_{i+1}) \in R$ *for all* w_i *of the sequence (except the last one if the sequence is finite).*

3. A loop branch *of* $\mathcal{T}$ *is a finite path* $w_0, w_1, \ldots, w_k, w_{k+1}$ *through* $\mathcal{T}$ *where* w_0 *is a root and for all* $i < k$ $w_{i+1} = w_i s_i$ *for some* $s_i \in S$. *The last state* w_{k+1} *is identical to a previous state, i.e.* $w_{k+1} = w_j$ *for a* $j \leq k$, *and it is called the* loop state *of the branch. The suffix path* $w_j, w_{j+1}, \ldots, w_k, w_{k+1}$ *is called the* loop *of the branch. We say that a path* π visits the loop branch *or simply* the loop *if* w_k, w_{k+1} *occurs in* π *(as a pair of consecutive states).*

4. *If* $\pi = w_0, \ldots, w_j, w_{j+1}, \ldots, w_k, w_{k+1}$ *is a loop branch, the set* $\text{open}_{\inf}(\pi, L)$ *is defined as:*

$$\text{open}_{\inf}(\pi, L) := \{C \cup D \mid C \cup D \in \text{open}(\pi, L)$$
$$\text{and } \forall i, (j \leq i \leq k \Rightarrow D \notin L(w_i))\}.$$

5. *The function* $\text{depth}_{\mathcal{T}} : W \to \mathbb{N}$ *is defined as follows:* 1. $\text{depth}_{\mathcal{T}}(w) := 0$ *for any root* w *of* $\mathcal{T}$. 2. $\text{depth}_{\mathcal{T}}(ws) := \text{depth}_{\mathcal{T}}(w) + 1$ *for any* $w, ws \in W$.

Remarks: 1. Note that a loop tree may contain several roots and may therefore represent several tree-like structures. 2. A loop branch is defined to contain the backward loop. Therefore a 'physical branch' can contain several loop branches that share a common prefix path (see Fig. 1). In particular, loops may also start at non-leaf nodes. 3. Obviously, $\text{open}_{\inf}(\pi, L)$ is a subset of $\text{open}(\pi, L)$. It denotes the eventualities of π which are not satisfied on the loop itself even if it is visited infinitely many times.

Proposition 1. *If* $w_j, w_{j+1}, \ldots, w_k, w_{k+1}$ *is a loop* ($w_{k+1} = w_j$) *and a path* π *visits it repeatedly (i.e. multiple occurrences of* w_k, w_{k+1} *on* π), *then obviously all other states of the loop* $w_{j+1}, \ldots, w_{k-1}$ *must occur in* π *between two occurrences of* w_k, w_{k+1}, *although not necessarily in a row.*

Definition 12. *Let* $\mathcal{T} = \langle W, S, L, R \rangle$ *be a loop tree. The subtree of* $\mathcal{T}$ *at* $w \in W$ *is a structure* $\mathcal{T}' = \langle W', S', L', R' \rangle$ *defined as follows:* 1. $S' := S$. 2. $W' := \{ww' \mid ww' \in W\}$. 3. $R' := \{(ww', ww'') \mid (ww', ww'') \in R\}$. 4. $L' := L \upharpoonright W'$. *We say that* $\mathcal{T}'$ *is an* isolated *subtree of* $\mathcal{T}$ *if* $(w', v) \notin R$ *for any* $w' \in W'$ *and* $v \in W \setminus W'$.

An isolated subtree is obviously a loop tree. Whether or not a subtree is isolated can be determined easily by checking the loop states of the loop branches that pass through the subtree's root.

Lemma 1. *Let* $\mathcal{T} = \langle W, S, L, R \rangle$ *be a loop tree and* $\mathcal{T}' = \langle W', S', L', R' \rangle$ *the subtree at* $w \in W$. *Then we have:* $\mathcal{T}'$ *is isolated iff*

$$\text{depth}_{\mathcal{T}}(w) \leq \min(\{\text{depth}_{\mathcal{T}}(w') \mid w' \text{is a loop state of a}$$
$$\text{loop branch of } \mathcal{T} \text{ containing } w\}).$$

Proof. If $\mathcal{T}'$ is isolated, then no loop branch of $\mathcal{T}$ containing w can have a loop state outside $\mathcal{T}'$. Since w is the root of $\mathcal{T}'$, the depth of a loop state must be greater or equal than the depth of w.

Conversely, if the depth of a loop state is greater or equal than the depth of w, then it must belong to $\mathcal{T}'$ since a branch may only loop back on itself. Therefore $\mathcal{T}'$ must be isolated.

Definition 13.

1. A *pre-Hintikka-tree* is a loop tree $\mathcal{T} = \langle W, S, L, R \rangle$ where L fulfills the properties (PC0-3) and (LC) for all paths through $\mathcal{T}$.
2. A *Hintikka-tree* for a formula A is a pre-Hintikka-tree $\mathcal{T} = \langle W, S, L, R \rangle$ with the additional property that there exists an infinite path $\pi = w_0, w_1, \ldots$ through $\mathcal{T}$ with $A \in L(w_0)$ and $open(\pi, L) = \emptyset$.

Proposition 2. *Let $\pi = w_j, w_{j+1}, \ldots, w_k, w_{k+1}$ be the loop ($w_{k+1} = w_j$) of a pre-Hintikka-tree $\mathcal{T} = \langle W, S, L, R \rangle$. Then we have: If an eventuality $C\mathcal{U}D$ is in $open_{\inf}(\pi, L)$, then $C\mathcal{U}D$ and $X(C\mathcal{U}D)$ are in $L(w_i)$ for all i with $j \leq i \leq k$.*

The following lemma states that the open eventualities of a path depend in a simple way on the unfulfilled eventualities of single loop branches.

Lemma 2. *Let π be an infinite path through the pre-Hintikka-tree $\mathcal{T} = \langle W, S, L, R \rangle$ and $\pi_1, \ldots, \pi_m$ be the loops of $\mathcal{T}$ that are visited infinitely many times by π. Then we have:*

$$open(\pi, L) = \bigcap_{i=1\ldots m} open_{\inf}(\pi_i, L).$$

Proof. $\supset$: Let $C\mathcal{U}D$ be in $\bigcap_{i=1\ldots m} open_{\inf}(\pi_i, L)$. There is a point in time after which only the loops $\pi_1, \ldots, \pi_m$ and, therefore, only states from $\pi_1, \ldots, \pi_m$ are visited. If $C\mathcal{U}D \in open_{\inf}(\pi_i, L)$, then D is not in any state of π_i, and by Proposition 2 we know that $C\mathcal{U}D$ is in each state of π_i. Therefore $C\mathcal{U}D$ must be in $open(\pi, L)$.

$\subset$: Let $C\mathcal{U}D$ be in $open(\pi, L)$. Then there is a state s in π such that for any future state s' the formula D is not in $L(s')$ but $X(C\mathcal{U}D)$ is in $L(s')$. This implies that for any state s'' from $\pi_1, \ldots, \pi_m$ the formula D is not in $L(s'')$ and $X(C\mathcal{U}D)$ is in $L(s'')$ since by Proposition 1 all these states are visited by π after s. Therefore $C\mathcal{U}D$ is in $open_{\inf}(\pi_i, L)$ for all π_i ($i = 1\ldots m$).

Theorem 2. *There is a Hintikka structure for a formula A iff there exists a Hintikka-tree for A.*

Proof. The direction from right to left is obvious. If $\mathcal{T} = \langle W, S, L, R \rangle$ is a Hintikka-tree for A then simply choose a path $\pi = w_0 w_1 \ldots$ through $\mathcal{T}$ with $A \in L(w_0)$ and $open(\pi, L) = \emptyset$. $\langle \pi, L \rangle$ is then a Hintikka structure for A.

For the direction from left to right assume that $\mathcal{H} = \langle S, L \rangle$ is a Hintikka structure for A with $S = s_0 s_1 \ldots s_i s_{i+1} \ldots$. First, we introduce an equivalence relation $\sim$ on the elements of S: $s_i \sim s_j$ iff $L(s_i) \cap cl(A) = L(s_j) \cap cl(A)$. The equivalence class of s_i is denoted by $[s_i]$. We construct a Hintikka-tree $\mathcal{T} = \langle W, S', L', R \rangle$ for A in the following way (w, w', w'' may be the empty word):

1. $S' := S/\sim$.
2. W and R are defined inductively:

(a) $[s_0]$ is an element of W.
(b) If $w[s_i]$ is an element of W, then we distinguish two cases:
   i. If $w[s_i]$ contains a state equivalent to s_{i+1}, that is, if $w[s_i] = w'[s_j]w''$ and $s_j \sim s_{i+1}$, then $(w[s_i], w'[s_j])$ is in R (a loop).
   ii. Otherwise $w[s_i][s_{i+1}]$ belongs to W and $(w[s_i], w[s_i][s_{i+1}])$ is in R.
3. The labeling L' is defined as $L'(w[s_i]) := L(s_i) \cap cl(A)$.

The structure $\mathcal{T}$ is obviously a loop tree. S' is finite since $cl(A)$ is finite, L' satisfies (PC0-3) and (LC), and by the construction there is a path $\pi = w_0, w_1, \ldots$ through $\mathcal{T}$ (corresponding to $s_0 s_1 \ldots$) with $w_0 = [s_0]$, $A \in L'(w_0)$ and $open(\pi, L') = \emptyset$. Therefore $\mathcal{T}$ is a Hintikka-tree for A.

4 The Calculus PLTL$_T$

We present a Tableau-like calculus for PLTL that is complete and correct with respect to the PLTL semantics. It operates on so-called prestates which contain the full information needed to decide satisfiability of formulae in negation normal form.

In the following we use Γ and Σ for finite sets of formulae in negation normal form, and Λ for sets of *literals* (and possibly constants). We also write A, Γ for the set $\{A\} \cup \Gamma$, and Γ, Σ for the union $\Gamma \cup \Sigma$, and $\mathsf{X}\Gamma$ is used for the set $\{\mathsf{X} A \mid A \in \Gamma\}$.

For lists we have the following conventions: We use $*$ for the concatenation of lists and $[]$ for the empty list. If M is a list, then we write $len(M)$ for the length of M and $M[i]$ for the ith element of M ($1 \leq i \leq len(M)$). If M is a list of tuples, then we write $M[i]_j$ to denote the projection to the jth element of $M[i]$.

Definition 14. *A* prestate *is a triple* $(\Gamma, Save, Res)$, *also written as* $\Gamma \mid Save \mid Res$ *where:*

1. *Γ is a finite set of formulae in negation normal form.*
2. *Save is a structure to store history information. It is a pair (Ev, Br), also written as $Ev\,;\,Br$, where Ev is a set of formulae in negation normal form representing the currently satisfied eventualities, Br is a list of pairs (Γ', Ev') representing the current branch, and Γ' and Ev' correspond to the Γ and Ev parts of previous prestates.*
3. *Res is a structure to store partial result information. It is a pair (n, uev), where n is a natural number indicating the 'earliest' prestate reachable by the current one, and uev is a set of eventuality formulae in negation normal form. It represents the <u>un</u>fulfilled <u>ev</u>entualities of the current branch.*

A prestate is said to be a state *if Γ is of the form $\Lambda, \mathsf{X}\Sigma$, that is, if Γ consists only of elementary formulae.*

According to the above definition, $\Gamma \mid Ev\,;\,Br \mid (n, uev)$ is the extended notion for an abstract prestate. To focus on the locally relevant parts of a prestate, we use '$\ldots$' for the 'unimportant' parts (e.g. $\Gamma \mid \ldots \mid \ldots$). If '$\ldots$' appears at the same position in the numerator and the denominator(s) of a rule, then we mean that the corresponding parts are the same.

Definition 15. *The Tableau calculus $PLTL_T$ is defined as follows:*
a) *Terminal rules:*

$$\text{false}, \Gamma \mid Ev \, ; \, Br \mid (len(Br), \{\text{false}\}) \quad (\text{false})$$

$$P, \neg P, \Gamma \mid Ev \, ; \, Br \mid (len(Br), \{\text{false}\}) \quad (\text{contr})$$

$$\Lambda, \mathsf{X} \, \Sigma \mid Ev \, ; \, Br \mid (k, uev) \quad (\text{loop})$$

where in (loop) there exists an i, $1 \le i \le len(Br)$, such that:
1. $\Lambda, \mathsf{X} \, \Sigma = Br[i]_1$.
2. $k = i - 1$ and $uev = \{C \, \mathcal{U} \, D \mid C \, \mathcal{U} \, D \in \Sigma \text{ and } D \notin (\cup_{j=i+1}^{len(Br)} Br[j]_2 \cup Ev)\}$.

b) *α-rules:*

$$\frac{\alpha, \Gamma \mid \ldots \mid \ldots}{\alpha_1, \alpha_2, \Gamma \mid \ldots \mid \ldots} \, (\alpha)$$

c) *β-rules:*

$$\frac{A \vee B, \Gamma \mid \ldots ; \, Br \mid (n, uev)}{A, \Gamma \mid \ldots ; \, Br \mid (n_1, uev_1) \quad B, \Gamma \mid \ldots ; \, Br \mid (n_2, uev_2)} \, (\vee)$$

$$\frac{C \, \mathcal{U} \, D, \Gamma \mid Ev \, ; \, Br \mid (n, uev)}{D, \Gamma \mid Ev \cup \{D\} \, ; \, Br \mid (n_1, uev_1) \quad C, \mathsf{X}(C \, \mathcal{U} \, D), \Gamma \mid Ev \, ; \, Br \mid (n_2, uev_2)} \, (\mathcal{U})$$

where in ($\vee$) and ($\mathcal{U}$):
1. $n = \min(n_1, n_2)$.
2. ($m := len(Br) - 1$).

$$uev = \begin{cases} \emptyset & \text{if } uev_1 = \emptyset \text{ or } uev_2 = \emptyset, \\ \{\text{false}\} & \text{if } n_1 > m \text{ and } n_2 > m \text{ (and } uev_1 \ne \emptyset, uev_2 \ne \emptyset), \\ uev_1 & \text{if } n_1 \le m \text{ and } n_2 > m \text{ (and } uev_2 \ne \emptyset), \\ & \text{or if } uev_2 = \{\text{false}\}, \\ uev_2 & \text{if } n_1 > m \text{ and } n_2 \le m \text{ (and } uev_1 \ne \emptyset), \\ & \text{or if } uev_1 = \{\text{false}\}, \\ uev_1 \cap uev_2 & \text{otherwise.} \end{cases}$$

d) *Nexttime rule:*

$$\frac{\Lambda, \mathsf{X} \, \Sigma \mid Ev \, ; \, Br \mid \ldots}{\Sigma \mid \emptyset \, ; \, Br * ((\Lambda, \mathsf{X} \, \Sigma), Ev) \mid \ldots} \, (\mathsf{X})$$

In order to ensure termination, the α- and β-rules and the nexttime rule are restricted to prestates that are not instances of a terminal rule. We call α in (α), $A \vee B$ in ($\vee$), and $C \, \mathcal{U} \, D$ in ($\mathcal{U}$) the decomposed formula of the respective rule.

Remark 1.

- The main difference to a modal calculus is the result part which is synthesized bottom-up (from children to parents). It is needed because a single branch need not be 'open' or 'closed'; it may be 'open' in connection with some other branches.
- (loop): The sequence $Br[i]_1, \ldots, Br[len(Br)]_1, (\Lambda, \mathsf{X} \, \Sigma)$ corresponds to the loop of a branch. uev is defined to be the set $open_{\text{inf}}(., .)$ of eventualities that are not satisfied on this loop branch (see proposition 2).

- (β-rules): A β-rule corresponds to a branching of the tree. n is set to the minimum depth of the states to which branches of the two subtrees can loop back. uev is the minimal set of eventualities that are left open by any infinite path visiting only loops of the subtree below this β node.
- (Nexttime rule): The current state and the eventualities that are satisfied by the current state are appended to the branch Br.
- Note that the sets Γ, Λ, and Σ may be empty. For instance, if Σ is empty in the numerator of (X), we obtain the following fragment of a tableau which ends in a basically empty instance of (loop). On the right the corresponding model is shown.

$$\cfrac{\cfrac{\Lambda \mid Ev\,;\, Br \mid \ldots}{\emptyset \mid \emptyset\,;\, Br * (\Lambda, Ev) \mid \ldots}\;(\mathsf{X})}{\emptyset \mid \emptyset\,;\, Br * (\Lambda, Ev) * (\emptyset, \emptyset) \mid (len(Br) + 1, \emptyset)\ \ \text{(loop)}}\;(\mathsf{X})$$

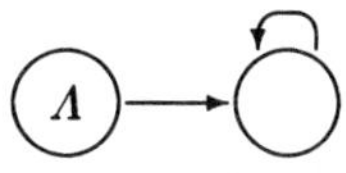

Definition 16. *A* tableau for a prestate *ps is a tree of prestates with root ps and where the sons of a node (prestate) correspond to an application of a* PLTL_T *rule to the node. We say that the tableau is* expanded, *if each leaf node is an instance of a terminal rule.*

Let A be a formula and n the number of subformulae of A. Then it is clear that any tableau for $A \mid \ldots \mid \ldots$ is finite. There are $2^{O(n)}$ many subsets of $cl(A) \cup cl(\overline{A})$. Each Γ of a prestate $\Gamma \mid \ldots \mid \ldots$ is such a subset, and since the terminal rules must be applied whenever they can be applied, the number of different prestates on each branch is finite. Therefore, the total number of prestates in any tableau for $A \mid \ldots \mid \ldots$ is finite and any expansion will eventually terminate.

Proposition 3.

a) For every formula A there is an $n \in \mathbb{N}$ and a set $uev \subseteq$ Fml such that there is an expanded tableau for $A \mid \emptyset\,;\, Br \mid (n, uev)$.

b) If in a tableau the set uev of a prestate is empty , then the set uev of the root of the tableau is also empty.

Example 1. We show the essential branch of a tableau for the satisfiable property $\mathsf{G}\,\mathsf{F}\,p \wedge \mathsf{G}\,\mathsf{F}\,\neg p$ (recall that $\mathsf{F}\,p$ can be written as $\mathsf{true}\,\mathcal{U}\,p$). The α- and β-rules are applied until we reach a state with only elementary formulae. The currently decomposed formula is in parentheses. It is left to the reader to fill in the missing *Save* and *Res* parts.

$$\begin{array}{ccl}
 & (\mathsf{G}\,\mathsf{F}\,p \wedge \mathsf{G}\,\mathsf{F}\,\neg p) \mid \ldots \mid \ldots & \\
\hline
 & (\mathsf{G}\,\mathsf{F}\,p), \mathsf{G}\,\mathsf{F}\,\neg p \mid \ldots \mid \ldots & \\
\hline
 & \mathsf{F}\,p, \mathsf{X}\,\mathsf{G}\,\mathsf{F}\,p, (\mathsf{G}\,\mathsf{F}\,\neg p) \mid \ldots \mid \ldots & \\
\hline
 & (\mathsf{F}\,p), \mathsf{X}\,\mathsf{G}\,\mathsf{F}\,p, \mathsf{F}\,\neg p, \mathsf{X}\,\mathsf{G}\,\mathsf{F}\,\neg p \mid \ldots \mid \ldots & \\
\hline
 & p, \mathsf{X}\,\mathsf{G}\,\mathsf{F}\,p, (\mathsf{F}\,\neg p), \mathsf{X}\,\mathsf{G}\,\mathsf{F}\,\neg p \mid \{p\}\,;\, . \mid \ldots & \mathrm{Sub}_1 \\
\hline
p, \neg p, \ldots & p, \mathsf{X}\,\mathsf{G}\,\mathsf{F}\,p, \mathsf{X}\,\mathsf{F}\,\neg p, \mathsf{X}\,\mathsf{G}\,\mathsf{F}\,\neg p \mid \{p\}\,;\, . \mid \ldots & \\
\hline
 & (\mathsf{G}\,\mathsf{F}\,p), \mathsf{F}\,\neg p, \mathsf{G}\,\mathsf{F}\,\neg p \mid \ldots \mid \ldots & (\mathsf{X}) \\
\hline
 & \mathsf{F}\,p, \mathsf{X}\,\mathsf{G}\,\mathsf{F}\,p, \mathsf{F}\,\neg p, (\mathsf{G}\,\mathsf{F}\,\neg p) \mid \ldots \mid \ldots &
\end{array}$$

$$\begin{array}{c}
\cfrac{\text{Sub}_2 \quad \cfrac{(\mathsf{F}p), \mathsf{XGF}p, \mathsf{F}\neg p, \mathsf{XGF}\neg p\,|\,\ldots\,|\,\ldots}{\mathsf{XF}p, \mathsf{XGF}p, (\mathsf{F}\neg p), \mathsf{XGF}\neg p\,|\,\ldots\,|\,\ldots}}{\boxed{\mathsf{XF}p, \mathsf{XGF}p, \neg p, \mathsf{XGF}\neg p\,|\,\{\neg p\}\,;\,.\,|\,\ldots}}\;(\mathsf{X}) \quad \text{Sub}_3
\end{array}$$

$$\cfrac{\cfrac{\mathsf{F}p, (\mathsf{GF}p), \mathsf{GF}\neg p\,|\,\ldots\,|\,\ldots}{\mathsf{F}p, \mathsf{XGF}p, (\mathsf{GF}\neg p)\,|\,\ldots\,|\,\ldots}}{(\mathsf{F}p), \mathsf{XGF}p, \mathsf{F}\neg p, \mathsf{XGF}\neg p\,|\,\ldots\,|\,\ldots}$$

$$\cfrac{p, \mathsf{XGF}p, (\mathsf{F}\neg p), \mathsf{XGF}\neg p\,|\,\{p\}\,;\,.\,|\,\ldots \quad\quad \text{Sub}_4}{p, \neg p, \ldots \quad p, \mathsf{XGF}p, \mathsf{XF}\neg p, \mathsf{XGF}\neg p\,|\,\ldots\,|\,(0, \emptyset) \quad (\text{loop})}$$

The highlighted prestates above the (X)'s are the *states* of the tableau; the first one (at 'state' depth 0) satisfies p and the second one satisfies $\neg p$. The essential branch ends in an instance of (loop), where in the *Res* part the 0 refers to the depth 0 of the first state and the $\emptyset$ indicates that all eventualities (the only candidate to check stems from $\mathsf{XF}\neg p$) are satisfied on this loop. $\text{Sub}_{1\ldots 4}$ stand for other branches in the expanded tableau.

The corresponding model is very simple: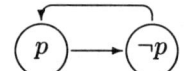

Definition 17. *The* loop tree $\mathcal{T} = \langle W, S, L, R\rangle$ *for an expanded tableau is defined in the following way:*

1. S *is the set of all states (not prestates!) of the tableau and the set of leaf nodes which are not instances of* (loop).
2. W *is the set of paths (in terms of S) to the elements of S in the tableau.*
3. R:
   (a) (w, ws) *is in R for all $w, ws \in W$.*
   (b) (ws, ws) *is in R if s is an instance of* (false) *or* (contr). *That is, we draw a loop to the last state itself if it is inconsistent.*
   (c) *If $w \in W$ is a path to a state which is the last state before an instance of* (loop) *in the tableau, then w must be of the form $w'sw''$ where s is the referenced loop state. Then $(w, w's)$ is in R.*
4. $L(ws)$ *is the set Γ if $s = \Gamma\,|\,\ldots\,|\,\ldots$ plus all the formulae that are decomposed in the tableau between w and ws.*

We could also (formally) omit the classical contradictions from the loop tree (and the states which have only contradictory prestates below), and we would obtain a pre-Hintikka-tree. However, the relevant information is always in the result part.

Lemma 3. *Let $\mathcal{T} = \langle W, S, L, R\rangle$ be the loop tree for an expanded tableau. Then we have for all $ws \in W$, $s = \Lambda, \mathsf{X}\Sigma\,|\,Ev\,;\,Br\,|\,(n, uev)$:*
a) L *fulfills (PC0-3) and (LC) for ws if s is not an instance of* (false) *or* (contr).
b) $\text{depth}_{\mathcal{T}}(ws) = len(Br)$.
c) $n = \min(\{\text{depth}_{\mathcal{T}}(v) \mid v \text{ is a loop state of a branch } \pi = \ldots ws \ldots\})$.
d) *The subtree of $\mathcal{T}$ at ws is isolated iff $n \geq len(Br)$.*

Proof. Follows from the definition of the calculus and Definition 17. The part d) follows from b) and c) and Lemma 1.

Note that the β-rule applications in the tableau are between two states. The lower state is at depth $len(Br)$. The conditions in the β-rules, however, control the result synthesis for the upper state which is at depth $len(Br) - 1$.

Theorem 3 (Correctness). *If A is a formula in negation normal form and if there exists an n such that there is a expanded tableau for $A\,|\,\emptyset\,;\,[\,]\,|\,(n,\emptyset)$, then A is satisfiable.*

Proof. (Sketch) We basically show that for any prestate $ps = \ldots\,|\,\ldots\,|\,(n,uev)$ in the tableau with $uev \neq \{\mathsf{false}\}$ there is a pre-Hintikka structure $\mathcal{H}_{ps} = \langle S, L \rangle$ for A with $open(S,L) = uev$. We represent the pre-Hintikka structure as a loop tree with one single branch π starting with the path that leads to ps. We proceed bottom-up, that is, by induction on the depth of the tableau subtree with root ps. The main case involves a linearization of two loops into a single one as depicted in Fig. 2 (the capital letters denote sections of the path). Lemma 2 ensures that the set of open eventualities is the intersection of the corresponding sets of the two loops, according to the condition in the β-rules.

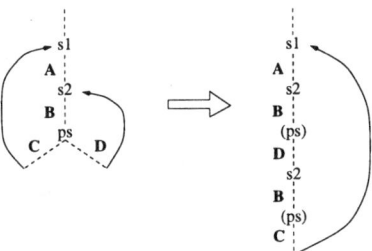

Fig. 2. Loop linearization.

Lemma 4. *Let $\mathcal{T} = \langle W, S', L', R \rangle$ be the loop tree for an expanded tableau for $A\,|\,\emptyset\ [\,]\,|\,(n_A, uev_A)$. Then we have: If there is a infinite path π through $\mathcal{T}$ with $open(\pi, L) = \emptyset$, and if L fulfills (PC0-3) and (LC) on π, then uev_A must be $\emptyset$.*

Proof. (Sketch) We use Lemma 2, 1 and Lemma 3 b),c) and proceed by induction on the depth of the subtree visited by π.

Theorem 4 (Completeness). *If a formula A in negation normal form is satisfiable, then there exists a tableau for $A\,|\,\emptyset\,;\,[\,]\,|\,(n,\emptyset)$ for some $n \in \mathbb{N}$.*

Proof. If A is satisfiable, there exists a complete Hintikka structure $\mathcal{H} = \langle S, L \rangle$ for A by Theorem 1. Let S be the sequence $s_0 s_1 \ldots$, and let $\mathcal{T} = \langle W, S', L', R \rangle$ be the loop tree for an expanded tableau for $A\,|\,\emptyset\ [\,]\,|\,(n, uev)$. We define inductively a map $\varphi : S \to W$ which provides us with a path $\pi = \varphi(s_0)\varphi(s_1)\ldots$ through $\mathcal{T}$ with the following properties:

(a) $A \in L'(\varphi(s_i))$ if $i = 0$.

(b) $L'(\varphi(s_i)) \subseteq L(s_i)$.

(c) If $C\mathcal{U}D \in L'(\varphi(s_i))$ and $D \in L(s_i)$, then $D \in L'(\varphi(s_i))$.

$i = 0$: First, A is in $L'(w)$ for every root $w \in W$ since the loop tree stems from a tableau for $A \mid \ldots \mid \ldots$, and A is also in $L(s_0)$, since $\mathcal{H}$ is a Hintikka structure for A. Second, there must exist a root $w_0 \in W$ with $L'(w_0) \subseteq L(s_0)$ since in the tableau there is a root state for each possible decomposition of A, and $L(s_0)$ must contain at least one set of decomposed formulae ($L(s_0)$ contains A and L fulfills (PC2) and (PC3)). Third, we can choose a w_0 so that for each $C\mathcal{U}D \in L'(w_0)$ with $D \in L(s_0)$ the decomposition $\{C\mathcal{U}D, D\}$ rather than $\{C\mathcal{U}D, \mathsf{X}(C\mathcal{U}D)\}$ is a subset of $L'(w_0)$. Set $\varphi(s_0) = w_0$.

$i \to i+1$: Assume that we have defined the map up to $\varphi(s_i)$. We define the sets $\mathit{next} := \{C \mid \mathsf{X}C \in L(s_i)\}$ and $\mathit{next'} := \{C \mid \mathsf{X}C \in L'(\varphi(s_i))\}$. We know that $\mathit{next} \subseteq L(s_{i+1})$ since L fulfills (LC), and that for every successor w of $\varphi(s_i)$ $\mathit{next'} \subseteq L'(w)$ (see the (X) rule of PLTL_T). Moreover, because of (b), we have $\mathit{next'} \subseteq \mathit{next}$. Again, since in the tableau there is a successor for each possible decomposition of $\mathit{next'}$, and since $L(s_{i+1})$ must contain at least one decomposition, there must exist a $w_{i+1} \in W$ so that (b) and (c) are fulfilled if we set $\varphi(s_{i+1})$ to w_{i+1}.

Obviously, L' fulfills (PC0-3) and (LC) on $\pi = \varphi(s_0)\varphi(s_1)\ldots$. Suppose now that there exists an eventuality $C\mathcal{U}D \in \mathit{open}(\pi, L')$. Then there exists a state $\varphi(s_i)$ so that $C\mathcal{U}D \in L'(\varphi(s_j))$ and $D \notin L'(\varphi(s_j))$ for all $j \geq i$. However, because of (b) and (c) this would mean that $C\mathcal{U}D$ is in $\mathit{open}(S, L)$ as well, which is a contradiction. Applying the previous lemma 4 concludes the proof of the theorem.

5 Conclusion

We have presented a new one-pass tableau calculus for PLTL which works, as most modal calculi, on trees rather than graphs. The representation is minimal but complete, that is, it can be used directly as the basis for a decision procedure without a second phase. It has inherent advantages compared to previous approaches: 1. Only one branch needs to be considered at any one time. This makes it into a natural candidate for parallelization. 2. A simple linearization of loops allows to actually extract linear models in a canonical way. Having the details of the eventuality checking incorporated in a formal way, the calculus is also a good starting point for theoretical investigations, for instance the verification of pruning techniques. These are certainly simpler to check when the underlying structure is a tree. A decision procedure based on PLTL_T has been implemented and tested and will be publicly available as a part of the Logics Workbench [10] version 1.1.

References

1. M. Browne, E. Clarke, D. Dill, and B. Mishra. Automatic verification of sequential circuits using temporal logic. *IEEE Transactions on Computers*, 35:1035–1044, December 1986.
2. C. Courcoubetis, M. Vardi, P. Wolper, and M. Yannakakis. Memory-efficient algorithms for the verification of temporal properties. In *Formal Methods in System Design*, volume 1, pages 275–288, 1992.
3. M. D'Agostino, D. Gabbay, R. Hähnle, and J. Posegga, editors. *Handbook of Tableau Methods*, chapter Tableau Methods for Modal and Temporal Logics. Kluwer, to appear. (currently available as technical report TR-ARP-15-95, Australian National University (ANU)).
4. E. Emerson. Temporal and modal logic. In J. v. Leeuwen, editor, *Handbook of Theoretical Computer Science. Volume B*, pages 995–1072. Elsevier, 1990.
5. M.J. Fischer and R.L. Ladner. Propositional dynamic logic of regular programs. *Journal of Computer and System Sciences*, 18:194–211, 1979.
6. M. Fitting. *Proof Methods for Modal and Intuitionistic Logics*. Reidel, Dordrecht, 1983.
7. R. Gerth, D. Peled, M. Vardi, and P. Wolper. Simple on-the-fly automatic verification of linear temporal logic. In P. Dembinski and M. Sredniawa, editors, *Protocol Specification Testing and Verification*, volume XV, pages 3–18. Chapman & Hall, 1996.
8. R. Goré. *Cut-free Sequent and Tableau Systems for Propositional Normal Modal Logic*. PhD thesis, Computer Laboratory, University of Cambridge, England, 1992.
9. G. Gough. Decision procedures for temporal logic. Technical Report UMCS-89-10-1, Department of Computer Science, University of Manchester, 1989.
10. A. Heuerding, G. Jäger, S. Schwendimann, and M. Seyfried. Propositional logics on the computer. In P. Baumgartner, R. Hähnle, and J. Posegga, editors, *Theorem Proving with Analytic Tableaux and Related Methods*, LNCS 918, pages 310–323, 1995.
11. A. Heuerding, M. Seyfried, and H. Zimmermann. Efficient loop-check for backward proof search in some non-classical propositional logics. In P. Miglioli, U. Moscato, D. Mundici, and M. Ornaghi, editors, *Tableaux 96*, LNCS 1071, pages 210–225, 1996.
12. Y. Kesten, Z. Manna, H. McGuire, and A.Pnueli. A decision algorithm for full propositional temporal logic. In *Computer Aided Verification*, LNCS 697, pages 4–35. Springer, 1993.
13. Z. Manna and A. Pnueli. *The Temporal Logic of Reactive and Concurrent Systems*. Springer Verlag, 1991.
14. K. L. McMillan. *Symbolic Model Checking*. Kluwer Academic Publishers, 1993.
15. A. Sistla and E. Clarke. The complexity of propositional linear temporal logic. *Journal of the Association for Computing Machinery*, 32(3):733–749, 1985.
16. R. Tarjan. Depth first search and linear graph algorithms. *SIAM Journal Computing*, 1(2):146–160, 1972.
17. P. Wolper. The tableau method for temporal logic: an overview. *Logique et Analyse*, 110-111:119–136, 1985.

Decision Procedures for Intuitionistic Propositional Logic by Program Extraction

Klaus Weich

Mathematisches Institut der Universität München, Theresienstr. 39,
D–80333 München, Germany; tel: +49 89 2394 4417;
email: weich@rz.mathematik.uni-muenchen.de

Abstract. We present two constructive proofs of the decidability of intuitionistic propositional logic by simultaneously constructing either a counter–model or a derivation. From these proofs, we extract two programs which have a sequent as input and return a derivation or a counter–model. The search tree of these algorithms is linearly bounded by the number of connectives of the input. Soundness of these programs follows from giving a correct construction of the derivations, similarly to Hudelmaier's work [7]; completeness from giving a correct construction of the counter–models, inspired by Miglioli, Moscato, and Ornaghi [8].

1 Introduction

Intuitionistic proofs can be considered as programs together with their verification. Consequently intuitionistic logic is a method for developing correct programs. To demonstrate the advantage of this approach, we construct two theorem provers for the propositional part by extracting them from a decidability proof.

Taking up Fitting's [2] completeness proof, Underwood [12] outlined how a decidability proof could be implemented in a formal system like NUPRL. For this, she proved that each set of sequents either contains a provable one or has a Kripke model so that each sequent is refuted at a certain node. The sequents themselves correspond to the nodes of a tableau. The proof is by induction on the number of formulas which can be added to the set of sequents, corresponding to a tableau rule. Underwood's extracted program has a sequent as input and returns a derivation or a counter–model. Her program uses a loop–check and tries to construct bottom–up a repetition–free derivation in a sequent calculus, similarly to the decision procedure already given by Gentzen [3].

To avoid a loop–check, contraction–free sequent calculi were introduced by Hudelmaier [6] and Dyckhoff [1], rediscovering the work of Vorob'ev [13]. The main idea of their completeness proof is that every derivation in a sequent calculus can be transformed so that every left premise of a left rule for implication (L-⊃ in our notation below) is either an axiom or the conclusion of a right rule (R-...). Formalising their proof would be a thankless and hard task. Fortunately, Miglioli, Moscato, and Ornaghi [8] gave an alternative completeness proof of a similar system by constructing a Kripke model. Hudelmaier developed his calculus further and gave an $O(n \log n)$–space algorithm [7].

In Sect. 4, we give a new proof of the decidability of the intuitionistic propositional logic. For this, we show that for every sequent there is a derivation or a counter–model. We simultaneously construct a derivation, taking up Hudelmaier's approach [7], or a counter–model, simplifying Migliogli, Moscato, and Ornaghi's idea [8]. The extracted algorithm has a sequent as input and returns a derivation or a counter–model. The height of the search tree of this algorithm is linearly bounded by the number of logical connectives of the input. Indeed, our algorithm restricted to derivability coincides with Hudelmaier's [7]. In addition, our algorithm terminates faster than the algorithms for refutability presented by Pinto and Dyckhoff [9] and Hudelmaier [4, 5], the search tree of which is only exponentially bounded.

In Sect. 5, we describe how the counter–models constructed during the search can be used to prune the search space.

In Sect. 6, we present an alternative proof of decidability. The algorithm described by this proof is completely different. Whereas the algorithm of Sect. 4 examines the premise A of implications $A \supset B$ in the left–hand side of the sequents, the one of Sect. 6 looks at the right–most conclusion P of the implications $A_0 \supset \ldots \supset A_n \supset P$ and selects this formula only if the right–hand side of the sequent coincides with P, similarly to the search strategy of PROLOG. This algorithm turns out to be faster with sequents which are hard on the algorithm presented in Sect. 4; and vice versa. Thus it might be useful to have several algorithms.

Since our proof is rather simple and elementary, formalising this approach seems promising. Indeed the $\supset$–fragment is formalised in the system MINLOG [10].

2 Notation

We use P, Q to denote atomic formulas and A, B, ... to denote formulas. Formulas are generated by atomic formulas, the falsum $\bot$, implications $A \supset B$, conjunctions $A \wedge B$, and disjunctions $A \vee B$. Negation $\neg A$ is treated as an abbreviation for $A \supset \bot$. Lists of formulas are denoted by Γ, Δ and sequents by $\Gamma \Rightarrow A$.

To avoid some parentheses, we write $A \supset B \supset C$ instead of $A \supset (B \supset C)$, $A \wedge B \supset C$ instead of $(A \wedge B) \supset C$, and $A \vee B \supset C$ instead of $(A \vee B) \supset C$.

2.1 Counter–Models

For an introduction to Kripke models, see e.g. Troelstra and van Dalen [11]. A Kripke model $\mathcal{K}$ is a tripel $(K, \leq, \Vdash)$, where K is a non–empty set of worlds denoted by k. We list the properties we need:

- $k \nVdash \bot$.
- $k \Vdash A \wedge B$ iff $k \Vdash A$ and $k \Vdash B$.
- $k \Vdash A \vee B$ iff $k \Vdash A$ or $k \Vdash B$.
- $k \Vdash A \supset B$ iff $k' \Vdash A$ implies $k' \Vdash B$ for all $k' \geq k$.

– If $k \Vdash A$ and $k' \geq k$ then $k' \Vdash A$ (monotonicity).

We extend the relation $\Vdash$ in a natural way to lists and write $k \Vdash \Gamma$ if $k \Vdash A$ for each $A \in \Gamma$.

Since $k \Vdash A$ cannot be assumed to be decidable, we have to specialise our semantics. A *Kripke tree* is a non–empty, finite, and finitely branching tree, the nodes of which are labelled by finite sets L of atomic formulas. We use $(L, \vec{\mathcal{K}})$ to denote such a node, where $\vec{\mathcal{K}}$ are its successors. A Kripke tree defines a tripel $(K, \leq, \Vdash)$ by taking K as the set of the nodes, $\leq$ the transitive and reflexive closure of the successor relation, and by $(L, \vec{\mathcal{K}}) \Vdash P$ iff $P \in L$. A Kripke tree is a *Kripke tree model* if the associated tripel $(K, \leq, \Vdash)$ is a Kripke model.

For a Kripke model $\mathcal{K} = (K, \leq, \Vdash)$, we write $\mathcal{K} \Vdash A$ if $k \Vdash A$ holds for each $k \in K$. For a Kripke tree model this is, by monotonicity, equivalent to "the root node forces A". Hence "a node $(L, \vec{\mathcal{K}})$ forces A" coincides with "a tree having the root $(L, \vec{\mathcal{K}})$ forces A". Furthermore, every node of a Kripke tree model is a Kripke tree model itself. Thus we identify a node $(L, \vec{\mathcal{K}})$ with a tree having that root.

A Kripke model $\mathcal{K}$ is a *counter–model* to $\Gamma \Rightarrow A$ if $\mathcal{K} \Vdash \Gamma$ and $\mathcal{K} \nVdash A$. In this case we call the sequent $\Gamma \Rightarrow A$ *refutable*.

We will later use the following lemmata to construct a Kripke tree model and to compute $(L, \vec{\mathcal{K}}) \Vdash A \supset B$ inductively.

Lemma 1. *$(L, \vec{\mathcal{K}})$ is a Kripke tree model iff for each $\mathcal{K} \in \vec{\mathcal{K}}$, $\mathcal{K}$ is Kripke tree model, and $\mathcal{K} \Vdash L$.* □

Lemma 2. *Let $(L, \vec{\mathcal{K}})$ be a Kripke tree model. $(L, \vec{\mathcal{K}}) \Vdash A \supset B$ if and only if $(L, \vec{\mathcal{K}}) \Vdash A$ implies $(L, \vec{\mathcal{K}}) \Vdash B$, and $\mathcal{K} \Vdash A \supset B$ for each $\mathcal{K} \in \vec{\mathcal{K}}$.* □

2.2 Derivations

The proofs in the following sections can easily be carried out by any notation for derivations for the intuitionistic logic. Instead of fixing a notation for derivations, we list the neccessary properties by giving some rules in Fig. 1. These rules have to be read constructively: for each instance of the rule, we can compute a derivation of the conclusion from derivations of all premises. In that case, we say a rule *preserves derivability*.

3 Invertibility

In this section, we will summarise some simple properties of propositional intuitionistic logic so that we can focus on the essence in the next section.

A rule is called *invertible* iff for each instance of the rule, the derivability of the conclusion implies that of all premises. It is well–known that most of the rules are invertible. Although it is not hard to prove this (by induction on derivations), we will not use invertibility. Instead, we will work with a semantic notation. We

$$\frac{}{\Gamma, P, \Delta \Rightarrow P}\text{ Ax} \qquad \frac{}{\Gamma, \bot, \Delta \Rightarrow A}\text{ Efq}$$

$$\frac{A, \Gamma \Rightarrow B}{\Gamma \Rightarrow A \supset B}\text{ R-}\supset \qquad \frac{\Gamma, A \supset B, \Delta \Rightarrow A \quad \Gamma, B, \Delta \Rightarrow C}{\Gamma, A \supset B, \Delta \Rightarrow C}\text{ L-}\supset$$

$$\frac{\Gamma \Rightarrow A \quad \Gamma \Rightarrow B}{\Gamma \Rightarrow A \wedge B}\text{ R-}\wedge \qquad \frac{\Gamma, A, B, \Delta \Rightarrow C}{\Gamma, A \wedge B, \Delta \Rightarrow C}\text{ L-}\wedge$$

$$\frac{\Gamma \Rightarrow A}{\Gamma \Rightarrow A \vee B}\text{ R-}\vee_l \quad \frac{\Gamma \Rightarrow B}{\Gamma \Rightarrow A \vee B}\text{ R-}\vee_r \qquad \frac{\Gamma, A, \Delta \Rightarrow C \quad \Gamma, B, \Delta \Rightarrow C}{\Gamma, A \vee B, \Delta \Rightarrow C}\text{ L-}\vee$$

$$\frac{\Gamma, \Delta \Rightarrow A \quad \Gamma, A, \Delta \Rightarrow B}{\Gamma, \Delta \Rightarrow B}\text{ cut} \qquad \frac{\Gamma, \Delta \Rightarrow B}{\Gamma, A, \Delta \Rightarrow B}\text{ weakening}$$

Fig. 1.

say a rule *preserves refutability* if for each instance, given a counter–model to one premise, we can compute a counter–model to the conclusion.

Invertibility and preservation of refutability have a close connection. We suppose soundness for a moment. As soon as we have proved completeness, invertibility will imply preservation of refutability. As soon as we have proved decidability, preservation of refutability will entail invertibility.

Each rule we introduce and which preserves refutability turns out to satisfy the following stronger property, which is very easy to verify.

Definition 3. *A rule* preserves counter–models *if for each instance, a counter–model to at least one premise is also a counter–model to the conclusion.*

Lemma 4. *The rules R-$\supset$, R-$\wedge$, L-$\wedge$, and L-$\vee$ preserve counter–models.* □

We will eliminate formulas of the form $\bot \supset B$ on the left–hand side of sequents. For this, we introduce the following rule:

$$\frac{\Gamma, \Delta \Rightarrow A}{\Gamma, \bot \supset B, \Delta \Rightarrow A}\text{ L-}\bot\supset$$

Lemma 5. *L-$\bot\supset$ preserves derivability and counter–models.* □

Furthermore, proofs become slightly shorter if we replace $B_0 \wedge B_1 \supset C$ by $B_0 \supset B_1 \supset C$ and $B_0 \vee B_1 \supset C$ by $B_0 \supset C$ and $B_1 \supset C$. For this, we take the following rules already introduced by Vorob'ev [13].

$$\frac{\Gamma, B_0 \supset B_1 \supset C, \Delta \Rightarrow A}{\Gamma, B_0 \wedge B_1 \supset C, \Delta \Rightarrow A}\text{ L-}\wedge\supset \qquad \frac{\Gamma, B_0 \supset C, B_1 \supset C, \Delta \Rightarrow A}{\Gamma, B_0 \vee B_1 \supset C, \Delta \Rightarrow A}\text{ L-}\vee\supset$$

Lemma 6. L-∧⊃ *and* L-∨⊃ *preserve derivability and counter–models.* □

The following linear degree w will be used in the proof of decidability.

Definition 7. *For formulas we define*

$w(P) := w(\bot) := 0$
$w(A \supset B) := 1 + w(A) + w(B)$
$w(A \wedge B) := 2 + w(A) + w(B)$
$w(A \vee B) := 3 + w(A) + w(B).$

For lists of formulas we define $w(B_1, \ldots, B_n) := w(B_1) + \ldots + w(B_n)$ *and for sequents* $w(\Gamma \Rightarrow A) := w(\Gamma) + w(A)$.

Each instance of any premise of the rules R-⊃, R-∧, R-∨, L-∨, L-⊥⊃, and L-∧⊃ has a smaller w–degree than the corresponding instance of the conclusion. This does not hold for L-∨⊃ if the instance of C is a composed formula. To avoid this, we use the following rule:

$$\frac{\Gamma, B \supset P, P \supset C, \Delta \Rightarrow A}{\Gamma, B \supset C, \Delta \Rightarrow A} \text{ L-S , provided } P \text{ does not occur in the conclusion.}$$

Lemma 8. L-S *preserves derivability and counter–models.*

Proof. For derivability see Hudelmaier [7]. Obviously, L-S preserves counter–models. □

Now combining L-∨⊃ and L-S

$$\frac{\dfrac{\Gamma, B_0 \supset P, B_1 \supset P, P \supset C, \Delta \Rightarrow A}{\Gamma, B_0 \vee B_1 \supset P, P \supset C, \Delta \Rightarrow A} \text{L-∨⊃}}{\Gamma, B_0 \vee B_1 \supset C, \Delta \Rightarrow A} \text{L-S}$$

we verify that each instance of the upper sequent has a smaller w–degree than the lower sequent.

At first sight, the rule L-S appears to increase the non–determinism by replacing one formula $B \supset C$ by two formulas $B \supset P$ and $P \supset C$, but, as it turns out, the second formula will only be considered if B is derived.

In general, the rules L-⊃ and R-∨ are neither invertible nor do they preserve refutability or counter–models. L-⊃ is known to be semi–invertible, i.e. for each instance, the derivability of the conclusion $\Gamma, A \supset B, \Delta \Rightarrow C$ implies that of the right premise $\Gamma, B, \Delta \Rightarrow C$. We will not use this property; instead, we will use the following property.

Lemma 9. *If* $\mathcal{K}$ *is a counter–model to* $\Gamma, B, \Delta \Rightarrow C$, *then also to* $\Gamma, A \supset B, \Delta \Rightarrow C$. □

In the case that the premise of an implication is atomic and is in the left–hand side already, we get the following rule(cf. Hudelmaier [6, 7] and Dyckhoff [1]).

$$\frac{\Gamma, C, \Delta \Rightarrow A}{\Gamma, P \supset C, \Delta \Rightarrow A} \text{ L-P⊃, provided } P \in \Gamma, \Delta$$

Corollary 10. L-$P\supset$ *preserves derivability and counter–models.*

Proof. By L-$\supset$ and Ax, we get that L-$P\supset$ preserves derivability. Lemma 9 says that L-$P\supset$ preserves counter–models. □

The above section can be summed up in the next lemma applying the following definition, which is an extension of Dyckhoff's [1].

Definition 11. *A sequent $\Gamma \Rightarrow A$ is called* irreducible *if Γ contains only atomic formulas[1], or formulas of the form $P \supset B$ where P is not in Γ, or else formulas of the form $(B_0 \supset B_1) \supset C$. Furthermore, we require that A is either a disjunction, or falsum, or else an atomic formula not in Γ. A sequent is called* reducible *if it is not irreducible.*

Thus a sequent $\Gamma \Rightarrow A$ is irreducible iff $\Gamma \Rightarrow A$ is neither an instance of a conclusion of one of the counter–models preserving rules introduced so far(L-S only if B is a disjunction and C is a composed formula), nor an instance of a conclusion of Ax and Efq. We will sometimes denote a sequent by $\mathcal{S}$.

Lemma 12. *For each reducible $\Gamma \Rightarrow A$ one can find $\mathcal{S}_1, \ldots, \mathcal{S}_n$ such that*
  (a) $w(\mathcal{S}_i) < w(\Gamma \Rightarrow A)$ for each i.
  (b) If each $\mathcal{S}_i$ is derivable, then so is $\Gamma \Rightarrow A$.
  (c) If $\mathcal{K}$ is a counter–model to at least one $\mathcal{S}_i$ then also to $\Gamma \Rightarrow A$.

Proof. Case $\Gamma = \Gamma_0, B_0 \wedge B_1, \Gamma_2$. Let $\mathcal{S}_1, \mathcal{S}_2$ be the left respectively right premise of L-$\wedge$ where $B_0 \wedge B_1$ is the principal formula. (a) is obvious. (b) holds since L-$\wedge$ preserves derivability, and (c) holds because L-$\wedge$ preserves counter–models.

Case $\Gamma = \Gamma_0, (B_0 \vee B_1) \supset C, \Gamma_2$. If C is a composed formula, then we proceed by combining L-$\vee\supset$ and L-S as described above. If C is atomic or $\bot$, we obtain our statement by L-$\vee\supset$.

The *remaining cases* follow similarly using the corresponding rules, all of which preserve counter–models. □

By a similar proof, we obtain a variant of the previous lemma, which will be motivated in Theorem 16.

Lemma 13. *Let A be a disjunction and B an arbitrary formula. For each reducible $\Gamma_1, \Gamma_2 \Rightarrow A$ one can find $\mathcal{S}_1, \ldots, \mathcal{S}_n$ such that*
  (a) The left–hand side of each $\mathcal{S}_i$ has a smaller w–degree than $\Gamma_1, A \supset B, \Gamma_2$.
  (b) If each $\mathcal{S}_i$ is derivable, then so is $\Gamma_1, A \supset B, \Gamma_2 \Rightarrow A$.
  (c) If $\mathcal{K}$ is a counter–model to at least one $\mathcal{S}_i$ then also to $\Gamma_1, A \supset B, \Gamma_2 \Rightarrow A$.
  (d) Each $\mathcal{S}_i$ is of the form $\Gamma_{1,i}, A \supset B, \Gamma_{2,i} \Rightarrow A$. □

---

[1] Note that $\bot$ is by definition not atomic.

4 Forward chaining

How can we deal with the implication in the context $\Gamma, A \supset B, \Delta \Rightarrow C$? One tactic is to search for a derivation of $\Gamma, A \supset B, \Delta \Rightarrow A$ and then to go on with $\Gamma, B, \Delta \Rightarrow C$. This can be called *forward chaining*. The problem is how to enforce termination without destroying completeness.

The first premise $\Gamma, A \supset B, \Delta \Rightarrow A$ has a particular form: the right–hand side A occurs as a premise of an implication on the left–hand side. If A is a composed formula, then, as observed by Hudelmaier, the problem can be reduced to a smaller one of the same form. For this, he introduced the rules $GI2{\to}$, $GI2{\land}$, and $GI2{\lor}$ in [7]. We rename $GI2{\to}$ to L2-$\supset\supset$ and $GI2{\land}$ to L2-$\land\supset$. In $GI2\lor$, the rule L-S is incorporated. We separate L-S and call the remaining part L2-$\lor\supset$.

$$\frac{A_0, \Gamma, A_1 \supset B, \Delta \Rightarrow A_1}{\Gamma, (A_0 \supset A_1) \supset B, \Delta \Rightarrow A_0 \supset A_1} \text{ L2-}\supset\supset$$

$$\frac{\Gamma, A_0 \supset B, \Delta \Rightarrow A_0 \quad \Gamma, A_1 \supset B, \Delta \Rightarrow A_1}{\Gamma, A_0 \land A_1 \supset B, \Delta \Rightarrow A_0 \land A_1} \text{ L2-}\land\supset$$

$$\frac{\Gamma, A_0 \supset B, A_1 \supset B, \Delta \Rightarrow A_i}{\Gamma, A_0 \lor A_1 \supset B, \Delta \Rightarrow A_0 \lor A_1} \text{ L2-}\lor\supset_i, \text{ where } i \in \{0,1\}$$

Lemma 14. *L2-$\supset\supset$ and L2-$\land\supset$ preserves derivability and counter–models, L2-$\lor\supset$ preserves derivability (only).*

Proof. To see that L2-$\supset\supset$ preserves derivability, use cut and a derivation of $A_0, (A_0 \supset A_1) \supset B \Rightarrow A_1 \supset B$. To see that L2-$\land\supset$ preserves derivability, use cut and a derivation of $(A_0 \supset B) \supset A_0, (A_1 \supset B) \supset A_1, A_0 \land A_1 \supset B \Rightarrow A_0 \land A_1$. To prove that L2-$\lor\supset$ preserves derivability, combine R-$\lor$ and L-$\lor\supset$.

The proof that L2-$\supset\supset$ and L2-$\land\supset$ preserve counter–models is easy and is therefore left to the reader. □

Note that L2-$\lor\supset$ is neither invertible nor refutability–preserving.

How can we deal with the implications $P \supset A$ where P is atomic? If P is in the left–hand side, we apply L-$P\supset$. What do we have to do if P is not in the left–hand side? Then we need not consider these formulas, as the following lemma shows. This was already observed by Vorob'ev [13], but we give a much simpler proof.

Lemma 15. *Let $\Gamma \Rightarrow A$ be an irreducible sequent. Let $\vec{\mathcal{K}}$ be Kripke tree models so that $\mathcal{K}' \Vdash \Gamma$ for each $\mathcal{K}' \in \vec{\mathcal{K}}$, and for each formula $(B_0 \supset B_1) \supset C$ in Γ there is a counter–model in $\vec{\mathcal{K}}$ to $\Gamma \Rightarrow B_0 \supset B_1$. Furthermore, if A is a disjunction $A_0 \lor A_1$, then suppose that there are counter–models $\mathcal{K}_0$, $\mathcal{K}_1$ in $\vec{\mathcal{K}}$ to $\Gamma \Rightarrow A_0$, $\Gamma \Rightarrow A_1$ respectively. Then $\mathcal{K} := (\{P : P \in \Gamma\}, \vec{\mathcal{K}})$ is a counter–model to $\Gamma \Rightarrow A$.*

Proof. As $\mathcal{K}' \Vdash \Gamma$ implies $\mathcal{K}' \Vdash \{P : P \in \Gamma\}$, Lemma 1 provides that $\mathcal{K}$ is a Kripke tree model.

Let $D \in \Gamma$. We have to show $\mathcal{K} \Vdash D$. Since $\Gamma \Rightarrow A$ is irreducible, D is either an atom, or an implication $B \supset C$ where B is either $B_0 \supset B_1$, or else an atom not in Γ. If D is an atomic formula, then $D \in \{P : P \in \Gamma\}$ and hence $\mathcal{K} \Vdash D$. If $D = B \supset C$, then by Lemma 2 and because of $\mathcal{K}' \Vdash \Gamma$, it is sufficient to show that $\mathcal{K} \Vdash B$ implies $\mathcal{K} \Vdash C$. We show $\mathcal{K} \nVdash B$. If B is an atom not in Γ, then $B \notin \{P : P \in \Gamma\}$ and hence $\mathcal{K} \nVdash B$. Otherwise $B = B_0 \supset B_1$ and there is a $\mathcal{K}'$ in $\vec{\mathcal{K}}$ such that $\mathcal{K}' \nVdash B$, implying $\mathcal{K} \nVdash B$ by monotonicity.

It remains to show $\mathcal{K} \nVdash A$. Since $\Gamma \Rightarrow A$ is irreducible, A is either an atom not in Γ, or $\bot$, or an disjunction $A_0 \vee A_1$. In the first case we have $A \notin \{P : P \in \Gamma\}$ and hence $\mathcal{K} \nVdash A$. If $A = \bot$, then $\mathcal{K} \nVdash A$ by definition. Otherwise $A = A_0 \vee A_1$. Here there are $\mathcal{K}_0, \mathcal{K}_1 \in \vec{\mathcal{K}}$ where $\mathcal{K}_i \nVdash A_i$. By monotonicity we get $\mathcal{K} \nVdash A_0$ and $\mathcal{K} \nVdash A_1$. Hence $\mathcal{K} \nVdash A_0 \vee A_1$ by definition. □

Particularly, if $\Gamma \Rightarrow P/\bot$ is irreducible and Γ does not contain formulas of the form $(B_0 \supset B_1) \supset C$, then $(\{P : P \in \Gamma\}, \varepsilon)$ is a counter-model to $\Gamma \Rightarrow P/\bot$. The previous lemma is the computational content of the refutation rules

$$\frac{\Gamma \nRightarrow B_1 \supset C_1 \quad \cdots \quad \Gamma \nRightarrow B_n \supset C_n}{\Gamma \nRightarrow P/\bot} \; n \geq 0$$

$$\frac{\Gamma \nRightarrow B_1 \supset C_1 \quad \cdots \quad \Gamma \nRightarrow B_n \supset C_n \quad \Gamma \nRightarrow A_0 \quad \Gamma \nRightarrow A_1}{\Gamma \nRightarrow A_0 \vee A_1} \; n \geq 0$$

provided that $\Gamma \Rightarrow P$, $\Gamma \Rightarrow \bot$, and $\Gamma \Rightarrow A_0 \vee A_1$ respectively, are irreducible, and where $(B_1 \supset C_1) \supset D_1, \ldots, (B_n \supset C_n) \supset D_n$ are all nested implications in Γ. Separating L2-$\supset\supset$ from the rule (11) given by Pinto and Dyckhoff [9], our rules coincide with a non-multi-succedent version of theirs. While Pinto and Dyckhoff work with the rules themselves, we work with the computational content of the rules. Thus we do not have to construct the counter-models in a second step, unlike the approach of Pinto and Dyckhoff.

Now we are ready to prove the decidability of the intuitionistic propositional logic. This will be done in part (i) of the following theorem. For the proof, we also need part (ii). Note that in (ii) the right-hand side of the sequent does not contribute to the w-degree.

Theorem 16. *Let n be any natural number.*
(i) Each $\Gamma \Rightarrow A$ with $w(\Gamma, A) \leq n$ has either a derivation or a counter-model.
(ii) Each $\Gamma_1, A \supset B, \Gamma_2 \Rightarrow A$ with $w(\Gamma_1, A \supset B, \Gamma_2) \leq n$ has either a derivation or a counter-model.

Proof. We proceed by simultaneous, progressive induction on n, i.e. let n be given and we are allowed to use the induction hypothesis for (i) and (ii) for all $m < n$. First we prove (i), then we prove (ii). We write IH for "induction hypothesis".

As for (i), we proceed by case analysis on whether (1) the sequent is reducible, or (2) the sequent is an instance of Efq or Ax, or (3) the IH(ii) on the left premise of an instance of a combination of L2-⊃⊃ and L-⊃ yields a derivation for at least one formula $(B_0 \supset B_1) \supset C$ of Γ, or (4) A is a disjunction and the IH(i) on at least one premise of R-∨ provides a derivation, or else the remaining case.

Case 1. $\Gamma \Rightarrow A$ is reducible. Let $S_1, \ldots, S_n$ be the sequents obtained from Lemma 12. The IH(i) on each S_i yields either a derivation for all S_i or a counter–model to at least one S_i. In the first case, we get a derivation of $\Gamma \Rightarrow A$ by (b) of Lemma 12. In the second case, we have already got a counter–model to $\Gamma \Rightarrow A$ according to (c) of Lemma 12.

Case 2. $\bot \in \Gamma$ or A is atomic and in Γ. Here we get a derivation by Efq or Ax respectively.

Case 3. For a partition $\Gamma = \Gamma_1, (B_0 \supset B_1) \supset C, \Gamma_2$ the IH(ii) on $B_0, \Gamma_1, B_1 \supset C, \Gamma_2 \Rightarrow B_1$ yields a derivation. Here we apply L2-⊃⊃ to obtain a derivation of $\Gamma_1, (B_0 \supset B_1) \supset C, \Gamma_2 \Rightarrow B_0 \supset B_1$. Moreover, the IH(i) on $\Gamma_1, C, \Gamma_2 \Rightarrow A$ yields a derivation or a counter–model. In the first case, L-⊃ provides a derivation of $\Gamma \Rightarrow A$. In the second case, we have already got a counter–model to that sequent by Lemma 9.

Case 4. $A = A_0 \vee A_1$ and the IH(i) on $\Gamma \Rightarrow A_0$ or on $\Gamma \Rightarrow A_1$ yields a derivation. Here, R-∨ provides a derivation of $\Gamma \Rightarrow A$.

Remaining case. For every partition $\Gamma = \Gamma_1, (B_0 \supset B_1) \supset C, \Gamma_2$ the IH(ii) yields a counter–model to $B_0, \Gamma_1, B_1 \supset C, \Gamma_2 \Rightarrow B_1$ and if $A = A_0 \vee A_1$, the IH(i) on both $\Gamma \Rightarrow A_0$ and $\Gamma \Rightarrow A_1$ also yields two counter–models. Let $\vec{\mathcal{K}}$ be all these counter–models. Since L2-⊃⊃ preserves counter–models, counter–models to $B_0, \Gamma_1, B_1 \supset C, \Gamma_2 \Rightarrow B_1$ are also counter–models to $\Gamma \Rightarrow B_0 \supset B_1$. Hence, by Lemma 15, $(\{P : P \in \Gamma\}, \vec{\mathcal{K}})$ is a counter–model to $\Gamma \Rightarrow A$. This completes the proof of (i).

As for (ii) we proceed by case analysis on the form of A.

Case A atomic or $\bot$. As we have already proved (i) for n, we can apply (i) to the sequent $\Gamma_1, A \supset B, \Gamma_2 \Rightarrow A$.

Case $A = A_0 \supset A_1$. The IH(ii) on $A_0, \Gamma_1, A_1 \supset B, \Gamma_2 \Rightarrow A_1$ yields a derivation or a counter–model. Since L2-⊃⊃ preserves derivability and counter–models, we obtain a derivation of $\Gamma_1, A_0 \wedge A_1 \supset B, \Gamma_2 \Rightarrow A_0 \wedge A_1$ or we have already got a counter–model to that sequent.

Case $A = A_0 \wedge A_1$. Similarly using L2-∧⊃.

Case $A = A_0 \vee A_1$. Here we have to consider similar cases as in (i).

Subcase 1. $\Gamma_1, \Gamma_2 \Rightarrow A$ is not irreducible. Let $S_1, \ldots, S_n$ be the sequents obtained from Lemma 13. The IH(ii) on each S_i yields either a derivation for all S_i or a counter–model to at least one S_i. We proceed as in (i) case 1.

Subcase 2. $\bot \in \Gamma$ or A is atomic and in Γ. Analogously to (i) case 2.

Subcase 3. For a formula $(C_0 \supset C_1) \supset D$ of Γ_1, Γ_2, the IH(ii) yields a derivation as in (i) case 3. We proceeds as in (i) case 3, replacing "IH(i)" by "IH(ii)".

Subcase 4a. B is atomic or $\bot$ and the IH(ii) on $\Gamma_1, A_0 \supset B, A_1 \supset B, \Gamma_2 \Rightarrow A_0$ or on $\Gamma_1, A_0 \supset B, A_1 \supset B, \Gamma_2 \Rightarrow A_1$ yields a derivation. Then L2-∨⊃ provides a derivation of the sequent under consideration.

Subcase 4b. B is a composed formula and the IH(ii) on $\Gamma_1, A_0 \supset P, A_1 \supset P, P \supset B, \Gamma_2 \Rightarrow A_0$ or on $\Gamma_1, A_0 \supset P, A_1 \supset P, P \supset B, \Gamma_2 \Rightarrow A_1$ where P is a new atomic formula yields a derivation. Using L2-∨⊃ and L-S provides a derivation.

In the *remaining subcase* we construct a counter–model as in (i), again replacing "IH(i)" by "IH(ii)". □

This proof describes an algorithm returning a derivation or a counter–model for a sequent $\Gamma \Rightarrow A$. The algorithm consists of two parts, say search(i) and search(ii). After applying bottom–up all rules preserving counter–models, if the new sequent is not an instance of the conclusion of the rules Ax or Efq, search(i) picks up all the formulas $(B_0 \supset B_1) \supset C$ one after the other and applies search(ii) to $B_0, \Gamma_1, B_1 \supset C, \Gamma_2 \Rightarrow B_1$ where $\Gamma = \Gamma_1, (B_0 \supset B_1) \supset C, \Gamma_2$. If a derivation is returned for one of these sequents, search(i) is called on $\Gamma_1, C, \Gamma_2 \Rightarrow A$. Otherwise, search(i) tries to apply R-∨; if this does not succeed either, a counter–model is returned.

Search(ii) applies L2-⊃⊃ and L2-∧⊃ until A is an atom or a disjunction $A_0 \vee A_1$. In the first case, search(i) is applied. In the second case, all nested implications are picked up as in search(i), and if this fails, L2-∨⊃ is tested (if B is a composed formula, only in combination with L-S). If this does not succeed either, a counter–model is returned.

Immediately, we see that the number of recursions is bounded by twice the w–degree of the input sequent. If we do not count the call of search(i) in search(ii) on the identical sequent, the number of recursions is even bounded by the w–degree of the input. Hence we obtain the following estimate.

Corollary 17. *If a sequent $\Gamma \Rightarrow A$ is refutable, then the algorithm described above returns a counter–model with height less than $w(\Gamma \Rightarrow A) + 1$.* □

While Hudelmaier [7], and Pinto and Dyckhoff [9] essentially proved completeness and termination for their calculi LF, LJT* respectively, allowing one to extract a decision algorithm for intuitionistic propositional logic, the present work starts off with a proof of decidability from which one can read off a decision procedure. However, when dropping the task of producing counter–models, this decision procedure coincides with that for LF. Furthermore, if we proved decidability using the computational contents of the rules given by Pinto and Dyckhoff, we would obtain an algorithm similar to that for LJT*, when dropping counter–models, and similar to CRIP, when dropping derivations.

5 Pruning the search tree

In contrast to Underwood's [12] approach, where the counter–model is constructed after the whole search fails, the present approach constructs local counter–models during the search. These local counter–models can be used to reduce the non–determinism. For example we consider (cf. Fig. 2) an irreducible sequent $(B_0 \supset B_1) \supset B_2, \Gamma, (D_0 \supset D_1) \supset D_2 \Rightarrow A$. To derive that sequent, we select $(B_0 \supset B_1) \supset B_2$ and search for a derivation of $(B_0 \supset B_1) \supset B_2, \Gamma, (D_0 \supset D_1) \supset D_2 \Rightarrow$

$B_0 \supset B_1$. Now L2-⊃⊃ reduces this to the search for $B_0, B_1 \supset B_2, \Gamma, (D_0 \supset D_1) \supset D_2 \Rightarrow B_1$. For simplicity, we assume that B_1 is atomic and that the sequent is irreducible. Furthermore, we assume that selecting each formula $(C_0 \supset C_1) \supset C_2$ of Γ yields a counter–model

$$\mathcal{K}_C \text{ to } B_0, B_1 \supset B_2, \Gamma, (D_0 \supset D_1) \supset D_2 \Rightarrow C_0 \supset C_1.$$

Finally we select $(D_0 \supset D_1) \supset D_2$ and apply L2-⊃⊃ bottom–up.

$$\cfrac{\cfrac{D_0, B_0, B_1 \supset B_2, \Gamma, D_1 \supset D_2 \Rightarrow D_1 \quad B_0, B_1 \supset B_2, \Gamma, D_2 \Rightarrow B_1}{\cfrac{B_0, B_1 \supset B_2, \Gamma, (D_0 \supset D_1) \supset D_2 \Rightarrow B_1}{} \quad B_2, \Gamma, D^\dagger \Rightarrow A}}{(B_0 \supset B_1) \supset B_2, \Gamma, (D_0 \supset D_1) \supset D_2 \Rightarrow A}$$

$\dagger \ D = (D_0 \supset D_1) \supset D_2$

Fig. 2.

We have to search for the sequents

(1a) $D_0, B_0, B_1 \supset B_2, \Gamma, D_1 \supset D_2 \Rightarrow D_1$,
(2a) $B_0, B_1 \supset B_2, \Gamma, D_2 \Rightarrow B_1$, and
(3a) $B_2, \Gamma, (D_0 \supset D_1) \supset D_2 \Rightarrow A$.

We assume that these sequents are irreducible. For the search we have to select each formula $(C_0 \supset C_1) \supset C_2$ of Γ again, unless we have a counter–model to

(1b) $B_0, B_1 \supset B_2, \Gamma, D_0, D_1 \supset D_2 \Rightarrow C_0 \supset C_1$,
(2b) $B_0, B_1 \supset B_2, \Gamma, D_2 \Rightarrow C_0 \supset C_1$,
(3b) $B_2, \Gamma, (D_0 \supset D_1) \supset D_2 \Rightarrow C_0 \supset C_1$, respectively.

Possibly $\mathcal{K}_C$ is a counter–model to one of these sequents; we only have to check if

(1c) $\mathcal{K}_C \Vdash D_0$,[2]
(2c) $\mathcal{K}_C \Vdash D_2$,
(3c) $\mathcal{K}_C \Vdash B_2$, respectively.

If we do not obtain a derivation of (1a) or (2a), we have to select each formula $(C_0 \supset C_1) \supset C_2$ of $\Gamma, (D_0 \supset D_1) \supset D_2$ in order to search for either a derivation of or a counter–model to $(B_0 \supset B_1) \supset B_2, \Gamma, (D_0 \supset D_1) \supset D_2 \Rightarrow C_0 \supset C_1$. Yet, $\mathcal{K}_C$ are counter–models to those sequents already!

In other steps of the algorithm, we can proceed similarly. In this way, we can prune the search tree. To do this, we have to pass the counter–models up, left, and down, which cannot be done by a sequent calculus.

---

[2] Since $\mathcal{K}_C \Vdash (D_0 \supset D_1) \supset D_2$ implies $\mathcal{K}_C \Vdash D_1 \supset D_2$.

Of course, in the worst case, we have to consider the whole tree in order to verify that a Kripke tree model forces a formula. Thus pruning the search tree in this way does not always reduce the runtime for every kind of pruning. But if the formula contains no $\supset$, we only have to consider the root of a Kripke tree.

Moreover, although an automatic theorem prover can handle most sequents very fast, an automatic theorem prover may exceed an acceptable runtime, since deciding intuitionistic propositional logic is PSPACE–hard. In this case an interactive prover may help the user by indicating which formulas he or she need not select. Here the additional runtime of checking whether a Kripke tree model forces a formula is almost always acceptable.

6 Backward Chaining

We will now give an alternative proof of decidability. The proof is much simpler if we restrict ourselves to the $\supset$-fragment. We will do so now. In this section, formulas are only generated by atomic formulas and implications. By abuse of notation, we use $\vec{A} \supset B$ to denote $A_1 \supset \ldots \supset A_n \supset B$, where the list $\vec{A}$ is possibly empty; in this case, we identify $\vec{A} \supset B$ with B. In the $\supset$-fragment, every formula B has the form $(\vec{A}_1 \supset Q_1) \supset \ldots \supset (\vec{A}_k \supset Q_k) \supset Q$ for $k \geq 0$. Q is called head of B. Since R-$\supset$ preseres derivability and counter–models, we only have to consider sequents where the right–hand side is atomic.

To derive a sequent $\Gamma \Rightarrow P$, we want to select a formula B from the context Γ only if the head of B is equal to P. We call this *backward chaining*.

We will use a generalisation of L2-$\wedge\supset$ and L2-$\supset\supset$:

$$\frac{\vec{A}_1, \Gamma_1, Q_1 \supset Q, \Gamma_2 \Rightarrow Q_1 \quad \ldots \quad \vec{A}_k, \Gamma_1, Q_k \supset Q, \Gamma_2 \Rightarrow Q_k}{\Gamma_1, (\vec{A}_1 \supset Q_1) \supset \ldots \supset (\vec{A}_k \supset Q_k) \supset Q, \Gamma_2 \Rightarrow Q} \text{ L3-}\supset^*$$

Note that in the case $k > 1$ or $k = 1$ and $\vec{A}_1 \neq \varepsilon$, each premise has a lower w–degree than the conclusion. Also note that, in the $\supset$-fragment, w coincides with the total number of $\supset$'s.

Lemma 18. *L3-$\supset^*$ preserves derivability and counter–models.*

Proof. To prove that L3-$\supset^*$ preserves derivability, we assume derivations of $\vec{A}_i, \Gamma_1, Q_i \supset Q, \Gamma_2 \Rightarrow Q_i$ for each i. By induction on the length of $\vec{A}_i$ and by L2-$\supset\supset$, we obtain derivations of $\Gamma_1, (\vec{A}_i \supset Q_i) \supset Q, \Gamma_2 \Rightarrow \vec{A}_i \supset Q_i$. By induction on k and by L2-$\wedge\supset$, we get a derivation of $\Gamma_1, (\vec{A}_1 \supset Q_1) \wedge \ldots \wedge (\vec{A}_k \supset Q_k) \supset Q, \Gamma_2 \Rightarrow Q$. Since $(\vec{A}_1 \supset Q_1) \supset \ldots \supset (\vec{A}_k \supset Q_k) \supset Q \Rightarrow (\vec{A}_1 \supset Q_1) \wedge \ldots \wedge (\vec{A}_k \supset Q_k) \supset Q$ is derivable, cut yields a derivation of $\Gamma_1, (\vec{A}_1 \supset Q_1) \supset \ldots \supset (\vec{A}_k \supset Q_k) \supset Q, \Gamma_2 \Rightarrow Q$.

To prove that L-$\supset^*$ preserves counter–models, one either proceeds in a similar way, or one proves it directly. □

Definition 19. *A formula B is a* properly nested implication *if B is of the form $(\vec{A}_1 \supset Q_1) \supset \ldots \supset (\vec{A}_k \supset Q_k) \supset Q$ where $k > 1$, or $k = 1$ and $\vec{A}_1 \neq \varepsilon$.*

If B is not a properly nested implication, i.e. $B = Q_1 \supset Q$, we do not reach our goal completely. But we can deal with these formulas by means of a simple loop–check.

Definition 20. *A* chain *in Γ from P_1 to P_n is a list of implications $P_1 \supset P_2, P_2 \supset P_3, \ldots, P_{n-1} \supset P_n$ where each formula $P_i \supset P_{i+1}$ is in Γ. The empty list is a chain in Γ from P to P, for every P.*

Lemma 21. *Suppose there is a chain in Γ from P_1 to P_n. If $\Gamma \Rightarrow P_1$ is derivable, then so is $\Gamma \Rightarrow P_n$.* □

Definition 22. *An atomic formula P_1 is called* significant *for $\Gamma \Rightarrow P$ if there is a chain from P_1 to P in Γ.*

Lemma 23. *Let $P_1 \supset P_2 \in \Gamma$. If P_2 is significant for $\Gamma \Rightarrow P_n$, then so is P_1.* □

Now we give an alternative construction of a counter–model.

Lemma 24. *Let a sequent $\Gamma \Rightarrow P$ and a list of Kripke tree models $\vec{\mathcal{K}}$ be given. Let L be the set of all atomic formulas significant for $\Gamma \Rightarrow P$. Suppose $L \cap \Gamma = \emptyset$ and $\mathcal{K}' \Vdash \Gamma$ for each $\mathcal{K}'$ in $\vec{\mathcal{K}}$. Furthermore, suppose that for each properly nested implication $(\vec{A}_1 \supset Q_1) \supset \ldots \supset (\vec{A}_k \supset Q_k) \supset Q$ in Γ where Q is significant for $\Gamma \Rightarrow P$, there is a $\mathcal{K}' \in \vec{\mathcal{K}}$ such that $\mathcal{K}'$ is a counter–model to $\Gamma \Rightarrow Q$. Then $\mathcal{K} := (\bigcap \pi_{\text{left}}(\vec{\mathcal{K}}) - L, \vec{\mathcal{K}})$ is a counter–model to $\Gamma \Rightarrow P$, where $\bigcap \pi_{\text{left}}(\vec{\mathcal{K}})$ denotes the intersection of the labels of the roots of $\vec{\mathcal{K}}$.*[3]

Proof. First note that $\mathcal{K}$ is a Kripke tree model by Lemma 1. Furthermore, we have $\mathcal{K} \not\Vdash P$, since $P \in L$. It remains to show $\mathcal{K} \Vdash \Gamma$. Assume $B \in \Gamma$.

Case B is atomic. By assumption $B \notin L$ and $\mathcal{K}' \Vdash B$, i.e. by definition $B \in \pi_{\text{left}}(\mathcal{K}')$ for each $\mathcal{K}'$. Hence $B \in \bigcap \pi_{\text{left}}(\vec{\mathcal{K}}) - L$.

Case $B = P_1 \supset P_2$. Due to $\mathcal{K}' \Vdash \Gamma$ for each $\mathcal{K}'$ and Lemma 2, we only have to show that $\mathcal{K} \Vdash P_1$ implies $\mathcal{K} \Vdash P_2$. Assume $\mathcal{K} \Vdash P_1$. Then by definition $P_1 \in \bigcap \pi_{\text{left}}(\vec{\mathcal{K}})$ and $P_1 \notin L$. The former is equivalent to $\mathcal{K}' \Vdash P_1$ for each $\mathcal{K}'$. By assumption $\mathcal{K}' \Vdash \Gamma$, particularly $\mathcal{K}' \Vdash P_1 \supset P_2$, it follows that $\mathcal{K}' \Vdash P_2$. Hence $P_2 \in \bigcap \pi_{\text{left}}(\vec{\mathcal{K}})$. The latter is equivalent to P_1 non–significant. By Lemma 23, P_2 is non–significant as well, i.e. $P_2 \notin L$.

Case $B = (\vec{A}_1 \supset Q_1) \supset \ldots \supset (\vec{A}_k \supset Q_k) \supset Q$ is a properly nested implication: By Lemma 2 we only have to show that $\mathcal{K} \Vdash \vec{A}_i \supset Q_i$ for each i implies $\mathcal{K} \Vdash Q$.

Subcase Q is significant. Then there is a $\mathcal{K}'$ so that $\mathcal{K}'$ is a counter–model to $\Gamma \Rightarrow Q$. Since $\mathcal{K}' \Vdash B$ and $\mathcal{K}' \Vdash \vec{A}_i \supset Q_i$ for each i using monotonicity, we get $\mathcal{K}' \Vdash Q$. This yields absurdity by $\mathcal{K}' \not\Vdash Q$ and hence $\mathcal{K} \not\Vdash \vec{A}_i \supset Q_i$.

Subcase Q is not significant, i.e. $Q \notin L$. Assume that $\mathcal{K} \Vdash \vec{A}_i \supset Q_i$ for each i. By monotonicity $\mathcal{K}' \Vdash \vec{A}_i \supset Q_i$ for each $\mathcal{K}'$ and for each i. By the assumption $\mathcal{K}' \Vdash \Gamma$, particularly $\mathcal{K}' \Vdash B$, it follows $\mathcal{K}' \Vdash Q$. Hence $Q \in \bigcap \pi_{\text{left}}(\vec{\mathcal{K}}) - L$, i.e. $\mathcal{K} \Vdash Q$ by definition. □

---

[3] To avoid running into infinite sets, we interpret $\bigcap \pi_{\text{left}}(\varepsilon)$ as the set of all atomic formulas occurring as sub–formulas in $\Gamma \Rightarrow A$. Here a more elegant approach would be to work with $(L, \vec{\mathcal{K}}) \Vdash P$ iff $P \notin L$ instead.

In the previous section, the labelling of the nodes was minimal in the sense that each label has to contain at least all atomic formulas in Γ.

By monotonicity the label must be included in $\bigcap \pi_{\text{left}}(\vec{\mathcal{K}})$. Furthermore, the label must not contain any significant atomic formula. Hence, now the labelling is maximal. The advantage of this labelling is that the trees are smaller, since there are fewer possibilities to enlarge the label. To see this, note that a Kripke tree model with each node having identical labels is equivalent to the leaf with the same label; hence enlarging the label is essential.

Now we give the alternative proof of decidability.

Theorem 25. *Each sequent $\Gamma \Rightarrow P$ has a derivation or a counter–model.*

Proof. We proceed by progressive induction on $w(\Gamma)$.

Case. There is a significant Q in Γ. Then there is a chain from Q to P in Γ by definition. Lemma 21 provides a suitable derivation.

Case. For a partition $\Gamma = \Gamma_1, B, \Gamma_2$ where $B = (\vec{A}_1 \supset Q_1) \supset \ldots \supset (\vec{A}_k \supset Q_k) \supset Q$ is a properly nested implication and where Q is significant, the induction hypothesis on $\vec{A}_i, \Gamma_1, Q_i \supset Q, \Gamma_2 \Rightarrow Q_i$ yields a derivation for each $1 \leq i \leq k$. Since L3-$\supset^*$ preserves derivability, we obtain a derivation of $\Gamma \Rightarrow Q$. Lemma 21 yields a derivation of $\Gamma \Rightarrow P$ as required.

Remaining case. Γ has no significant atom and for each partition $\Gamma = \Gamma_1, B, \Gamma_2$ where $B = (\vec{A}_1 \supset Q_1) \supset \ldots \supset (\vec{A}_k \supset Q_k) \supset Q$ is a properly nested formula and where Q is significant, the induction hypothesis yields a counter–model to $\vec{A}_i, \Gamma_1, Q_i \supset Q, \Gamma_2 \Rightarrow Q_i$ for an i. Since L3-$\supset^*$ preserves counter–models, this is also a counter–model to $\Gamma \Rightarrow Q$. Let $\vec{\mathcal{K}}$ be all these counter–models. By Lemma 24, $(\bigcap \pi_{\text{left}}(\vec{\mathcal{K}}) - L, \vec{\mathcal{K}})$ is a counter–model to $\Gamma \Rightarrow P$. □

Again we extract an algorithm from this proof. It successively picks up each properly nested formula $(\vec{A}_1 \supset Q_1) \supset \ldots \supset (\vec{A}_k \supset Q_k) \supset Q$ where Q is significant, and applies the algorithm recursively to $\vec{A}_i, \Gamma_1, Q_i \supset Q, \Gamma_2 \Rightarrow Q_i$ until all recursive calls are successful. In this case we get a derivation, otherwise we obtain a counter–model. Pruning as described in Sect. 5 is possible here as well.

Since treating the whole fragment would go beyond the limit of space, we have to omit this.

7 Conclusion

We have seen that developing an algorithm by extracting a program from a proof is not only possible, but even easier than proving that a given algorithm is sound and complete. The reason is that we work with the computational content of the rules rather than with their operational semantics. In that way, we presented a new aspect of Hudelmaier's work [7].

8 Implementation

The $\supset$–part of the proof of Theorem 16 is already implemented in the formal system MINLOG [10]. In this system, we can use the modified realisability method to extract an algorithm (see e.g. Troelstra and van Dalen [11]). The present version of MINLOG requires the normalisation of the proof for extracting the algorithm. This means all the recursions are unfolded. Since all the case distinctions are implemented as boolean recursions, the program term is very large and unreadable. In future versions this will no longer be the case. Then, it should not take too much effort to implement the whole proofs of Theorem 16 and 25 and to extract competitive algorithms.

Acknowledgements

Thanks for helpful comments are due to Roy Dyckhoff, Grigori Mints, Karl–Heinz Niggl, Wolfgang Zuber, the anonymous referees and others.

References

1. Roy Dyckhoff. Contraction–free sequent calculi for intuitionistic logic. *Journal of Symbolic Logic*, 57(3):795–807, 1992.
2. Melvin C. Fitting. *Intuitionistic Logic, Model Theory and Forcing*. North-Holland, Amsterdam, 1969.
3. Gerhard Gentzen. Untersuchungen über das logische Schließen. *Math. Zeitschrift*, (39), 1935.
4. Jörg Hudelmaier. Bicomplete calculi for intuitionistical propositional logic. Aufsatz, Universität Tübingen.
   http://www-pu.informatik.uni-tuebingen.de/logik/joerg/bicomp.ps.
5. Jörg Hudelmaier. A note on kripkean countermodels for intuitionistically unprovable sequents. Aufsatz, Universität Tübingen.
   http://www-pu.informatik.uni-tuebingen.de/logik/joerg/kripke.ps.Z.
6. Jörg Hudelmaier. A PROLOG program for intuitionistic logic. SNS–Bericht 88–28, Universität Tübingen, 1988.
7. Jörg Hudelmaier. An $O(n \log n)$–space decision procedure for intuitionistic propositional logic. *Journal of Logic and Computation*, 3(1):63–75, 1993.
8. Pierangelo Migliloli, Ugo Moscato, and Mario Ornaghi. How to avoid duplications in refutation systems for intuitionistic logic and Kuroda logic. Rapporto interno 99-93, Dipartimento di Scienze dell'Informazione, Università degli Studi di Milano, 1993.
9. Luis Pinto and Roy Dyckhoff. Loop–free construction of counter–models for intuitionistic propositional logic. In Behara, Fritsch, and Lintz, editors, *Symposia Gaussiana, Conf. A*, pages 225–232. De Gruyter, 1995.
10. Helmut Schwichtenberg. *Minlog–an interactive proof system*.
    http://www.mathematik.uni-muenchen.de/~logik/minlog_e.html.
11. Anne S. Troelstra and Dirk van Dalen. *Constructivism in Mathematics–an Introduction*, volume 1. North–Holland, 1988.
12. Judith Underwood. A constructive completeness proof for intuitionistic propositional calculus. Technical Report 90-1179, Cornell University, 1990. Also in *Proceedings of the Second Workshop on Theorem Proving with Analytic Tableaux and Related Methods*, April 1993, Marseille, France.
13. N. N. Vorob'ev. A new algorithm for derivability in the constructive propositional calculus. *Amer. Math. Soc. Transl.*, 94(2):37–71, 1970.

The FaCT System

Ian Horrocks

Medical Informatics Group, Department of Computer Science,
University of Manchester, Manchester M13 9PL, UK
horrocks@cs.man.ac.uk
http://www.cs.man.ac.uk/~horrocks

Abstract. FaCT is a Description Logic classifier which has been implemented as a test-bed for a highly optimised tableaux satisfiability (subsumption) testing algorithm. The correspondence between modal and description logics also allows FaCT to be used as a theorem prover for the propositional modal logics **K**, **KT**, **K4** and **S4**. Empirical tests have demonstrated the effectiveness of the optimised implementation and, in particular, of the dependency directed backtracking optimisation.

1 Introduction

FaCT [1] is a Description Logic (DL) classifier which has been implemented as a test-bed for a highly optimised tableaux satisfiability/subsumption testing algorithm. The underlying logic, $\mathcal{ALCH}_{R^+}$, is a superset of the $\mathcal{ALC}$ DL, and this means that FaCT can be used as a theorem prover for the propositional modal logic $\mathbf{K}_{(m)}$ (**K** with multiple modalities) by exploiting the well known correspondence between the two logics [17]. Because $\mathcal{ALCH}_{R^+}$ supports transitive relations, FaCT can also be used as a prover for $\mathbf{K4}_{(m)}$, and it extends the range of logics it can deal with to include $\mathbf{KT}_{(m)}$ and $\mathbf{S4}_{(m)}$ by embedding formulae in $\mathbf{K}_{(m)}$ and $\mathbf{K4}_{(m)}$ respectively.

In order to make the FaCT system usable in realistic DL applications, a wide range of optimisation techniques are used in the implementation of the $\mathcal{ALCH}_{R^+}$ satisfiability testing algorithm. Although some of these techniques were designed to take advantage of the structure of a DL knowledge base (KB), and the repetitive nature of the satisfiability problems encountered when classifying a KB, some of the optimisations are also effective in improving FaCT's performance with respect to single satisfiability problems.

2 Description Logics and Modal Logics

Description Logics support the logical description of concepts and roles (relationships) and their combination, using a variety of operators, to form more complex descriptions. The $\mathcal{ALC}$ DL [18] allows descriptions to be formed using

---

[1] Fast Classification of Terminologies.

standard logical connectives as well as both universally and existentially quantified relational operators: if C is a concept and R is a role then an $\mathcal{ALC}$ concept expression is of the form $C \mid \top \mid \bot \mid \neg C \mid C \sqcap D \mid C \sqcup D \mid \exists R.C \mid \forall R.C$. A Tarski style model theoretic semantics is used to interpret expressions [3].

Table 1 shows how propositional $\mathbf{K}_{(m)}$ formulae correspond to $\mathcal{ALC}$ concept expressions. Note that the modal operators $\Box$ and $\Diamond$ correspond to $\exists R.C$ and $\forall R.C$ expressions, with different roles corresponding to distinct modalities or accessibility relations. Standard modal $\mathbf{K}$ ($\mathbf{K}_{(1)}$) has only one modality, so modal $\mathbf{K}$ formulae correspond to $\mathcal{ALC}$ concept expressions containing a single role. The correspondence can be extended to $\mathbf{K4}_{(m)}$ simply by making all roles transitive.

Table 1. The correspondence between modal $\mathbf{K}_{(m)}$ and $\mathcal{ALC}$

$\mathbf{K}_{(m)}$	$\mathcal{ALC}$	$\mathbf{K}_{(m)}$	$\mathcal{ALC}$
True	$\top$	False	$\bot$
ϕ	C	$\neg\phi$	$\neg C$
$\phi \land \varphi$	$C \sqcap D$	$\phi \lor \varphi$	$C \sqcup D$
$\Box_i \phi$	$\forall R_i.C$	$\Diamond_i \phi$	$\exists R_i.C$

FaCT also supports $\mathbf{KT}_{(m)}$ and $\mathbf{S4}_{(m)}$ by embedding formulae in $\mathbf{K}_{(m)}$ and $\mathbf{K4}_{(m)}$: $\Box_i \phi$ becomes $\phi \land \Box_i \phi$ and $\Diamond_i \phi$ becomes $\phi \lor \Diamond_i \phi$.

3 The $\mathcal{ALCH}_{R^+}$ Tableaux Algorithm

The tableau algorithm for $\mathcal{ALCH}_{R^+}$ is extended from an algorithm for the $\mathcal{ALC}_{R^+}$ DL described in [16]. The full algorithm, along with a proof of its soundness and correctness, is given in [14].

The main features of the algorithm are:

1. it uses a "single pass" tableau construction and search method as is usual in DL tableaux algorithms where logics generally have the finite model property;
2. transitive roles are dealt with simply by propagating $\Box_i \phi$ terms along i relations;
3. termination is ensured by "blocking"—checking for cycles in the tableau construction [7, 1].

4 Optimisations

To improve the performance of the $\mathcal{ALCH}_{R^+}$ satisfiability testing algorithm, a range of optimisations have been employed. These include lexical normalisation and encoding, semantic branching search and dependency directed backtracking.

4.1 Normalisation and Encoding

In DL terminologies, large and complex concepts are seldom described monolithically, but are built up from a hierarchy of named concepts whose descriptions are less complex. The tableaux algorithm can take advantage of this structure by trying to find contradictions between concept names before substituting them with their definitions and continuing with the tableau expansion: we will call this strategy *lazy unfolding*. In fact it has been shown (in the Kris system) that a significant improvement in performance can be obtained simply by not deleting names when they are lazily unfolded [2]. This is because obvious contradictions can often be detected earlier by comparing names rather than unfolded definitions.

FaCT takes this optimisation to its logical conclusion by lexically normalising and encoding all formulae and, recursively, their sub-formulae, so that:

1. All formulae are named; e.g., $\Diamond_i(\phi \wedge \varphi)$ would be encoded as $\Diamond_i \Phi$, where $\Phi = \phi \wedge \varphi$.
2. All formulae are in a standard form; e.g., all $\Diamond$ formulae are converted to $\Box$ formulae, so $\Diamond_i \phi$ would be normalised to $\neg \Box_i \neg \phi$. The encoded sub-formulae in conjunctions and disjunctions are also sorted.

Adding normalisation (step 2) allows lexically equivalent formulae to be recognised and identically encoded; it can also lead to the detection of formulae which are trivially satisfiable or unsatisfiable.

4.2 Semantic Branching Search

Standard tableaux algorithms use an inherently inefficient search technique for the non-deterministic expansion of disjunctive formulae—they choose an unexpanded disjunction and check the different tableaux obtained by adding each of the disjuncts [11]. As the alternative branches of the search are not disjoint, there is nothing to prevent the recurrence of unsatisfiable disjuncts.

FaCT deals with this problem by using a semantic branching technique adapted from the Davis-Putnam-Logemann-Loveland procedure (DPL) commonly used to solve propositional satisfiability (SAT) problems [8, 10]. Instead of choosing an unexpanded disjunction, a single disjunct ϕ is chosen from the set of unexpanded disjunctions, and the two possible tableaux obtained by adding either ϕ or $\neg \phi$ are then searched.

During the DPL search, FaCT also performs boolean constraint propagation (BCP) [9], a technique which maximises deterministic expansion, and thus pruning of the search via contradiction detection, before performing non-deterministic expansion. BCP works by deterministically expanding disjunctions which present only one expansion possibility, and detecting a contradiction when there is a disjunction which no longer has any expansion possibilities. In effect, BCP applies the inference rule $\frac{\neg \phi, \phi \vee \varphi}{\varphi}$ to disjunctive formulae encountered in the tableau expansion, or in other words, performs some localised propositional resolution.

4.3 Dependency Directed Backtracking

Inherent unsatisfiability concealed in sub-formulae can lead to large amounts of unproductive backtracking search known as thrashing. For example, expanding the formula $(\phi_1 \vee \varphi_1) \wedge \ldots \wedge (\phi_n \vee \varphi_n) \wedge \Diamond_i(\phi \wedge \varphi) \wedge \Box_i \neg \phi$ could lead to the fruitless exploration of 2^n possible expansions of $(\phi_1 \vee \varphi_1) \wedge \ldots \wedge (\phi_n \vee \varphi_n)$ before the inherent unsatisfiability of $\Diamond_i(\phi \wedge \varphi) \wedge \Box_i \neg \phi$ is discovered.

This problem is addressed by adapting a form of dependency directed backtracking called *backjumping*, which has been used in solving constraint satisfiability problems [5]. Backjumping works by labeling formulae with a dependency set indicating the branching choices on which they depend. When a contradiction is discovered, the dependency sets of the contradictory formulae can be used to identify the most recent branching point where exploring an alternative branch might alleviate the cause of the contradiction. The algorithm can then jump back over intervening branching points *without* exploring any alternative branches. A similar technique was employed in the HARP theorem prover [15].

5 Performance

FaCT's performance as a modal logic theorem prover has been tested using both randomly generated formulae, a test method described in [12] and derived from a widely used procedure for testing SAT decision procedures [10], and a corpus of carefully designed benchmark formulae [13].

FaCT performs well in tests using randomly generated formulae [14], but its advantages are more clearly demonstrated by the benchmark formulae, and in particular by the provable formulae.[2] This is because the hardness of these formulae often derives from hidden unsatisfiability, a phenomenon which rarely occurs in the randomly generated formulae where hardness is simply a feature of the problem size. Figure 1, for example, shows CPU time plotted against problem size for 2 classes of formulae from the **K** benchmark suite, *k-dum-p* and *k-grz-p*. FaCT's performance is compared with that of the Crack DL [6], the KSAT $\mathbf{K_{(m)}}$ theorem prover [11] and the Kris DL [4]. The performance of FaCT with the backjumping optimisation disabled is also shown, indicated in the graphs by FaCT*. All the systems use compiled Lisp code (Allegro CL 4.3), and the tests were performed on a Sun Ultra 1 with a 147 MHz CPU and 64MB of RAM.

It can be seen from the graphs that FaCT not only outperforms the other systems, but that it exhibits a completely different qualitative performance: solution times for all the other systems increase exponentially with increasing formula size, whereas those for FaCT increase only very slowly. Extrapolating the results for the other systems suggests that FaCT would be several orders of magnitude faster for the largest problems in the test sets. The results for FaCT* demonstrate that, for these formulae, backjumping accounts for FaCT's performance advantage.

---

[2] Note that a formula is proved by showing that its negation is unsatisfiable.

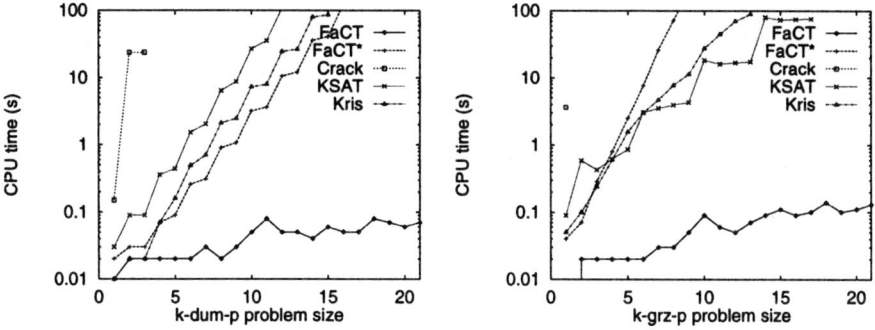

Fig. 1. Solving **K** satisfiability problems

6 Conclusion

Although it was designed for subsumption testing in a DL classifier, FaCT's optimised $\mathcal{ALCH}_{R+}$ satisfiability testing algorithm also performs well as a propositional modal logic theorem prover, and enables FaCT to outperform the other systems with which it has been compared. Backjumping has been shown to be particularly effective, changing both the quantitative and the qualitative performance of the algorithm for some classes of hard unsatisfiable formulae.

References

1. F. Baader, M. Buchheit, and B. Hollunder. Cardinality restrictions on concepts. *Artificial Intelligence*, 88(1-2):195–213, 1996.
2. F. Baader, E. Franconi, B. Hollunder, B. Nebel, and H.-J. Profitlich. An empirical analysis of optimization techniques for terminological representation systems or: Making KRIS get a move on. In B. Nebel, C. Rich, and W. Swartout, editors, *Principals of Knowledge Representation and Reasoning: Proceedings of the Third International Conference (KR'92)*, pages 270–281. Morgan-Kaufmann, 1992. Also available as DFKI RR-93-03.
3. F. Baader, H.-J. Heinsohn, B. Hollunder, J. Muller, B. Nebel, W. Nutt, and H.-J. Profitlich. Terminological knowledge representation: A proposal for a terminological logic. Technical Memo TM-90-04, Deutsches Forschungszentrum für Künstliche Intelligenz GmbH (DFKI), 1991.
4. F. Baader and B. Hollunder. KRIS: Knowledge representation and inference system. *SIGART Bulletin*, 2(3):8–14, 1991.
5. A. B. Baker. *Intelligent Backtracking on Constraint Satisfaction Problems: Experimental and Theoretical Results*. PhD thesis, University of Oregon, 1995.
6. P. Bresciani, E. Franconi, and S. Tessaris. Implementing and testing expressive description logics: a preliminary report. In Gerard Ellis, Robert A. Levinson, Andrew Fall, and Veronica Dahl, editors, *Knowledge Retrieval, Use and Storage for Efficiency: Proceedings of the First International KRUSE Symposium*, pages 28–39, 1995.

7. M. Buchheit, F. M. Donini, and A. Schaerf. Decidable reasoning in terminological knowledge representation systems. *Journal of Artificial Intelligence Research*, 1:109–138, 1993.
8. M. Davis, G. Logemann, and D. Loveland. A machine program for theorem proving. *Communications of the ACM*, 5:394–397, 1962.
9. J. W. Freeman. *Improvements to propositional satisfiability search algorithms*. PhD thesis, Department of Computer and Information Science, University of Pennsylvania, Philadelphia, PA, USA, 1995.
10. J. W. Freeman. Hard random 3-SAT problems and the Davis-Putnam procedure. *Artificial Intelligence*, 81:183–198, 1996.
11. F. Giunchiglia and R. Sebastiani. Building decision procedures for modal logics from propositional decision procedures—the case study of modal K. In Michael McRobbie and John Slaney, editors, *Proceedings of the Thirteenth International Conference on Automated Deduction (CADE-13)*, number 1104 in Lecture Notes in Artificial Intelligence, pages 583–597. Springer, 1996.
12. F. Giunchiglia and R. Sebastiani. A SAT-based decision procedure for $\mathcal{ALC}$. In L. C. Aiello, J. Doyle, and S. Shapiro, editors, *Principals of Knowledge Representation and Reasoning: Proceedings of the Fifth International Conference (KR'96)*, pages 304–314. Morgan Kaufmann, November 1996.
13. A. Heuerding and S. Schwendimann. A benchmark method for the propositional modal logics k, kt, s4. Technical report IAM-96-015, University of Bern, Switzerland, October 1996.
14. I. Horrocks. *Optimising Tableaux Decision Procedures for Description Logics*. PhD thesis, University of Manchester, 1997.
15. F. Oppacher and E. Suen. HARP: A tableau-based theorem prover. *Journal of Automated Reasoning*, 4:69–100, 1988.
16. U. Sattler. A concept language extended with different kinds of transitive roles. In G. Görz and S. Hölldobler, editors, *20. Deutsche Jahrestagung für Künstliche Intelligenz*, number 1137 in Lecture Notes in Artificial Intelligence, pages 333–345. Springer Verlag, 1996.
17. K. Schild. A correspondence theory for terminological logics: Preliminary report. In *Proceedings of the 12th International Joint Conference on Artificial Intelligence (IJCAI-91)*, pages 466–471, 1991.
18. M. Schmidt-Schauß and G. Smolka. Attributive concept descriptions with complements. *Artificial Intelligence*, 48:1–26, 1991.

Implementation of Proof Search in the Imperative Programming Language Pizza

Christian Urban

Computer Laboratory, University of Cambridge,
Gonville and Caius College, Cambridge CB2 1TA, UK.
Christian.Urban@cl.cam.ac.uk

Abstract. Automated proof search can be easily implemented in logic programming languages. We demonstrate the technique of success continuations, which provides an equally simple method for encoding proof search in imperative programming languages. This technique is exemplified by developing an interpreter for the calculus G4ip in the language Pizza.
Keywords: Success Continuations, G4ip, Pizza

1 Introduction

A sequent-calculus formulation of a logic is a convenient starting-point for automating proof search because the corresponding inference rules are 'local' operations on proofs. A sequent can be proved by applying inference rules until one reaches axioms; or can make no further progress in which case one must backtrack or even abandon the search. This proof method is a simple depth-first strategy; it is preferred over a less efficient breadth-first strategy. However, this method requires the mechanism of choice points in order to facilitate the backtracking. Logic programming languages provide substantial support for depth-first proof search and therefore simplify considerably the implementation of a proof search engine. Unfortunately these languages have a rather limited support for user interfaces, which makes them unsuitable for larger applications. Imperative programming languages, on the other hand, provide a rich environment for user interfaces, but seem to need a significant overhead of code for an implementation of a proof search engine. We illustrate in the paper an implementation technique using success continuations which provides a simple encoding of proof search. The technique is not new: it was introduced in [Car84] with a rather technical illustration in LISP. Later an excellent paper [EP91] appeared which implements a full-fledged interpreter for λProlog in SML.

In the paper we focus on intuitionistic logic, for which Gentzen's calculus LJ is a standard formalisation. Unfortunately a proof method for LJ using a depth-first search strategy cannot guarantee termination. Some modifications can be made to the inference rules of LJ without loss of soundness and completeness. As a result an efficient depth-first proof search procedure can be designed for the propositional fragment of intuitionistic logic. The name G4ip has been assigned

$$\overline{A, \Gamma \vdash A} \; Axiom \qquad \frac{\Gamma \vdash B \quad \Gamma \vdash C}{\Gamma \vdash B \& C} \; \&\text{-R} \qquad \frac{\Gamma \vdash B_i}{\Gamma \vdash B_1 \vee B_2} \; \vee\text{-R}_i$$
(A being atomic)

$$\overline{false, \Gamma \vdash G} \; false\text{-L} \qquad \frac{B, \Gamma \vdash C}{\Gamma \vdash B \supset C} \; \supset\text{-R} \qquad \frac{B, C, \Gamma \vdash G}{B \& C, \Gamma \vdash G} \; \&\text{-L}$$

$$\frac{B, \Gamma \vdash G \quad C, \Gamma \vdash G}{B \vee C, \Gamma \vdash G} \; \vee\text{-L} \qquad \frac{B, A, \Gamma \vdash G}{A \supset B, A, \Gamma \vdash G} \; \supset\text{-L}_1 \qquad \frac{C \supset (D \supset B), \Gamma \vdash G}{(C \& D) \supset B, \Gamma \vdash G} \; \supset\text{-L}_2$$
(A being atomic)

$$\frac{C \supset B, D \supset B, \Gamma \vdash G}{(C \vee D) \supset B, \Gamma \vdash G} \; \supset\text{-L}_3 \qquad \frac{D \supset B, \Gamma \vdash C \supset D \quad B, \Gamma \vdash G}{(C \supset D) \supset B, \Gamma \vdash G} \; \supset\text{-L}_4$$

Fig. 1. The Inference Rules of G4ip.

to the corresponding calculus in [TS96]. This calculus is also known as LJT which has been studied thoroughly in [Dyc92]. The inference rules of G4ip are given in Fig. 1. G4ip has the pleasant property that a depth-first proof search does not loop and that it is sound and complete in terms of provable theorems.

In the paper we demonstrate the implementation technique using the imperative programming language Pizza. Pizza is a strict extension of the object-oriented programming language Java such that Pizza programs can be translated into Java code or can be compiled into ordinary Java Byte Code (see [OW97] for a technical introduction into Pizza). We make use of the following two new language constructs:

- higher-order functions, i.e. functions may be passed as parameters or returned from methods;
- class cases and pattern matching: this allows much simpler and more readable code.

These features are not directly present in Java, but Pizza makes them accessible by translating them into Java. The higher-order functions are essential for our implementation. The success continuations are functions passed as parameters or returned as values. Pizza provides the programmer with access to the same extensive libraries for graphic and network applications as Java. The paper assumes some familiarity with Java or Pizza.

2 Design of the Proving Method

In this section we are concerned with the abstract design of the proof search. Each step in the depth-first search is a reversed application of an inference rule. The inference rule analyses one formula, called the principal formula, of the sequent being proved. The construction of proofs, however, is not deterministic;

it does not determine the choice of the principal formula. All formulae in a sequent are possible candidates and a naive proof method has to explore them all. A widely used optimisation is the search for uniform proofs [MNPS91]. Uniform proofs are constrained on the choice of the principal formula, i.e., whenever sequents have non-atomic goal formulae then only those inference rules which analyse these goal formulae are chosen. This optimised search strategy can be employed to logics for which uniform proofs are complete. However for G4ip, this optimisation cannot be made. For example, the sequent $p \vee q \vdash p \vee q$ has only one proof which is non-uniform. Therefore, we shall implement a proof method which first enumerates all formulae from the left-hand side of the sequent as being principal and applies corresponding left rules. Subsequently the goal formula is chosen and a right rule is applied. The sequents of G4ip are specified with multisets of formulae on the left-hand side. Accordingly, the enumeration of formulae as being principal can be done in any order. We have avoided making some optimisations in favour of a clear illustration of the success continuation technique. A λProlog implementation of the outlined proof method has been provided in order to compare the technique of success continuations and more traditional implementations using logic programming (see Sect. 5). A similar implementation was presented in [HM94] using Lolli, a logic programming language based on linear logic. This implementation differs from ours in the inactive parts of the sequents (i.e. Γ) which are treated implicitly.

3 The Representation of Formulae and Sequents

Amongst the new language features of Pizza are class cases and pattern matching, which provide a very pleasant syntax for algebraic data types. The formulae of G4ip are specified by the following grammar:

$$F ::= \ false \ | \ A \ | \ F \& F \ | \ F \vee F \ | \ F \supset F$$

(where A is taken from a set of atomic formulae). The class cases allow a straightforward implementation of this specification; it is analogous to the SML implementation of λProlog's formulae in [EP91]. The class of formulae is given below. On the right-hand side two examples illustrate the use of this class:

```
public class Form {
    case False();
    case Atm(String c);
    case And(Form c1,Form c2);
    case Or(Form c1,Form c2);
    case Imp(Form c1,Form c2);
}
```

$p \supset p$ is represented as:
Imp(Atm("p"),Atm("p"))

$a \vee (a \supset false)$ is represented as:
Or(Atm("a"),Imp("a",False()))

The class cases of Pizza also support an implementation of formulae specified by a mutually recursive grammar. This is required, for example, when implementing hereditary Harrop formulae.

The sequents of G4ip, which have the form $\Gamma \vdash G$, are represented by means of the class below. The left-hand sides are specified by multisets of formulae. Therefore, we do not need to worry about the order in which the formulae occur.

```
public class Sequent {
    Form     G;
    Context  Gamma;
    public Sequent(Context _Gamma, Form _G) {...}
}
```

We have a constructor for generating new sequents during proof search. `Context` is a class which represents multisets; it is a simple extension of the class Vector available in the Java libraries. `Context` provides methods for adding elements to a multiset (`add`), taking out elements from a multiset (`removeElement`) and testing the membership of an element in a multiset (`includes`).

4 The Implementation of the Proof Search

In a first attempt we could implement the choice of a principal formula and the application of an inference rule in a recursive style. Suppose we have a method *prove*. This method receives a sequent $F_1, \ldots, F_{n-1} \vdash F_n$ as its only argument. In *prove* we could enumerate all formulae of the sequent as being principal and apply a corresponding inference rule, say R_i. This produces for each F_i some new sequents to be proved:

$$\frac{Premise_1 \ldots Premise_m}{F_1, \ldots, F_{n-1} \vdash F_n} R_i$$

We perform *prove* again by calling it recursively with $Premise_1$ to $Premise_m$ as arguments. However this simple method intermingles the separate concepts of proof obligations (which must be proved) and choice points (which can be tried out). To make this first attempt work we need to make some non-trivial modifications. A simpler method is to add another argument to the method *prove*. This additional argument will be an anonymous function, which is permitted by Pizza. The method *prove* is now of the form *prove(sequent,sc)*. Somewhat simplified the first argument is the leftmost premise ($Premise_1$) and the second argument sc, which is the *success continuation*, represents the other proof obligations ($Premise_2$ to $Premise_m$). In case we succeed in proving the first premise we attempt to prove the other premises.

The inference rules of G4ip fall into three groups: inference rules with a single premise, inference rules with two premises and inference rules without premises (e.g. *Axiom*). Suppose we have called *prove* with a sequent s and a success continuation sc. The inference rules of the first group manipulate s obtaining s' and call *prove* again with the new sequent s' and the current success continuation sc (Steps 2-3 and 4-5 in Fig. 2). The inference rules of the second group have

$$\frac{\dfrac{}{3\ p,q\vdash p}\ Axiom}{\dfrac{}{2\ p\&q\vdash p}}\ \&\text{-L}\quad \frac{\dfrac{}{5\ p,r\vdash p}\ Axiom}{\dfrac{}{4\ p\&r\vdash p}}\ \&\text{-L}$$
$$\frac{}{1\ (p\&q)\vee(p\&r)\vdash p}\ \vee\text{-L}$$

6	is		
5	$p,r\vdash p$	is	
4	$p\&r\vdash p$	is	
3	$p,q\vdash p$	$p\&r\vdash p$	is
2	$p\&q\vdash p$	$p\&r\vdash p$	is
1	$(p\&q)\vee(p\&r)\vdash p$	is	

Fig. 2. An Example for the Technique of Success Continuations.

two premises, s_1 and s_2. These rules call *prove* with s_1 and the new success continuation *prove(s_2,sc)* (Step 1-2 in Fig. 2). The third group of inference rules only invoke the success continuation if the rule is applicable (Steps 3-4 and 5-6 in Fig. 2).

We are going to give a detailed description of the code for the rules: &-L, ∨-R_i, ∨-L and *Axiom*. For lack of space the code of the other rules is omitted. The method *prove* enumerates all formulae as being principal and two switch statements select a corresponding rule depending on the form and the occurrence of the principal formula. The &-L rule is in the first group; it modifies the sequent being proved and calls *prove* again with the current success continuation. The code is as follows:[1]

```
case And(Form B, Form C):    //&-L
    prove(new Sequent(Gamma.add(B,C),G),sc); break;
```

The ∨-R_i rule is an exception in the first group. It breaks up a goal formula of the form $B_1\vee B_2$ and proceeds with one of its components. Since we do not know in advance which component leads to a proof we have to try both. Therefore this rule acts as a choice point, which is encoded by a recursive call of *prove* for each case.

```
case Or(Form B1,Form B2):    //∨-Ri
    prove(new Sequent(Gamma,B1),sc);
    prove(new Sequent(Gamma,B2),sc); break;
```

The ∨-L rule falls into the second group where the current success continuation is modified. It calls *prove* with the first premise $(B,\Gamma\vdash G)$ and wraps up the success continuation with the new proof obligation $(C,\Gamma\vdash G)$. The construction `fun()->void {...}` defines an anonymous function: the new success continuation. In case the sequent $B,\Gamma\vdash G$ can be proved, this function is invoked.

---

[1] Gamma stands for the multiset of formulae on the left-hand side of a sequent excluding the principal formula; G stands for the goal formula of a sequent; B and C stand for the two components of the principal formula.

```
case Or(Form B,Form C):    //∨-L
  prove(new Sequent(Gamma.add(B),G),
      fun()->void {prove(new Sequent(Gamma.add(C),G),sc);}
  ); break
```

The *Axiom* rule falls into the third group. It first checks if the principal formula (which is an atom) matches with the goal formula and then invokes the current success continuation *sc* in order to prove all remaining proof obligations.

```
case Atm(String c):    //Axiom
  if (G instanceof Atm) { if (G.c.compareTo(c) == 0) { sc(); }
  } break;
```

The proof search is started with an initial success continuation *is* (cf. Fig. 2). This initial success continuation is invoked when a proof has been found. In this case we want to give some response to the user. An example for the initial success continuation *is* could be as follows:

```
public void initial_sc() { System.out.println("Provable!"); }
```

Suppose we attempt to start the proof search with $prove(p, p \vdash p, is)$. We would find that the prover responds twice with "Provable!" because it finds two proofs. In our implementation this problem is avoided by encoding the proof search as a thread. Whenever a proof is found, the initial success continuation displays the proof and suspends the thread. The user can decide to resume with the proof search or abandon the search.

5 Conclusion

We have adapted the technique of success continuations presented in [Car84] and [EP91] and provided an implementation of G4ip in the imperative programming language Pizza. This imperative language provides substantial support for user interfaces and network applications—more than current logic programming languages. Our implementation of G4ip cannot be considered as optimal in terms of speed. A much more efficient (but less clear) proof-search algorithm for G4ip has been implemented by Dyckhoff in Prolog. Similar ideas could be encoded in our Pizza implementation; but our point was not the efficiency but the clarity of the implementation using success continuations. The technique is applicable wherever backtracking is required. We compared the code of our implementation with a similar implementation of the naive proving strategy in λProlog: the ratio of code is approximately 2 to 1. This result is partly due to the fact that we had to implement a class for multisets. In a future version of Java, we could have accessed a package in the library. The technique of success continuation can also be applied to a first-order calculus as shown in [EP91], but the required mechanism of substitution needs to be implemented separately.

The code of the implementations and some accompanying information are available under the address http://www.cl.cam.ac.uk/~cu200/Prover/. It includes an applet which can be executed on a Java-capable browser.

Acknowledgements: I am very grateful for Dr Roy Dyckhoff's constant encouragement and many comments on my work. I thank Dr Gavin Bierman who helped me to test the prover applet and commented on the paper. The work was supported by a scholarship from the German Academic Exchange Service.

References

[Car84] M. Carlsson. On Implementing Prolog in Functional Programming. *New Generation Computing*, pages 347–359, 1984.

[Dyc92] R. Dyckhoff. Contraction-Free Sequent Calculi for Intuitionistic Logic. *Journal of Symbolic Logic*, 57(3):795–807, September 1992.

[EP91] C. Elliott and F. Pfenning. A Semi-Functional Implementation of a Higher-Order Logic Programming Language. In P. Lee, editor, *Topics in Advanced Language Implementation*, pages 289–352. MIT Press, 1991.

[HM94] J. Hodas and D. Miller. Logic Programming in a Fragment of Intuitionistic Linear Logic. *Information and Computation*, 110(2):327–365, 1994.

[MNPS91] D. Miller, G. Nadathur, F. Pfenning, and A. Scedrov. Uniform Proofs as a Foundation for Logic Programming. *Annals of Pure Applied Logic*, 51:125–157, 1991.

[OW97] M. Odersky and P. Wadler. Pizza into Java: Translating Theory into Practice. In *Proc. 24th ACM Symposium on Principles of Programming Languages*, January 1997.

[TS96] A. S. Troelstra and H. Schwichtenberg. *Basic Proof Theory*. Cambridge Tracts in Theoretical Computer Science. Cambridge University Press, 1996.

p-SETHEO: Strategy Parallelism in Automated Theorem Proving*

Andreas Wolf

Technische Universität München, Institut für Informatik, D-80290 Munich
wolfa@informatik.tu-muenchen.de

Abstract. Automated theorem provers use search strategies. Unfortunately, there is no unique strategy which is uniformly successful on all problems. A combination of more than one strategy increases the chances of success. Limitations of resources such as time or the number of available processors enforce efficient use of these resources by partitioning them adequately among the involved strategies. One of the problems to be solved in the context of resource scheduling is an optimization problem. We describe this problem and discuss the prototypical theorem prover p-SETHEO.

Introduction. A search problem is typically solved by applying a *uniform* search procedure. In automated deduction, different search strategies may have a strongly different behavior on a given problem. In general, it cannot be decided in advance which strategy is the best for a given problem. This motivates the *competitive* use of different strategies. In order to be successful with such an approach, the set of strategies must satisfy the following two intuitively given conditions.

1. For a given set of problems, the function $f(t) = \frac{|s(t)|}{t}$, where $s(t)$ is the set of problems solved within time t is sub-linear, i.e., with each new time interval less new problems are solved.
2. The strategies must be *complementary* w.r.t. a given problem set and a given time limit, i.e., they have to solve different sets of problems, or, at least, the sets of solved problems (w.r.t. a given problem set) must differ "significantly".

If both conditions are satisfied, then a competitive use of different strategies can be more successful than the best single strategy. The first condition is typically satisfied in automated theorem proving whereas the second condition has not been in the focus of automated deduction research so far. The success of p-SETHEO shows that it is worthwhile to develop methods for achieving complementary strategies, which is left to future research.

Related Approaches. The method introduced here differs significantly from a *partitioning* of the search space which is done, for instance, in PARTHEO [8].

---

* This work is supported by the Deutsche Forschungsgemeinschaft within the Sonderforschungsbereich 342, subproject A5.

Partitioning *guarantees* that no part of the search space is explored more than once. However, partitioning has the main disadvantage that completeness can only be guaranteed if all agents are reliable. In contrast to partitioning, strategy parallelism retains completeness as long as one agent is reliable.

A combination of different strategies is used, e.g., within the TeamWork concept of DISCOUNT [1], where a combination of several completion strategies for unit equality problems periodically exchanges intermediate results. The clause diffusion concept of AQUARIUS [2] uses a resolution based prover with cooperating agents on splitted databases. A third approach is the Nagging concept [6]. Here, dependent subtasks will be sent by a master process to the naggers, which try to solve them and report on their success. The results will be integrated into the main proof attempt.

Strategy Parallelism. The selection of more than one search strategy in combination with techniques to partition the available resources (time and processors) with respect to the actual task defines a new parallelization method, which we call *strategy parallelism*. (Distributed) competitive agents traverse similar search spaces, at least in different order. Whenever an agent finds a solution, all other agents are stopped. With this method it is intended that the strategies traverse the search space in such a manner that, in practice, the repeated consideration of the same parts can be avoided. In the simple form of strategy parallelism which is discussed here, there is no interaction between the competitive agents. This enables the combination of even completely different search paradigms, for example, resolution and model elimination.

An Optimization Problem. A combination of strategies increases the chances of success. Limitations of resources such as time or processors enforce efficient use of these resources by partitioning them adequately among the involved strategies. This leads to the following *strategy allocation problem*.

GIVEN a set $F = \{f_1, \ldots, f_n\}$ of objects (formulae or training problems), a set $S = \{s_1, \ldots, s_m\}$ of functions (admissible strategies) $s_i : F \to \mathbb{N}^+ \cup \{\infty\}^1$ ($1 \leq i \leq m$), and nonnegative integers t (time limit) and p (number of available processors).

FIND ordered pairs $(t_1, p_1), \ldots, (t_m, p_m)$ (strategy i will be scheduled for time t_i on processor p_i) of nonnegative integers such that

$$\sum_{\{i : p_i = j\}} t_i \leq t \text{ for all } j = 1, \ldots, p, \text{ and } | \bigcup_{i=1}^{m} \{f : s_i(f) \leq t_i\} | \text{ is maximal}^2, \text{ i.e.,}$$

assign resources to the strategies such that a maximal number of problems from the training set can be solved. The decision variant[3] of the problem is in NP: a given satisfying allocation can be verified in polynomial time. Unfortunately, the decision variant of the problem is already *strongly NP-complete*[4] for a single

---

[1] ∞ is added because of the strategy s_i may be incomplete, or the problem cannot be solved in the given time.
[2] $s(f)$ is the time the strategy s needs to solve the formula f.
[3] Guess a resource allocation such that more than k problems can be solved.
[4] Recognizable by providing a polynomial reduction of the *minimum cover problem*.

processor. Therefore, in practice the determination of an optimal solution for the full problem will be not possible, at least not on larger sets and with classical methods. One reasonable possibility is to use a gradient procedure. This indeed was done to select p-SETHEO's strategies.

Implementation. In order to investigate the potential of strategy parallelism in practice, we have evaluated the method on different strategies of the sequential model elimination prover SETHEO [5]. We implemented the PVM based [4] p-SETHEO system. p-SETHEO can be configured very easily by an ASCII file containing information about the usable hosts, the maximal number of simultaneously running strategies per processor, and the resource allocation for the selected strategies. Currently, all contained strategies are variants of SETHEO which have been obtained by modifying the parameter settings. Because of the generic layout of the p-SETHEO controlling mechanism, new strategies and even new theorem provers can be integrated very easily. The implementation uses PERL, PVM, C, and shell tools in approximately 1000 lines of code (excluding SETHEO). The parallelization model of p-SETHEO works as follows:

1. Select a set of triples of strategies, computation times, and assigned processors.
2. Perform the preprocessing steps needed for all selected strategies (reordering, equality treatment, lemma generation etc.)
3. Start all prover strategies. The first prover finding a proof stops all other processes.

Figure 1 shows the parallel flow of processes for an example p-SETHEO configuration.

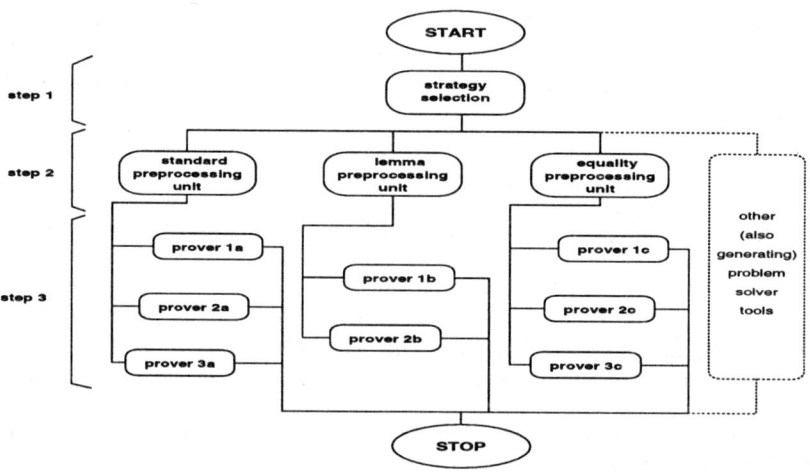

Fig. 1. Parallelization model of p-SETHEO (example).

Experimental Results. *A simple TPTP configuration.* A proof for the performance of the strategy parallel approach is the combination of four arbitrarily selected strategies on four processors: iterative deepening on tableau depth with and without folding-up, and the iterative deepening with the weighted depth bound also with and without folding-up[5]. The result of this experiment is shown in Table 1. The best single strategy solves within 310% of time only 70% of

configuration	solutions	%	time	%	work	%	work/solution	%
-dr	151	66	106128	377	106128	94	703	143
-dr with -foldup	159	69	91561	325	91561	81	576	118
-wdr	122	53	128587	456	128587	114	1054	215
-wdr with -foldup	161	70	87321	310	87321	78	542	111
p-SETHEO	230	100	28168	100	112672	100	490	100

Table 1. Combination of four strategies on a TPTP subset: the 230 tasks from TPTP, which can be solved by at least one strategy within 1000 seconds but which are solved by at most two strategies within less than 20 seconds. p-SETHEO combines all strategies. Time and work are given in seconds. The time limit is 1000 seconds.

the problems p-SETHEO solves. If we consider parallel work instead of time, p-SETHEO solves significantly more problems with nearly the same costs; so the best single strategy needs 111% of the work per solution p-SETHEO needs.

Application of the strategy allocation problem. As training set F we selected the 420 problems in clausal form of the TPTP problem library [7] that have been used for the CADE-14 theorem prover competition. For a given set of 20 strategies S (parameter settings of SETHEO), the solution times $s(f)$, for any $f \in F, s \in S$ were computed with a time limit T of 300 seconds per problem. An approximative solution of the mentioned strategy allocation problem for one processor and 300 seconds CPU time leads to a strategy schedule, which was tested on the whole TPTP and compared with the best single strategy. The result is shown in Figure 2.

Assessment. Many issues of importance for strategy parallelism have not been discussed here: How can a set of strategies be obtained that solve as many problems as fast as possible and have sets of solved problems that differ as much as possible? Often, strategies are only successful for a certain class of problems. For example, unit equality problems need different treatment than problems without equality. If such features can be identified, the selection of strategies can be made more specific and hence more successful. The success of the selected strategies depends on the given training set. How do we obtain a training set which is representative for the considered domain of problems? The number of

---

[5] For details on the parameter settings see [5].

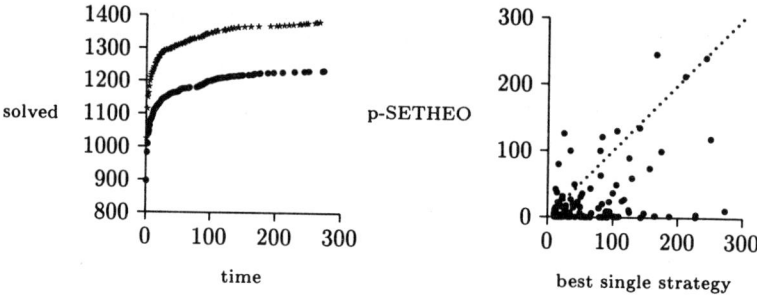

Fig. 2. In the left plot, the • marks the number of problems solved by the best single strategy within a certain time, the ⋆ shows the problems solved by p-SETHEO. The right plot compares proof times of p-SETHEO with the best single strategy. The points lying underneath the dotted line represent all problems that p-SETHEO solves in less time than the best single strategy.

sensible strategies which are successful and differ as much as possible seems to be bounded. This restricts the scalability of strategy parallelism to large platforms of parallel processors. Can we find a systematic method for producing as many successful and differing strategies as we want? Probably such a method must contain randomized elements.

We have investigated the problems and perspectives of strategy parallelism. As our experiments show, even with a very simple strategy allocation algorithm and a non-optimized set of admissible strategies, one can significantly increase the performance of theorem provers. While, in theorem proving, often the system developer or advanced user can tune his system to a given problem by using his experience, this is not possible in practice if the theorem prover is integrated into a larger proof environment like ILF [3]. In this case the configuration should be done automatically, and strategy parallelism is a good alternative.

References

1. J. Avenhaus et al. Discount: A system for distributed equational deduction. RTA-6, 1995.
2. M. P. Bonacina, J. Hsiang. Distributed deduction by clause diffusion: The Aquarius prover. LNAI 722, Springer, 1993.
3. B. I. Dahn et al. Integration of Automated and Interactive Theorem Proving in ILF. CADE-14, 1997.
4. A. Geist et al. PVM: Parallel Virtual Machine. MIT Press, 1994.
5. M. Moser et al. Setheo and E-Setheo. The CADE-13 Systems. JAR, 18(2), 1997.
6. D. B. Sturgill, A. M. Segre. A Novel Asynchronous Parallelism Scheme for First-Order Logic. CADE-12, 1994.
7. G. Sutcliffe et al. The TPTP Problem Library. CADE-12, 1994.
8. Ch. Suttner, J. Schumann. Parallel Automated Theorem Proving. PPAI, Elsevier, 1993.

Author Index

Baldoni, M.	44	Langsteiner, H.	156
Balsiger, P.	25,35	Leach, J.	202
Baumgartner, P.	60	Levy, M.	31
Beckert, B.	33,77,93		
Berezin, S.	18	Martelli, A.	44
Bowman, H.	108	Martin, P.J.	202
Bruijn, N.G. de	1	Massacci, F.	217
Bundy A.	10	Monz, C.	232
		Murray, N.	172
Cerrito, S.	124		
Cialdea Mayer, M.	124	Negri, S.	247
Clarke, E.	18		
		Patel-Schneider, P.F.	27
Egly, U.	141	Pitt, J.	38
		Plato, J. von	247
Fermüller, C.	156		
		Rijke, M. de	232
Gabbay, D.	77	Rosenthal, E.	172
Gavilanes, A.	202		
Giordano, L.	44	Schmidt, R.A.	36,187
Goré, R.	33	Schmitt, S.	262
		Schwendimann, S.	35,277
Hähnle, R.	172		
Hartmer, U.	93	Thompson, S.	108
Heuerding, A.	25,35	Tompits, H.	141
Horrocks, I.	27,307		
Hustadt, U.	36,187	Urban, C.	313
		Weich, K.	292
Janssen, G.L.J.M.	40	Weidenbach, C.	36
		Wolf, A.	320
Komara, J.	42		
Kreitz, C.	262	Voda, P.J.	42

Springer and the environment

At Springer we firmly believe that an international science publisher has a special obligation to the environment, and our corporate policies consistently reflect this conviction.

We also expect our business partners – paper mills, printers, packaging manufacturers, etc. – to commit themselves to using materials and production processes that do not harm the environment. The paper in this book is made from low- or no-chlorine pulp and is acid free, in conformance with international standards for paper permanency.

Lecture Notes in Artificial Intelligence (LNAI)

Vol. 1224: M. van Someren, G. Widmer (Eds.), Machine Learning: ECML-97. Proceedings, 1997. XI, 361 pages. 1997.

Vol. 1227: D. Galmiche (Ed.), Automated Reasoning with Analytic Tableaux and Related Methods. Proceedings, 1997. XI, 373 pages. 1997.

Vol. 1228: S.-H. Nienhuys-Cheng, R. de Wolf, Foundations of Inductive Logic Programming. XVII, 404 pages. 1997.

Vol. 1229: G. Kraetzschmar, Distributed Reason Maintenance for Multiagent Systems. XIV, 296 pages. 1997.

Vol. 1236: E. Maier, M. Mast, S. LuperFoy (Eds.), Dialogue Processing in Spoken Language Systems. Proceedings, 1996. VIII, 220 pages. 1997.

Vol. 1237: M. Boman, W. Van de Velde (Eds.), Multi-Agent Rationality. Proceedings, 1997. XII, 254 pages. 1997.

Vol. 1244: D. M. Gabbay, R. Kruse, A. Nonnengart, H.J. Ohlbach (Eds.), Qualitative and Quantitative Practical Reasoning. Proceedings, 1997. X, 621 pages. 1997.

Vol. 1249: W. McCune (Ed.), Automated Deduction – CADE-14. Proceedings, 1997. XIV, 462 pages. 1997.

Vol. 1257: D. Lukose, H. Delugach, M. Keeler, L. Searle, J. Sowa (Eds.), Conceptual Structures: Fulfilling Peirce's Dream. Proceedings, 1997. XII, 621 pages. 1997.

Vol. 1263: J. Komorowski, J. Zytkow (Eds.), Principles of Data Mining and Knowledge Discovery. Proceedings, 1997. IX, 397 pages. 1997.

Vol. 1266: D.B. Leake, E. Plaza (Eds.), Case-Based Reasoning Research and Development. Proceedings, 1997. XIII, 648 pages. 1997.

Vol. 1265: J. Dix, U. Furbach, A. Nerode (Eds.), Logic Programming and Nonmonotonic Reasoning. Proceedings, 1997. X, 453 pages. 1997.

Vol. 1285: X. Jao, J.-H. Kim, T. Furuhashi (Eds.), Simulated Evolution and Learning. Proceedings, 1996. VIII, 231 pages. 1997.

Vol. 1286: C. Zhang, D. Lukose (Eds.), Multi-Agent Systems. Proceedings, 1996. VII, 195 pages. 1997.

Vol. 1297: N. Lavrač, S. Džeroski (Eds.), Inductive Logic Programming. Proceedings, 1997. VIII, 309 pages. 1997.

Vol. 1299: M.T. Pazienza (Ed.), Information Extraction. Proceedings, 1997. IX, 213 pages. 1997.

Vol. 1303: G. Brewka, C. Habel, B. Nebel (Eds.), KI-97: Advances in Artificial Intelligence. Proceedings, 1997. XI, 413 pages. 1997.

Vol. 1307: R. Kompe, Prosody in Speech Understanding Systems. XIX, 357 pages. 1997.

Vol. 1314: S. Muggleton (Ed.), Inductive Logic Programming. Proceedings, 1996. VIII, 397 pages. 1997.

Vol. 1316: M.Li, A. Maruoka (Eds.), Algorithmic Learning Theory. Proceedings, 1997. XI, 461 pages. 1997.

Vol. 1317: M. Leman (Ed.), Music, Gestalt, and Computing. IX, 524 pages. 1997.

Vol. 1319: E. Plaza, R. Benjamins (Eds.), Knowledge Acquisition, Modelling and Management. Proceedings, 1997. XI, 389 pages. 1997.

Vol. 1321: M. Lenzerini (Ed.), AI*IA 97: Advances in Artificial Intelligence. Proceedings, 1997. XII, 459 pages. 1997.

Vol. 1323: E. Costa, A. Cardoso (Eds.), Progress in Artificial Intelligence. Proceedings, 1997. XIV, 393 pages. 1997.

Vol. 1325: Z.W. Raś, A. Skowron (Eds.), Foundations of Intelligent Systems. Proceedings, 1997. XI, 630 pages. 1997

Vol. 1328: C. Retoré (Ed.), Logical Aspects of Computational Linguistics. Proceedings, 1996. VIII, 435 pages. 1997.

Vol. 1342: A. Sattar (Ed.), Advanced Topics in Artificial Intelligence. Proceedings, 1997. XVIII, 516 pages. 1997.

Vol. 1348: S. Steel, R. Alami (Eds.), Recent Advances in AI Planning. Proceedings, 1997. IX, 454 pages. 1997.

Vol. 1359: G. Antinou, A. Ghose, M. Truszczynski (Eds.), Learning and Reasoning with Complex Representations. Proceedings, 1996. X, 283 pages. 1998.

Vol. 1360: D. Wang (Ed.), Automated Deduction in Geometry. Proceedings, 1996. VII, 235 pages. 1998.

Vol. 1365: M.P. Singh, A. Rao, M.J. Wooldridge (Eds.), Intelligent Agents IV. Proceedings, 1997. XII, 351 pages. 1998.

Vol. 1371: I. Wachsmuth, M. Fröhlich (Eds.), Gesture and Sign-Language in Human-Computer Interaction. Proceedings, 1997. XI, 309 pages. 1998.

Vol. 1374: H. Bunt, R.-J. Beun, T. Borghuis (Eds.), Multimodal Human-Computer Communication. VIII, 345 pages. 1998.

Vol. 1387: C. Lee Giles, M. Gori (Eds.), Adaptive Processing of Sequences and Data Structures. Proceedings, 1997. XII, 434 pages. 1998.

Vol. 1394: X. Wu, R. Kotagiri, K.B. Korb (Eds.), Research and Development in Knowledge Discovery and Data Mining. Proceedings, 1998. XVI, 424 pages. 1998.

Vol. 1397: H. de Swart (Ed.), Automated Reasoning with Analytic Tableaux and Related Methods. Proceedings, 1998. X, 325 pages. 1998.

Vol. 1398: C. Nédellec, C. Rouveirol (Eds.), Machine Learning: ECML-98. Proceedings, 1998. XII, 420 pages. 1998.

Lecture Notes in Computer Science

Vol.Vol. 1359: G. Antinou, A. Ghose, M. Truszczynski (Eds.), Learning and Reasoning with Complex Representations. Proceedings, 1996. X, 283 pages. 1998. (Subseries LNAI).

1360: D. Wang (Ed.), Automated Deduction in Geometry. Proceedings, 1996. VII, 235 pages. 1998. (Subseries LNAI).

Vol. 1361: B. Christianson, B. Crispo, M. Lomas, M. Roe (Eds.), Security Protocols. Proceedings, 1997. VIII, 217 pages. 1998.

Vol. 1362: D.K. Panda, C.B. Stunkel (Eds.), Network-Based Parallel Computing. Proceedings, 1998. X, 247 pages. 1998.

Vol. 1363: J.-K. Hao, E. Lutton, E. Ronald, M. Schoenauer, D. Snyers (Eds.), Artificial Evolution. XI, 349 pages. 1998.

Vol. 1364: W. Conen, G. Neumann (Eds.), Coordination Technology for Collaborative Applications. VIII, 282 pages. 1998.

Vol. 1365: M.P. Singh, A. Rao, M.J. Wooldridge (Eds.), Intelligent Agents IV. Proceedings, 1997. XII, 351 pages. 1998. (Subseries LNAI).

Vol. 1367: E.W. Mayr, H.J. Prömel, A. Steger (Eds.), Lectures on Proof Verification and Approximation Algorithms. XII, 344 pages. 1998.

Vol. 1368: Y. Masunaga, T. Katayama, M. Tsukamoto (Eds.), Worldwide Computing and Its Applications — WWCA'98. Proceedings, 1998. XIV, 473 pages. 1998.

Vol. 1370: N.A. Streitz, S. Konomi, H.-J. Burkhardt (Eds.), Cooperative Buildings. Proceedings, 1998. XI, 267 pages. 1998.

Vol. 1371: I. Wachsmuth, M. Fröhlich (Eds.), Gesture and Sign-Language in Human-Computer Interaction. Proceedings, 1997. XI, 309 pages. 1998. (Subseries LNAI).

Vol. 1372: S. Vaudenay (Ed.), Fast Software Encryption. Proceedings, 1998. VIII, 297 pages. 1998.

Vol. 1373: M. Morvan, C. Meinel, D. Krob (Eds.), STACS 98. Proceedings, 1998. XV, 630 pages. 1998.

Vol. 1374: H. Bunt, R.-J. Beun, T. Borghuis (Eds.), Multimodal Human-Computer Communication. VIII, 345 pages. 1998. (Subseries LNAI).

Vol. 1375: R. D. Hersch, J. André, H. Brown (Eds.), Electronic Publishing, Artistic Imaging, and Digital Typography. Proceedings, 1998. XIII, 575 pages. 1998.

Vol. 1376: F. Parisi Presicce (Ed.), Recent Trends in Algebraic Development Techniques. Proceedings, 1997. VIII, 435 pages. 1998.

Vol. 1377: H.-J. Schek, F. Saltor, I. Ramos, G. Alonso (Eds.), Advances in Database Technology – EDBT'98. Proceedings, 1998. XII, 515 pages. 1998.

Vol. 1378: M. Nivat (Ed.), Foundations of Software Science and Computation Structures. Proceedings, 1998. X, 289 pages. 1998.

Vol. 1379: T. Nipkow (Ed.), Rewriting Techniques and Applications. Proceedings, 1998. X, 343 pages. 1998.

Vol. 1380: C.L. Lucchesi, A.V. Moura (Eds.), LATIN'98: Theoretical Informatics. Proceedings, 1998. XI, 391 pages. 1998.

Vol. 1381: C. Hankin (Ed.), Programming Languages and Systems. Proceedings, 1998. X, 283 pages. 1998.

Vol. 1382: E. Astesiano (Ed.), Fundamental Approaches to Software Engineering. Proceedings, 1998. XII, 331 pages. 1998.

Vol. 1383: K. Koskimies (Ed.), Compiler Construction. Proceedings, 1998. X, 309 pages. 1998.

Vol. 1384: B. Steffen (Ed.), Tools and Algorithms for the Construction and Analysis of Systems. Proceedings, 1998. XIII, 457 pages. 1998.

Vol. 1385: T. Margaria, B. Steffen, R. Rückert, J. Posegga (Eds.), Services and Visualization. Proceedings, 1997/1998. XII, 323 pages. 1998.

Vol. 1386: T.A. Henzinger, S. Sastry (Eds.), Hybrid Systems: Computation and Control. Proceedings, 1998. VIII, 417 pages. 1998.

Vol. 1387: C. Lee Giles, M. Gori (Eds.), Adaptive Processing of Sequences and Data Structures. Proceedings, 1997. XII, 434 pages. 1998. (Subseries LNAI).

Vol. 1388: J. Rolim (Ed.), Parallel and Distributed Processing. Proceedings, 1998. XVII, 1168 pages. 1998.

Vol. 1389: K. Tombre, A.K. Chhabra (Eds.), Graphics Recognition. Proceedings, 1997. XII, 421 pages. 1998.

Vol. 1391: W. Banzhaf, R. Poli, M. Schoenauer, T.C. Fogarty (Eds.), Genetic Programming. Proceedings, 1998. X, 232 pages. 1998.

Vol. 1393: D. Bert (Ed.), B'98: Recent Advances in the Development and Use of the B Method. Proceedings, 1998. VIII, 313 pages. 1998.

Vol. 1394: X. Wu. R. Kotagiri, K.B. Korb (Eds.), Research and Development in Knowledge Discovery and Data Mining. Proceedings, 1998. XVI, 424 pages. 1998. (Subseries LNAI).

Vol. 1396: G. Davida, M. Mambo, E. Okamoto (Eds.), Information Security. Proceedings, 1997. XII, 357 pages. 1998.

Vol. 1397: H. de Swart (Ed.), Automated Reasoning with Analytic Tableaux and Related Methods. Proceedings, 1998. X, 325 pages. 1998. (Subseries LNAI).

Vol. 1371: I. Wachsmuth, M. Fröhlich (Eds.), Gesture and Sign-Language in Human-Computer Interaction. Proceedings, 1997. XI, 309 pages. 1998. (Subseries LNAI).